Advanced Energy Storage Technologies and Their Applications (AESA)

Special Issue Editors

Rui Xiong
Hailong Li
Joe (Xuan) Zhou

MDPI • Basel • Beijing • Wuhan • Barcelona • Belgrade

MDPI

Special Issue Editors
Rui Xiong   Hailong Li
Beijing Institute of Technology   Mälardalens University
China   Sweden
Joe (Xuan) Zhou
Kettering University
USA

Editorial Office
MDPI AG
St. Alban-Anlage 66
Basel, Switzerland

This edition is a reprint of the Special Issue published online in the open access journal *Energies* (ISSN 1996-1073) from 2016–2017 (available at: http://www.mdpi.com/journal/energies/special_issues/AESA2017).

For citation purposes, cite each article independently as indicated on the article page online and as indicated below:

Lastname, F.M.; Lastname, F.M. Article title. *Journal Name*. **Year**. *Article number, page range.*

First Edition 2018

ISBN 978-3-03842-544-1 (Pbk)
ISBN 978-3-03842-545-8 (PDF)

Table of Contents

About the Special Issue Editors

Rui Xiong is the deputy director of Beijing Engineering Research Center for Electric Vehicles and an associate professor of National Engineering Laboratory for Electric Vehicles, Beijing Institute of Technology (BIT), China. He received the M.Sc. degree in vehicle engineering and the Ph.D. degree in mechanical engineering from BIT in 2010 and 2014, respectively. He conducted scientific research as a joint Ph.D. student at the University of Michigan, Dearborn, USA, between 2012 and 2014.

Since 2014, he has been an Associate Professor in the Department of Vehicle Engineering, BIT, China. During 2015 and 2017, he was a visiting Associate Professor in the Faculty of Science, Engineering and Technology, Swinburne University of Technology, Australia. He has authored more than 100 peerreviewed articles and hold five patents. His research interests mainly include electrical/hybrid vehicles, energy storage, and battery management system.

He received the Excellent Doctoral Dissertation from BIT in 2014, the first prize of Chinese Automobile Industry Science and Technology Progress Award in October 2015 and the second prize of National Defense Technology Invention Award in December 2016. He received Best Paper Awards from the journal Energies and International conferences four times. He is an Associate Editor of *IEEE Access and Energy, Ecology and Environment* (E3). He is serving as the Editorial Board of the *Journal of Cleaner Production and Energies, subject assistant editor of Applied Energy*. He has been the conference chair of the 2017 International Symposium on Electric Vehicles hold in Stockholm, Sweden.

Hailong Li is a Senior Lecturer and Associate Professor in the department of energy engineering and Future Energy Centre of Mälardalen University, Sweden. Before started working at MDH in 2010, he worked as a research scientist in SINTEF Energy Research, the biggest research institute in Norway. He is also a visiting professor at Shandong University, Tianjin University of Commerce, and China University of Petroleum-Beijing.

Dr. Li has published more than 100 scientific papers, including 75 peer reviewed journal papers and more than 50 peer reviewed conference papers, resulting in an H-Index of 25 (by Google Scholar). His research interests mainly lie in the mitigation of the climate change and the efficient conversion and utilization of energy. Currently, Dr Li is serving as the Assistant Editor of Applied Energy. He has been the member of scientific committee, organizing committee, track director and session chair of International Conference on Applied Energy (ICAE) since 2012 and Applied Energy Symposium and Forum—Low carbon cities and urban energy systems (CUE) since 2015.

Joe (Xuan) Zhou is an assistant professor in the department of Electrical and Computer Engineering, Kettering University, USA. He received B.S. degree in Metal Material Manufacturing from Taiyuan University of Technology, Shanxi, China 2002, M.S. degree in Materials Physics from Xi'an Jiaotong University, Shaanxi, China 2005 and the Ph.D. degree from University of Michigan, Dearborn, USA in Automotive System Engineering in 2012.

He was a visiting professor in Beijing Institute of Technology, Beijing, and scientist at Csquared Innovations, Michigan. He has published more than 30 papers, authored one book chapter and holds two patents. He is serving as a guest editor for one SCI-indexed journal, reviewers for four journals. He is also active member of professional societies such as IEEE and ECS. His research interests include battery design and manufacturing, battery modeling and control on electric vehicles etc. Dr. Zhou was recipient of the Outstanding New Research Award of Kettering University in 2016 and the 2014 Kettering Faculty Research Fellowship for the work on the development of battery for renewable energy storage.

Preface to "Advanced Energy Storage Technologies and Their Applications (AESA)"

The depletion of fossil fuels, the increase of energy demands, and the concerns over climate change are the major driving forces for the development of renewable energy, such as solar, wind and wave energy. However, the intermittency of renewable energy has hindered its large-scale deployment, which, therefore, has necessitated the development of advanced energy storage technologies. The use of large-scale energy storage can effectively improve the efficiency of energy resource utilization, and increase the adoption of variable renewable resources, the energy access, and the end-use sector electrification (e.g., electrification of transport sector).

To unlock the world's creativity, remove the existing barriers and encourage international cooperation, an international joint research association entitled "Advanced Energy Storage and Applications (AESA)" has been created. It aims to provide a platform for presenting the latest research outcomes on the technology development of large-scale energy storage and empower both academia and industry to explore success through this platform.

This special issue, published in the journal of Energies, is a milestone achievement of AESA, which includes 22 outstanding contributions (20 original research papers and 2 reviews) across the world. The topics cover a wide range including lithium-ion battery, electric vehicles (EVs), and thermal energy storage. New models have been proposed for parameter verification, state of charge and peak power estimation, equalization, and capacity decay predication. Battery self-heating-related research have also been reported. As one of the major applications of Lithium-ion battery, EVs represent another popular topic. The energy distribution of the power sources, energy optimization strategies and automatic control techniques were investigated comprehensively. In addition, latest progresses about superconducting magnetic, latent thermal, and compressed air energy storages have also been reported.

Finally, we wish to express our deep gratitude to all the authors and reviewers who have significantly contributed to this special issue. Sincere thanks also go to the editorial team of MDPI and Energies for giving the opportunity to publish this book and helping in all possible ways, especially Dr. Terry Zhang for his endless support.

Rui Xiong, Hailong Li and Joe (Xuan) Zhou

Special Issue Editors

energies

MDPI

Editorial

Advanced Energy Storage Technologies and Their Applications (AESA2017)

Rui Xiong [1,2,*], **Hailong Li** [3,4] **and Xuan Zhou** [5]

[1] National Engineering Laboratory for Electric Vehicles, Department of Vehicle Engineering, School of Mechanical Engineering, Beijing Institute of Technology, Beijing 100081, China

[2] Faculty of Science, Engineering and Technology, Swinburne University of Technology, John Street, Hawthorn, VIC 3122, Australia

[3] Energy Technology, School of Business, Society and Technology Mälardalens University, Box 883, 72123 Västerås, Sweden; hailong.li@mdh.se

[4] Tianjin Key Laboratory of Refrigeration Technology, School of Mechanical Engineering, Tianjin University of Commerce, Tianjin 300134, China

[5] Electrical and Computer Engineering Department, Kettering University, 1700 University Avenue, Flint, MI 48504, USA; xzhou@kettering.edu

* Correspondence: rxiong@bit.edu.cn; Tel.: +86-10-6891-4070

Received: 24 August 2017; Accepted: 7 September 2017; Published: 9 September 2017

Abstract: This editorial summarizes the performance of the special issue entitled Advanced Energy Storage Technologies and Applications (AESA), which is published in MDPI's Energies journal in 2017. The special issue includes a total of 22 papers from four countries. Lithium-ion battery, electric vehicle, and energy storage were the topics attracting the most attentions. New methods have been proposed with very sound results.

Keywords: lithium-ion battery; electric vehicle; energy storage

---

To reduce the usage of fossil fuel and ease air pollution, many countries have put huge efforts to promote the development of electric vehicles. Lithium-ion batteries are the main power sources of electric vehicles, and have been the research focus in both industry and academia [1,2].

This special issue has focused on advanced energy storage technologies and their applications, which covers all kinds of energy storage and application fields, such as:

(1) Novel energy storage materials and topologies;
(2) Application in electrical/hybrid driven system and electrical/hybrid vehicles;
(3) Next generation energy storage devices, systems, or techniques;
(4) Large-scale energy storage system modeling, simulation and optimization, including testing and modeling ageing processes;
(5) Advanced energy storage management systems, including advanced control algorithms and fault diagnosis/online condition monitoring for energy storage systems;
(6) Business model for the application and deployment of energy storage;
(7) Lifecycle analysis, repurposing, and recycling.

After peer-reviewing, papers in high scientific quality and innovativeness were accepted. A total of twenty-two papers were accepted, with the following geographical distribution of authors:

(1) China (18).
(2) USA (2).
(3) Germany (1).

(4)   Italy (1).

The lithium-ion battery has been investigated broadly, including equivalent circuit modeling and parameter estimation [3,4], state of charge and peak power estimation [5,6], battery pack equalization [7], and battery capacity decay [8]. Recently, battery heating-related characteristics have been a research focus. Zhu et al. [9] investigated an impedance-based temperature estimation method considering the electrochemical non-equilibrium with short-term relaxation time for facilitating vehicular application. Hong et al. [10] developed a thermal runaway prognosis scheme for battery systems in electric vehicles based on the big data platform and entropy method. The low-temperature preheating techniques of lithium-ion batteries were investigated in [11,12].

The driving performance of electric vehicles (EVs) is highly dependent on the energy distribution of the power sources and the electronics' reliability. Energy optimization strategies and automatic control techniques were investigated in [13–15]. Ding et al. [16] investigated the impact of silicon carbide (SiC) metal oxide semiconductor field effect transistors (MOSFETs) on the dynamic performance of permanent magnet synchronous motor (PMSM) drive systems. In another paper, Ding et al. [17] investigated the impact of SiC on the powertrain systems in EVs.

Other energy storage forms have also been investigated aside from lithium-ion batteries or DC-DC, including superconducting magnetic energy storage (SMES) [18], latent thermal energy storage (LTS) [19], and compressed air energy storage [20]. Refs. [21,22] investigated the design of pump-turbines.

Two reviews are presented in this special issue. Lanahan et al. [23] analyzed recent case studies—numerical and field experiments—seen by borehole thermal energy storage (BTES) in space heating and domestic hot water capacities, coupled with solar thermal energy. Benato et al. [24] offered a wide overview on the large-scale electrochemical energy projects installed in the high-voltage Italian grid. Detailed descriptions of energy (charge/discharge times of about 8 h) and power intensive (charge/discharge times ranging from 0.5 h to 4 h) installations were presented with some insights into the authorization procedures, safety features, and ancillary services.

Acknowledgments: Guest editors would like to express their sincerest gratitude to Energies' in-house editor and reviewers for their wonderful work and effort. Without their support, the efficient handling of all receive manuscripts, it would not have been possible to publish this special issue. Sincere gratitude is also expressed to the joint support by the National Natural Science Foundation of China under Grant No. 51507012 and Beijing Nova Program under Grant No. Z171100001117063.

Conflicts of Interest: The authors declare no conflict of interest.

References

1.   Xiong, R.; Zhang, Y.; He, H.; Zhou, X.; Pecht, M. A double-scale, particle-filtering, energy state prediction algorithm for lithium-ion batteries. *IEEE Trans. Ind. Electron.* **2017**. [CrossRef]
2.   Xiong, R.; Yu, Q.Q.; Wang, L.Y.; Lin, C. A novel method to obtain the open circuit voltage for the state of charge of lithium ion batteries in electric vehicles by using H infinity filter. *Appl. Energy* **2017**. [CrossRef]
3.   Chen, D.; Jiang, J.; Li, X.; Wang, Z.; Zhang, W. Modeling of a Pouch Lithium Ion Battery Using a Distributed Parameter Equivalent Circuit for Internal Non-Uniformity Analysis. *Energies* **2016**, *9*, 865. [CrossRef]
4.   Yang, J.; Xia, B.; Shang, Y.; Huang, W.; Mi, C. Improved Battery Parameter Estimation Method Considering Operating Scenarios for HEV/EV Applications. *Energies* **2017**, *10*, 5. [CrossRef]
5.   Zhang, C.; Jiang, J.; Zhang, L.; Liu, S.; Wang, L.; Loh, P.C. A Generalized SOC-OCV Model for Lithium-Ion Batteries and the SOC Estimation for LNMCO Battery. *Energies* **2016**, *9*, 900. [CrossRef]
6.   Jiang, B.; Dai, H.; Wei, X.; Zhu, L.; Sun, Z. Online Reliable Peak Charge/Discharge Power Estimation of Series-Connected Lithium-Ion Battery Packs. *Energies* **2017**, *10*, 390. [CrossRef]
7.   Shang, Y.; Zhang, Q.; Cui, N.; Zhang, C. A Cell-to-Cell Equalizer Based on Three-Resonant-State Switched-Capacitor Converters for Series-Connected Battery Strings. *Energies* **2017**, *10*, 206. [CrossRef]
8.   Wu, X.; Wang, T. Optimization of Battery Capacity Decay for Semi-Active Hybrid Energy Storage System Equipped on Electric City Bus. *Energies* **2017**, *10*, 792. [CrossRef]

9. Zhu, J.; Sun, Z.; Wei, X.; Dai, H. Battery Internal Temperature Estimation for LiFePO4 Battery Based on Impedance Phase Shift under Operating Conditions. *Energies* **2017**, *10*, 60. [CrossRef]

10. Hong, J.; Wang, Z.; Liu, P. Big-Data-Based Thermal Runaway Prognosis of Battery Systems for Electric Vehicles. *Energies* **2017**, *10*, 919. [CrossRef]

11. Wu, X.; Chen, Z.; Wang, Z. Analysis of Low Temperature Preheating Effect Based on Battery Temperature-Rise Model. *Energies* **2017**, *10*, 1121. [CrossRef]

12. Zhang, C.; Jin, X.; Li, J. PTC Self-heating Experiments and Thermal Modeling of Lithium-ion Battery Pack in Electric Vehicles. *Energies* **2017**, *10*, 572. [CrossRef]

13. Liu, Y.; Li, J.; Ye, M.; Qin, D.; Zhang, Y.; Lei, Z. Optimal Energy Management Strategy for a Plug-in Hybrid Electric Vehicle Based on Road Grade Information. *Energies* **2017**, *10*, 412. [CrossRef]

14. Lei, Z.; Cheng, D.; Liu, Y.; Qin, D.; Zhang, Y.; Xie, Q. A Dynamic Control Strategy for Hybrid Electric Vehicles Based on Parameter Optimization for Multiple Driving Cycles and Driving Pattern Recognition. *Energies* **2017**, *10*, 54. [CrossRef]

15. Sun, J.; Xing, G.; Zhang, C. Data-Driven Predictive Torque Coordination Control during Mode Transition Process of Hybrid Electric Vehicles. *Energies* **2017**, *10*, 441. [CrossRef]

16. Ding, X.; Du, M.; Cheng, J.; Chen, F.; Ren, S.; Guo, H. Impact of Silicon Carbide Devices on the Dynamic Performance of Permanent Magnet Synchronous Motor Drive Systems for Electric Vehicles. *Energies* **2017**, *10*, 364. [CrossRef]

17. Ding, X.; Cheng, J.; Chen, F. Impact of Silicon Carbide Devices on the Powertrain Systems in Electric Vehicles. *Energies* **2017**, *10*, 533. [CrossRef]

18. Wang, X.; Yang, J.; Chen, L.; He, J. Application of Liquid Hydrogen with SMES for Efficient Use of Renewable Energy in the Energy Internet. *Energies* **2017**, *10*, 185. [CrossRef]

19. Kuboth, S.; König-Haagen, A.; Brüggemann, D. Numerical Analysis of Shell-and-Tube Type Latent Thermal Energy Storage Performance with Different Arrangements of Circular Fins. *Energies* **2017**, *10*, 274. [CrossRef]

20. Safaei, H.; Aziz, M.J. Thermodynamic Analysis of Three Compressed Air Energy Storage Systems: Conventional, Adiabatic, and Hydrogen-Fueled. *Energies* **2017**, *10*, 1020. [CrossRef]

21. Wang, Z.; Zhu, B.; Wang, X.; Qin, D. Pressure Fluctuations in the S-Shaped Region of a Reversible Pump-Turbine. *Energies* **2017**, *10*, 96. [CrossRef]

22. Liu, L.; Zhu, B.; Bai, L.; Liu, X.; Zhao, Y. Parametric Design of an Ultrahigh-Head Pump-Turbine Runner Based on Multiobjective Optimization. *Energies* **2017**, *10*, 1169. [CrossRef]

23. Lanahan, M.; Tabares-Velasco, P.C. Seasonal Thermal-Energy Storage: A Critical Review on BTES Systems, Modeling, and System Design for Higher System Efficiency. *Energies* **2017**, *10*, 743. [CrossRef]

24. Benato, R.; Bruno, G.; Palone, F.; Polito, R.M.; Rebolini, M. Large-Scale Electrochemical Energy Storage in High Voltage Grids: Overview of the Italian Experience. *Energies* **2017**, *10*, 108. [CrossRef]

Article

Parametric Design of an Ultrahigh-Head Pump-Turbine Runner Based on Multiobjective Optimization

Linhai Liu [1], Baoshan Zhu [1,*], Li Bai [2], Xiaobing Liu [2] and Yue Zhao [3]

[1] Department of Thermal Engineering, State Key Laboratory of Hydro Science and Engineering, Tsinghua University, Beijing 100084, China; liu-lh14@mails.tsinghua.edu.cn

[2] School of Energy and Power Engineering, Xihua University, Chengdu 610039, China; baili7023yjj@163.com (L.B.); liuxb@mail.xhu.edu.cn (X.L.)

[3] Harbin Institute of Large Electrical Machinery, Harbin 150040, China; zhaoyue1967@sina.com

* Correspondence: bszhu@mail.tsinghua.edu.cn; Tel.: +86-10-6279-6797

Received: 19 April 2017; Accepted: 4 August 2017; Published: 8 August 2017

Abstract: Pumped hydro energy storage (PHES) is currently the only proven large-scale energy storage technology. Frequent changes between pump and turbine operations pose significant challenges in the design of a pump-turbine runner with high efficiency and stability, especially for ultrahigh-head reversible pump-turbine runners. In the present paper, a multiobjective optimization design system is used to develop an ultrahigh-head runner with good overall performance. An optimum configuration was selected from the optimization results. The effects of key design parameters—namely blade loading and blade lean—were then investigated in order to determine their effects on runner efficiency and cavitation characteristics. The paper highlights the guidelines for application of inverse design method to high-head reversible pump-turbine runners. Middle-loaded blade loading distribution on the hub, back-loaded distribution on the shroud, and large positive blade lean angle on the high pressure side are good for the improvement of runner power performance. The cavitation characteristic is mainly influenced by the blade loading distribution near the low pressure side, and large blade lean angles have a negative impact on runner cavitation characteristics.

Keywords: ultrahigh-head pump-turbine; multiobjective optimization; blade loading; blade lean

1. Introduction

Benefits of pumped hydro energy storage (PHES) on electrical system operations are prominent. The flexible generation of PHES can provide upregulation and downregulation in power systems. Furthermore, PHES enable quick start and the provision of spinning and standing reserves. Interest in this technology has been renewed because of the increase in variable renewable energy, such as wind power [1,2]. In recent years, higher head and larger capacity PHES stations have been developed in order to reduce the construction costs [3].

The pump-turbine is a key component in PHES stations. It usually takes only one runner functioning as pump or turbine. Therefore, pump and turbine efficiencies should be guaranteed for the runners during water pumping and electricity generation. Furthermore, the cavitation performance and operation stability have to be improved for both operating conditions. It is difficult to develop a pump-turbine runner with high overall performance because the targets affect each other and sometimes conflict in its two operations [4,5].

The pump-turbine runners are usually designed from pump mode, and then verified with turbine mode [1,5], given that the requirements for pump operation are difficult to meet, and the relatively good performance can be maintained when pumps operate as turbines [6,7]. The runners are more like centrifugal pump impellers in shape, rather than Francis turbine runners. Furthermore, pump-turbine runners with higher working heads possess more prolonged flow channels. Low efficiency and bad cavitation characteristics are the main challenges in the development of ultrahigh-head pump turbines, especially the runners.

Computational fluid dynamics (CFD) has been widely used in the development of the pump-turbine runner [8,9]. The profile of the runner can be modified by changing the design parameters on the basis of internal flow analysis [10,11]. However, this CFD flow analysis cannot directly propose a blade configuration with favorable flow pattern. Moreover, the direct CFD-based modification technique is considerably time consuming and requires intensive experience. With the development of design theory and computer technology, three-dimensional (3D) inverse design methods have been increasing in popularity for turbomachinery in the past 30 years [12–14]. In the so-called inverse design methods, the geometry of the blades is unknown and it can be directly calculated according to the design specifications. The main advantage of the inverse design methods is the closer relationship between the design parameters and the hydrodynamic flow field. However, no direct relationship can be given between geometric parameters and runner performances. Accordingly, trial and error in flow analyses and model tests is still necessary.

More systematic approaches, such as optimization techniques, have been applied in the design of turbomachinery [9,15]. Optimal design associated to turbomachinery is a multiobjective and difficult problem by its nature. Gradient-based optimization methods have been successfully applied in the foil design [16,17]. It is known that gradient techniques are efficient in terms of convergence rate, but do not guarantee production of the global optimum. On the other hand, multiobjective evolutionary algorithms (MOEAs) have gained increasing popularity over the past two or three decades [18–20]. These population-based methods mimic the evolution of species and the survival of the fittest, and comparted to the gradient-based optimization techniques, they offer advantages, such as good approximations to optimal sets of solutions, generating multiple trade-off solutions in a single iteration [18,21]. Recently, a multiobjective optimization design strategy has been used to develop pump-turbine runners [22,23]. The strategy has been built by combining 3D design method, CFD analysis, design of experiment (DoE), response surface methodology (RSM), and multiobjective genetic algorithm (MOGA). A middle-high-head turbine runner with high efficiency and stability has been designed by using this strategy [23]. Because of its simplicity, its ease of use and its suitability to be coupled with specialized numerical tools, for instance CFD techniques, the strategy can be widely used in the development of fluid machines.

In this study, a parametric design study of an ultra-head pump-turbine runner is carried out based on multiobjective optimization. First, the multiobjective optimization design system was introduced and an ultrahigh-head pump-turbine runner was designed. The runner with high overall performance was obtained. Then, the impact of blade loading and stacking conditions on the runner performance was assessed, where the runners are optimally described using the inverse design method and their performance was estimated with CFD analyses. The main aim is to offer a guideline for the design ultrahigh-head pump-turbine runners by means of comparisons and analyses of design parameters on the runners' performances.

2. Optimization Design System

Figure 1 shows the flow chart of the design strategy used in this study. The design approach was based on the coupling of the parameterization of the blade shape with a 3D inverse design method to produce the blade geometry, DoE to reduce the number of calculation times, CFD analysis to estimate the objective functions, RSM to correlate the design parameters with the objectives, and MOGA to search the Pareto front for the trade-off design [22,23].

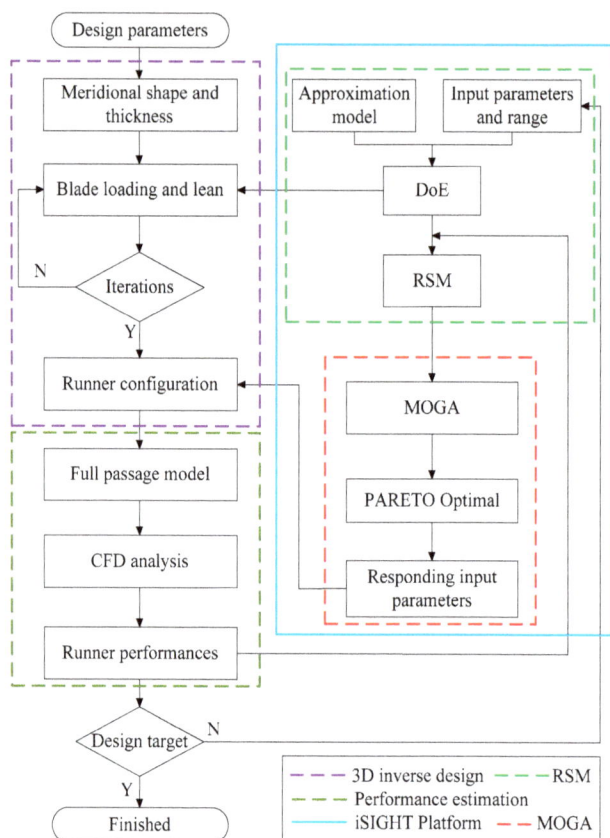

Figure 1. Procedures for multiobjective optimization.

2.1. 3D Inverse Design

The 3D design software TURBODesign 5.2 was used to parametrically describe the runner shape [12,15]. The flow through the runner is considered as water at normal temperature. When TURBODesign 5.2 is used for design, the flow is simplified to steady and inviscid, and the blades are represented by sheets of vorticity. Strength of the vorticity is determined by a circumferentially averaged velocity torque $r\overline{V}_\theta$, defined as

$$r\overline{V}_\theta = \frac{B}{2\pi}\int_0^{\frac{2\pi}{B}} rV_\theta d\theta \tag{1}$$

It is referred to as the "blade loading", here is the blade number.

For the incompressible potential flow, blade pressure distribution can be expressed as follows [12,15].

$$p^+ - p^- = \frac{2\pi}{B}\rho W_{bl}\frac{\partial\left(r\overline{V}_\theta\right)}{\partial m} \tag{2}$$

where subscripts + and − represent either side of the blades, ρ is the water density, W_{bl} is the relative velocity on the blade surface, and m is in the direction of streamlines in the meridional plane.

Equation (2) shows the direct relationship of $\partial(r\overline{V}_\theta)/\partial m$ with the difference between pressure on the upper and lower surfaces of the blade. The other important input specification is the stacking condition. This condition specifies the blade lean at the high pressure side (HPS) of the pump-turbine runner blades as shown in Figure 2, which affects the wrap angle of the blades [12,15,24].

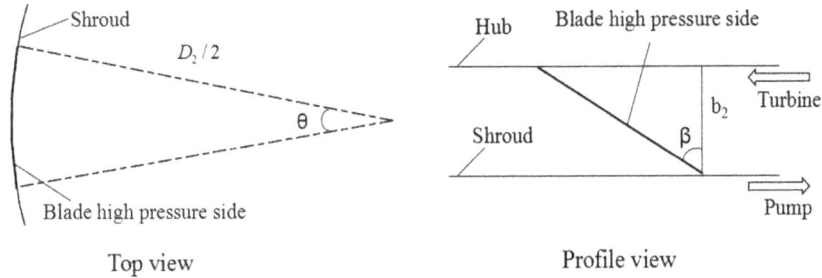

Figure 2. Blade lean at high pressure side.

2.2. CFD Analyses

The widely used commercial code ANSYS CFX 15.0 was used to conduct the CFD analyses. CFD analyses were conducted for two purposes: one was the estimation of objective functions and the other was validation and analyses of optimization results. The accuracy of the objective functions is important for the optimization process. Thus, as shown in Figure 3, 3D, turbulent, and steady flow simulations were performed for the full passage pump-turbine using the Reynolds-Averaged Navier–Stokes (RANS) equations [22,23]. For steady flow simulations, the RANS equations can be expressed as

$$\frac{\partial V_i}{\partial x_i} = 0 \tag{3}$$

$$\frac{\partial(V_i V_j)}{\partial x_j} = -\frac{1}{\rho}\frac{\partial p}{\partial x_i} + \nu\frac{\partial^2 V_i}{\partial x_i \partial x_j} + \frac{\partial(-\overline{V_i' V_j'})}{\partial x_j} \tag{4}$$

where V is the velocity, p is the pressure, ρ is the density, and ν is the kinematic viscosity, respectively. The Reynolds stresses are modeled according to the turbulent viscosity hypothesis as $-\overline{V_i' V_j'} = \nu_t\left(\frac{\partial V_i}{\partial x_j} + \frac{\partial V_j}{\partial x_i}\right) - \frac{2}{3}k\delta_{ij}$, here k is the turbulent kinematic energy, and δ_{ij} is the Dirac Delta function. The turbulence model is an important factor for CFD. For turbomachinery, performance parameters like efficiency and cavitation can be predicted with reasonable accuracy by solving the RANS equations with advanced turbulence models, such as standard $k - \varepsilon$, and renormalization group (RNG) $k - \varepsilon$ [10,22,23,25]. In this study, RNG $k - \varepsilon$ turbulence model was used for the closure of the RANS equations with the standard wall function method since it is economical and robust for predicting steady calculation with acceptable accuracy [25].

The computational domain includes spiral casing, stay vanes, guide vanes, runner, and draft tube as shown in Figure 3. The frozen rotor model was used at interfaces between the stationary and rotating components. No-slip wall conditions were set for stationary and rotating parts. Inlet and outlet boundaries were set as follows: static pressure zero was set at the inlet and the flow discharge ($Q_m = 0.284\,\text{m}^3/\text{s}$, listed in Table 1) was set at the outlet under pump mode; the flow discharge ($Q_m = 0.305\,\text{m}^3/\text{s}$, listed in Table 2) was set at the inlet and the static pressure was set at the outlet under turbine mode. Stochastic fluctuations of the velocities with a 5.0% free stream turbulent intensity were adopted as the mass flow rate was specified. ANSYS ICEM and TurboGrid were used for mesh generation. Hexahedral meshes were mainly used except in the volute tongue with tetrahedral meshes because of its complicated structure.

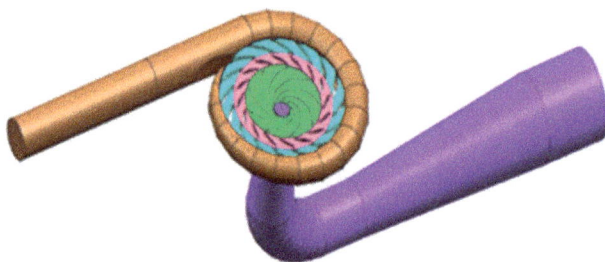

Figure 3. Whole flow passage model.

2.3. Optimization Strategy

The RSM model was used to describe the approximate relationships between the optimization targets and input design parameters. The second-order polynomial function was used in this study.

$$\hat{y}_i = \beta_0 + \sum_{j=1}^{m} \beta_i^j x_i + \sum_{j=1}^{m} \beta_{ii}^j x_i^2 + \sum_{i \neq j} \beta_{ik}^j x_i x_k \tag{5}$$

where $\hat{y}_i$ is the target, x_i and x_k are input parameters, β_0, β_i, β_{ii}, and β_{ik} can be determined by following the principle of least square regression with the help of a set of sample points in the design space.

The distribution of sample points in design space has significant influence on the accuracy of RSM model. The Latin hypercube sampling method was used in DoE, wherein the sample points are equiprobable, random, and orthogonally distributed in the design space. As the quadratic approximation model Equation (5) is used, the least number S of sample points should be

$$S \geq (N+1)(N+2)/2 \tag{6}$$

where N is the number of input variables selected.

When the RSM between the optimization targets and inputs was generated, the multiobjective optimization was then implemented with modified non-dominated sorted genetic algorithm (NSGA-II). In NSGA-II, the fast non-dominated sorting and crowding technique is adopted. NSGA-II is suitable for the optimization design of the pump-turbine runners with a reduction in computation complexity and an improvement in elitist strategy.

All the utilized software was integrated into the iSIGHT platform as shown in Figure 1. The optimization design process began with the selection of input parameters. After the variation ranges on the input parameters were determined, different combinations of the input parameters were used to generate a number of runner configurations with TURBODesign5.2. Then, runner performances were estimated under different operating conditions by using ANSYS CFX 15.0 and the RSM model between the optimization targets was generated. The CFD calculations were time consuming. Finally, NSGA-II was implemented on the RSM model and the optimal solutions were determined. It was unnecessary to regenerate the runners and estimate their performance in this optimization process. The optimal solutions could be obtained in a short time.

3. Design of the Ultrahigh-Head Pump Turbine Runner

3.1. Design Specifications

The specific design parameters were based on Yangjiang PHES station located in the Guangdong Province of China [26]. In turbine mode, the rated head was $H_r = 659.0$ m, and the maximum and minimum net head were $H_{max} = 693.85$ m and $H_{min} = 624.66$ m, respectively. In pump mode, the maximum and minimum heads were $H_{max} = 712.46$ m and $H_{max} = 652.11$ m, respectively. The rated capacity of the reversible synchronous motor was $P_r = 444.44$ MW and its rotational speed was $n_r = 500$ rpm.

In order to conduct the model tests on a standard test rig [27], scaled pump-turbine runners were designed. The design parameters are shown in Table 1, where H_m, Q_m, and n_m are the design head, design flow discharge, and rotational speed of the model runners, respectively. The number of the blades is $B = 9$. Figure 4 shows the meridional shape of the blades, which were derived on the basis of the centrifugal pump and one-dimensional flow calculation. The main geometrical parameters of the runner are high-pressure side (HPS) diameter D_2, HPS width b_2, low-pressure side (LPS) shroud diameter D_{1s}, and LPS hub diameter D_{1h}. The value of these main parameters is given in Table 2.

Table 1. Design parameters of a model pump-turbine.

Mode	H_m/m	Q_m/m^3	n_m/rpm
Pump	59.40	0.284	1200
Turbine	59.31	0.305	1200

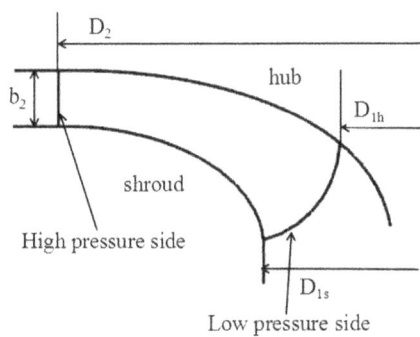

Figure 4. Meridional blade shape.

Table 2. Geometric parameters for meridional blade shape.

Parameter	b_2/m	D_{1h}/m	D_{1s}/m	D_2/m
Value	0.042	0.132	0.250	0.540

3.2. Optimization Settings

As the description in Section 2.1, blade loading and blade stacking are the most important parameters in determining the blade shape [12,15,24]. Blade loading distributions are usually given along the hub and shroud streamlines. The blade loading between the hub and shroud is determined by using linear interpolation. As shown in Figure 5, along each streamline, three-segment distribution was adopted. Four parameters—namely, connection point locations *NC* and *ND*, slope of the linear line *SLOPE*, and loading at the low pressure edge *DVRT*—were used to control the distribution curve.

Blade stacking specifies the blade lean angle θ at the HPS of the blade as shown in Figure 2. The rake angle β in Figure 2 is given as

$$\beta = \arctan\left[\left(\theta \cdot \frac{D_2}{2}\right)/b_2\right] \tag{7}$$

The stacking condition was imposed linearly along the HPS of the blade in this study.

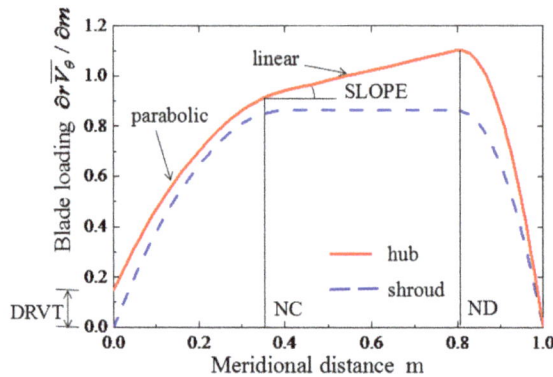

Figure 5. Blade loading distributions.

As shown in Figure 5, a total of eight variables are necessary for the control of the blade loading distribution. When the blade loading and blade lean angle are identified as optimization parameters, there are nine variables. The sensitivity on the runners' shape and the range of these variables are tested by the trial designs of the runners [22,23]. In this study, extensive trial designs were made by using the design software TURBODesign 5.2 to check whether blades with a reasonable shape could be obtained. With these trial designs, three variables were fixed at $NC_h = 0.7$, $ND_h = 0.8$, and $ND_s = 0.8$, and the variation range of the other input variables were determined as shown in Table 3.

The optimization targets were set as the runner efficiencies η_{mP} and η_{mT} at the pump design point and turbine rated point. In pump-turbines, the cavitation performance in pump mode is usually worse than that in turbine mode. During the design, as the cavitation requirement is satisfied in pump mode, it can be satisfied in turbine mode. Therefore, the lowest pressure p_{low} on the blade at the pump design point was also set as an optimization target. The optimization was made to maximize the runner efficiencies, η_{mP} and η_{mT}, and to increase pressure p_{low}. Considering these three objective functions were likely conflicting, MOGA was employed to find a number of trade-off solutions.

Table 3. Variation range of input parameters.

Optimized Inputs	Parameters	Range
	NC_s	$0.7 \sim 0.8$
	$SLOP_h$	$-1.0 \sim 2.0$
Blade loading	$SLOP_s$	$-2.0 \sim 2.0$
	$DRVT_h$	$-0.2 \sim 0.2$
	$DRVT_s$	$-0.2 \sim 0.2$
Blade lean angle	θ	$-20.0° \sim 20.0°$

The efficiencies η_{mP} and η_{mT} are defined in Equations (8) and (9), respectively.

$$\eta_{mP} = \frac{\rho g H_{mP} Q_{mP}}{M_{mP} \omega_{mP}} \tag{8}$$

$$\eta_{mT} = \frac{M_{mT} \omega_{mT}}{\rho g H_{mT} Q_{mT}} \tag{9}$$

where Q_m and H_m are the discharge and head given in Table 1, ω_m is the angular velocity. Momentum M_m acting on the runner was calculated through CFD analyses introduced in Section 2.2.

In order to generate the quadratic RSM model as shown by Equation (5), 40 different runner geometries were provided by using TURBODesign 5.2. Therefore, 80 CFD calculations were conducted. The parameters setting for NSGA-II is shown in Table 4.

Table 4. Parameter settings for NSGA-II.

Parameters	Value
Population size	100
Number of generations	100
Crossover probability	0.9
Crossover distribution index	10
Mutation distribution index	20
Initialization mode	Random

3.3. Optimization Results

Figure 6 shows the optimization results. There are a total of 10,000 different optimized runners as shown in Figure 6a. The original 40 sample runners produced in DoE are also shown in Figure 6a. These samples were random, equiprobable, and orthogonally distributed. So that a high accuracy could be obtained as the RSM model shown in Equation (5) was used. The trade-off relationship between pump efficiency, turbine efficiency, and minimum pressure at the blade surface is indicated in the Pareto front surface in Figure 6b.

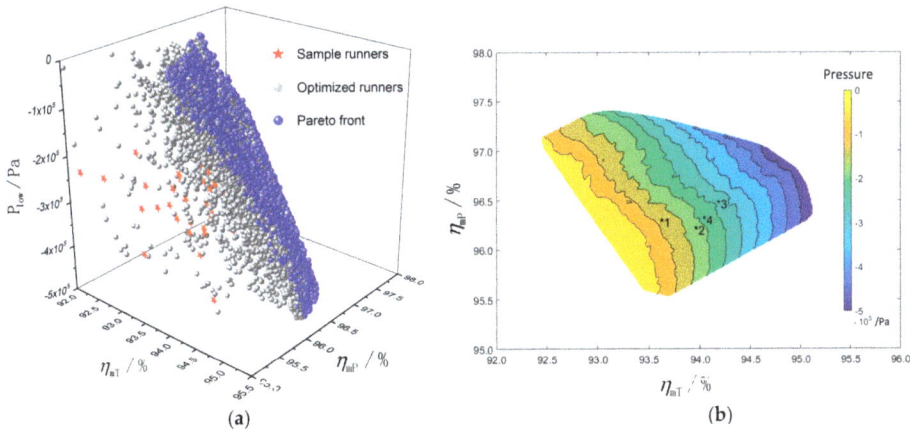

Figure 6. Optimization results. (**a**) Optimized runners; and (**b**) Pareto front surface.

Four runner configurations on the Pareto front in Figure 6b—denoted by 1, 2, 3, and 4—are selected for further detailed study. These runners were selected with an artificial screening method, in which a limited range was set to each optimization target, and the runners satisfying the target conditions were selected. Table 5 shows the performance comparisons calculated by CFD and estimated by RSM. The initial baseline runner was also reported. The CFD results shown in Table 5 were obtained from the redesigned runners by using optimized blade loading and blade lean. As shown in Table 5, there are some differences in the objective functions between the RSM estimation and CFD calculation. When the RSM model approach expressed in Equation (5) is used in the optimization procedure, the response surface is an approximation of runner performance predicted by CFD analyses. Simulation-based objective functions are inherently noisy, which is the typical problem in the numerical optimization process [18,21]. Therefore, it is necessary to develop robust and efficient optimization methodologies that can afford satisfactory designs even for limited computational resources.

Table 5. Comparison of the selected runners and the initial runner.

Runner	Mode	Runner Efficiency η/%		Low Pressure on Blade Surface p_{min}/Pa	
		RSM	CFD	RSM	CFD
1	pump	96.29	95.28	−72,461.6	−178,610
	turbine	93.61	93.25	-	-
2	pump	96.21	95.91	−247,114.3	−282,009
	turbine	93.94	93.04	-	-
3	pump	96.47	95.66	−302,416.2	−276,037
	turbine	94.16	93.42	-	-
4 (Preferred runner)	pump	96.31	96.43	−248,645.4	−263,678
	turbine	94.02	93.14	-	-
Initial runner	pump	-	95.86	-	−326,890
	turbine	-	92.45	-	-

Runner 4 is recommended as the preferred runner through comprehensive consideration of runner efficiencies and minimum pressure on blade surface. As shown in Figure 7, the blade loading distributions of the initial runner on the hub and shroud are both back-loaded, while the optimized blade loading distributions are middle-loaded on the hub and back-loaded on the shroud for preferred runner 4. The blade lean is $\theta = 0°$ for the initial runner, and $\theta = -2.0°$ for the preferred runner 4 at the HPS, respectively.

Compared the preferred runner 4 with the initial runner, the runner's efficiencies are increased about 0.6% and 0.7% under pump mode and turbine mode, respectively. At the same time, the runner cavitation performance is greatly increased by raising the minimum pressure on the blade surface. Figure 8 shows the comparison of the shapes among the preferred and initial runners. The preferred runner has a negative blade lean angle $\theta = -2.0°$, and rake angle $\beta = -12.65°$, whereas the initial runner has no lean on the HPS. Near the low pressure side (LPS), the blade cross-sections are in distorted shape in the preferred runner.

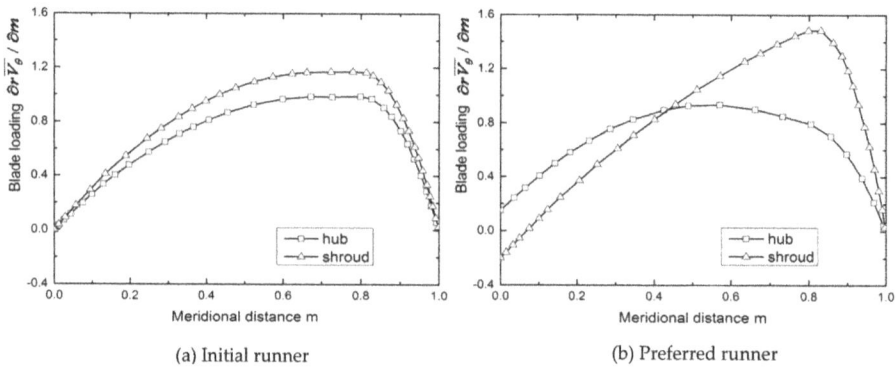

(a) Initial runner

(b) Preferred runner

Figure 7. Blade loading distributions of the initial runner and the preferred runner.

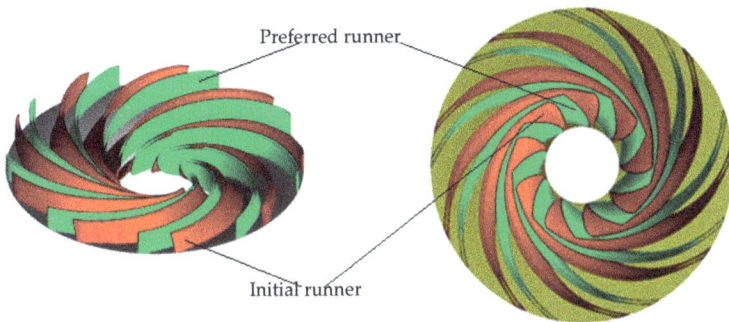

Figure 8. Comparison of the blades for two runners.

4. Parametric Effects on the Runner Performance

With the optimization, the runner with good overall performance could be developed as shown in Table 5. As discussed in Section 3.3, there were some differences in the performance estimated by RSM model and CFD prediction. In order to assess the impact of the main design parameters on the runner performances and increase the quantitative credibility of the optimized results, besides runners 1–4, more runners (A–H) were selected from the optimized results as shown in Figure 9. These runners were redesigned using the optimized design parameters and numerically simulated with CFD. Table 6 shows main design parameters and the CFD calculated performances for runners A–H, as well as the initial runner and the preferred runner.

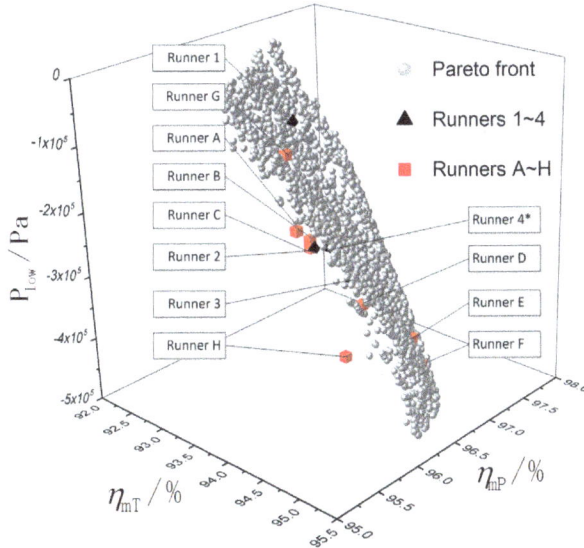

Figure 9. Selected runners and Pareto front surface.

Table 6. Design parameters and performances of the runners.

Runner	Design Parameters		Performance		
	$\partial(r\overline{V}_\theta)/\partial m$	$\theta,\ \beta/^\circ$	$\eta_{mT}/\%$	$\eta_{mP}/\%$	p_{low}/Pa
Initial runner	Figure 7a	0.0, 0.0	92.45	95.86	−326,890
Preferred runner	Figure 7b	−2.0, −12.65	93.14	96.43	−263,678
A	Figure 12a	−2.0, −12.65	93.09	95.53	−286,616
B	Figure 12b	−2.0, −12.65	92.82	95.89	−282,646
C	Figure 12c	−3.0, −18.60	93.15	95.88	−388,540
D	Figure 12d	0.0, 0.0	93.34	96.10	−230,361
E	Figure 7b	19.0, 64.87	93.96	95.96	−336,242
F	Figure 7b	20.0, 65.89	94.08	96.45	−341,955
G	Figure 7b	−19.0, −64.87	92.97	96.03	−360,081
H	Figure 7b	−18.0, −63.66	93.14	95.41	−405,167

4.1. Effects of Blade Loading

The blade lean angles for runners A–D are $\theta_A = \theta_B = -2.0^\circ$, $\theta_C = -3.0^\circ$, and $\theta_D = 0.0^\circ$, while the blade lean angles for the initial runner and the preferred runner are $\theta_i = 0.0^\circ$ and $\theta_P = -2.0^\circ$, respectively. Figure 10 shows the comparisons of the shapes among runners A–D and the preferred runner. For runners A, B, and the preferred runner, their blade shapes are similar near HPS. As shown in Figure 10b, the blade shapes are a little different near HPS for runners C, D, and the preferred runner because of a slightly different blade lean. Near LPS, the blade shapes of runner D and the preferred runner are similar, and the blades tilt more to the turbine rotation direction than the other runners.

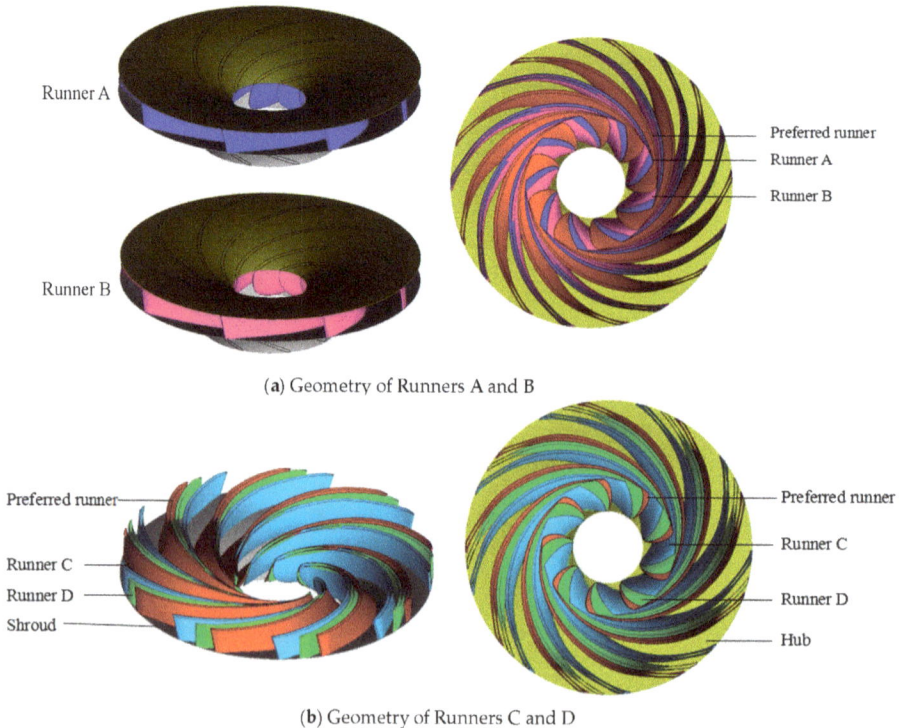

(**a**) Geometry of Runners A and B

(**b**) Geometry of Runners C and D

Figure 10. Blade configuration comparisons.

According to Table 6, the preferred runner and runner D have a higher efficiency in both turbine and pump mode. Furthermore, the minimum pressure at the blade surface is lower for these two runners. Figure 11 shows the pressure distributions on the blade suction surface under pump mode for different runners. Smaller low pressure zones on the blade suction side in pump mode show that the preferred runner and runner D have better cavitation characteristics.

Figure 12 shows the blade loading for runners A–D. Blade loading distributions are aft-loaded on the hub and shroud for runners A, B, and C, similar to the initial runner in Figure 7a. For runner D, blade loading distributions are middle-loaded on the hub and aft-loaded on the shroud, similar with the preferred runner shown in Figure 7b. The preferred runner has same blade lean angle with runners A and B, meanwhile runner D has the same blade lean angle as the initial runner. Therefore, the performance improvement for the preferred runner and runner D is mainly provided by the blade loading distribution. Synthetically considering the effects on efficiency and cavitation, it is recommended to design the runner to be middle-loaded on hub and back-loaded on shroud for blade loading distributions.

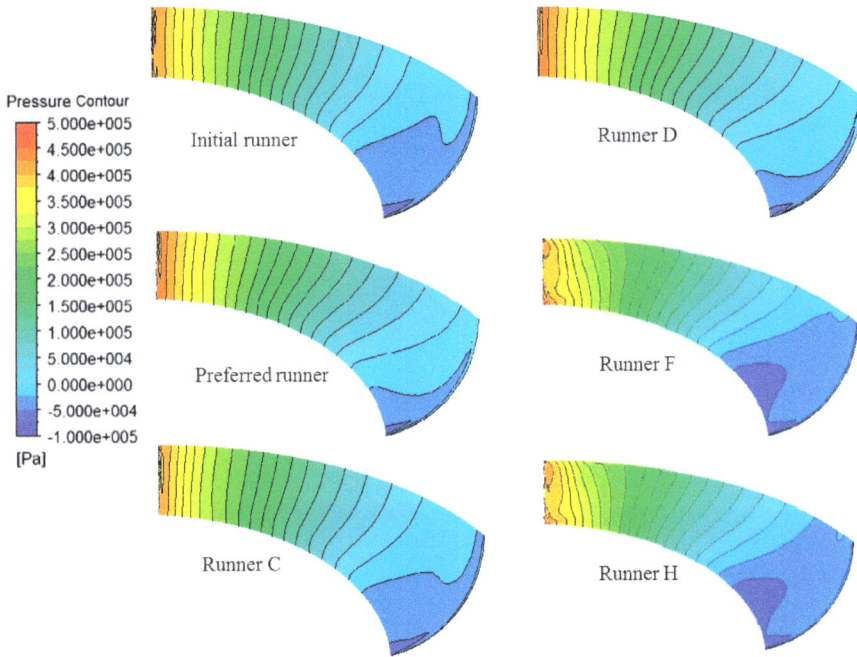

Figure 11. Pressure distribution on suction surface for different runners.

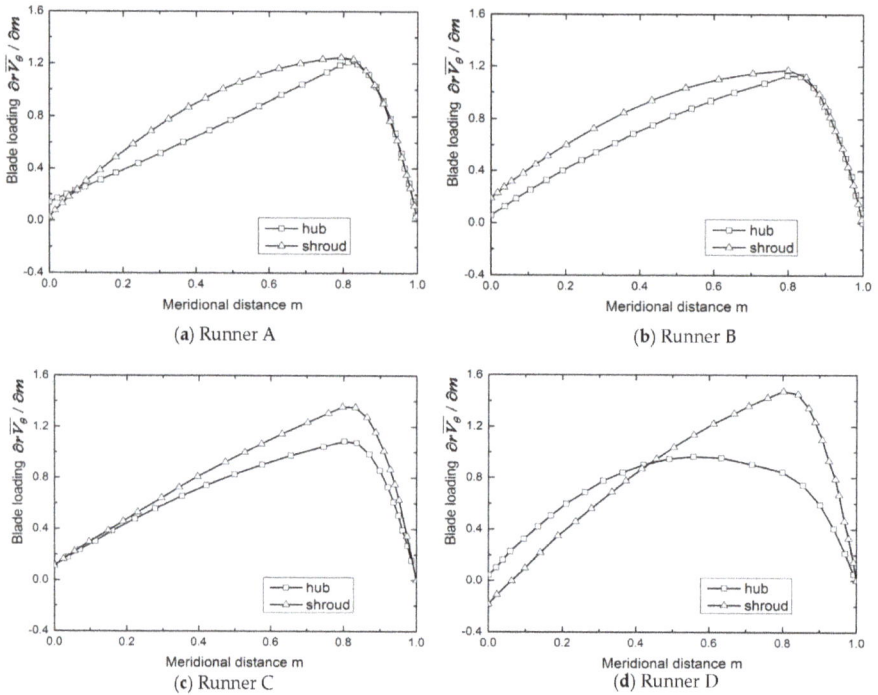

Figure 12. Blade loading distributions for runners A–D.

4.2. Effects of the Blade Lean

More runners marked by E–H were investigated. For these four runners, the blade loading distributions were almost the same with the preferred runner and runner D, while the blade lean angles changed greatly, $\theta_E = 19.0°$, $\theta_F = 20.0°$, $\theta_G = -19.0°$, and $\theta_H = -18.0°$, respectively. Figure 13 shows the shapes of these four runners. It can found that large positive or negative blade lean angles significantly change the spatial shape of the blades from shroud to hub.

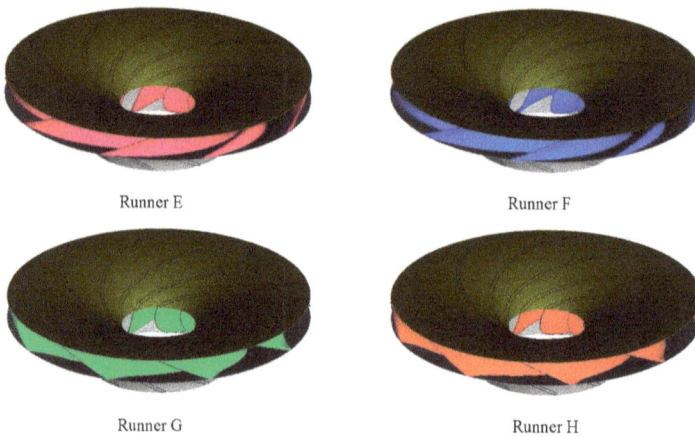

Runner E

Runner F

Runner G

Runner H

Figure 13. Geometry comparison for runners E–H.

Table 6 shows that runners E and F with large positive blade lean angle have a higher efficiency than the preferred runner and the initial runner in both turbine mode and pump mode. Runners G and H with large negative blade lean angles retain a relative high efficiency in pump mode, but the efficiency in turbine mode decreases. For all these four runners, the minimum pressure at the blade surface is lower than that of the initial runner and the preferred runner. It is clearly shown in Figure 11 that low pressure zones on the blade suction surface for runners F and H are larger than those for the initial and preferred runner.

The cavitation characteristics of the runner mainly depend on the blade shape near the runner's LPS. The blade loading for the preferred runner and runners D–H are almost the same, so that the large blade lean on the HPS induces the blade shape change near the runner's LPS, and deteriorates the runner's cavitation characteristics. Therefore, the large blade lean on the HPS is not recommended to be used for the ultra-head reversible pump-turbine runner considering cavitation characteristics.

5. Conclusions

In the present paper, a multiobjective optimization design strategy is briefly presented. The design approach is a combination of 3D inverse design to parameterize the blade geometry, CFD for flow analysis, DoE to reduce the number of calculation times, RSM to correlate the design parameters with the objectives, and MOGA to search the trade-off design. The strategy is used to develop an ultrahigh-head pump-turbine runner. Based on the trade-offs among the optimized targets, a runner is recommended from the optimized runners. Compared to the initial runner, the preferred runner's efficiency under turbine mode is increased by about 0.7% and the pump efficiency by about 0.6%, while the runner's cavitation is greatly promoted.

The hydrodynamic performance characteristics of the pump-turbine correlate strongly with the design parameters. Based on the optimization, the effects of blade loading and blade lean on the runners' geometry and performance are studied. It is suggested that middle-loaded blade loading distribution on the hub, and back-loaded distribution on the shroud—as shown in Figure 7b—are good for the improvement for the runner efficiencies under two operating modes. On the shroud, the blade loading should be reduced near the LPS because the cavitation is most likely to occur in this zone. The large positive blade lean angle on the high pressure side can increase the runner efficiency under turbine mode. However, large blade lean angles may induce drop on the lowest pressure, and deteriorate the cavitation characteristics.

For the large capacity pump-turbine unit, besides the efficiency and cavitation performances, the operation stability for both operating conditions should be guaranteed [3,28]. Under pump mode, instabilities with cavitation in the hump region limit the normal operating range of the unit. Under turbine mode, pressure fluctuations mainly determine smooth operations for the unit. The flow field is converted into a fully separated unsteady state in these cases. Therefore, enlarging the present strategy to consider the unsteady characteristics of the pump-turbine would be valuable.

Acknowledgments: The present work was partially supported by the National Natural Science Foundation of China (Grant No. 51679122).

Author Contributions: Linhai Liu made the computational simulations and prepared the first draft of the paper. Baoshan Zhu planned the study project and revised the paper. Li Bai and Xiaobin Liu made some simulations and checked the calculation data. Yue Zhao and Baoshan Zhu conceived and directed this study.

Conflicts of Interest: The authors declare no conflict of interest.

Nomenclature

b_2	High pressure side width
B	Number of blades
D_{1H}	Hub diameter for low pressure side
D_{1S}	Shroud diameter for low pressure side

D_2	High pressure side diameter
$DRVT$	Blade loading at leading edge
H	Head height
H_{max}	Maximum head height
H_{min}	Minimum head height
H_r	Rated head height
k	Turbulence kinetic energy
m	Percentage meridional distance
M	Torque acting on runner
n	Revolution speed
NC	Fore connection point on blade loading distribution curve
ND	Aft connection point on blade loading distribution curve
P_r	Rated output power
Q	Discharge
r	Radius or radial direction
$SLOPE$	Slope of the middle line on blade loading distribution curve
V	Tangentially velocity
W_{mbl}	Relative velocity on blade surface
x	Input parameter or Cartesian coordinate
y	Optimization targets
β	Blade rake angle or coefficients in polynomial for RSM
η	Unit efficiency
θ	Blade lean angle
ρ	Fluid density
ω	Angular velocity
ν_t	Turbulent kinematic viscosity

Superscripts

	Circumferential average
$+$	Upper side of blade
$-$	Lower side of blade
$'$	Fluctuation

Subscripts

bl	Blade surface
H	Hub
m	Model unit or meridional direction
P	Pump mode
S	Shroud
T	Turbine mode
θ	Tangential component

References

1. Mei, Z.Y. *Power Generation Technology of Pumped Storage Power Station*; China Machine Press: Beijing, China, 2000; p. 47. (In Chinese)
2. Hino, T.; Lejeune, A. Pumped storage hydropower developments. *Compr. Renew. Energy* **2012**, *6*, 405–434.
3. Zuo, Z.G.; Liu, S.H.; Sun, Y.K.; Wu, Y.L. Pressure fluctuations in the vaneless space of high-head pump-turbines—A review. *Renew. Sustain. Energy Rev.* **2015**, *41*, 965–974. [CrossRef]
4. Pugliese, F.; De Paola, F.; Fontana, N.; Giugni, M.; Marini, G. Experimental characterization of two pumps as turbines for hydropower generation. *Renew. Energy* **2016**, *99*, 180–187. [CrossRef]
5. Chen, C.; Zhu, B.; Singh, P.M.; Choi, Y.D. Design of a pump-turbine based on the 3D inverse design method. *KSFM J. Fluid Mach.* **2015**, *18*, 5–10. [CrossRef]
6. Derakhshan, S.; Nourbakhsh, A. Theoretical, numerical and experimental investigation of centrifugal pumps in reverse operation. *Exp. Therm. Fluid Sci.* **2008**, *32*, 1620–1627. [CrossRef]

7.   Frosina, E.; Buono, D.; Senatore, A. A Performance prediction method for pumps as turbines (PAT) using a computational fluid dynamics (CFD) modeling Approach. *Energies* **2017**, *10*, 103. [CrossRef]

8.   Kerschberger, P.; Gehrer, A. Hydraulic development of high specific-speed pump-turbines by means of an inverse design method, numerical flow-simulation (CFD) and model testing. *IOP Conf. Ser. Earth Environ. Sci.* **2010**, *12*, 012039. [CrossRef]

9.   Schleicher, W.C.; Oztekin, A. Hydraulic design and optimization of a modular pump-turbine runner. *Energy Convers. Manag.* **2015**, *93*, 388–398. [CrossRef]

10. Olimstad, G.; Nielson, T.; Borresen, B. Dependency on runner geometry for reversible-pump turbine characteristics in turbine mode of operation. *J. Fluids Eng.* **2012**, *134*, 1428–1435. [CrossRef]

11. Yin, J.L.; Wang, D.Z.; Wei, X.Z.; Wang, L.Q. Hydraulic improvement to eliminate S-shaped curve in pump turbine. *J. Fluids Eng.* **2013**, *135*, 071105. [CrossRef]

12. Zangeneh, M. A compressible three-dimensional design method for radial and mixed flow turbomachinery blades. *Int. J. Numer. Meth. Fluids* **1991**, *13*, 599–624. [CrossRef]

13. Peng, G. A practical combined computation method of mean through-flow for 3D inverse design of hydraulic turbomachinery blades. *J. Fluids Eng.* **2005**, *127*, 1183–1190. [CrossRef]

14. Tan, L.; Cao, S.; Wang, Y.; Zhu, B. Direct and inverse iterative design method for centrifugal pump impellers. *Proc. Inst. Mech. Eng. Part A* **2012**, *226*, 764–775. [CrossRef]

15. Bonaiuti, D.; Zangeneh, M. On the coupling of inverse design and optimization techniques for the multiobjective, multipoint design of turbomachinery blades. *J. Turbomach.* **2009**, *131*, 21014–21029. [CrossRef]

16. Hua, J.; Kong, F.; Liu, P.Y.; Zingg, D. Optimization of long-endurance airfoils. In Proceedings of the AIAA-2003-3500, 21st AIAA Applied Aerodynamics Conference, Orlando, FL, USA, 23–26 June 2003.

17. Secanell, M.; Suleman, A. Numerical evaluation of optimization algorithms for low-Reynolds number aerodynamics shape optimization. *AIAA J.* **2005**, *10*, 2262–2267. [CrossRef]

18. Zitler, E.; Thiele, L. Multiobjective evolution algorithm: A comparative case study and the strength Pareto approach. *IEEE Trans. Evolut. Comput.* **1999**, *3*, 257–271. [CrossRef]

19. Lee, D.S.; Gonzalez, L.F.; Periaus, J.; Srinivas, K. Robust design optimization using multi-objective evolution algorithms. *Comput. Fluids* **2008**, *37*, 565–583. [CrossRef]

20. Oyama, A.; Okabe, Y.; Shimoyama, K.; Fujii, K. Aerodynamics multiobjective design exploration of a flapping airfoil using a Navier-Stokes solver. *J. Aeros. Comp. Inf. Com.* **2009**, *6*, 256–270. [CrossRef]

21. Koziel, S.; Yang, X.S. Computational optimization, methods and algorithm. In *Studies in Computational Intelligence*; Springer: Berlin, Germany, 2011; Volume 356.

22. Wang, X.H.; Zhu, B.S.; Tan, L.; Zhai, J.; Cao, S.L. Development of a pump-turbine runner based on multiobjective optimization. *IOP Conf. Ser. Earth Environ. Sci.* **2014**, *22*. [CrossRef]

23. Zhu, B.; Wang, X.; Tan, L.; Zhou, D.; Zhao, Y.; Cao, S. Optimization design of a reversible pump–turbine runner with high efficiency and stability. *Renew. Energy* **2015**, *81*, 366–376. [CrossRef]

24. Zangeneh, M.; Goto, A.; Harada, H. On the design criteria for suppression of secondary flows in centrifugal and mixed flow impellers. *J. Turbomach.* **1998**, *120*, 723–735. [CrossRef]

25. Ding, H.; Visser, F.C.; Jiang, Y.; Furmanczyk, M. Demonstration and validation of a 3D CFD simulation tool predicting pump performance and cavitation for industrial applications. *J. Fluids Eng.* **2011**, *133*, 011101. [CrossRef]

26. Huang, L.C. The introduction of Yangjiang pumped storage power plant. In *The Collected Works for the Pumped Storage Power Plants Construction in 2005*; China Electric Power Press: Beijing, China, 2005; pp. 414–417. (In Chinese)

27. International Electrotechnical Commission (IEC). *Hydraulic Turbines, Storage Pumps and Pump-Turbines—Model Acceptance Tests*; IEC Standard 60193; International Electrotechnical Commission (IEC): Geneva, Switzerland, 1999.

28. Wang, Z.; Zhu, B.; Wang, X.; Qin, D. Pressure fluctuations in the S-shaped region of a reversible pump-turbine. *Energies* **2017**, *10*, 96. [CrossRef]

energies

MDPI

Article

Analysis of Low Temperature Preheating Effect Based on Battery Temperature-Rise Model

Xiaogang Wu [1,2,*], Zhe Chen [1] and Zhiyang Wang [1]

[1] College of Electrical and Electronics Engineering, Harbin University of Science and Technology, Harbin 150000, China; 18946092365@163.com (Z.C.); m18346559815@163.com (Z.W.)
[2] State Key Laboratory of Automotive Safety and Energy, Tsinghua University, Beijing 100084, China
* Correspondence: xgwu@hrbust.edu.cn

Received: 13 May 2017; Accepted: 27 July 2017; Published: 1 August 2017

Abstract: It is difficult to predict the heating time and power consumption associated with the self-heating process of lithium-ion batteries at low temperatures. A temperature-rise model considering the dynamic changes in battery temperature and state of charge is thus proposed. When this model is combined with the ampere-hour integral method, the quantitative relationship among the discharge rate, heating time, and power consumption, during the constant-current discharge process in an internally self-heating battery, is realized. Results show that the temperature-rise model can accurately reflect actual changes in battery temperature. The results indicate that the discharge rate and the heating time present an exponential decreasing trend that is similar to the discharge rate and the power consumption. When a 2 C discharge rate is selected, the battery temperature can rise from $-10\,°C$ to $5\,°C$ in 280 s. In this scenario, power consumption of the heating process does not exceed 15% of the rated capacity. As the discharge rate gradually reduced, the heating time and power consumption of the heating process increase slowly. When the discharge rate is 1 C, the heating time is more than 1080 s and the power consumption approaches 30% of the rated capacity. The effect of discharge rate on the heating time and power consumption during the heating process is significantly enhanced when it is less than 1 C.

Keywords: lithium ion battery; low temperature preheating; temperature-rise model; heating time; power consumption

---

1. Introduction

Lithium batteries have become the main source of power for electric vehicles because of the advantages they offer, such as reduced pollution, a long life cycle, high energy density, and good power performance [1]. However, the performance of lithium batteries at low temperatures is poor. When the temperature decreases, the ohmic, polarization, and total internal resistance of batteries increase [2]. For example, the ohmic resistance of a charging $LiFePO_4$ battery at $-5\,°C$ is five times that at room temperature [3]. When the temperature is below $-10\,°C$, there is a significant drop in battery capacity, as well as a loss in power [4]. Battery charging is also more difficult than discharging in this environment. In this case, if the battery is forced to charge, lithium deposits and dendrites will appear on its negative electrode, which cause an internal short circuit [5]. So far, it has been difficult to solve the low-temperature performance problem of lithium batteries through the use of innovative materials [6]. Therefore, it is often necessary to heat the battery to a suitable operating temperature before using the battery in low temperature conditions.

At present, methods for heating batteries in low temperature environments are divided primarily into external heating and internal heating. Wang Facheng et al. [7] used a heating wire to heat air at the inlet of an air duct of a battery box, and subsequently heat batteries through air convection. Hyun-Sik Song et al. [8] also achieved battery heating by way of air convection. The above heating method can make the battery temperature rise rapidly to the appropriate temperature and the battery performance is improved significantly at low temperatures. However, this method causes unnecessary energy loss in the heating process, and the energy utilization of techniques that heat by way of air convection is low. Zhang Chengning et al. [9] heat batteries using a wide-line metal film. Comparing with that it is almost not able to discharge prior to heating, the battery can subsequently release 50% of the stored electric energy after heating.

Liu Cunshan et al. [10] established a low-temperature heating model for power batteries, and compared the effect of a positive temperature coefficient (PTC) heater and an electrothermal film heater. The electrothermal film heating mode does not affect the heat dissipation of the battery and has insulating performance at some degree. However, the power batteries used in electric vehicles are composed of a plurality of cells, which are arranged closely together, in series and in parallel [11]. In the external heating mode, battery cells are not uniformly heated, which causes a rapid rise in local temperature. As a result, battery consistency deteriorates and the life of the battery pack is greatly shortened. In more severe cases, the deterioration in battery consistency causes failure of isolated cells, resulting in serious accidents. Compared to the external heating methods, the main advantage of internal heating is the use of heat generated by internal resistance in the charging/discharging process. The internal heating methods are characterized by high energy efficiency and can achieve uniform battery heating. Yan Ji et al. [12] simulated a battery pack equivalent to two groups of cells, which, at a certain frequency, are alternately charged and discharged for battery heating after DC/DC boost, ultimately getting the ideal temperature rise effect. The mutual pulse heating consumes little battery power and is free of convective heat transfer system. However, it appears that the current used in this process is too large. In addition, the charging voltage of the battery in the heating process may reach 4.5 V, which is significantly higher than the charging cut-off voltage and increases the possibility of the formation of lithium dendrites. Zhang Jianbo et al. [5] established a frequency domain model for a lithium-ion battery, which had a rated capacity of 3.1 A·h, and proposed the use of sinusoidally alternating currents for internal heating. The battery can be heated from $-20\,^\circ$C to 5 $^\circ$C within 15 min and the temperature distribution remains essentially uniform. However, the heating process is accompanied by large transient voltages. The maximum battery voltage recorded experimentally is 4.5 V. If an appropriate AC amplitude and frequency cannot be selected in practical applications, the battery may continue to be in a state of over-voltage, causing some damage. Zhao Xiaowei et al. [13] proposed the use of a large current pulse for heating a 3.2 V, 12 A·h lithium-iron phosphate battery. The charge and discharge cut-off voltages were 2.1 V and 3.6 V respectively. The heating process comprised a total of 18 charge and discharge cycles. In the final realization, the battery temperature rises from $-10\,^\circ$C to 3 $^\circ$C. Ruan Haijun et al. [14] heated batteries with a high-frequency alternating current, using a constant polarization voltage as a boundary condition. Ultimately, the battery temperature can be raised from $-15\,^\circ$C to 5.6 $^\circ$C in 338 s. The constant polarization voltage is managed for battery heating to achieve a good tradeoff between short heating time and less damage to battery lifetime based on an electro-thermal coupled model. However, as the study only proved that there was no significant capacity decay in the battery after 30 repeated internal heating tests, the overall health of the battery, if the test is repeated more than 30 times, cannot be ascertained. Although pulsed heating can effectively heat batteries, alleviating the impact of low temperatures, larger charge pulse amplitudes result in stronger polarization of the anode surface, leading to the formation of lithium dendrites [15].

The main reason for the failure of lithium batteries is the generation of lithium dendrites during the charging process in low-temperature environments [16]. The lithium metal precipitates on the graphite anode surface at low temperatures or during charging at a high rate, and further reacts with the electrolyte. As a result, both available electrolyte and lithium ions are lost, and the battery volume changes, leading to poor contact between active substances and the current collector [17]. The embedding of both electrolyte and lithium ions accelerates the peeling of graphite particles. The corrosion of both the collector and the adhesive reduces battery capacity [18], eventually causing permanent damage to the battery. Though the discharging capacity of lithium batteries decreases and the discharging platform voltage drops, discharging in low temperature conditions does not cause permanent damage to the battery.

On the basis of the foregoing, this study develops a method to internally preheat lithium-ion batteries at low temperatures by way of constant-current discharging. This indicates that the temperature generated by internal resistance during battery discharging is used to heat the battery in a low temperature environment. Besides, it is difficult to predict the heating time and power consumption associated with the self-heating process of lithium-ion batteries at low temperatures. A temperature-rise model considering the dynamic changes in both battery temperature and state of charge (SOC) is thus proposed. When this model is combined with the ampere-hour integral method, the quantitative relationship among the discharge rate, heating time, and power consumption, during the progress of constant-current discharging for internally self-heating battery, is realized. Further, the problem of predicting the heating time and power consumption of the self-heating at low temperature is solved in this paper.

2. The Temperature-Rise Model

The Thevenin model is used to analyze the discharging process. As shown in Figure 1, R_r represents the ohmic resistance, U_r is the voltage on R_r, C_p and R_p represents the polarization capacitance and polarization resistance respectively, U_p is the voltage on C_p and R_p, U_{OCV} is the open circuit voltage, E is the terminal voltage, $I$is the discharging current. In this paper, R_{total} is equivalent to the combination of R_r, C_p and R_p, which is annotated as R in the temperature-rise model.

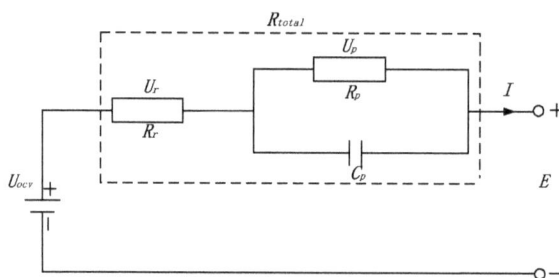

Figure 1. Thevenin model

Heat generated by a battery can be divided into irreversible heat and reversible heat. The irreversible heat includes Joule heat and concentration polarization heat. The reversible heat, also known as reaction heat, refers to energy that is released or absorbed in the electrochemical reaction to maintain the energy balance of the reaction. Referring to [19], the simplified heat generation equation used in this paper can be expressed as (1):

$$Q_t = Q_J + Q_r = I(E - U_{OCV}) + IT\frac{\partial U_{OCV}}{\partial T} \tag{1}$$

$$Q_J = I(E - U_{OCV}) = I^2 R, \tag{2}$$

where, I is the operating current of the battery (positive for charge, negative for discharge), E is the battery voltage, U_{OCV} is the open circuit voltage, Q_t is the total heat generation power. Q_J is the irreversible heat generation power, which represents the sum of both the heat generated by ohmic resistance when current flows and the heat generated by concentration difference through material transfer in the battery. Q_r is the reversible entropic heat or reaction heat, which depends on the direction of current and the sign of the entropy coefficient. The entropy potential is greatly influenced by the state of charge (SOC) and varies with different chemical compositions [20]. The difference between the battery terminal voltage and the open circuit voltage results from the voltage generated by internal resistance when current flows [21]. Therefore, the irreversible heat can be expressed as Equation (2), where R is the equivalent internal resistance of the battery.

Battery temperature is influenced by heat generation, heat conduction, and thermal diffusion [22]. In addition to internal heat production, the battery also distributes heat to the exterior when it works at a low temperatures. There are two main approaches for heat loss: convection and heat radiation. Thermal radiation is very small compared to thermal convection and is therefore ignored [23]. The heat dissipation can be expressed by (3):

$$Q_{dis} = -hA(T - T_\infty),\tag{3}$$

where h is the equivalent heat transfer coefficient, A is the surface area of the battery, T is the battery temperature, and T_∞ is the ambient temperature. Therefore, the heat balance equation can be obtained as the following equation:

$$mc\frac{dT}{dt} = Q_J + Q_r + Q_{dis} = I^2R + IT\frac{\partial U_{OCV}}{\partial T} - hA(T - T_\infty),\tag{4}$$

where m is the mass of battery and c is the specific heat capacity. From Equation (4) we can see that the total heat generated by the battery is influenced by current, resistance, entropy potential, the equivalent heat transfer coefficient and battery temperature. One can yield that a greater current and resistance lead to greater heat generation. Conversely, a greater equivalent transfer coefficient and battery temperature results in more heat dissipation. As a result, the total heat generated is reduced. The battery temperature-rise model developed in this paper will take into account changes in the resistance and entropy coefficient during the process of battery heating so as to guarantee accuracy.

According to Equation (4), we can get the linear differential equation relating to battery temperature in Equation (5).

$$\frac{dT(t)}{dt} = (\frac{I\frac{\partial V_{OCV}}{\partial T}}{mc} - \frac{hA}{mc})T(t) + \frac{I^2R}{mc} + \frac{hAT_\infty}{mc}.\tag{5}$$

Equation (5) can be rewritten in discrete-time. The relevant expression, shown in Equation (6), is deduced, using the Laplace transform as,

$$sT(s) - T(t_0) = (\frac{I\frac{\partial V_{OCV}}{\partial T}}{mc} - \frac{hA}{mc})T(s) + (\frac{I^2R}{mc} + \frac{hAT_\infty}{mc})\frac{1}{s},\tag{6}$$

where, t_0 is the initial time and t is the current time. Under periodic sampling conditions, $t_0 = kT_0$, $t = (k+1)T_0$, and $k = 0,1,2,3...$, Equation (6) can be rewritten as:

$$sT(s) - T(kT_0) = (\frac{I\frac{\partial V_{OCV}}{\partial T}}{mc} - \frac{hA}{mc})T(s) + (\frac{I^2R}{mc} + \frac{hAT_\infty}{mc})\frac{1}{s}\tag{7}$$

Upon further rearrangement, we can get Equation (8) as,

$$T(s) = \frac{T(kT_0)}{s + \frac{hA-I\frac{\partial V_{OCV}}{\partial T}}{mc}} + \frac{1}{s(s + \frac{hA-I\frac{\partial V_{OCV}}{\partial T}}{mc})}\frac{I^2R + hAT_\infty}{mc}\tag{8}$$

Equation (9) is obtained from Equation (8) by the inverse Laplace transform

$$T((k+1)T_0) = e^{-\frac{hA-I\frac{\partial V_{OCV}}{\partial T}}{mc}t}T(kT_0) + \frac{mc}{hA - I\frac{\partial V_{OCV}}{\partial T}}(1 - e^{-\frac{hA-I\frac{\partial V_{OCV}}{\partial T}}{mc}t})\frac{I^2R + hAT_\infty}{mc}.\tag{9}$$

3. Model Validation

3.1. Model Parameter Acquisition

The battery tested in this study was a commercial 18650 lithium-ion battery, which has a rated capacity of 2.6 A·h. The cathode of the battery is $Li_xNiCoAlO_2$, and the anode is graphite. The specifications are shown in Table 1.

Table 1. Battery parameters.

Parameters	Symbol	Value
Mass	m	45 g
Surface area	A	4.287×10^{-3} m^2
Capacity	Q	2.6 A·h
Voltage	V_{rate}	3.63 V
Upper cut-off voltage	V_{up}	4.2 V
Lower cut-off voltage	V_{low}	2.75 V

The experimental set-up is shown in Figure 2. The temperature sensor is attached to the battery, and the side surface of the battery tested in this experiment is covered by an insulating film. The battery temperature data measured by the temperature sensor is sent to the computer via the battery temperature measuring device. And the computer controls the battery to charge and discharge via the Arbin battery tester. Detailed parameters of the battery tester and temperature chamber are shown in Table 2.

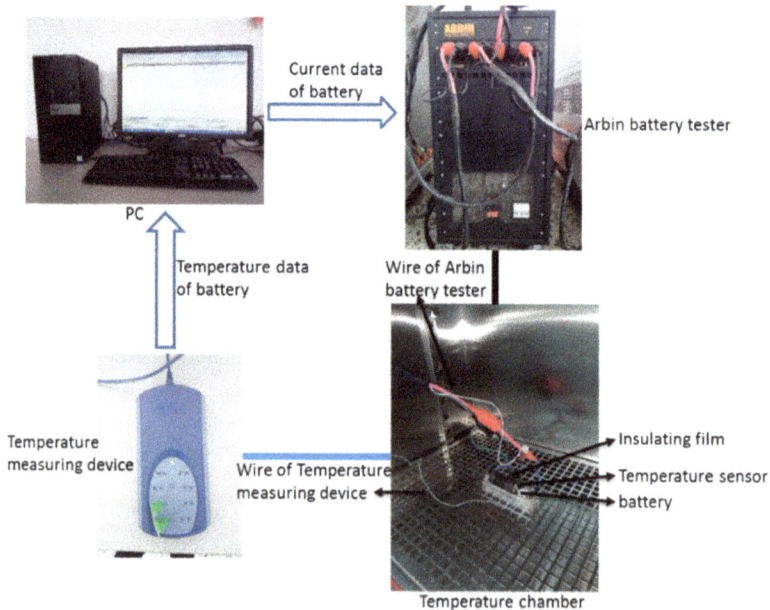

Figure 2. Experimental set-up.

Table 2. Equipment parameters.

Arbin Battery Tester	Voltage range: 0 V–5 V Current range: 0 A–50 A Voltage accuracy: full-range ± 0.05% FSR Current accuracy: full-range ± 0.01% FSR Number of channels: 4
Temperature Chamber	Temperature range: −50 °C~150 °C Temperature error:<0.5 °C Chamber volume: 0.5 m × 0.5 m × 0.6 m

The batteries are tested by the hybrid pulse power characteristic (HPPC) rule [24] to obtain the relation between internal resistance and SOC at different temperatures. The schematic of the HPPC test is shown in Figure 3, the battery is excited by a charging pulse and a discharging pulse at a certain SOC, and the pulse width is set to 10 s. After that, the battery is discharged to the next SOC point. The result of HPPC test with 10% SOC intervals at 25 °C is shown in Figure 4. The equations of charging ohmic resistance R_r^c, charging total resistance R_{total}^c, discharging ohmic resistance R_r^d and discharging total resistance R_{total}^d are as follows:

$$R_r^c = \frac{U_2 - U_1}{I_c} \tag{10}$$

$$R_r^d = \frac{U_5 - U_4}{I_d} \tag{11}$$

$$R_{total}^c = \frac{U_3 - U_1}{I_c} \tag{12}$$

$$R_{total}^d = \frac{U_6 - U_4}{I_d} \tag{13}$$

Above all, U_1, U_2, U_3, U_4, U_5, U_6, are the terminal voltage at point 1, 2, 3, 4, 5, 6. In addition, I_c, I_d are charging current and discharging current respectively.

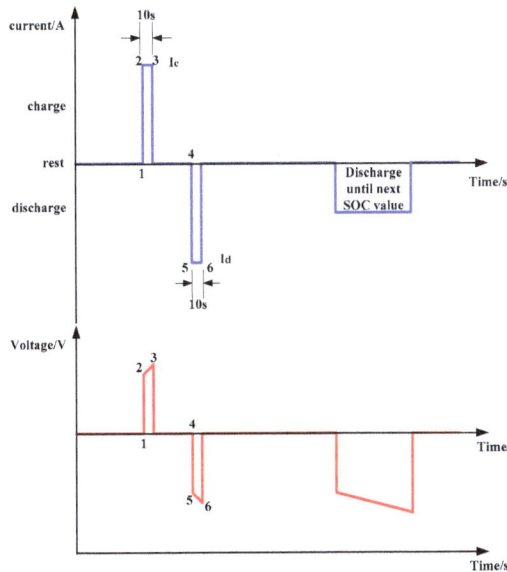

Figure 3. The schematic of the HPPC test.

25

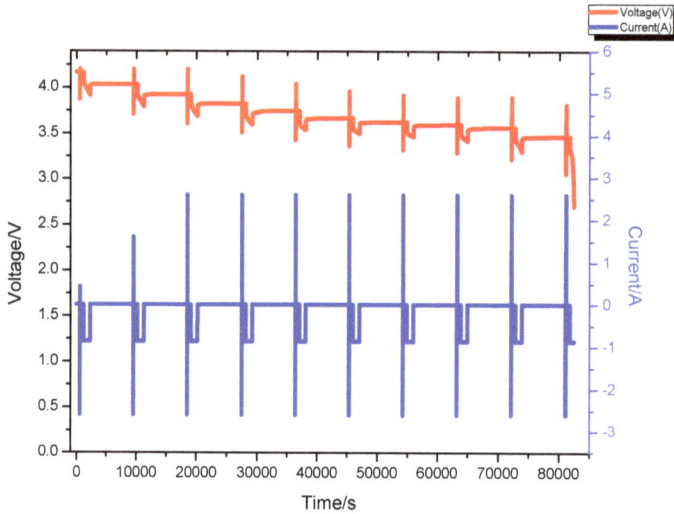

Figure 4. The result of HPPC test at 25 °C.

The results of HPPC tests at different temperatures are shown in Figure 5. The battery is excited by a pulse of 2.6 A under different conditions, and the voltage of the battery will exceed the range of cut-off voltage at both low SOC and low temperature. As a result, the data of resistance at both low SOC and low temperature are missing. Besides, the battery is excited by the mode of constant current—constant voltage at high SOC, preventing the voltage of battery from exceeding the upper cut-off voltage. In this paper, the SOC of the battery is defined as the ratio of residual capacity to the rated capacity. According to Figure 5, the internal resistance gradually increased with decreasing temperature. The battery resistance is effectively stable when the SOC is between 50% and 90%. The resistance increases when the SOC is less than 50% or more than 90%.

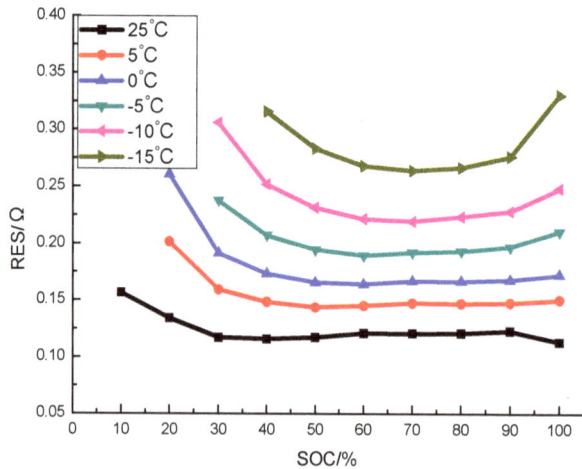

Figure 5. Battery internal resistance curves at different temperatures.

The relationship between the open circuit voltage (OCV) and SOC is significant for describing the basic performance of a battery. SOC-OCV curves vary with different types of batteries [25]. The SOC-OCV curves obtained by the battery test system are shown in Figure 6.

The open circuit voltage of the battery is mainly affected by the SOC and the temperature. The open circuit voltage increases gradually with increasing SOC, and decreases gradually with decreasing temperature. The effect of SOC on the open circuit voltage is significantly greater than that of temperature. The fluctuation of open circuit voltage caused by the variation of 5 °C in temperature does not exceed 5 mV.

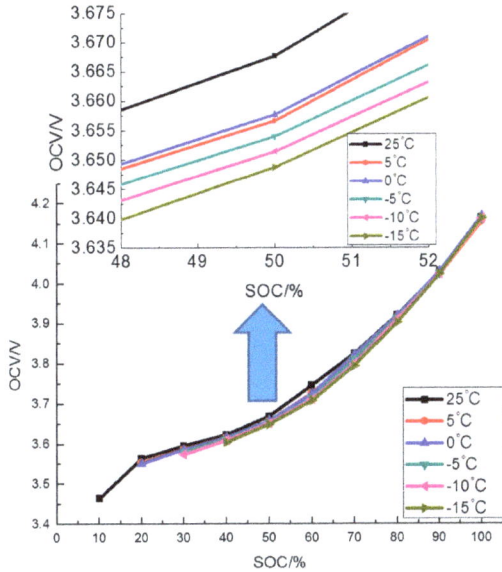

Figure 6. Curves of battery open circuit voltage at different temperatures.

The entropy coefficient is an important parameter for estimating the reaction heat. Firstly, the open circuit voltages of the battery should be measured at different temperatures and SOC points. Through the analysis of the data measured, different open circuit voltage corresponding to different temperatures is obtained at a certain SOC. Referring to [26], a linear function of the temperature and the OCV at a specified SOC is fitted by the least square method. The slope of the derived function is used as the entropy coefficient at the defined SOC. The entropy coefficient at 50% SOC is shown in Figure 7. The above fitting methods were implemented at different SOC points. An entropy coefficient curve with 10% SOC intervals was obtained, as shown in Figure 8. The entropy coefficient is more than zero when the SOC is within a 20–90% range but less than zero when the SOC exceeds 90%. The value of the entropy coefficient is small, which is always in the range of −0.4 to 1.6 mV/°C. According to the reaction heat equation, which is $IT\frac{\partial U_{OCV}}{\partial T}$, the small value of the entropy coefficient implies that the contribution of reaction heat is limited. This also shows that most of the heat is generated by Joule heat, and reaction heat contributes less to the temperature rise of the battery.

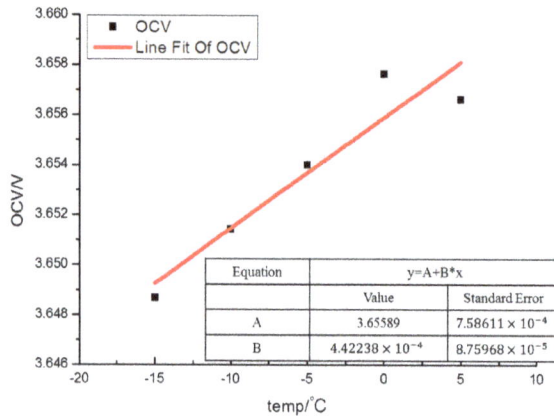

Figure 7. Entropy coefficient for SOC = 50%.

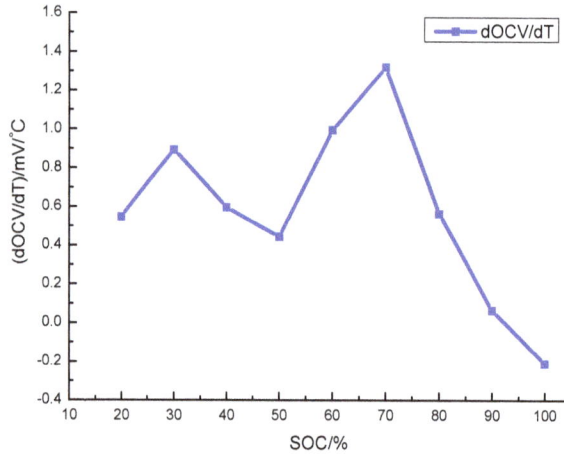

Figure 8. Entropy coefficient curve for varying SOCs.

The heat dissipation in the heating process can be expressed by the equivalent heat transfer coefficient. The equivalent heat transfer coefficient is an important parameter in the energy conservation model, which can affect the accuracy of the battery temperature raise model. In practical application, hundreds of individual batteries are connected in series to compose a battery pack, which are put in the battery box of an electric vehicle. The battery box has an insulating effect on the batteries. In order to simulate the actual environment of the battery box in an electric vehicle and reduce the heat dissipation of the battery at low temperatures, the side surface of the battery tested in this experiment is covered by an insulating film which is a thin sponge with stickiness [27]. As a result, the equivalent heat transfer coefficient will be smaller due to the insulating film. The equivalent heat transfer coefficient of the battery is obtained by the temperature gradient calculated during the battery cooling process. The energy conservation equation is shown in Equation (14).

$$mc\frac{dT}{dt} = -hA(T - T_\infty), \tag{14}$$

where $m = 45$ g, $c = 1.72$ J/g·K [5], and $T_\infty = -10\,°C$. If h is a constant value, a solution to Equation (14) is:

$$\ln(T - T_\infty) = -\frac{hA}{mC_p}t + con. \tag{15}$$

Equation (15) shows that there is a linear function between $\ln(T - T_\infty)$ and time, and the equivalent heat transfer coefficient can be determined from the slope of the curve of $\ln(T - T_\infty)$ with t [28]. The battery temperature and $\ln(T - T_\infty)$ when the battery cools down are shown in Figure 9. Figure 9b shows the linear relationship with time, resulting in an equivalent heat transfer coefficient of 5.035 W/m²·K.

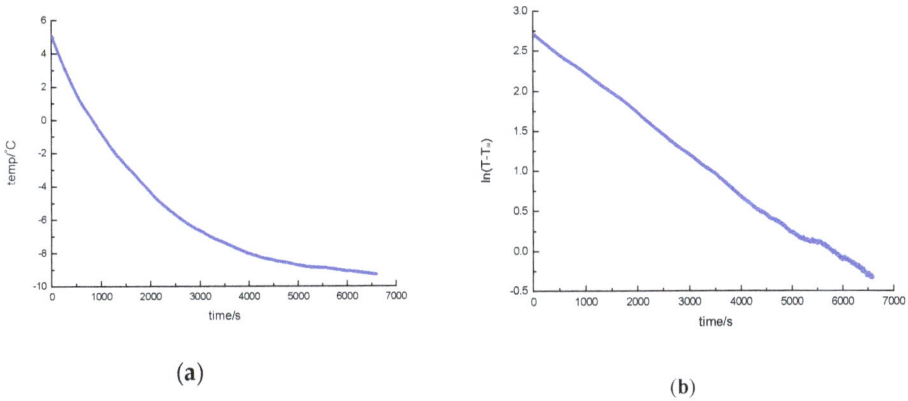

(a)

(b)

Figure 9. Curves of (**a**) battery temperature; (**b**) $\ln(T - T_\infty)$, when the battery cools down.

3.2. Temperature-Rise Model Validation

The fluctuation of both battery temperature and SOC is relatively large in the process of battery discharge for self-heating. This paper establishes a temperature-rise model which takes into account the dynamic characteristic of the battery temperature and SOC. Tests analyzing the discharge process for self-heating at low temperature were carried out. The discharge rates selected are 1 C, 1.5 C, and 2 C. The ambient temperature is $-10\,°C$, the target temperature is 5 °C [5], and the initial SOC of the tested battery is 80%. The experimental results are compared with the simulation results, and the accuracy of the temperature rise model is verified by the error between the actual temperature and the simulation.

Plots of the predicted temperature, actual temperature and the error between these values are shown in Figure 10. The predicted temperature obtained from the temperature-rise model is essentially identical to the actual temperature of the battery. The maximum error between the predicted temperature and the actual temperature does not exceed 1 °C during the process of self-heating, which is the same as [26]. Hence, it can be demonstrated that the temperature-rise model established for ICR18650 batteries in this paper is highly accurate.

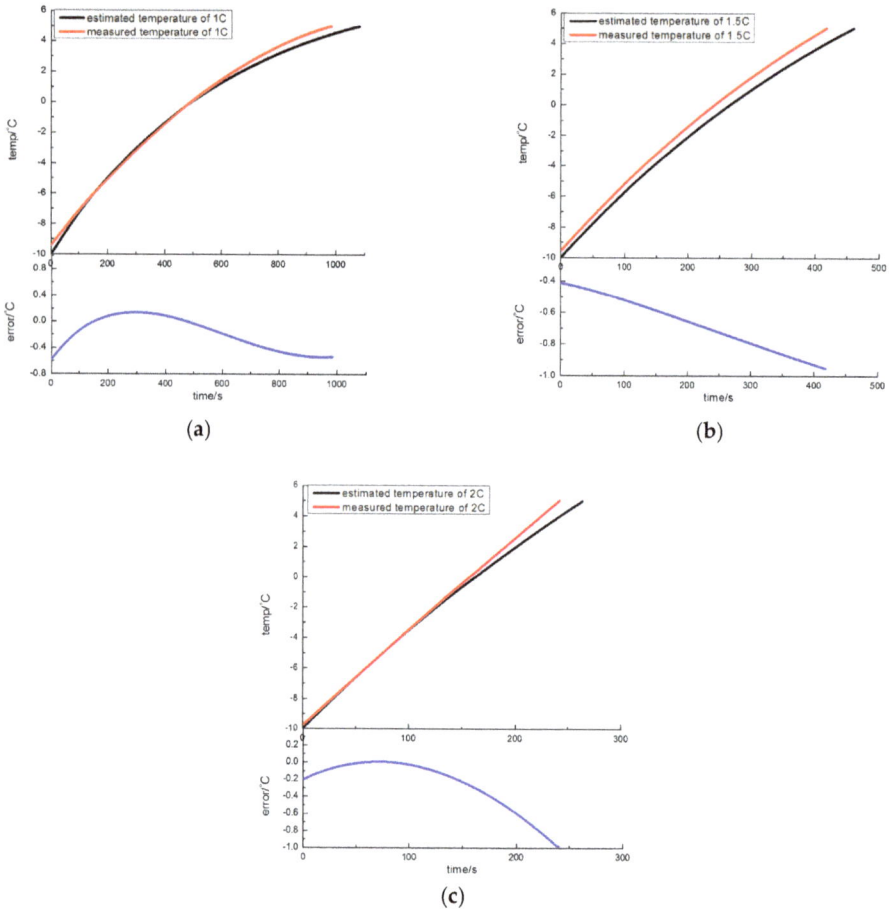

Figure 10. Comparing the estimated temperature with the measured temperature at (**a**) 1 C; (**b**) 1.5 C; (**c**) 2 C discharge rates.

4. Calculation Results and Analysis

According to the temperature-rise model developed in this paper, the time required for heating the battery from the ambient temperature to the target temperature at different discharge rates is obtained, as shown in Figure 11. The curve is fitted by the least squares method to obtain the function of the battery discharge rate and the heating time, which is shown in Equation (12), where x is the discharge rate and y is the heating time in seconds.

$$y = 3.74227 \times 10^{-9} e^{-\frac{x}{0.0484}} + 5.35283 \times 10^{-8} e^{-\frac{x}{0.06003}} + 8293.17524 e^{-\frac{x}{0.43995}} + 182.07697. \quad (16)$$

As can be seen from Figure 11, the battery temperature can be raised from $-10\,^\circ$C to $5\,^\circ$C in 280 s when the discharge rate is 2 C. When the discharge rate decreases, the heating time gradually increases in response. The heating time is 1080 s when the discharge rate is 1 C. The effect of current discharge on the heating time is significantly enhanced when the discharge rate is less than 1 C. As the discharge rate continues to decrease, the heating time rapidly increases. The heating time is more than 2640 s when the discharge rate is 0.8 C, which is far longer than the reasonable heating time in actual applications.

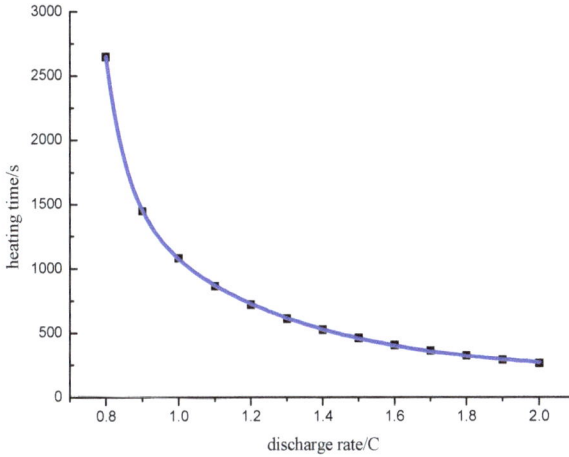

Figure 11. Time required for heating the battery from ambient temperature to the target temperature at different discharge current rates.

Further, the power consumption of the self-heating process can be calculated by combining the battery temperature-rise model with the ampere-hour integral method [29]. The ampere-hour integral equation is shown as Equation (17).

$$SOC_t = SOC_0 - \int_0^t \frac{I}{Q} dt, \tag{17}$$

where SOC_0 is the initial SOC of the battery, SOC_t is the SOC at time t, I is the discharge current of the battery, and Q is the rated capacity of the battery. Additionally, $\int_0^t \frac{I}{Q} dt$ is defined as the power consumption in this paper. The power consumption of the battery during heating at different discharge current rates is shown in Figure 12. The curve is fitted utilizing the least squares method to further obtain the function of the battery discharge rate and power consumption shown in Equation (18), where x is the discharge rate and z is the total variation of SOC during the heating process, i.e., the power consumption.

$$z = 2.54425 \times 10^{-6} e^{-\frac{x}{0.04527}} + 86035.57986 e^{-\frac{x}{0.06101}} + 0.93799 e^{-\frac{x}{0.59824}} + 0.11457. \tag{18}$$

According to Figure 12, the power consumption of the battery at a 2 C discharge rate is less than 15% of the rated capacity. As the discharge rate gradually reduced, the power consumption increases slowly. The power consumption of the heating process is 30% of the rated capacity when the discharge rate is 1 C. The effect of discharge rate on power consumption is significantly enhanced when it is less than 1 C. When the discharge rate is 0.8 C, the power consumption of the heating process is 60% of the rated capacity, which is twice the value at 1 C. Therefore, the discharge rate should be selected in the range of 1 C–2 C in applying the constant-current discharge method to heating a battery at low temperature.

Energies **2017**, *10*, 1121

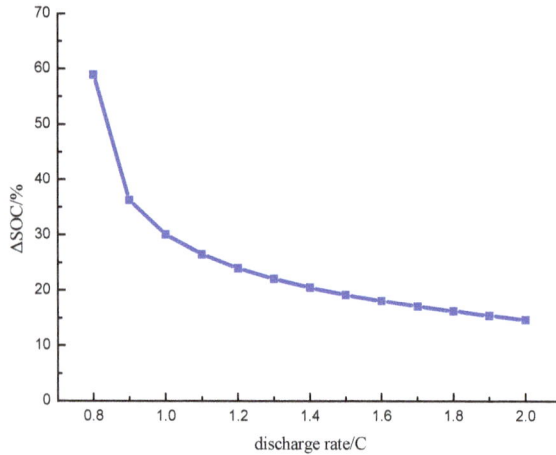

Figure 12. Power consumption of the battery heated from ambient temperature to the target temperature at different discharge current rates.

5. Conclusions

A temperature-rise model considering the dynamic fluctuation in battery temperature and SOC is proposed, and it is possible to predict the battery temperature during the progress of battery self-heating at low temperature. Tests in which the battery was heated from −10 °C to 5 °C were conducted at different discharge rates. The results show that the temperature-rise model can accurately reflect actual variation in battery temperature. The maximum error between the predicted temperature and actual temperature is less than 1 °C during the process of battery self-heating.

When the temperature-rise model developed in this paper is combined with the ampere-hour integral method, the quantitative relationship among the discharge rate, the heating time, and the power consumption during the self-heating process is realized. The difficulty in predicting the heating time and power consumption during the self-heating process is thus addressed. The results indicate that the discharge rate and the heating time present an exponential decreasing trend and it is similar with the discharge rate and the power consumption. When a 2 C discharge rate is selected for constant-current discharging to the internal heating battery, the battery temperature can rise from −10 °C to 5 °C in 280 s. In this case, the power consumption of the self-heating process does not exceed 15% of the rated capacity. As the discharge rate gradually reduced, the heating time and power consumption of the heating process increased slowly. When the discharge rate was 1 C, the heating time exceeded 1080 s, and the power consumption reached 30% of the rated capacity. The effect of discharge rate on the heating time and power consumption during the self-heating process is significantly enhanced when the discharge rate is less than 1 C. When the discharge rate is 0.8 C, the power consumption of self-heating process is 2.45 times that at 1 C, and the heating time is twice that at 1 C. Therefore, the discharge current rate should be selected in the range of 1 C–2 C in applying the constant-current discharge method to battery self-heating. The method of self-heating is suitable for heating the lithium-ion battery which is fully charged at low temperature before the normal operation.

Acknowledgments: This work was supported by the State Key Laboratory of Automotive Safety and Energy under Project No. KF16062 and Science Funds for the Young Innovative Talents of HUST, No. 201503.

Author Contributions: Xiaogang Wu, Zhe Chen and Zhiyang Wang designed the simulations and experiments; Xiaogang Wu analyzed the data; Zhe Chen wrote the paper; Zhiyang Wang polished the paper.

Conflicts of Interest: The authors declare no conflict of interest.

References

1. Sun, F.; Xiong, R.; He, H. A systematic state-of-charge estimation framework for multi-cell battery pack in electric vehicles using bias correction technique. *Appl. Energy* **2016**, *162*, 1399–1409. [CrossRef]

2. Li, Z.; Han, X.; Lu, L.; Ouyang, M. Temperature Characteristics of Power LiFePO$_4$ Batteries. *Chin. J. Mech. Eng.* **2011**, 115–120. [CrossRef]

3. Lin, C.; Li, B.; Chang, G.; Xu, S. Experimental study on internal resistance of LiFePO$_4$ batteries under different ambient temperatures. *Chin. J. Power Sources* **2015**, *39*, 22–25.

4. Cho, H.; Choi, W.; Go, J.; Bae, S.; Shin, H. A study on time-dependent low temperature power performance of a lithium-ion battery. *J. Power Sources* **2012**, *198*, 273–280. [CrossRef]

5. Zhang, J.; Ge, H.; Li, Z.; Ding, Z. Internal heating of lithium-ion batteries using alternating current based on the heat generation model in frequency domain. *J. Power Sources* **2015**, *273*, 1030–1037. [CrossRef]

6. Lei, Z.; Zhang, C.; Li, J.; Fan, G.; Lin, Z. Preheating method of lithium-ion batteries in an electric vehicle. *J. Mod. Power Syst. Clean Energy* **2015**, *3*, 289–296. [CrossRef]

7. Wang, F.C.; Zhang, J.Z.; Wang, L.F. Design of electric air-heated box for batteries in electric vehicles. *Chin. J. Power Sources* **2013**, *37*, 1184–1187.

8. Song, H.; Jeong, J.; Lee, B.; Shin, D.H.; Kim, B.; Kim, T.; Heo, H. Experimental Study on the Effects of Pre-Heating a Battery in a Low-Temperature Environment. In Proceedings of the 2012 IEEE Vehicle Power and Propulsion Conference (VPPC), Seoul, Korea, 9–12 October 2012; pp. 1198–1201.

9. Zhang, C.N.; Lei, Z.G.; Dong, Y.G. Method for Heating Low-Temperature Lithium Battery in Electric Vehicle. *Trans. Beijing Inst. Technol.* **2012**, *32*, 921–925.

10. Liu, C.; Zhang, H. Research on heating method at low temperature of electric vehicle battery. *Chin. J. Power Sources* **2015**, *39*, 1645–1647, 1701.

11. Xiong, R.; Sun, F.; Chen, Z.; He, H. A data-driven multi-scale extended Kalman filtering based parameter and state estimation approach of lithium-ion olymer battery in electric vehicles. *Appl. Energy* **2014**, *113*, 463–476. [CrossRef]

12. Ji, Y.; Wang, C.Y. Heating strategies for Li-ion batteries operated from subzero temperatures. *Electrochim. Acta* **2013**, *107*, 664–674. [CrossRef]

13. Zhao, X.W.; Zhang, G.Y.; Yang, L.; Qiang, J.X.; Chen, Z.Q. A new charging mode of Li-ion batteries with LiFePO$_4$/C composites under low temperature. *J. Therm. Anal. Calorim.* **2011**, *104*, 561–567. [CrossRef]

14. Ruan, H.; Jiang, J.; Sun, B.; Zhang, W.; Gao, W.; Wang, L.Y.; Ma, Z. A rapid low-temperature internal heating strategy with optimal frequency based on constant polarization voltage for lithium-ion batteries. *Appl. Energy* **2016**, *177*, 771–782. [CrossRef]

15. Fan, J.; Tan, S. Studies on Charging Lithium-Ion Cells at Low Temperatures. *J. Electrochem. Soc.* **2006**, *153*, A1081–A1092. [CrossRef]

16. Ouyang, M.; Chu, Z.; Lu, L.; Li, J.; Han, X.; Feng, X.; Liu, G. Low temperature aging mechanism identification and lithium deposition in a large format lithium iron phosphate battery for different charge profiles. *J. Power Sources* **2015**, *286*, 309–320. [CrossRef]

17. Verma, P.; Maire, P.; Novák, P. A review of the features and analyses of the solid electrolyte interphase in Li-ion batteries. *Electrochim. Acta* **2010**, *55*, 6332–6341. [CrossRef]

18. Vetter, J.; Novák, P.; Wagner, M.R.; Veit, C.; Möller, K.C.; Besenhard, J.O.; Winter, M.; Wohlfahrt-Mehrens, M.; Vogler, C.; Hammouche, A. Ageing mechanisms in lithium-ion batteries. *J. Power Sources* **2005**, *147*, 269–281. [CrossRef]

19. Liu, G.; Ouyang, M.; Lu, L.; Li, J.; Han, X. Analysis of the heat generation of lithium-ion battery during charging and discharging considering different influencing factors. *J. Therm. Anal. Calorim.* **2014**, *116*, 1001–1010. [CrossRef]

20. Viswanathan, V.V.; Choi, D.; Wang, D.; Xu, W.; Towne, S.; Williford, R.E.; Zhang, J.; Liu, J.; Yang, Z. Effect of entropy change of lithium intercalation in cathodes and anodes on Li-ion battery thermal management. *J. Power Sources* **2010**, *195*, 3720–3729. [CrossRef]

21. Zhang, Y.; Xiong, R.; He, H.; Shen, W. Lithium-Ion Battery Pack State of Charge and State of Energy Estimation Algorithms Using a Hardware-in-the-Loop Validation. *IEEE Trans. Power Electr.* **2017**, *32*, 4421–4431. [CrossRef]

22. Li, X.; Zhang, J.; Wei, Y.; Liu, Y.; Ding, S. Analysis of Specific Heat of Lithium-ion Power Battery. *J. Mater. Sci. Eng.* **2014**, *32*, 908–912.

23. Ruan, H.; Jiang, J.; Sun, B.; Wu, N.; Shi, W.; Zhang, Y. Stepwise Segmented Charging Technique for Lithium-ion Battery to Induce Thermal Management by Low-Temperature Internal Heating. In Proceedings of the 2014 IEEE Conference and Expo Transportation Electrification Asia-Pacific (ITEC Asia-Pacific), Beijing, China, 31 August–3 September 2014.

24. Hunt, G. Freedom CAR Battery Test Manual for Power-Assist Hybrid Electric Vehicles. In *DOE/ID-11069*; Idaho National Engineering & Environmental Laboratory: Idaho Falls, ID, USA, 2003.

25. Wu, X.G.; Mei, Z.Y.; Hu, C.; Zhu, C.B.; Sun, J.L. Temperature Percormance Comparative Analysis of Different Power Batteries. In Proceedings of the 2016 IEEE Vehicle Power and Propulsion Conference (VPPC), Hangzhou, China, 17–20 October 2016.

26. Sun, J.; Wei, G.; Pei, L.; Lu, R.; Song, K.; Wu, C.; Zhu, C. Online Internal Temperature Estimation for Lithium-Ion Batteries Based on Kalman Filter. *Energies* **2015**, *8*, 4400–4415. [CrossRef]

27. Zhang, J.; Wu, B.; Li, Z.; Huang, J. Simultaneous estimation of thermal parameters for large-format laminated lithium-ion batteries. *J. Power Sources* **2014**, *259*, 106–116. [CrossRef]

28. Ge, H.; Huang, J.; Zhang, J.; Li, Z. Temperature-Adaptive Alternating Current Preheating of Lithium-Ion Batteries with Lithium Deposition Prevention. *J. Electrochem. Soc.* **2016**, *163*, 290–299. [CrossRef]

29. Chen, C.; Xiong, R.; Shen, W. A lithium-ion battery-in-the-loop approach to test and validate multi-scale dual H infinity filters for state of charge and capacity estimation. *IEEE Trans. Power Electr.* **2017**, *99*, 1. [CrossRef]

energies

MDPI

Article

Thermodynamic Analysis of Three Compressed Air Energy Storage Systems: Conventional, Adiabatic, and Hydrogen-Fueled

Hossein Safaei and Michael J. Aziz *

Harvard John A. Paulson School of Engineering and Applied Sciences, Pierce Hall, 29 Oxford Street, Cambridge, MA 02138, USA; Hossein.Safaei@gmail.com
* Correspondence: maziz@harvard.edu; Tel.: +1-617-495-9884

Received: 10 May 2017; Accepted: 27 June 2017; Published: 18 July 2017

Abstract: We present analyses of three families of compressed air energy storage (CAES) systems: conventional CAES, in which the heat released during air compression is not stored and natural gas is combusted to provide heat during discharge; adiabatic CAES, in which the compression heat is stored; and CAES in which the compression heat is used to assist water electrolysis for hydrogen storage. The latter two methods involve no fossil fuel combustion. We modeled both a low-temperature and a high-temperature electrolysis process for hydrogen production. Adiabatic CAES (A-CAES) with physical storage of heat is the most efficient option with an exergy efficiency of 69.5% for energy storage. The exergy efficiency of the conventional CAES system is estimated to be 54.3%. Both high-temperature and low-temperature electrolysis CAES systems result in similar exergy efficiencies (35.6% and 34.2%), partly due to low efficiency of the electrolyzer cell. CAES with high-temperature electrolysis has the highest energy storage density (7.9 kWh per m^3 of air storage volume), followed by A-CAES (5.2 kWh/m^3). Conventional CAES and CAES with low-temperature electrolysis have similar energy densities of 3.1 kWh/m^3.

Keywords: compressed air energy storage (CAES); adiabatic CAES; high temperature electrolysis; hydrogen storage; thermodynamics

1. Introduction

Large penetrations of wind and solar energies challenge the reliability of the electricity grid, due to their intermittency and uncertainty. Storage technologies are being developed to tackle this challenge. Compressed air energy storage (CAES) is a relatively mature technology with currently more attractive economics compared to other bulk energy storage systems capable of delivering tens of megawatts over several hours, such as pumped hydroelectric [1–3]. CAES stores electrical energy as the exergy of compressed air. Figure 1 is a simplified schematic of a CAES plant. Electricity is supplied by the grid to run the air compressors and charge the storage system. Waste heat is released during the compression phase. Air is stored for later use—often in an underground cavern. During the discharge phase, compressed air is combusted with a fuel, and expanded in a turbine (expander) to regenerate electricity. Currently, there are two commercial CAES plants in operation: Huntorf in Germany (since 1978) and McIntosh in USA (since 1991) [4]. Moreover, there are some smaller projects in operations or in construction and planning phases, most notably General Compression's 2 MW, 300 MWh project in Texas, USA and SustainX's 1.5 MW, 1 MWh project in New Hampshire, USA [5].

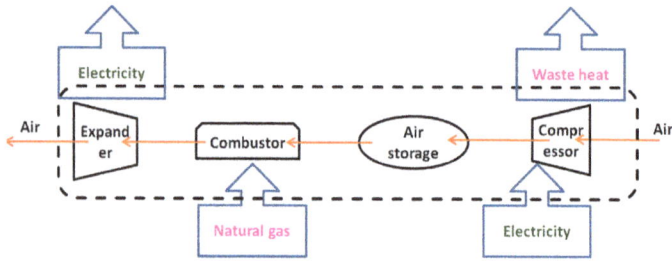

Figure 1. Schematic of a generic conventional compressed air energy storage (CAES) system.

The prospects for the conventional CAES technology are poor in low-carbon grids [2,6–8]. Fossil fuel (typically natural gas) combustion is needed to provide heat to prevent freezing of the moisture present in the expanding air [9]. Fuel combustion also boosts the work output in comparison to solely harnessing the energy stored in the compressed air.

We develop analytical models to assess the thermodynamics of two strategies to make CAES greenhouse gas (GHG) emissions-free. Both utilize the temperature increase from the air compression process to eliminate the need for gas combustion. This heat is generated during the charging phase. Because of its low temperature and correspondingly low exergy, the compression heat is rejected to the ambient environment in the conventional CAES setup. This heat could, in principle, be stored to heat the expanding air provided that the temperature of this stored heat is high enough. The primary method to achieve such high temperatures is to increase the operating pressure of the compressors and to eliminate intercooling between compression stages (i.e., adiabatic compression). This, however, poses practical challenges due to high operating pressures and temperatures of the compressors (e.g., metallurgical limits on compressor blades).

Physical storage of the compression heat is the core of the Adiabatic CAES (A-CAES) concept—the first carbon-free CAES system we investigate. Chemical storage of the compression heat in the form of hydrogen, and combustion of hydrogen instead of natural gas during the discharge phase is the second strategy we analyzed. Hydrogen can be produced via electrolysis of steam at high temperatures (HTE) or water at low (ambient) temperatures (LTE). The HTE concept benefits from the lower electricity demand of the electrolysis process at higher temperatures. Utilizing the high-temperature heat of compression lowers the electricity demand of hydrogen production in the CAES-HTE system. This saving is achieved at the expense of higher electricity demand of the air compressor which, in CAES-HTE, operates at higher pressures with limited or no cooling. The CAES-LTE concept is comparable to the conventional CAES system (diabatic compression with the use of coolers between compressor stages). However, hydrogen is produced onsite with a low-temperature electrolyzer.

Electrolysis of steam (HTE) instead of water (LTE) requires more thermal but less electrical energy. Figure 2 illustrates the theoretical energy requirements as functions of the electrolysis reaction temperature (see Appendix A for details). In a high-temperature electrolyzer, steam is disassociated in the cathode to produce hydrogen and O^{2-}, while O^{2-} is oxidized in the anode to produce oxygen. The theoretical total energy demand of the electrolysis process (change in enthalpy, ΔH) equals the electricity demand (reversible work, i.e., change in the Gibbs free energy, ΔG) plus the heat demand of the reaction (change in entropy multiplied by the reaction temperature, $\Delta S \times T$) from a source of at least as high a temperature as the reaction temperature. While the total energy demand (enthalpy change) of electrolyzing steam increases at higher temperatures, its electricity demand decreases. The savings in electricity consumption of the electrolyzer come at the expense of its higher heat load. Therefore, electrolysis of steam instead of water could be particularly attractive when electricity supply is constrained and high-temperature heat is abundant. Note that the actual electricity demand of the electrolyzers will be higher than the theoretical value (ΔG) because the electrolyzer cell efficiency is less than 100%.

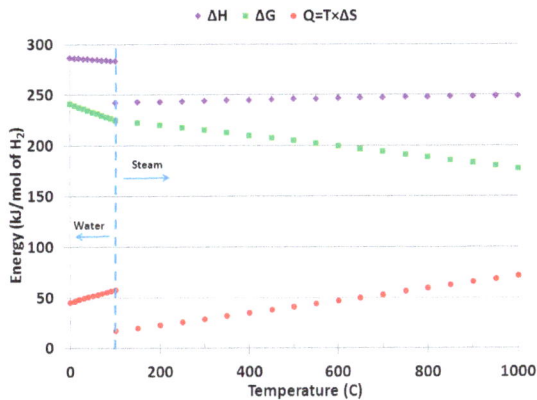

Figure 2. Theoretical energy demand for electrolysis as a function of the reaction temperature.

Existing literature has studied the thermodynamics of conventional CAES [10–15], A-CAES [16–22], and combustion of hydrogen instead of natural gas to fuel conventional CAES [23]. Moreover, alternative advanced CAES designs have also been studied such as isothermal CAES, CAES paired with cogeneration of heat and power, CAES with humidification, and trigeneration CAES systems [24–29].

With this paper, we introduce the concepts of CAES-HTE and CAES-LTE, and provide a comparative thermodynamic analysis of these approaches against A-CAES and conventional CAES. We also assess the sensitivity of our results to two key design parameters: the storage pressure of compressed air and the maximum discharge temperature of the high-pressure compressor. CAES-HTE can potentially be an alternative to A-CAES as a zero-carbon energy storage system that makes use of the otherwise wasted heat of compression. A-CAES stores it as high-temperature thermal energy whereas CAES-HTE stores it as chemical energy. This paper explores whether the use of the compression heat at sufficiently high temperatures could reduce the electricity demand of hydrogen production enough to make the efficiency of CAES-HTE competitive with A-CAES. CAES-LTE is analyzed to provide the most direct baseline for CAES-HTE.

Based on our analysis, A-CAES scored the highest storage efficiency (69.6%) followed by conventional CAES (54.3%), CAES-HTE (35.6%, assuming an electrolyzer efficiency of 50%), and CAES-LTE (34.2%, assuming an electrolyzer efficiency of 50%). CAES-HTE has the highest energy storage density (7.9 kWh per m^3 of storage volume) compared with A-CAES (5.2 kWh/m^3). Conventional CAES and CAES-LTE have similar energy intensities (3.1 kWh/m^3). The conventional CAES system modeled here uses natural gas at a rate of 3.97 GJ per MWh of gross (total) electricity generated. This corresponds to 15.27 GJ per MWh of net or incremental electricity (difference between electricity released and stored) delivered by the plant. Other technical figures of merit are introduced and evaluated as well.

2. Materials and Methods

We use an analytical model to compare the thermodynamics of the conventional CAES, A-CAES, CAES-HTE, and CAES-LTE systems. Our general strategy is applying the First and Second Laws of thermodynamics to the individual system components and modeling air as an ideal gas with temperature-independent specific heat values. We quantify the mass, energy, and exergy flows into and out of the storage facility. The schematics of the modeled conventional CAES, A-CAES, CAES-HTE and CAES-LTE systems are illustrated in Figures 3–6.

We also assess the sensitivity of our results to two key design parameters: the storage pressure of compressed air and the maximum discharge temperature of the high-pressure compressor.

The modeled compressed air storage systems use both electrical energy (to compress air and possibly to generate hydrogen) and heating energy provided by natural gas (only conventional CAES). We use three metrics to compare their energy use: heat rate, work ratio, and roundtrip exergy efficiency (storage efficiency). The heat rate is defined as the external heating fuel (natural gas here) consumed per unit of gross (total) electricity generated by the storage plant (GJ/MWh, on a lower heating value basis, LHV). The heat rate of A-CAES, CAES-HTE, and CAES-LTE is zero as they do not use an external fuel (i.e., natural gas). We also report the heat rate based on the net (incremental) electricity delivered by the storage plant. This metric assists in comparing conventional CAES with conventional gas turbines to manage intermittency of wind and solar in low-carbon grids.

The work ratio quantifies the amount of electrical energy consumed by the compressor (and also the electrolyzer, when applicable) per unit of gross electrical energy generated by the expander. The roundtrip exergy efficiency is the ratio of the exergy delivered (i.e., turbine work) to the exergy provided to the storage plant. The input exergy is the summation of the compression work, the LHV exergy of natural gas (conventional CAES), and the electricity consumed by the electrolyzer (CAES-HTE and CAES-LTE).

One of the critiques of using compressed air to store electricity at scale is its low exergy density. Here, we define exergy density of the storage facility as the ratio of the delivered exergy (i.e., expansion work) to the volume of the air storage cavern. Exergy density is especially important when the storage medium is scarce.

We define the emissions intensity as the ratio of the GHG emissions from natural gas consumption to the gross electricity supply by the storage plant (i.e., total electricity delivered). This variable is zero for all systems studied except conventional CAES—the only configuration in which a fossil fuel is burned. As we do for the heat rate, we express the emissions intensity of conventional CAES based on both the gross and the net electricity delivered.

This section summarizes our general modeling assumptions and simplifications. See Appendix A and nomenclature for the full thermodynamic analysis and the list of symbols.

We model one complete charge and discharge cycle at full load of the compressor and expander (i.e., no part-load operations). We treat air as an ideal gas with temperature-independent specific heat. We ignore the fuel mass and treat the mixture of air and fuel as pure air. Equations (1)–(5) show the general ideal gas formulae we use. The ambient environment (subscript 0) is set at the standard ambient temperature and pressure of 25 °C and 101 kPa. This condition is the reference state for calculating the internal energy, enthalpy, entropy, and exergy throughout our analysis.

$$m = \frac{P\,V}{R\,T} \qquad \text{mass of the air present in cavern} \qquad (1)$$

$$h = (T - T_0)C_p \qquad \text{specific enthalpy of air} \qquad (2)$$

$$u = (C_v T) - (C_p T_0) \qquad \text{specific internal energy of air} \qquad (3)$$

$$s = C_p \ln\left(\frac{T}{T_0}\right) - R \ln\left(\frac{P}{P_0}\right) \qquad \text{specific entropy of air} \qquad (4)$$

$$\psi = (h - h_0) - (s - s_0)T_0 \qquad \text{specific stream exergy of air} \qquad (5)$$

The air storage cavern has a fixed volume. Its pressure varies between a minimum (P_{em}) and a maximum (P_{fl}) during the charge and discharge processes. In order to maintain its mechanical integrity and to ensure high-enough flow rates for the discharging air, the cavern is not fully discharged in practice. The air mass remaining in the storage at the end of the discharge phase (when all the "working air" has been withdrawn) is called the "cushion air". We model the cavern as adiabatic. Raju et al. [10], Steta [20], and Xia et al. [30] studied heat transfer between the stored air and the cavern wall, which is beyond the scope of our work. The rate of heat transfer depends on several factors such as residence time of air in the cavern and its temperature, rock properties, cavern size and shape.

Raju et al. estimated the rate of heat loss at the Huntorf CAES plant in the order of few percent of the compressor power.

The coolers (heat exchangers) following each compression stage are assigned a fixed approach temperature, T_{ac}. This is defined as the difference between the temperature of the cooling fluid (e.g., water) entering the cooler ($T^{CL}_{in, coolant}$, set at T_0) and that of the cooled compressed air leaving the compressor cooler ($T^{CL}_{out, air}$). This implies the inlet temperature of the cavern and the output of all the compressor coolers are fixed and equal to $T_{ac} + T_0$ (see Equation (6)).

$$T^{CN}_{in} = T^{CL}_{out,air} = T_{ac} + T_0 \quad \text{inlet temperature of cavern and discharge of compressor coolers} \quad (6)$$

The discharge temperature of the combustion chambers (T^{cc}_{out}) is maintained at a fixed value. The expander has two stages. The high-pressure (HP) and low-pressure (LP) stages, which have equal but variable expansion ratios (XR) and determined according to the instantaneous pressure of the cavern (Equation (7)).

$$\text{XR}_{HP} = \text{XR}_{LP} = \sqrt{\frac{P_0}{P_{CN}}} = \sqrt{\text{XR}} \quad \text{instantaneous expansion ratio} \quad (7)$$

The temperature of the air stream leaving the storage plant during the discharge process (T_{et}) is constrained to be fixed and constant. Following Osterle [13], an imaginary final heat exchanger (FHX) is placed at the exhaust of the storage plant to account for the exergy loss by the exhaust stream to the ambient environment. This heat exchanger cools down the expanded air from T_{et} to the ambient temperature.

Heat flows (Q) are reckoned to be positive if they enter the system (e.g., heat added in the combustor). Work done by the system on the surroundings has a positive sign (e.g., expansion work).

As shown in the Appendix A, the First and Second Laws of thermodynamics are applied to each system component to quantify the work, heat, and exergy fluxes during the charge and discharge processes. Once these are determined, the roundtrip exergy efficiency ($\eta_{storage}$), work ratio (WR), heat rate (HR), emissions intensity (GI_{plant}), and exergy density (ρ) of the storage plant are calculated by applying Equations (8)–(12).

$$\eta_{storage} = \frac{W_{TB}}{-W_{CM} + X_{NG} - W_{electrolysis}} \quad (8)$$

$$\text{WR} = \frac{-W_{CM} - W_{electrolysis}}{W_{TB}} \quad (9)$$

$$\text{HR} = \left(\frac{m_{NG} \, \text{LHV}_{NG}}{W_{TB}} \right) \left(\frac{3.6 \, \text{GJ}}{\text{MWh}} \right) \quad (10)$$

$$GI_{plant} = (\text{HR})(GI_{NG}) \quad (11)$$

$$\rho = \frac{W_{TB}}{V} \quad (12)$$

2.1. Modeling Conventional CAES

In the conventional CAES system we modeled (Figure 3), air is compressed in a three-stage compressor (CM) and then stored in the cavern (CN). Each compression stage is followed by a cooler (CL) to reduce the compression work of the succeeding stage and to reduce the volumetric requirement of air storage by increasing the density of the stored air. The compression heat is released to the ambient environment.

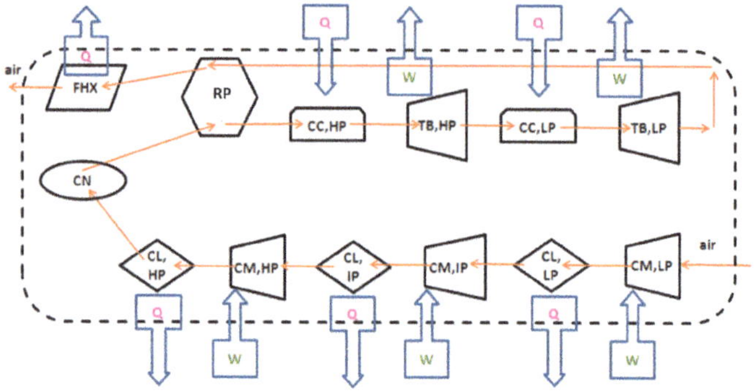

Figure 3. Schematic of the conventional CAES system. The compression train (CM) is composed of three stages: low-, intermediate-, and high-pressure (LP, IP, and HP). The expansion train (TB) is made of high- and low-pressure stages (HP and LP). "Q" and "W" represent heat and work interactions between the system and the surroundings. The air leaving the cavern is preheated in the recuperator by the hot air leaving the low pressure turbine (internal heat transfer). "0" indicates ambient condition. CL, CN, CC, RP, and FHX stand for the cooler, cavern, combustor, recuperator, and final exhaust heat exchanger.

During the discharge phase, air is first preheated in a recuperator (RP). It is then combusted with natural gas (NG) to generate work in the expanders (turbines, TB). We modeled a two-stage expander. In the recuperator, the exhaust of the low-pressure turbine preheats the air leaving the cavern and entering the high-pressure combustor to reduce the fuel demand.

The compressor has three stages: low (LP), intermediate (IP), and high pressure (HP). All stages have variable but equal compression ratios throughout the charging process. The compression ratios (CR) vary according to the instantaneous pressure of the cavern.

2.2. Modeling A-CAES

Figure 4 illustrates the A-CAES system we analyzed. The compression heat is stored in two thermal energy storage systems (TS1 and TS2). Coolers between compression stages are eliminated in A-CAES to increase the discharge temperature of the compressors. We therefore, model a two-stage (LP and HP) compressor. Only one cooler (heat exchanger, between the TS2 discharge and the cavern inlet) exists and cools the compressed air prior to storage.

The expansion train of A-CAES is made up of two stages (HP and LP). The withdrawn compressed air is heated in TS1 and TS2 before expanding and generating electricity (combustors are eliminated). No recuperator is considered. This is owing to the low discharge temperature of the LP expander. The discharge stream is cooled to the ambient temperature in the final exhaust heat exchanger (FHX). TS1 absorbs heat from the air leaving the low-pressure compressor and provides heat to the compressed air entering the high-pressure expander. TS2 interacts with the high-pressure compressor and the low-pressure expander. Refer to Appendix A for more details.

Similar to the analysis of conventional CAES, the temperature of the air entering the cavern is set as constant. The intake temperatures of the expanders (i.e., exhaust of TS1 and TS2) are constrained to be constant. However, their values are dictated by the amount of heat stored during the charging phase. Note that the inlet temperatures of the expanders in the conventional CAES system were constant as well, but their values were a preset design parameter, satisfied by variable combustion rates. The TS1 and TS2 units are modeled as isobaric and adiabatic.

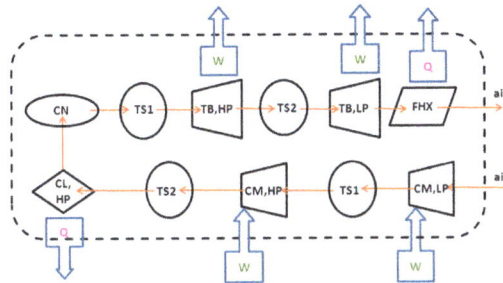

Figure 4. Schematic of the A-CAES system. All of the low pressure (LP) and a portion of the high pressure (HP) compression heat are stored in two thermal storage facilities (TS1 and TS2). The same heat storage units release the heat to the withdrawn compressed air prior to expansion during the discharge phase. "Q" and "W" represent heat and work interactions between the system and the surroundings.

The maximum exit temperature of the high-pressure compressor is a preset parameter, which will be varied in the sensitivity analysis section. Our rationale for this design constraint is the following. The exit temperature of the compressor is a key parameter for determining the exergy supplied to and stored in thermal storage. This consequently impacts the temperature of the air entering the expanders. Moreover, there are technical constraints such as the stress on and the fatigue of compressor blades driven by the maximum exit temperature of the compressor [31,32].

Once a full charge and discharge cycle is modeled, the overall performance of A-CAES are characterized with Equations (8)–(12). The heat rate and GHG emissions intensity of A-CAES are zero as no fuel is consumed.

2.3. Modeling CAES-HTE and CAES-LTE

As illustrated in Figure 5, the compressor of CAES-HTE is made up of two stages similar to A-CAES to increase the temperature of the compression heat stored, in contrast to our CAES-LTE and conventional CAES models with three stage compressors. A heat exchanger cools the exhaust stream of the LP compressor to a constant temperature before entering the HP compressor. The maximum discharge temperature of the HP compressor is a design parameter and is preset, similar to the A-CAES model. Whereas using a one stage compressor could generate higher temperature heat, the operating temperature of the compressor would be in excess of 1000 °C, compared to the 500–700 °C range considered in the literature for practical reasons (e.g., mechanical integrity of compressor blades) [31,32]. The heat absorbed from the exhaust stream of the HP compressor is used to make steam, to heat up the steam to the constant temperature of the electrolyzer, and to provide the heating energy required for the electrolysis process. A heat exchanger follows the electrolyzer to further cool the compressed air to a fixed temperature before entering the cavern. The generated hydrogen is stored to burn and heat the air during the discharge phase. We choose not to consider any physical storage of heat, similar to conventional CAES and in contrast to A-CAES. This is due to the relatively low temperature of air upon giving its heat to the HTE system. The discharge phase of CAES-HTE is identical to that of conventional CAES with the distinction that hydrogen (produced during charging), instead of natural gas, fuels the combustors.

The simulated CAES-LTE system (Figure 6) has a similar configuration as the CAES-HTE system, with the difference that a low-temperature electrolyzer (LTE) is used instead of a HTE. Moreover, a three-stage compression train is used, similar to conventional CAES. This is because there is no need to produce high-temperature heat. Increasing the number of compression stages in CAES-LTE reduces the compression work compared to CAES-HTE.

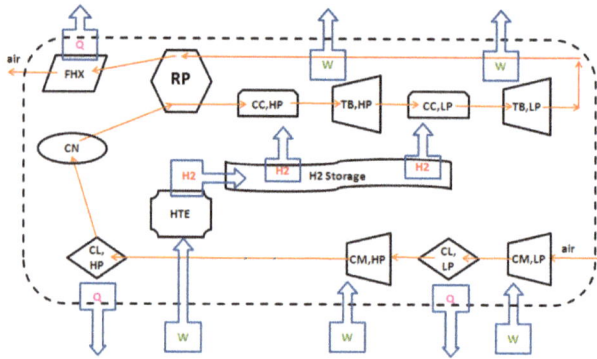

Figure 5. Schematic of the CAES-HTE system simulated. HTE stands for high temperature electrolyzer. "Q" and "W" represent heat and work interactions between the system and the surroundings. The air leaving the cavern is preheated in the recuperator by the hot air leaving the low pressure turbine (internal heat transfer). The air leaving the high pressure compressor provides the heat demand of the HTE (internal heat transfer).

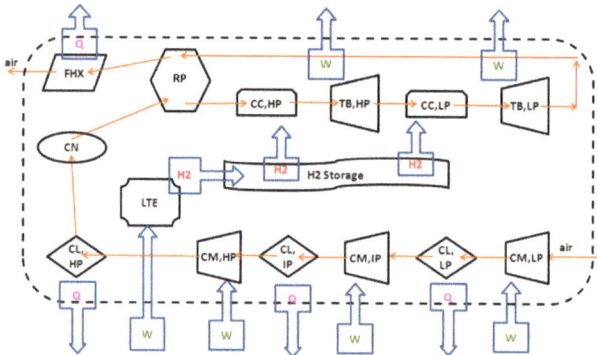

Figure 6. Schematic of the CAES-LTE system simulated. LTE stands for low temperature electrolyzer. "Q" and "W" represent heat and work interactions between the system and the surroundings. The air leaving the cavern is preheated in the recuperator by the hot air leaving the low pressure turbine (internal heat transfer).

Analysis of CAES-LTE is similar to that of CAES-HTE. The difference is that there is no explicit constraint on the maximum temperature of the air leaving the HP compressor. This is because the utilization of a three stage compressor and coolers results in temperature below the metallurgical limit of the turbines blades, similar to the conventional CAES design.

The heat rate and GHG emissions intensity of both CAES-HTE and CAES-LTE are zero because they burn hydrogen rather than natural gas. The storage efficiency and work ratio of the systems are quantified by Equations (8) and (9). CAES-HTE and CAES-LTE have higher work ratios compared to conventional CAES because of the work load of the electrolyzer.

3. Results

This section provides a series of numerical examples based on the analytical models developed in Section 2 and in Appendix A. The main question being addressed is whether the chemical storage of the high-temperature heat of compression as hydrogen (CAES-HTE) is thermodynamically superior to the physical storage of the heat (A-CAES). The sensitivity of the results to the storage pressure and discharge temperature of the high-pressure compressor are discussed as well. We also discuss the thermodynamics of conventional CAES and CAES-LTE to benchmark performance of CAES-HTE and A-CAES. The Appendix A section also includes details on the temperature range of each system component over the storage cycle.

3.1. Thermodynamic Comparison of A-CAES and CAES-HTE

Table 1 lists the parameters used to compare the thermodynamics of A-CAES and CAES-HTE in the base case scenario. We consulted the design parameters of the two existing commercial CAES plants (Huntorf and McIntosh [10,15,33–36]), as well as literature on design of A-CAES systems [15,16,20,31,32,36] to choose these values. In the sensitivity analysis section, we discuss the impact of two key design parameters, cavern storage pressure and discharge temperature of the HP compressor. The simulation results are tabulated in Table 2. Refer to the nomenclature for the list of symbols.

Table 1. Inputs for analysis of the A-CAES and CAES-HTE systems in the base case. The conventional CAES and CAES-LTE systems use the same values with the main exception that the maximum storage pressure is 7 MPa instead of 10 MPa.

Parameter	Value	Parameter	Value	Parameter	Value
γ	1.4	V	0.56 Mm^3	LHV$_{H2}$	120 MJ/kg
C_p	1.006 $\frac{kJ}{kg \cdot K}$	R	0.287 $\frac{kJ}{kg \cdot K}$	x_{H2}	114 MJ/kg
$P_{CN,fl}$	10 MPa	T_{ac}	30 °C	$T_{in}^{TB,LP}$	850 °C
$P_{CN,em}$	5 MPa	η_{TB}	85%	T_{et}	130 °C
η_{CM}	85%	$T_{in,coolant}^{CL}$	25 °C	$T_{in}^{TB,HP}$	530 °C
$T_{out}^{CM,HP,Max}$	600 °C	$T_{out}^{TS2,ch}$	100 °C		

Table 2. Simulation results for the A-CAES and CAES-HTE systems simulated in the base case scenario.

Variable	A-CAES	CAES-HTE (100% Efficient Electrolzer)	CAES-HTE (50% Efficient Electrolzer)
W_{CM} (TJ)	15.22	15.22	15.22
$Q_{electrolysis}$ (TJ)	-	7.96	7.96
$W_{electrolysis}$ (TJ)	-	14.82	29.65
Q_{TS} (TJ)	13.62	-	-
W_{TB} (TJ)	10.58	15.99	15.99
Q_{CC} (TJ)	-	17.59	17.59
$\eta_{storage}$ (%)	69.5	53.2	35.6
WR	1.44	1.88	2.81
ρ(kWh/m^3)	5.2	7.9	7.9

The compression work to fully charge the cavern for both A-CAES and CAES-HTE is the same (15.22 TJ). This is because the compressors in both systems are identical (equal pressure, temperature, and mass of working air). The high temperature electrolyzer of the CAES-HTE system uses 14.83 TJ to produce enough hydrogen for combustion during the discharge phase. The electrolyzer has a heating load of 7.96 TJ to generate steam and electrolyze it. This heat is supplied by the thermal energy dissipated from the compressors. Therefore, about 52% of the compression work is recovered and used in the HTE to produce hydrogen. In the adiabatic system, about 89% of the compression work is physically stored (13.62 GJ). The remainder of heat is released to the ambient environment. Therefore, the A-CAES system recovers and utilizes a higher portion of the energy supplied to the compressor (compression work).

Because the temperature of the expanding air is higher in the CAES-HTE configuration (due to the combustion of hydrogen), the generated work (15.99 TJ) is larger than that of the A-CAES system

(10.58 TJ). The heat load of CAES-HTE combustors is 17.59 TJ. This is while the heat transferred to the expanding air in the A-CAES system is 13.62 TJ (equal to the stored heat).

The A-CAES system is thermodynamically more efficient than the CAES-HTE system based on our analysis. Physical storage of the compression heat leads to an overall storage efficiency of 69.5% (A-CAES) compared to 35.6% for its chemical storage in the form of hydrogen (CAES-HTE, assuming a 50% efficient electrolyzer). The simulated CAES-HTE system uses 29.65 TJ more electricity (to produce hydrogen in the electrolyzer) compared to A-CAES (both systems have the same compression work of 15.22 TJ). The CAES-HTE system however, produces 5.41 TJ of additional work because it combusts hydrogen and the expanding air has a higher temperature. Therefore, for each unit of excess electrical energy used by the CAES-HTE system, roughly 0.18 unit of excess electrical energy is generated. This relatively large gap lowers the overall efficiency of CAES-HTE.

The performance of CAES-HTE would be less attractive in the real world due to the inefficiencies of the electrolyzer itself. The energy efficiency of the electrolysis process can be defined as the ratio of the theoretical electricity demand (ΔG) to the actual use. No commercial large-scale HTE facility currently exists to our knowledge. The efficiency of the laboratory-scale systems are reported as about 50% (at operating temperatures of about 850 °C [37]). To estimate the upper bound of the CAES-HTE performance, we also run the simulation with an ideal electrolyzer (100% efficient). This would halve the work load of the electrolyzer to 14.82 TJ. This decrease in energy consumption improves the roundtrip efficiency of the CAES-HTE system to 53.2% from 35.6%.

The storage requirements of hydrogen are likely to degrade the performance of CAES-HTE too. One would need to compress the hydrogen for storage and use later during the discharge phase in CAES-HTE. We have ignored this additional work load in our analysis.

The hydrogen-based system benefits, however, from a higher exergy density (7.9 kWh/m^3) compared to A-CAES (5.2 kWh/m^3). This is because the CAES-HTE system stores energy both as mechanical energy (compressed air) and as chemical energy (hydrogen). Thus A-CAES would require 52% more cavern volume to generate the same amount of work in our analysis.

3.2. Sensitivity of A-CAES and CAES-HTE to Exit Temperature of the HP Compressor

We treat the maximum exit temperature of the HP compressor ($T_{out}^{CM,HP,Max}$) as a key design parameter for A-CAES and CAES-HTE. Tables 3 and 4 present the sensitivity of the results to this parameter (all other parameters are similar to the base case scenario, Table 1).

Table 3. Sensitivity of the CAES-HTE results to the maximum exit temperature of the HP compressor ($T_{out}^{CM,HP,Max}$). Results are based on an ideal electrolyzer.

Variable	$T_{out}^{CM,HP,Max}$ (°C)			
	500	600	700	800
W_{CM} (TJ)	14.20	15.22	16.24	17.25
$Q_{electrolysis}$ (TJ)	7.24	7.96	8.70	9.45
$W_{electrolysis}$ (TJ)	15.19	14.83	14.45	14.07
T_{HTE} (°C)	460	555	650	745
$\eta_{storage}$ (%)	54.4	53.2	52.1	51.1

For CAES-HTE, setting a higher exit temperature for the HP compressor translates to a higher inlet temperature for this compressor and a lower cooling load for the LP cooler (because the compression ratio is constant). This provides more thermal energy for the electrolysis process, which reduces its electricity demand to produce the same amount of hydrogen. Nevertheless, as the maximum discharge temperature of the high pressure compressor ($T_{out}^{CM,HP,Max}$) increases from 500 °C to 800 °C, less cooling by the LP cooler increases the work load of the HP compressor (because of the higher inlet temperature of the HP compressor). As shown in Table 3, the net effect of a higher compression work and a lower electrolysis work is an increase in the exergy demand of CAES-HTE to charge the cavern. The total input exergy (summation of work of compressor and electrolyzer) increases from 29.39 TJ to 31.32 TJ. The discharge phase is insensitive to these changes in our model. Storage efficiency of the CAES-HTE system drops from 54.4% to 51.1% across this range. Results are presented for a CAES-HTE system with an ideal (100% efficient) electrolyzer.

Table 4. Sensitivity of the A-CAES performance to the maximum discharge temperature of the high-pressure compressor ($T_{out}^{CM,HP,Max}$).

Variable	$T_{out}^{CM,HP,Max}$ (°C)			
	500	600	700	800
W_{CM} (TJ)	14.20	15.22	16.24	17.25
Q_{TS} (TJ)	12.60	13.62	14.63	15.65
W_{TB} (TJ)	10.43	10.58	10.61	10.88
$\eta_{storage}$ (%)	73.4	69.5	65.4	63.1
$T_{out}^{TB,HP,Min}$ (°C)	54	26	−8	−31
$T_{out}^{TB,LP,Min}$ (°C)	142	181	217	259

In the A-CAES system, raising $T_{out}^{CM,HP,Max}$ increases the compression work too. This is because less cooling is done in the low pressure cooler and consequently, the inlet temperature to the HP compressor is elevated. At the same time, more heat is stored in TS2 from the air leaving the HP compressor. The turbine's total work increases by about 4% (from 10.43 to 10.88 TJ) as $T_{out}^{CM,HP,Max}$ is raised from 500 to 800 °C. This small increase in the expansion work despite a much higher (24%, from 12.60 to 15.65 TJ) increase in the total thermal energy stored (Q_{TS}) occurs because the thermal energy stored in TS1 (Q_{TS1}) and, consequently, the heat given to the air entering the HP turbine and the work generated by the HP turbine, ought to decrease to allow higher temperatures for the intake and thus discharge of the high pressure compressor. Note that TS1 precedes the HP compressor and HP expander (see Figure 4).

The net effect of increasing $T_{out}^{CM,HP,Max}$ is lowering the storage efficiency of A-CAES. Its efficiency decreases from 73.4% to 63.1% when $T_{out}^{CM,HP,Max}$ increases from 500 to 800 °C. Comparing Table 3 with Table 4 shows that storage efficiency of A-CAES is more sensitive to the temperature of air stream leaving the compressor, compared with that of CAES-HTE.

An important design consideration in our model is the discharge temperature of the expander. This variable needs to remain above the freezing point of water to avoid mechanical damage to the expanders. Referring to Table 4, the exit temperature of the HP turbine drops as $T_{out}^{CM,HP,Max}$ increases. Because less heat can be stored in TS1 and then released to the compressed air entering the high pressure expander. This temperature drops below the freezing point when $T_{out}^{CM,HP,Max}$ reaches 700 °C in our analysis. Designing A-CAES in the real world would need to include a detailed analysis to optimize the performance of the plant and avoid freezing concerns. For instance, although they are constrained to be equal in our model, the HP expander can be designed to have a lower expansion ratio than the LP expander. This will raise and lower the discharge temperatures of the HP and LP expanders, respectively.

3.3. Sensitivity of A-CAES and CAES-HTE to Storage Pressure

The storage pressure of air is our second key design parameter. The sensitivity of the CAES-HTE and A-CAES results to the maximum storage pressure are shown in Tables 5 and 6. All parameters are from Table 1. The maximum cavern pressure is varied in the range of 7–12 MPa, compared to 10 MPa in the base case. The minimum storage pressure is kept at 5 MPa in all cases.

For the CAES-HTE system, the compression work and hydrogen demand increase at higher storage pressures. This is because more air needs to be stored and heated. The operating temperature of the electrolyzer slightly decreases at higher pressures since we keep the maximum discharge temperature of the high-pressure compressor fixed. Higher cavern pressures translate to higher compression ratios. Keeping the $T_{out}^{CM,HP,Max}$ constant requires a lower inlet temperature for the HP compressor at higher cavern pressures. Therefore, TS1 (preceding the HP compressor) needs to absorb more heat from the compressed air leaving the LP compressor and entering the HP compressor. A lower discharge temperature for the HP compressor decreases the temperature of the air entering the electrolyzer, and thus the electrolysis reaction temperature. This temperature drops from 578 to 544 °C as the maximum storage pressure of air ($P_{CN,fl}$) is raised from 7 to 12 MPa for a CAES-HTE system with an ideal electrolyzer.

Table 5. Sensitivity of the CAES-HTE model to maximum storage pressure. Results are based on an ideal electrolyzer.

Variable	P_{fl} (MPa)			
	7	8	10	12
W_{CM} (TJ)	5.88	8.93	15.22	21.73
$Q_{electrolysis}$ (TJ)	3.15	4.75	7.96	21.15
$W_{electrolysis}$ (TJ)	5.72	8.69	14.82	22.79
T_{HTE} (°C)	578	569	555	544
W_{TB} (TJ)	6.18	9.39	15.99	22.79
Q_{CC} (TJ)	6.82	10.35	17.60	25.03
$\eta_{storage}$ (%)	53.3	53.3	53.2	53.1
ρ (kWh/m^3)	3.1	4.7	7.9	11.3

Table 6. Sensitivity of the A-CAES model to the maximum storage pressure.

Variable	P_{fl} (MPa)			
	7	8	10	12
W_{CM} (TJ)	5.88	8.93	15.22	21.73
Q_{TS} (TJ)	5.24	7.97	13.62	19.48
W_{TB} (TJ)	3.90	6.00	10.58	15.36
$\eta_{storage}$ (%)	66.3	67.2	69.5	70.7
$\rho_{STORAGE}$ (kWh/m^3)	1.9	3.0	5.2	7.6
$T_{out}^{TB,HP,Min}$ (°C)	5	11	26	33

The expansion work increases at higher cavern pressures because more compressed air is handled, and at higher pressures. The net impact of higher cavern pressures on the storage efficiency is negligible (slightly negative). Increased work loads for the compressor and the HTE cancel out the higher expansion work. The exergy density of the cavern increases ~2.7 times as the maximum storage pressure increases from 7 to 12 MPa. Therefore, increasing the cavern pressure substantially improves the exergy density of the plant while it marginally degrades the storage efficiency. In the real world, however, the storage efficiency is likely to degrade more compared to the scenario pictured here. For example, we have assumed a fixed isentropic efficiency for the compressors whereas their efficiency is likely to degrade at higher compression ratios [38].

For the A-CAES system, higher cavern pressures translate to higher compression work as well. At the same time, more waste heat recovery opportunities are available. The expansion work also increases as a larger mass of air and at a higher pressure is expanded. The net effect of higher compression work, recovered heat, and expansion work is positive on the storage efficiency of A-CAES. It rises from 66.3% to 70.7% as the cavern pressure is lifted from 7 to 12 MPa. The exergy density of the cavern at 12 MPa is almost 4 times that of 7 MPa, as more air is stored in the same cavern, and at higher pressures. Finally, the exit temperature of the HP expander is also raised, due to more stored heat despite the higher expansion ratios. This is beneficial in addressing the concerns with freezing of vapor in the expanding air and damaging the turbine blades.

3.4. Thermodynamics of Conventional CAES and CAES-LTE

In Table 7, we present the results for thermodynamics of conventional CAES and CAES-LTE systems. For the most part, we use the same input parameters as for A-CAES and CAES-HTE (Table 1), such as air storage temperature, minimum cavern pressure, and discharge temperature of the final heat exchanger. This is to benchmark the performance of A-CAES and CAES-HTE. The primary difference is that the compression train of CAES and CAES-LTE is made up of three stages. This is because there is no need to generate high temperature heat in these designs. Therefore, more intercooling can be performed to lower the compression work, similar to the McIntosh and Huntorf CAES plants. Moreover, the maximum storage pressure of the cavern is set at 7 MPa instead of 10 MPa, again as high temperature heat is not needed. As the number of compression stages increases and their compression ratio drops, energy losses during the compression (charging) phase decrease. This is because the compression process gets closer to an isothermal instead of an adiabatic process.

Table 7. Simulation results for the conventional CAES and CAES-LTE systems.

Variable	CAES-LTE (100% Efficient Electrolzer)	CAES-LTE (50% Efficient Electrolzer)	CAES
W_{CM} (TJ)	4.56	4.56	4.56
$Q_{electrolysis}$ (TJ)	1.38	1.38	-
$W_{electrolysis}$ (TJ)	6.47	13.48	-
W_{TB} (TJ)	6.18	6.18	6.18
Q_{CC} (TJ)	6.82	6.82	6.82
$\eta_{storage}$ (%)	54.7	34.2	54.3
WR	1.83	2.92	0.74
ρ (kWh/m^3)	3.1	3.1	3.1

The storage efficiency of the CAES-LTE system with a 50% efficient electrolyzer is 34.2%, which is comparable to that of the CAES-HTE system. This indicates that the lower electricity demand of the electrolyzer in CAES-HTE system is offset by its higher compression work. Using an ideal (100% efficient) electrolyzer instead of a 50% efficient electrolyzer leads to an overall storage efficiency of 54.7% for CAES-LTE.

The efficiency of the conventional CAES system is 54.3%, which is lower than that of A-CAES (69.5%) and similar to the hydrogen-fueled CAES systems with ideal electrolyzers (53.2% for HTE and 54.7% for LTE). The conventional CAES system has the lowest work ratio (0.74) because it burns natural gas with a heat rate of 3.97 GJ per MWh of gross electricity generated, or 15.27 GJ per MWh of net electricity.

4. Discussion

Our analysis shows that the A-CAES system has the highest exergy storage efficiency, followed by conventional CAES, and then the hydrogen based CAES systems. High exergy losses in electrolyzers constitute a key contributor to the overall low storage efficiency of CAES-HTE and CAES-LTE.

Current literature has identified A-CAES as a potentially important component of low carbon grids with large penetration of renewable energies from an economic point of view. This paper builds on the same premise and provides further insight into thermodynamic performance and competitiveness of A-CAES.

The economics of conventional CAES are likely to be more attractive compared to the other systems studied here unless significant GHG emissions restrictions are in place. The emissions intensity of the conventional CAES system modeled is 262 kgCO$_2$e/MWh of gross electricity generated whereas the other three systems emit no greenhouse gases. The emissions of the electricity consumed to charge these plants are not included for this calculation. However, if emissions per unit of net rather than gross electricity generated by the CAES plant (generation minus consumption, equal to $1 - WR$) is considered, the corresponding emissions intensity is 1008 kgCO$_2$e/MWh. In other words, the conventional CAES plant would emit 1 metric ton of CO$_2$e per incremental MWh of electricity it adds to the grid supply. This is almost 50% higher than that of a simple gas combustion turbine (679 kgCO$_2$e/MWh, using an efficiency of 35% and heat rate of 10.26 GJ/MWh), which competes with storage for filling in the gaps in the supply of intermittent renewables. This high emissions level highlights the shortcoming of the conventional CAES systems in carbon-constrained grids.

Assessing the competitiveness of these storage technologies to support integration of renewable energies into low-carbon grids requires a comprehensive analysis, including both thermodynamics and the economics of practical implementation. Precise thermodynamic assessment of these systems in the real world calls for complex numerical analyses due to their complexities, which is beyond the scope of this paper. Here we offer a few insights into the thermodynamic and economic trade-offs of these systems in the real-world.

Our thermodynamic analysis indicates that prospects for hydrogen-based CAES systems are likely weaker than those of A-CAES due to the lower storage efficiency. Even assuming an ideal electrolyzer leads to storage efficiencies in the lower 50% range as the high end for the CAES-HTE and CAES-LTE configurations studied here, compared to around 70% for A-CAES. Using a currently more realistic electrolyzer efficiency of 50% lowers the overall efficiency of the hydrogen-fueled systems to the mid 30% range. Although thermal losses would decrease the efficiency of A-CAES, they would not be as significant as the electrolyzer losses.

Capital and operating costs of these CAES systems are different, with conventional CAES currently being the most mature and inexpensive for large scale adoption, in the absence of tight emissions restrictions. The design and operation of A-CAES plants are complicated by the need for high-pressure and high-temperature compressors, thermal stores, and high-pressure turbines [31]. In contrast, the engineering and economic complications of high-temperature electrolyzers and hydrogen storage and combustion complicate the CAES-HTE systems. The design and operations of a CAES-LTE system would be simpler because can operate at pressures and temperatures of conventional CAES systems.

Acknowledgments: Hossein Safaei thanks Professor David Keith at Harvard University for the invaluable discussions on CAES systems. Research at Harvard University was partly supported by the Ziff Fund for Energy and Environment.

Author Contributions: H.S. developed the thermodynamic model in collaboration with M.J.A. H.S. and M.J.A. wrote the paper together.

Conflicts of Interest: The authors declare no conflict of interest. The founding sponsors had no role in the design of the study; in the collection, analyses, or interpretation of data; in the writing of the manuscript, and in the decision to publish the results.

Nomenclature

A-CAES	Adiabatic compressed air energy storage
CAES	Compressed air energy storage
CAES-HTE	Compressed air energy storage paired with a high-temperature electrolyzer
CAES-LTE	Compressed air energy storage paired with a low-temperature electrolyzer
CC	Combustor
Ch	Charge process
CL	Cooler (heat exchanger) following compressor stages
C_{Lnt}	Latent heat of evaporation (kJ/kg)
CM	Compressor
CN	Cavern for air storage
coolant	Cold stream of compressor cooler
C_p, C_v	Specific heat at constant pressure and volume ($\frac{kJ}{kg \cdot K}$)
CR	Compression ratio
dch	Discharge process
E	Voltage applied to electrolyzer (volts)
em	State of depleted cavern
et	Exhaust stream of the storage plant

FHX	Heat exchanger located at exhaust of the plant
fl	State of fully charged cavern
G	Gibbs free energy (kJ)
GI	GHG emissions intensity of plant or fuel (kgCO$_2$e/MWh or kgCO$_2$e/GJ, respectively)
h	Specific enthalpy (kJ/kg)
Δh_f^0	Standard enthalpy of formation (kJ/kg)
HP	High-pressure equipment
HR	Heat rate (GJ/MWh)
HTE	High temperature electrolysis/ electrolyzer
in	Inlet conditions
IP	Intermediate-pressure equipment
ist	Isentropic process
IX	Exergy loss (kJ)
LHV	Lower heating value (kJ/kg)
LP	Low-pressure equipment
LTE	Low-temperature electrolysis/ electrolyzer
m	Mass of air or fuel (kg)
M	Molar mass (kg/kMole)
n	Molar coefficient
NG	Natural gas
out	Outlet conditions
P	Pressure (kPa)
Q	Thermal energy (kJ)
R	Ideal gas constant ($\frac{kJ}{kg \cdot K}$)
RP	Recuperator
s	Specific entropy ($\frac{kJ}{kg \cdot K}$)
s^0	Standard entropy ($\frac{kJ}{kg \cdot K}$)
S	Entropy (kJ)
sns	Sensible heat (kJ/kg)
T	Temperature
T_{ac}	Approach temperature
TB	Turbine (expander)
TS	Thermal energy storage unit
u	Specific internal energy (kJ/kg)
U	Internal energy (kJ)
V	Volume of air storage (m^3)
W	Work (kJ)
WR	Work ratio
x	Specific exergy (kJ/kg)
X	Exergy (kJ)
XR	Expansion ratio of turbine
ρ	Exergy density of cavern (kWh/m^3)
γ	Specific heat ratio of air
η	Efficiency (%)
ψ	Exergy of air stream (kJ/kg)
0	Standard conditions

Appendix A

We here discuss the details of the analytical models developed to assess the thermodynamics of the compressed air energy storage systems.

We use the following general assumptions and simplifications. One complete charge and discharge cycle is analyzed (without part-load operation). Air is modeled as an ideal gas with temperature-independent specific heat. Mass of the fuel is assumed negligible compared to the compressed air and the mixture of air and fuel is treated as pure air. Equations (A1)–(A5) list the general ideal gas formulae used. The ambient environment (subscript 0) is at standard conditions ($P_0 = 101$ kPa and $T_0 = 25\,°C$). This condition is also the reference state for calculating the internal energy, enthalpy, entropy, and exergy in our analysis.

$$m = \frac{P\,V}{R\,T} \tag{A1}$$

$$h = (T - T_0)C_p \tag{A2}$$

$$u = (C_v\,T) - (C_p\,T_0) \tag{A3}$$

$$s = C_p \ln\left(\frac{T}{T_0}\right) - R\,\ln\left(\frac{P}{P_0}\right) \tag{A4}$$

$$\psi = (h - h_0) - (s - s_0)T_0 \tag{A5}$$

In our model, similar to Huntorf and McIntosh CAES plants, the cavern has a fixed volume. Cavern's pressure varies between a minimum pressure (P_{em}) and a maximum (P_{fl}). In order to maintain its mechanical integrity and to ensure high-enough flow rates of the withdrawn air, the cavern is not fully depleted. The air mass remaining in the cavern at the end of the discharge phase (when all the "working air" has been withdrawn to generate electricity) is called the "cushion air".

The cavern and thermal storage units are modeled as adiabatic. The coolers (heat exchangers) following each compressor stage are constrained to have a fixed approach temperature (T_{ac}), defined as the difference between temperature of the cooled compressed air leaving the heat exchanger ($T^{CL}_{out,air}$) and temperature of the coolant ($T^{CL}_{in,\,coolant}$). This implies the inlet temperature of the cavern and the discharge of all the compressor coolers is constant through the charging process (Equation (A6)).

$$T^{CN}_{in} = T^{CL}_{out,air} = T_{ac} + T^{CL}_{in,\,coolant} = T_{ac} + T_0 \tag{A6}$$

During the discharge phase, we constrain the exit temperature of the combustors at a fixed value (e.g., through controlling fuel combustion). The expander (turbine) has two stages; high-pressure (HP) and low-pressure (LP). These stages are constrained to have equal but variable expansion ratios (Equation (A7)), according to the instantaneous pressure of the cavern.

$$XR_{HP} = XR_{LP} = \sqrt{\frac{P_0}{P_{CN}}} = \sqrt{XR} \tag{A7}$$

Temperature of the air stream leaving the storage plant (T_{et}) is set to be fixed. An imaginary heat exchanger (FHX) is placed at the exhaust of the storage plant to account for the exergy loss of the exhaust stream to the ambient environment. This heat exchanger cools the air leaving the LP turbine or recuperator, from T_{et} down to the ambient temperature (see Figure 3).

The heat flows (Q) are reckoned to be positive if they enter the system (e.g., heat added in the combustor, Q_{CC}) and negative if they leave the system (e.g., heat dissipated in the compressor coolers, Q_{HC}). The work (W) done by the system on the surroundings has a positive sign (e.g., expansion work, W_{TB}) whereas the work done on the system has a negative sign (e.g., compression work, W_{CM}).

A.1. Thermodynamic Modeling of Conventional CAES

In the conventional CAES system (Figure 3), air is compressed in a multi-stage compressor (CM) and then stored in the cavern (CN). Each compression stage is followed by a heat exchanger (cooler, CL) to reduce compression work of the succeeding stage and to reduce the volumetric requirements for air storage. The compression heat (heat absorbed by the coolers) is rejected to the ambient environment.

We model the compressor of conventional CAES system to have three stages: low-pressure (LP), intermediate-pressure (IP), and high-pressure (HP). All the stages have variable but equal compression ratios and a fixed isentropic efficiency throughout the entire charging process. The compression ratio (CR) varies to match the instantaneous pressure of the cavern (P_{CN}) (see Equation (A8)).

$$CR_{HP} = CR_{IP} = CR_{LP} = \sqrt[3]{\frac{P_{CN}}{P_0}} = \sqrt[3]{CR} \tag{A8}$$

During the discharge phase, air is preheated in a recuperator. It is then combusted with natural gas (NG) and generates work in the expanders (TB). We model a two-stage expander. In the recuperator, the exhaust of the low-pressure (LP) turbine preheats the air entering the high-pressure (HP) combustor.

The two existing commercial CAES plants (Huntorf and McIntosh) have a similar configuration as the one modeled here, except that the Huntorf facility does not utilize a recuperator.

A.1.1. Charge Phase of Conventional CAES

At the beginning of each charging phase, the initial temperature (T_{em}) and pressure (P_{em}) of the cavern are known from the previous storage cycle. Therefore, the mass of the cushion air is calculated by applying the ideal gas equation of state. The relationship between changes in the mass of air present in the cavern ($dm_{air,\ ch}$) and its instantaneous pressure (P_{ch}) is found by applying the First Law of thermodynamics to the control volume of the cavern (Equation (A9)).

$$dQ - dW = dU_{CV} - h_{air,\ in}\ dm_{air,\ ch} \tag{A9}$$

Since the cavern is adiabatic and has a fixed volume, $dQ = dW = 0$. Using Equations (A1)–(A3), the above equation is transformed into Equation (A10).

$$dm_{air,ch} = \frac{dP_{ch}\ V}{R\ \gamma\ T_{in}^{CN}} \tag{A10}$$

Because the inlet temperature of the cavern (T_{in}^{CN}) is fixed and known, Equation (A10) is integrated to find the total mass of working air ($m_{air,ch}$) as the cavern is charged and its pressure raises from P_{em} to P_{fl} (Equation (A11)). Once the mass of air in the fully charged cavern ($m_{air,fl}$) is determined (Equation (A12)), its temperature ($T_{CN,\ fl}$) is calculated by applying the ideal gas equation of state (Equation (A13)).

$$m_{air,ch} = \frac{\left(P_{fl} - P_{em}\right)\ V}{R\ \gamma\ T_{in}^{CN}} \tag{A11}$$

$$m_{air,fl} = m_{air,em} + m_{air,ch} \tag{A12}$$

$$T_{CN,\ fl} = \frac{T_{em}\ P_{fl}}{P_{em} + \left(P_{fl} - P_{em}\right)\frac{T_{em}}{\gamma\ T_{in}^{CN}}} \tag{A13}$$

The compression work for charging the cavern is quantified by applying the First Law of thermodynamics to each compression stage and summing them up (Equation (A14)). Work of the low-pressure compressor is formulated in Equation (A15). A similar equation is applicable to the intermediate-pressure and high-pressure compressors. The inlet temperature of each stage is fixed and known. The instantaneous discharge temperature of each compressor stage is determined by applying the isentropic compression formulae (see Equation (A16) for the LP compressor).

$$W_{CM} = \int_{P_{em}}^{P_{fl}} (dW_{CM,LP} + dW_{CM,IP} + dW_{CM,HP}) \tag{A14}$$

$$dW_{CM,LP} = C_p \left(T_{in}^{CM,LP} - T_{out}^{CM,LP} \right) dm_{air,ch} \tag{A15}$$

$$T_{out}^{CM,LP} = T_{in}^{CM,LP} - \frac{T_{in}^{CM,LP} - T_{out,\,ist}^{CM,LP}}{\eta_{CM}} \tag{A16}$$

$$T_{out,\,ist}^{CM,LP} = T_{in}^{CM,LP} \, CR_{LP}^{(\gamma-1)/\gamma}$$

The total compression heat is determined by applying the First Law of thermodynamics to each compressor cooler and integration (Equation (A17)). As a case in point, Equation (A18) quantifies an increment of heat dissipated by the cooler of the HP compressor. Note the inlet temperature of each cooler equals the discharge temperature of the preceding compression stage. The discharge temperature of the coolers is set to be fixed (Equation (A6)).

$$Q_{CL} = \int_{P_{em}}^{P_{fl}} (dQ_{CL,LP} + dQ_{CL,IP} + dQ_{CL,HP}) \tag{A17}$$

$$dQ_{CL,LP} = C_p \left(T_{out}^{CL,LP} - T_{in}^{CL,LP} \right) dm_{air,ch} \tag{A18}$$

Once the initial (*em*) and final (*fl*) states of the cavern over the charging phase in addition to the compression work are determined, the change in internal energy (ΔU_{ch}), entropy (ΔS_{ch}), and exergy (ΔX_{ch}) of the cavern as well as the exergy loss (IX_{ch}) are calculated with Equations (A19)–(A22).

$$\Delta U_{ch} = m_{air,fl} \, u_{fl} - m_{air,em} \, u_{em} \tag{A19}$$

$$\Delta S_{ch} = m_{air,fl} \, s_{fl} - m_{air,em} \, s_{em} \tag{A20}$$

$$\Delta X_{ch} = \Delta U_{ch} - T_0 \, \Delta S_{ch} \tag{A21}$$

$$IX_{ch} = -\Delta X_{ch} - W_{CM} + m_{air,ch} \, \psi_{air,ch} \tag{A22}$$

since air entering the system is at ambient conditions, $\psi_{air,ch} = 0$.

A.1.2. Discharge Phase of Conventional CAES

Similar to the charging process, we use the First Law of thermodynamics to find the relationship between the instantaneous temperature and changes in the mass and pressure of the air present in the cavern during the discharge phase (Equation (A23)).

$$dm_{air,dch} = \frac{dP_{CN} \, V}{R \, \gamma \, T_{CN}} \tag{A23}$$

Using Equation (A23) and the state equation for ideal gas, mass of air left in the cavern at the end of the discharge process (cushion air, $m_{air,em}$) and total mass of air withdrawn (working air, $m_{air,dch}$) are calculated via Equations (A24) and (A25).

$$\frac{dm_{air,dch}}{m_{air,dch}} = \frac{dP_{CN}}{\gamma \, P_{CN}}$$

$$m_{air,em} = m_{air,fl} \left(\frac{P_{em}}{P_{fl}} \right)^{\frac{1}{\gamma}} \tag{A24}$$

$$m_{air,dch} = m_{air,fl} - m_{air,em} = \frac{V}{R} \left(\frac{P_{em}}{T_{em}} + \frac{P_{fl} - P_{em}}{\gamma \, T_{in}^{CN}} \right) \left(1 - \left(\frac{P_{em}}{P_{fl}} \right)^{\frac{1}{\gamma}} \right) \tag{A25}$$

Now that mass of the cushion air is known (Equation (A24)), the state equation is applied to determine the instantaneous and final temperature of the compressed air in the cavern (Equations (A26) and (A27)).

$$T_{CN,\,dch} = T_{fl} \left(\frac{P_{CV}}{P_{fl}} \right)^{(\gamma-1)/\gamma} \tag{A26}$$

$$T_{em} = T_{fl} \left(\frac{P_{em}}{P_{fl}} \right)^{(\gamma-1)/\gamma} \tag{A27}$$

The initial temperature of air in the fully discharged cavern is set to T_0 (i.e., temperature of the air in cavern at the beginning of the very first cycle). This temperature eventually reaches asymptotic limits after many cycles, regardless of the initial temperature we choose. Simulation is run until this asymptotic limit is reached and all the results are reported then.

Temperature of the air entering the low- and high-pressure turbines ($T_{in}^{TB,LP}$ and $T_{in}^{TB,HP}$) is set fixed and constant in the analysis. However, the discharge temperature varies according to the instantaneous expansion ratio, which is itself a function of the instantaneous cavern pressure. The First Law of thermodynamics is applied to each expander stage to find the work generated (Equations (A28) and (A29)). The instantaneous discharge temperature of the high-pressure expander is quantified by Equation (A30), based on the isentropic expansion formulae. A similar set of equations can be written for the low-pressure turbine. We use a fixed isentropic efficiency for all stages.

$$W_{TB} = \int_{P_{fl}}^{P_{em}} (dW_{TB,HP} + dW_{TB,LP}) \tag{A28}$$

$$dW_{TB,HP} = C_p \left(T_{in}^{TB,HP} - T_{out}^{TB,HP} \right) dm_{air,dch} \tag{A29}$$

$$T_{out}^{TB,HP} = T_{in}^{TB,HP} + \eta_{TB} \left(T_{out,\,ist}^{TB,HP} - T_{in}^{TB,HP} \right) \tag{A30}$$

$$T_{out,\,ist}^{TB,HP} = T_{in}^{TB,HP} \, XR_{HP}^{(\gamma-1)/\gamma}$$

Once the instantaneous exit temperature of the high-pressure expander ($T_{out}^{TB,HP}$) is quantified, total heat added in the low-pressure combustor is determined by applying the First Law of thermodynamics (Equation (A31)). Note that $T_{in}^{CC,LP} = T_{out}^{TB,HP}$ and $T_{out}^{CC,LP} = T_{in}^{TB,LP}$.

$$Q_{CC,LP} = \int_{P_{fl}}^{P_{em}} C_p \left(T_{out}^{CC,LP} - T_{in}^{CC,LP} \right) dm_{air,dch} \tag{A31}$$

The instantaneous temperature of air entering the high-pressure combustor is expressed as a function of the cavern's pressure by applying the First Law of thermodynamics to the recuperator (Equation (A32)). Similar to the low-pressure combustor, the heat added in the high-pressure combustor is quantified by applying the First Law of thermodynamics (see Equation (A33)). Note that $T_{in}^{CC,HP} = T_{out,coolant}^{RP}$ and $T_{out}^{CC,HP} = T_{in}^{TB,HP}$. T_{et} is the fixed temperature of the exhaust stream of the plant, which is leaving the recuperator (RP) and entering the final heat exchanger (FHX).

$$T_{in}^{CC,HP} = T_{CN,\,dch} + T_{out}^{TB,LP} - T_{et} \tag{A32}$$

$$Q_{CC,HP} = \int_{P_{fl}}^{P_{em}} C_p \left(T_{out}^{CC,HP} - T_{in}^{CC,HP} \right) dm_{air,dch} \tag{A33}$$

Once the heat added in the combustor is known, the total mass and exergy of the fuel (natural gas) are quantified (Equations (A34) and (A35)).

$$m_{NG} = \frac{Q_{LP,CC} + Q_{HP,CC}}{LHV_{NG}} \tag{A34}$$

$$X_{NG} = m_{NG}\, x_{NG} \tag{A35}$$

The First Law of thermodynamics is applied to quantify the heat recovered in the recuperator and the heat dissipated in the final heat exchanger (Equations (A36) and (A37)).

$$Q_{RP} = \int_{P_{fl}}^{P_{em}} C_p \left(T_{et} - T_{out}^{TB,LP} \right) dm_{air,dch} \tag{A36}$$

$$Q_{FHX} = \int_{P_{fl}}^{P_{em}} C_p \left(T_0 - T_{et} \right) dm_{air,dch} \tag{A37}$$

Finally, the change in internal energy, entropy, and exergy of the cavern as well as exergy lost over the discharge process are calculated (Equation (A38) to Equation (A41)). The exergy of air stream leaving the storage plant ($\psi_{air,dch}$) is zero, similar to the exergy of air entering the plant ($\psi_{air,ch}$). This is because both air streams are at ambient conditions.

$$\Delta U_{dch} = m_{air,em}\, u_{em} - m_{air,fl}\, u_{fl} \tag{A38}$$

$$\Delta S_{dch} = m_{air,em}\, s_{em} - m_{air,fl}\, s_{fl} \tag{A39}$$

$$\Delta X_{dch} = \Delta U_{dch} - T_0\, \Delta S_{dch} \tag{A40}$$

$$IX_{dch} = -\Delta X_{dch} - W_{TB} + X_{NG} - m_{air,dch}\, \psi_{air,dch} \tag{A41}$$

A.1.3. Roundtrip Analysis of Conventional CAES

Once the work, heat, and exergy fluxes during the charge and discharge processes are quantified, the storage efficiency, work ratio and heat rate of the storage plant are calculated (Equation (A42) to Equation (A44)). The GHG emissions intensity of the plant and cavern exergy density are determined by using Equations (A45) and (A46).

$$\eta_{storage} = \frac{W_{TB}}{-W_{CM} + X_{NG} - W_{electrolysis}} \tag{A42}$$

$$WR = \frac{-W_{CM} - W_{electrolysis}}{W_{TB}} \tag{A43}$$

$$HR = \left(\frac{m_{NG} LHV_{NG}}{W_{TB}} \right) \left(\frac{3.6\ \text{GJ}}{\text{MWh}} \right) \tag{A44}$$

$$GI_{storage} = (HR)\,(GI_{NG}) \tag{A45}$$

$$\rho = \frac{W_{TB}}{V} \tag{A46}$$

We use an emissions intensity of 66 kgCO$_2$e/GJ for natural gas to account for upstream emissions in addition to combustion emissions [2].

Figure A1 illustrates the temperature range for each system component over the full charge and discharge cycle. In parentheses by each component are the minimum and maximum temperatures in °C as the plant goes through a full charging and discharge cycle. For instance (39,70) at the discharge of the cavern indicates the temperature of the fully discharged cavern (39 °C at 5 MPa) and fully charged cavern (70 °C at 7 MPa). If only one number is shown (e.g., discharge of the HP combustor), then the temperature is constant throughout the cycle.

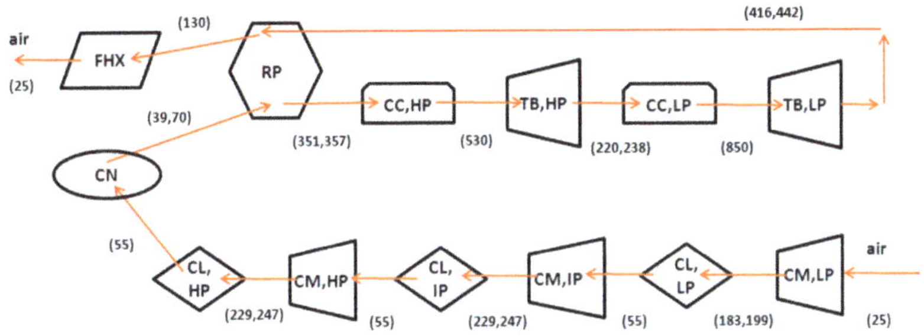

Figure A1. Temperature range of the conventional CAES system in the base case scenario. Values in each bracket show minimum and maximum temperature. The cavern operates between 5 and 7 MPa. All compression stages and expansion stages have equal pressure ratios.

A.2. Thermodynamic Modeling of Adiabatic CAES (A-CAES)

Figure 4 illustrates schematics of the A-CAES system simulated in this paper. The compression heat is stored in two thermal energy store facilities (TS1 and TS2). The coolers (heat exchangers) between the compressor stages are eliminated in order to increase the temperature of the compression heat available for thermal energy storage. Only one cooler is used, between TS2 and cavern, to lower temperature of the air prior to storage in the cavern.

The compressors have two stages, low-and high-pressure. The LP and HP stages have variable but equal compression ratios. The instantaneous compression ratio in the A-CAES system is therefore, given by Equation (A47).

$$CR_{HP} = CR_{LP} = \sqrt[2]{\frac{P_{CN}}{P_0}} = \sqrt[2]{CR} \tag{A47}$$

The expansion train of A-CAES is made up of two stages as well, LP and HP. The withdrawn air is heated in TS1 and TS2 before expansion and power generation (no combustor exists in A-CAES). The final heat exchanger (FHX) cools the discharge stream of the low-pressure expander to the ambient temperature.

The TS1 unit absorbs heat from the air leaving the low-pressure compressor and provides heat to air entering the high-pressure expander. TS2 interacts with the high-pressure compressor and the low-pressure expander. TS2 stores heat at higher temperatures compared to TS1 since it is in contact with the high temperature air leaving the HP compressor (see Figure A2). Therefore, in our design, TS2 is set to release its high temperature heat to the compressed air entering the LP compressor so that the temperature of air leaving the turbine remains above freezing, to avoid damaging turbine blades (the temperature and amount of heat stored in TS1 are not sufficient for this purpose).

The temperatures of the air entering the expanders (i.e., exhaust of TS1 and TS2) during the discharge phase are constrained to be constant. However, their values are dictated by amount of the heat stored. Note that the inlet temperatures of expander in the conventional CAES system were constant as well but their values were a design parameter (i.e., preset). The thermal energy storage units (TS1 and TS2) are modeled isobaric and adiabatic.

A.2.1. Charge Phase of A-CAES

Equations (A6) and (A10) remain applicable to determine the temperature and mass of compressed air entering the cavern over the charging period. Similarly, Equations (A11)–(A13) quantify changes in the mass and temperature of compressed air in the cavern.

The compression work of A-CAES is calculated by applying the First Law of thermodynamics to the LP and HP compressors and summing them up (Equation (A48)). The work and discharge temperature of the LP compressor are quantified by Equations (A49) and (A50). Similar formulae can be written for the HP compressor.

$$W_{CM} = \int_{P_{em}}^{P_{fl}} (dW_{CM,LP} + dW_{CM,HP}) \tag{A48}$$

$$dW_{CM,LP} = C_p \left(T_{in}^{CM,LP} - T_{out}^{CM,LP} \right) dm_{ch} \tag{A49}$$

$$T_{out}^{CM,LP} = T_{in}^{CM,LP} - \frac{T_{in}^{CM,LP} - T_{out,\,ist}^{CM,LP}}{\eta_{CM}} \tag{A50}$$

$$T_{out,\,ist}^{CM,LP} = T_{in}^{CM,LP} \, CR_{LP}^{(\gamma-1)/\gamma}$$

The maximum exit temperature of the high-pressure compressor is a preset parameter, which will be varied in the sensitivity analysis section. Our rationale for this design constraint is the following. The exit temperature of the compressor is a key parameter for determining the exergy supplied to and stored in thermal storage. This consequently impacts the temperature of the air entering the expanders during the discharge phase. Moreover, there are technical constraints such as the stress on and the fatigue of compressor blades driven by the maximum temperature of the compressor.

Thermal storage units (TS1 and TS2) are constrained to have a constant discharge temperature over the charging period. The constant exit temperature of TS2 (succeeding the HP compressor) is preset ($T_{out}^{TS2,ch}$). The constant exit temperature of TS1 (preceding the HP compressor) is, however, dictated by the value chosen for maximum discharge temperature of the HP compressor ($T_{out}^{CM,HP,\,Max}$). Solving the system of equations composed of Equations (A51)–(A53) finds the fixed inlet temperature of the high-pressure compressor ($T_{in}^{CM,HP}$) that limits its maximum exit temperature to the preset value of $T_{out}^{CM,HP,\,Max}$. Note that $T_{in}^{CM,HP}$ is equal to $T_{out}^{TS1,ch}$.

$$CR_{HP,\,Max} = \sqrt[2]{\frac{P_{fl}}{P_0}} = \sqrt[2]{CR_{Max}} \tag{A51}$$

$$T_{out}^{CM,HP,Max} = T_{in}^{CM,HP} + \frac{T_{out,\,ist}^{CM,HP,Max} - T_{in}^{CM,HP}}{\eta_{CM}} \tag{A52}$$

$$T_{out,\,ist}^{CM,HP,Max} = T_{in}^{CM,HP} (CR_{HP,Max})^{\frac{\gamma-1}{\gamma}} \tag{A53}$$

Equation (A54) gives the total heat stored in TS2 during the charging phase. The inlet temperature of TS2 is equal to the exit temperature of the high-pressure compressor and is a function of the instantaneous compression ratio (see Equations (A55) and (A56)). The exit temperature of TS2 ($T_{out}^{TS2,ch}$) is constrained to be constant throughout the charging process. Equations (A57)–(A59) show similar formulae for TS1 absorbing and storing the heat from compressed air leaving the LP compressor.

$$Q_{TS2,ch} = \int_{P_{em}}^{P_{fl}} C_p \left(T_{out}^{TS2,ch} - T_{out}^{CM,HP} \right) dm_{air,ch} \tag{A54}$$

$$T_{out}^{CM,HP} = T_{in}^{CM,HP} - \frac{T_{in}^{CM,HP} - T_{out,\,ist}^{CM,HP}}{\eta_{CM}} \tag{A55}$$

$$T_{out,\,ist}^{CM,HP} = T_{in}^{CM,HP} (CR_{HP})^{\frac{\gamma-1}{\gamma}} \tag{A56}$$

$$Q_{TS1,ch} = \int_{P_{em}}^{P_{fl}} C_p \left(T_{out}^{TS1,ch} - T_{out}^{CM,LP} \right) dm_{air,ch} \tag{A57}$$

$$T_{out}^{CM,LP} = T_{in}^{CM,LP} - \frac{T_{in}^{CM,LP} - T_{out,\,ist}^{CM,LP}}{\eta_{CM}} \tag{A58}$$

$$T_{out,\,ist}^{CM,LP} = T_{in}^{CM,LP} \left(CR_{LP}\right)^{\frac{\gamma-1}{\gamma}} \tag{A59}$$

Exergy of the heat stored in TS1 and TS2 is similarly quantified (Equations (A60) and (A61)). The exergy of the air stream is calculated according to Equations (A2), (A4) and (A5) since the inlet and outlet temperature and pressure of TS1 and TS2 are known.

$$\Delta X_{TS1,ch} = \int_{P_{em}}^{P_{fl}} \left(\psi_{out}^{TS1,ch} - \psi_{out}^{CM,LP} \right) dm_{air,ch} \tag{A60}$$

$$\Delta X_{TS2,ch} = \int_{P_{em}}^{P_{fl}} \left(\psi_{out}^{TS2,ch} - \psi_{out}^{CM,HP} \right) dm_{air,ch} \tag{A61}$$

Air stream leaving TS2 is further cooled in the aftercooler prior to entering the cavern. Equation (A62) quantifies the heat rejected to the ambient in the aftercooler. Equation (A63) expresses the total heat that is dissipated by the compressor, equal to the heat stored by TS1 and TS2 plus the heat rejected to the ambient environment by the aftercooler

$$Q_{CL} = \int_{P_{em}}^{P_{fl}} C_p \left(T_{in}^{CN} - T_{out}^{TS2,ch} \right) dm_{air,ch} \tag{A62}$$

$$Q_{CM} = Q_{CL} + Q_{TS1,ch} + Q_{TS2,ch} \tag{A63}$$

Equations (A19)–(A21) remain applicable for quantifying the changes in the internal energy, entropy, and exergy of the air stored in the cavern of A-CAES. Exergy destroyed over the charging period is shown in Equation (A64).

$$IX_{ch} = -\Delta X_{ch} - W_{CM} + m_{ch}\,\psi_0 - \Delta X_{TS1,ch} - \Delta X_{TS1,ch} \tag{A64}$$

A.2.2. Discharge Phase of A-CAES

Changes in the mass and temperature of compressed air present in the cavern of A-CAES during the discharge period are expressed by Equations (A23)–(A27), similar to the conventional CAES model.

We assume perfect storage of heat, i.e., all the thermal energy stored during the charging phase is released back to the expanding air without any losses. The exit temperature of TS1 and TS2 is assumed to remain constant over the discharge period.

The First Law of thermodynamics is applied to perform a series of trials and errors in order to find the fixed (but unknown) exit temperature of TS1 so that the heat released by TS1 during the discharge period (Equation (A65)) becomes equal to the heat previously stored in it during the charge period (Equation (A57)).

$$Q_{TS1,dch} = \int_{P_{fl}}^{P_{em}} C_p \left(T_{out}^{TS1,dch} - T_{CN,\,dch} \right) dm_{air,dch} \tag{A65}$$

$T_{CN,dch}$ is the instantaneous temperature of air leaving the cavern (Equation (A26)).

Once the exit temperature of TS1 is calculated, work and discharge temperature of the high-pressure expander are found through applying the First Law of thermodynamics to the high-pressure expander (Equations (A66) and (A67)). The instantaneous expansion ratio is expressed by Equation (A7).

$$W_{TB,HP} = \int_{P_{fl}}^{P_{em}} C_p \left(T_{out}^{TS1,dch} - T_{out}^{TB,HP} \right) dm_{air,dch} \tag{A66}$$

$$T_{out}^{TB,HP} = T_{out}^{TS1,dch} + \eta_{TB} \left(T_{out,\,is}^{TB,HP} - T_{out}^{TS1,dch} \right) \tag{A67}$$

$$T_{out,\,ist}^{TB,HP} = \left(T_{out}^{TS1,dch}\right) XR_{HP}^{(\gamma-1)/\gamma}$$

Upon determining the intake temperature of TS2 during the discharge phase (equal to $T_{out}^{TB,HP}$), we again use the First Law of thermodynamics to express heat of TS2 during the discharge phase (Equation (A68)). The same trial and error procedure explained for TS1 is applied to Equations (A68) and (A54) to find the exit temperature of TS2.

$$Q_{TS2,dch} = \int_{P_{fl}}^{P_{em}} C_p \left(T_{out}^{TS2,dch} - T_{out}^{TB,HP}\right) dm_{air,dch} \tag{A68}$$

The changes in the exergy stock of TS1 and TS2 are quantified by Equations (A69) and (A70). Since we model the thermal storage systems as ideal (no losses), the exergy released by TS1 and TS2 is equal to the exergy stored during the charging phase.

$$\Delta X_{TS1,dch} = \int_{P_{em}}^{P_{fl}} \left(\psi_{out}^{TS1,dch} - \psi_{dch}^{CN}\right) dm_{air,dch} \tag{A69}$$

$$\Delta X_{TS2,dch} = \int_{P_{em}}^{P_{fl}} \left(\psi_{out}^{TS2,dch} - \psi_{out}^{TB,HP}\right) dm_{air,dch} \tag{A70}$$

Now that the inlet temperature of the low-pressure turbine is determined (equal to discharge temperature of TS2), the First Law of thermodynamics is used to determine the work of the low-pressure turbine (Equation (A71)) and its exit temperature (Equation (A72)).

$$W_{TB,LP} = \int_{P_{fl}}^{P_{em}} C_p \left(T_{out}^{TS2,dch} - T_{out}^{TB,LP}\right) dm_{air,dch} \tag{A71}$$

$$T_{out}^{TB,LP} = T_{out}^{TS2,dch} + \eta_{TB} \left(T_{out,\,ist}^{TB,LP} - T_{out}^{TS2,dch}\right) \tag{A72}$$

$$T_{out,\,ist}^{TB,LP} = \left(T_{out}^{TS2,dch}\right)\left(XR_{LP}^{(\gamma-1)/\gamma}\right)$$

Finally, we quantify the heat dissipated by the final exhaust heat exchanger via Equation (A73).

$$Q_{FHX} = \int_{P_{fl}}^{P_{em}} C_p \left(T_0 - T_{out}^{TB,LP}\right) dm_{air,dch} \tag{A73}$$

Equations (A38)–(A40) remain applicable to calculate the changes in the internal energy, entropy, and exergy of cavern over the discharge phase. Equation (A74) quantifies the total exergy loss during the discharge phase of A-CAES.

$$IX_{dch} = -\Delta X_{dch} - W_{TB} - m_{air,dch}\,\psi_{air,dch} - \Delta X_{TS1,dch} - \Delta X_{TS1,dch} \tag{A74}$$

A.2.3. Roundtrip Analysis of A-CAES

Equations (A42)–(A46) are applicable for the performance of the A-CAES system. Since no fuel is consumed, the heat rate and GHG emissions intensity of A-CAES are zero.

Similar to Figure A1, Figure A2 shows the temperature range for each A-CAES system components over the storage cycle.

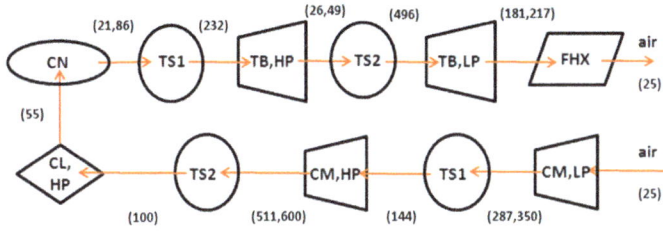

Figure A2. Temperature range of the A-CAES system in the base case scenario. Values in each bracket show minimum and maximum temperature. The cavern operates in 5–10 MPa range. The compressor and expander stages have equal pressure ratios.

A.3. Modeling CAES Paired with a High-Temperature Electrolyzer

The compressor of the CAES-HTE system simulated has two stages (similar to A-CAES). A cooler (heat exchanger) lowers the temperature of the exhaust stream of the LP compressor to a fixed value before entering the high-pressure compressor. The maximum discharge temperature of the high-pressure compressor is a design parameter and is preset, similar to A-CAES. The discharge of the HP compressor is fed into a high-temperature electrolyzer. Its heat is used to boil water, to heat up the generated steam to the constant temperature of the electrolyzer (T_{HTE}), and to provide the heating energy required for the electrolysis process.

A cooler follows the electrolyzer, lowering temperature of the compressed air to a fixed value ($T_{CN,\,in}$) before storage in the cavern. The generated hydrogen is stored separately to combust and heat the expanding air during the discharge phase. No physical storage of heat is performed, in contrast to the A-CAES configuration.

The discharge phase of CAES-HTE is identical to that of the conventional CAES in our model, with the distinction that hydrogen fuels the combustors instead of natural gas.

The heating loads of the high-pressure and low-pressure combustors during the discharge phase determine the amount of hydrogen fuel needed and consequently the energy demand of the electrolyzer during the charge phase. We therefore, discuss the discharging phase first and then the charging phase.

A.3.1. Discharge Phase of CAES-HTE

Equations (A23)–(A27) remain applicable for quantifying the changes in the mass and temperature of the compressed air remaining in the cavern during discharge. Similar to the conventional CAES system, the inlet temperatures of the low-pressure and high-pressure turbines ($T_{in}^{TB,LP}$ and $T_{in}^{TB,HP}$) are preset. Therefore, Equations (A28)–(A30) quantify the expansion work and the instantaneous exit temperature of the turbines (expanders). The heat load of the low-pressure and high-pressure combustors are determined by applying Equations (A31), (A33) and (A75). Finally, the mass and exergy of the hydrogen fuel needed are calculated via Equations (A76) and (A77).

$$Q_{CC} = Q_{LP,CC} + Q_{HP,CC} \tag{A75}$$

$$m_{H2} = \frac{Q_{CC}}{LHV_{H2}} \tag{A76}$$

$$X_{H2} = m_{H2}\, x_{H2} \tag{A77}$$

Equations (A38)–(A40) remain applicable for quantifying the changes in internal energy, entropy, and exergy of the compressed air present in the cavern. The exergy lost during the discharge phase is quantified by Equation (A78).

$$I_{dch} = -\Delta X_{dch} - W_{TB} + X_{H2}^{CC} - m_{air,dch}\, \psi_{air,dch} \tag{A78}$$

A.3.2. Charge Phase of CAES-HTE

Charging phase of CAES-HTE is similar to that of conventional CAES with the key difference that exhaust stream of the HP compressor transfers some of its heat to an electrolyzer to generate hydrogen and the rest to the ambient in an aftercooler. As with the A-CAES simulation, the maximum discharge temperature of the HP compressor is preset. This parameter dictates the exit temperature of the LP cooler (a constant but unknown parameter).

Equation (A9) to Equation (A13) remain valid for quantifying changes in the mass and temperature of compressed air in the cavern during the charge period. Work by the LP compressor and its discharge temperature are calculated by applying the First Law of thermodynamics to the low-pressure compressor (Equations (A79) and (A80)).

$$W_{CM,LP} = \int_{P_{em}}^{P_f} C_p \left(T_{in}^{CM,LP} - T_{out}^{CM,LP} \right) dm_{air,ch} \tag{A79}$$

$$T_{out}^{CM,LP} = T_{in}^{CM,LP} - \frac{T_{in}^{CM,LP} - T_{out,\, ist}^{CM,LP}}{\eta_{CM}} \tag{A80}$$

$$T_{out,\, ist}^{CM,LP} = T_{in}^{CM,LP} \, CR_{LP}^{(\gamma-1)/\gamma}$$

The next step after quantifying the inlet temperature of the LP cooler (equal to exit temperature of the LP compressor, Equation (A80)) is finding its discharge temperature, which is same as the inlet temperature of the HP compressor. This temperature is constrained to remain constant during the charging period. Its value, however, is dictated by another design parameter: the maximum exit temperature of the HP compressor. Similar to the A-CAES model, solving the system of equations of Equations (A51)–(A53) finds the inlet temperature of the high-pressure compressor ($T_{in}^{CM,HP}$, which is equal to $T_{out}^{CL,LP}$).

Once the inlet temperature of the HP compressor is known, its work and instantaneous exit temperature are determined by applying Equations (A81) and (A82).

$$W_{CM,HP} = \int_{P_{em}}^{P_{fl}} C_p \left(T_{in}^{CM,HP} - T_{out}^{CM,HP} \right) dm_{air,ch} \tag{A81}$$

$$T_{out}^{CM,HP} = T_{in}^{CM,HP} - \frac{T_{in}^{CM,HP} - T_{out,\, ist}^{CM,HP}}{\eta_{CM}} \tag{A82}$$

$$T_{out,\, ist}^{CM,HP} = T_{in}^{CM,HP} \, CR_{HP}^{(\gamma-1)/\gamma}$$

As illustrated in Figure 2, the energy demand of the electrolyzer depends on the reaction temperature. For the sake of simplicity, we assume a fixed electrolysis temperature, equal to the average exit temperature of the HP compressor over the charge process.

Since the instantaneous inlet temperature of the electrolyzer is known as a function of the compression ratio (equal to $T_{out}^{CM,HP}$, Equation (A82)), one can apply the First Law of thermodynamics to the electrolyzer to find its exit temperature (T_{out}^{HTE}), which results in the desired heat load of the electrolyzer ($Q_{HTE,\, total}$, quantified in the next paragraphs and expressed by Equation (A97)). A trial and error process is used (similar to the A-CAES system) to find T_{out}^{HTE}.

Total heat demand of the electrolyzer ($Q_{HTE,\, total}$) is dictated by the mass of hydrogen needed to fuel the combustors during the discharge process. This heating load is made up of four components: sensible heat load to bring water to 100 °C ($Q_{Sns,water}$), latent heat load to boil it ($Q_{Lnt,water}$), sensible heat load to bring the steam from 100 °C to the electrolysis temperature ($Q_{Sns,steam}$), and the heat required for the electrolysis process itself (Q_{HTE}).

In a high-temperature electrolyzer, steam is disassociated in the cathode to produce hydrogen and O^{2-} while O^{2-} gets oxidized in the anode producing oxygen (Equations (A83)–(A85)).

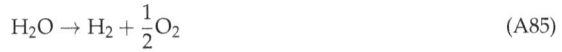

$$H_2O + 2e \rightarrow H_2 + O^{2-} \tag{A83}$$

$$O^{2-} \rightarrow \frac{1}{2}O_2 + 2e \tag{A84}$$

$$H_2O \rightarrow H_2 + \frac{1}{2}O_2 \tag{A85}$$

One mole of hydrogen and half a mole of oxygen are generated per mole of water. The relationship between mass of hydrogen and water are shown in Equation (A86).

$$m_{H_2O} = m_{H_2} \left(\frac{M_{H_2O}}{M_{H_2}} \right) \left(\frac{n_{H_2O}}{n_{H_2}} \right) \tag{A86}$$

The sensible heat load of water, sensible heat load of steam, and the latent heat are quantified by Equations (A87)–(A89).

$$Q_{Sns,water} = m_{H_2O} \, C_p \, (T_{boil} - T_0) \tag{A87}$$

$$Q_{Sns,steam} = m_{H_2O}^{HTE} \, C_p \, (T_{HTE} - T_{boil}) \tag{A88}$$

$$Q_{Ltn,water} = m_{H_2O}^{HTE} \, C_{Lnt} \tag{A89}$$

For the electrolysis process itself, the total energy requirement (ΔH), the change in entropy (ΔS), and the electricity demand (change in the Gibbs free energy, ΔG) are determined by Equations (A90)–(A92) as a function of the reaction temperature (T).

$$\Delta H_{HTE}(T) = n_{H_2} \, \Delta h_{H_2}(T) + n_{O_2} \, \Delta h_{O_2}(T) - n_{H_2O} \, \Delta h_{H_2O}(T) \tag{A90}$$

$$\Delta S_{HTE}(T) = n_{H_2} \, \Delta s_{H_2}(T) + n_{O_2} \, \Delta s_{O_2}(T) - n_{H_2O} \, \Delta s_{H_2O}(T) \tag{A91}$$

$$\Delta G_{HTE}(T) = \Delta H_{HTE}(T) - T_{HTE} \, \Delta S_{HTE}(T) \tag{A92}$$

The changes in specific enthalpy (Δh) and entropy (Δs) of the reactant and products are expressed by Equations (A93) and (A94), respectively. Symbols Δh_f^0 and s^0 represent the standard specific enthalpy of formation and the standard specific entropy. Specific heat (c_p) is itself a function of temperature. The values for the standard enthalpy of formation, standard entropy, and specific heat of water and steam are shown in Table A1.

$$\Delta h(T) = \Delta h_f^0 + \int_{T_0}^{T_{HTE}} C_p(T) \, dT \tag{A93}$$

$$\Delta s(T) = s^0 + \int_{T_0}^{T_{HTE}} \frac{C_p(T)}{T} \, dT \tag{A94}$$

We therefore, use Table A1 and Equations (A90)–(A94) to quantify the heat and work load of the electrolyzer, as shown in Equations (A95) and (A96).

$$Q_{HTE} = T_{HTE} \, \Delta S_{HTE}(@T_{HTE}) \tag{A95}$$

$$W_{HTE} = \Delta G_{HTE}(@T_{HTE}) \tag{A96}$$

Finally, the total heat load of the electrolyzer ($Q_{HTE, \, total}$) is determined by Equation (A97). This variable is used to find the discharge temperature of the electrolyzer (T_{out}^{HTE}) through the trial and error procedure explained earlier.

$$Q_{HTE, \, total} = Q_{Sns,water} + Q_{Lnt,water} + Q_{Sns,steam} + Q_{HTE} \tag{A97}$$

Table A1. Thermodynamic properties used to quantify energy demand of electrolysis [39,40].

Component	Δh_f^0 (kJ/mol)	s^0 (kJ/kmolK)	C_p (kJ/kmolK)	C_{Lnt} (kJ/mol)	n
H_2 (gas)	0	131	$27.28 + 0.00326\,T + 50,000/T^2$	-	1
O_2 (gas)	0	205	$29.96 + 0.00418\,T - 167,000/T^2$	-	0.5
HO_2 (liquid)	-285.83	70	75.44	40.7	1
HO_2 (gas)	-241.82	189	$30 + 0.01071\,T + 33,000/T^2$	-	1

The First Law of thermodynamics is applied to each cooler to determine the heat dissipated by the LP and HP coolers (Equations (A98)–(A100)). Note that the discharge temperature of the HP cooler is preset to the approach temperature plus the ambient temperature ($T_{ac} + T_0$), same as the conventional CAES model. The intake temperature of the high-pressure compressor ($T_{in}^{CM,HP}$) is equal to the exit temperature of the LP cooler. The total compression heat is equal to the heat dissipated in the compressor coolers (Equation (A98)) plus the heat load of the electrolyzer (Equation (A97)), as shown in Equation (A101).

$$Q_{CL} = \int_{P_{em}}^{P_{fl}} (dQ_{CL,LP} + dQ_{CL,HP}) \tag{A98}$$

$$dQ_{CL,LP} = C_p \left(T_{in}^{CM,HP} - T_{out}^{CM,LP} \right) dm_{air,ch} \tag{A99}$$

$$dQ_{CL,HP} = C_p \left(T_{CN,in} - T_{out}^{HTE} \right) dm_{air,ch} \tag{A100}$$

$$Q_{CM} = Q_{CL} + Q_{HTE,\,total} \tag{A101}$$

Equations (A19)–(A21) remain valid for quantifying the changes in the internal energy, entropy, and exergy of the compressed air in the cavern. The exergy lost over the charging process of CAES-HTE is calculated by Equation (A102).

$$IX_{ch} = m_{ch}\,\psi_0 + X_{H2} - \Delta X_{ch} - W_{CM} - W_{HTE} \tag{A102}$$

A.3.3. Roundtrip Performance of CAES-HTE

Equations (A42)–(A46) remain applicable for expressing the roundtrip performance of the CAES-HTE system. Since no fuel is consumed, the heat rate and GHG emissions intensity of HTE-CAES are zero.

The temperature range of the components of the CAES-HTE system in the base case are illustrated in Figure A3.

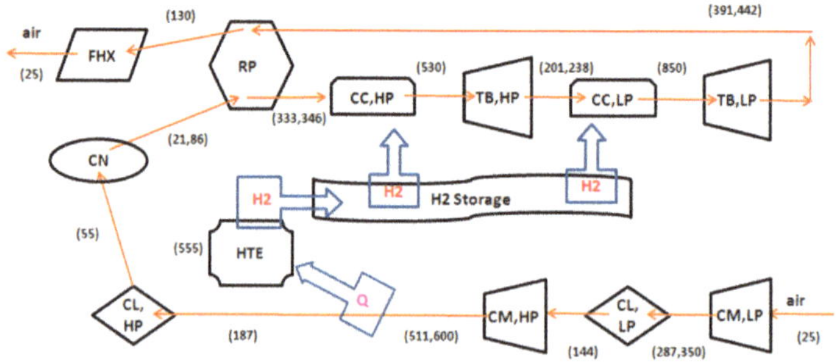

Figure A3. Temperature range of the CAES-HTE system in the base case scenario. Values in each bracket show minimum and maximum temperature. The cavern operates in 5–10 MPa range. The compressor and expander stages have equal pressure ratios.

A.4. Modeling CAES Paired with a Low-Temperature Electrolyzer (CAES-LTE)

The CAES-LTE system modeled is similar to the CAES-HTE system in the sense that hydrogen is produced onsite to fuel combustors during the discharge phase. However, water instead of steam is electrolyzed in a low-temperature electrolyzer.

Similar to the conventional CAES system and in contrast to CAES-HTE, we do not impose any constraint on the maximum discharge temperature of the high-pressure compressor ($T_{out}^{CM,HP,Max}$). Additionally, the exit temperature of the compressor coolers is constant and preset to the ambient temperature plus the approach temperature, similar to the conventional CAES model.

We model a three-stage compressor for CAES-LTE since there is no need for producing high-temperature heat in this system (electrolysis occurs at the ambient temperature). Using three stages reduces the work requirements of the compressor, as discussed for conventional CAES. The compression ratio in the CAES-LTE system is given by Equation (A8).

Equations (A42)–(A46) are used to characterize performance of the CAES-LTE system. Similar to the CAES-HTE system, the CAES-LTE design does not consume natural gas, therefore its heat rate and GHG emissions are zero.

Similar to Figure A3, Figure A4 shows the temperature range for each CAES-LTE system components in the base case scenario.

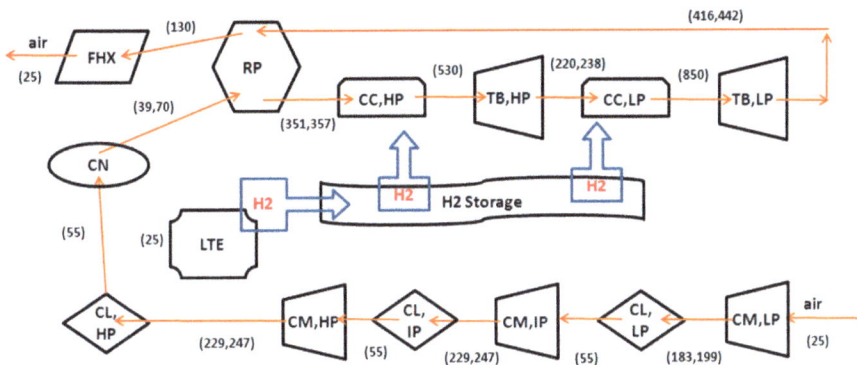

Figure A4. Temperature range of the CAES-LTE system in the base case scenario. Values in each bracket show minimum and maximum temperature. The cavern operates in 5–7 MPa range. The different stages of the compression and expansion trains have equal pressure ratios.

References

1. Kintner-Meyer, M.; Balducci, P.; Colella, W.; Elizondo, M.; Jin, C.; Nguyen, T.; Viswanathan, V.; Zhang, Y. *National Assessment of Energy Storage for Grid Balancing and Arbitrage: Phase 1, WECC*; Pacific Northwest National Laboratory: Richland, WA, USA, 2012.
2. Safaei, H.; Keith, D.W. How much bulk energy storage is needed to decarbonize electricity? *Energy Environ. Sci.* **2015**, *8*, 3409–3417. [CrossRef]
3. Schoenung, S. *Energy Storage Systems Cost Update: A Study for the DOE Energy Storage Systems Program*; Sandia National Laboratories: Livermore, CA, USA, 2011.
4. Safaei, H.; Keith, D.W.; Hugo, R.J. Compressed air energy storage (CAES) with compressors distributed at heat loads to enable waste heat utilization. *Appl. Energy* **2013**, *103*, 165–179. [CrossRef]
5. John, J.S. SustainX to Merge with General Compression, Abandon Above-Ground CAES Ambitions. Available online: https://www.greentechmedia.com/articles/read/sustainx-to-merge-with-general-compression-abandon-above-ground-caes-ambiti (accessed on 4 June 2017).
6. Fertig, E.; Apt, J. Economics of compressed air energy storage to integrate wind power: A case study in ERCOT. *Energy Policy* **2011**, *39*, 2330–2342. [CrossRef]

7.   Drury, E.; Denholm, P.; Sioshansi, R. The value of compressed air energy storage in energy and reserve markets. *Energy* **2011**, *36*, 4959–4973. [CrossRef]

8.   Greenblatt, J.B.; Succar, S.; Denkenberger, D.C.; Williams, R.H.; Socolow, R.H. Baseload wind energy: Modeling the competition between gas turbines and compressed air energy storage for supplemental generation. *Energy Policy* **2007**, *35*, 1474–1492. [CrossRef]

9.   Akhil, A.A.; Huff, G.; Currier, A.B.; Kaun, B.C.; Rastler, D.M.; Chen, S.B.; Cotter, A.L.; Bradshaw, D.T.; Gauntlett, W.D. *DOE/EPRI 2013 Electricity Storage Handbook in Collaboration with NRECA*; Sandia National Laboratories: Albuquerque, NM, USA, 2013.

10.   Raju, M.; Khaitan, S.K. Modeling and simulation of compressed air storage in caverns: A case study of the Huntorf plant. *Appl. Energy* **2012**, *89*, 474–481. [CrossRef]

11.   Kim, Y.M.; Favrat, D. Energy and exergy analysis of a micro-compressed air energy storage and air cycle heating and cooling system. *Energy* **2009**, *35*, 213–220. [CrossRef]

12.   Buffa, F.; Kemble, S.; Manfrida, G.; Milazzo, A. Exergy and exergoeconomic model of a ground-based CAES plant for peak-load energy production. *Energies* **2013**, *6*, 1050–1067. [CrossRef]

13.   Osterle, J.F. The thermodynamics of compressed air exergy storage. *J. Energy Resour. Technol.* **1991**, *113*, 7–11. [CrossRef]

14.   Skorek, J.; Banasiak, K. Thermodynamic analysis of the compressed-air energy storage systems operation. *Inzynieria Chemiczna i Procesowa* **2006**, *27*, 187–200.

15.   Succar, S.; Williams, R.H. *Compressed Air Energy Storage, Theory, Resources, and Applications for Wind Power*; Princeton University: Princeton, NJ, USA, 2008.

16.   Zhang, Y.; Yang, K.; Li, X.; Xu, J. The thermodynamic effect of air storage chamber model on advanced adiabatic compressed air energy storage system. *Renew. Energy* **2013**, *57*, 469–478. [CrossRef]

17.   Grazzini, G.; Milazzo, A. Thermodynamic analysis of CAES/TES systems for renewable energy plants. *Renew. Energy* **2008**, *33*, 1998–2006. [CrossRef]

18.   Wolf, D.; Budt, M. LTA-CAES: A low-temperature approach to adiabatic compressed air energy storage. *Appl. Energy* **2014**, *125*, 158–164. [CrossRef]

19.   Hartmann, N.; Vohringer, O.; Kruck, C.; Eltrop, L. Simulation and analysis of different adiabatic compressed air energy storage plant configurations. *Appl. Energy* **2012**, *93*, 541–548. [CrossRef]

20.   Steta, F.D.S. *Modeling of an Advanced Adiabatic Compressed Air Energy Storage (AA-CAES) Unit and an Optimal Model-Based Operation Strategy For Its Integration Into Power Markets*; Swiss Federal Institute of Technology: Zurich, Switzerland, 2010.

21.   Arsie, I.; Marano, V.; Nappi, G.; Rizzo, G. A model of a hybrid power plant with wind turbines and compressed air energy storage. In Proceedings of the ASME Power Conference, Chicago, IL, USA, 5–7 April 2005; pp. 987–1000.

22.   Bullough, C.; Gatzen, C.; Jakiel, C.; Koller, M.; Nowi, A.; Zunft, S. Advanced adiabatic compressed air energy storage for the integration of wind energy. In Proceedings of the European Wind Energy Conference (EWEC 2004), London, UK, 22–25 November 2004.

23.   Khaitan, S.K.; Raju, M. Dynamics of hydrogen powered CAES based gas turbine plant using sodium alanate storage system. *Int. J. Hydrog. Energy* **2012**, *37*, 18904–18914. [CrossRef]

24.   Bidini, G.; Grimaldi, C.N.; Postrioti, L. Thermodynamic analysis of hydraulic air compressor-gas turbine power plants. *Proc. Inst. Mech. Eng. Part A J. Power Energy* **1997**, *211*, 429–437. [CrossRef]

25.   Enis, B.M.; Lieberman, P.; Rubin, I. Operation of hybrid wind-turbine compressed-air system for connection to electric grid networks and cogeneration. *Wind Eng.* **2003**, *27*, 449–459. [CrossRef]

26.   Lund, H.; Salgi, G. The role of compressed air energy storage (CAES) in future sustainable energy systems. *Energy Convers. Manag.* **2009**, *50*, 1172–1179. [CrossRef]

27.   Kim, Y.M.; Lee, J.H.; Kim, S.J.; Favrat, D. Potential and evolution of compressed air energy storage: Energy and exergy analyses. *Entropy* **2012**, *14*, 1501–1521. [CrossRef]

28.   Li, Y.; Wang, X.; Li, D.; Ding, Y. A trigeneration system based on compressed air and thermal energy storage. *Appl. Energy* **2012**, *99*, 316–323. [CrossRef]

29.   Najjar, Y.S.H.; Jubeh, N.M. Comparison of performance of compressed-air energy-storage plant with compressed-air storage with humidification. *Proc. Inst. Mech. Eng. Part A J. Power Energy* **2006**, *220*, 581–588. [CrossRef]

30. Xia, C.; Zhou, Y.; Zhou, S.; Zhang, P.; Wang, F. A simplified and unified analytical solution for temperature and pressure variations in compressed air energy storage caverns. *Renew. Energy* **2015**, *74*, 718–726. [CrossRef]

31. Jakiel, C.; Zunft, S.; Nowi, A. Adiabatic compressed air energy storage plants for efficient peak load power supply from wind energy: The European project AA-CAES. *Int. J. Energy Technol. Policy* **2007**, *5*, 296–306. [CrossRef]

32. Zunft, S.; Jakiel, C.; Koller, M.; Bullough, C. Adiabatic compressed air energy storage for the grid integration of wind power. In Proceedings of the Sixth international workshop on large-scale integration of wind power and transmission networks for offshore windfarms, Delft, The Netherlands, 26–28 October 2006.

33. Weber, O. Huntorf air storage gas turbine power plant. *Brown Boveri Rev.* **1975**, *62*, 332–337.

34. *Compressed Air Energy Storage State-of-Science*; Electric Power Research Institute (EPRI): Palo Alto, CA, USA, 2009.

35. Knoke, S. Compressed air energy storage (CAES). In *Handbook of Energy Storage for Transmission or Distribution Applications*; Eckroad, S., Ed.; The Electric Power Research Institute (EPRI): Palo Alto, CA, USA, 2002.

36. Wright, S. *Reference Design Description and Cost Evaluation for Compressed Air Energy Storage Systems*; Electric Power Research Institute (EPRI): Palo Alto, CA, USA, 2011; p. 104.

37. Gupta, R.B. *Hydrogen Fuel: Production, Transport, and Storage*; CRC Press: Boca Raton, FL, USA, 2009.

38. Boyce, M.P. Principles of operation and performance estimation of centrifugal compressors. In Proceedings of the Twenty-Second Turbomachinery Symposium, Houston, TX, USA, 14–16 September 1993; pp. 161–177.

39. Liu, M.; Bo, Y.; Xu, J.; Jing, C. Thermodynamic analysis of the efficiency of high-temperature steam electrolysis system for hydrogen production. *J. Power Sources* **2008**, *177*, 493–499.

40. Shin, Y.; Park, W.; Chang, J.; Park, J. Evaluation of the high temperature electrolysis of steam to produce hydrogen. *Int. J. Hydrogen Energy* **2007**, *32*, 1486–1491. [CrossRef]

energies

MDPI

Article

Big-Data-Based Thermal Runaway Prognosis of Battery Systems for Electric Vehicles

Jichao Hong [1,2], **Zhenpo Wang** [1,2,*] and **Peng Liu** [1,2,*]

1   National Engineering Laboratory for Electric Vehicles, Beijing Institute of Technology, Beijing 100081, China; qdbithong@163.com
2   Beijing Co-Innovation Center for Electric Vehicles Lecturer, Beijing 100081, China
*   Correspondence: wangzhenpo@bit.edu.cn (Z.W.); bitliupeng@bit.edu.cn (P.L.)

Academic Editor: Hailong Li
Received: 15 May 2017; Accepted: 26 June 2017; Published: 4 July 2017

Abstract: A thermal runaway prognosis scheme for battery systems in electric vehicles is proposed based on the big data platform and entropy method. It realizes the diagnosis and prognosis of thermal runaway simultaneously, which is caused by the temperature fault through monitoring battery temperature during vehicular operations. A vast quantity of real-time voltage monitoring data is derived from the National Service and Management Center for Electric Vehicles (NSMC-EV) in Beijing. Furthermore, a thermal security management strategy for thermal runaway is presented under the Z-score approach. The abnormity coefficient is introduced to present real-time precautions of temperature abnormity. The results illustrated that the proposed method can accurately forecast both the time and location of the temperature fault within battery packs. The presented method is flexible in all disorder systems and possesses widespread application potential in not only electric vehicles, but also other areas with complex abnormal fluctuating environments.

Keywords: thermal runaway; battery systems; big data platform; National Service and Management Center for Electric Vehicles

1. Introduction

Battery systems are critical components that strongly influence the driving performance and cost-effectiveness of electric vehicles (EVs). The travel distance, acceleration performance, and security requirements of EVs cannot be satisfied by the energy density and power density of the single-cell. Therefore, the cells need to be assembled into a small battery module according to certain forms, and battery systems can be composed of a number of battery modules in series or parallel to satisfy the driving requirement of EVs [1]. Thermal runaway may occur with extreme phenomena, such as battery leakage, smoking, or gas venting in the event the heating rate exceeds the dissipation rate. In recent years, a spectrum of fatal fire accidents has shown the great threat to system safety and durability. Generally, thermal runaway occurs when an exothermic reaction gets out of control, which is interpreted as the reaction rate increasing due to the temperature increasing, and causes a further increase in temperature and, hence, a further increase in the reaction rate. In some serious cases, thermal runaway possibly results in an explosion [2]. Battery degradation and failure are strongly dependent on the abnormality in cell temperature. Furthermore, to maintain the healthy state of the battery, thermal management strategies are employed in electric vehicles [3].

A preeminent battery thermal management system (BTMS) is necessary and essential because extreme temperatures affect the driving performance and safety of EVs. In some extreme cases, thermal runaway might trigger fires and explosions if the battery temperature gets out of the safety scope. The effectiveness of a BTMS depends on the design of the battery system and the operating conditions. Daowd et al. [4] proposed an intelligent battery management system (BMS), including a battery pack charging and discharging control, with a battery pack thermal management system. Finally, an experimental setup was implemented for the validation of the proposed balancing system. Panchal et al. [5,6] presented in situ measurements of the heat generation rate for a prismatic lithium-ion battery and a lithium-ion pouch cell (20 Ah capacity) at 1C, 2C, 3C, and 4C discharge rates and 5 °C, 15 °C, 25 °C, and 35 °C boundary conditions (BCs). The results show that the highest rate of heat generation was found to be 91 W for the 4C discharge rate and 5 °C BC, while the minimum value was 13 W measured at a 1C discharge rate and 35 °C BC. This illustrated that the increase in the discharge rate and the discharge current caused a consistent increase in the heat generation rate for an equal depth of discharge points. A model was developed using the neural network approach and the predicted heat generation rate demonstrates an identical behavior with experimental results from this model. Lan et al. [7] developed a novel design of BTMS based on aluminum mini-channel tubes and applied it to a single prismatic lithium-ion cell under different discharge rates. To investigate the thermal performance of a lithium-ion battery pack, Qian et al. [8] established a three-dimensional numerical model using a type of liquid cooling method based on mini-channels and cold-plates. Though simplified approaches, Mastali et al. [9] developed the simplified electrochemical multi-particle model and homogenous pseudo-two-dimensional model to decrease the computational time; the speed and simplicity of three-dimensional electrochemical-thermal models are still of concern. The second type of model is the equivalent circuit model (ECM), where the battery is regarded as a mass point [10,11]. Therefore, they are suitable to be implanted in the battery management system (BMS) for the state of charge (SOC) or the state of health (SOH) estimation [12–15]. Lin et al. [16] and Forgez et al. [17] added lumped-parameter thermal models to the ECM to predict the thermal characteristics of the cell, which made the model more comprehensive. The results showed this method could effectively control the battery temperature at a 5 °C discharge and the temperature uniformity was obviously improved. Through the studies mentioned in the literature, apart from a few studies monitoring temperature changes through the temperature sensor, no effective and systematic theory or method concerns the accurate and timely temperature fault detection and early detection and warning of thermal runaway during real operation.

Meanwhile, in order to maintain higher energy density, the size and complexity of the battery cell is growing, which leads to a potential temperature imbalance and a risk of various battery faults. So many fault diagnosis methodologies have been presented to reveal the thermal runaway of battery systems. For external short-circuit detection, Xiong et al. [18] extracted the OCV-SOC relationship from any existing current-voltage measurements by using an H infinity filter within several seconds. The results show that the estimated OCV can result in accurate SoC estimation with a maximum error of 1%. Seo et al. [19] proposed a high accuracy model-based switching model method (SMM) to detect the internal short circuit (ISCr) in the lithium-ion battery, which helps the battery management system to fulfill early detection of the ISCr. Zhang et al. [20] proposed a novel method to perform online and real-time capacity fault diagnosis for a parallel-connected battery group (PCBG) and the fault simulation and validation results demonstrate that the proposed methods have good accuracy and reliability. Due to the inconsistent and varied characteristics of lithium-ion battery cells, Chen et al. [21] and Liu et al. [22] proposed the multi-scale dual H infinity filters and model-based sensor fault diagnosis method, which can significantly reduce the computation work and retains good model accuracy. Bai et al. [23] applied a combined power generation system (CPGS) to achieve a reliable evaluation of a distribution network with micro-grids combined with fault duration. In addition, many model-based diagnostic algorithms, such as extended Kalman, were presented to diagnose thermal faults in lithium-ion batteries [24–27], and the simulation and experimental studies were demonstrated to illustrate the effectiveness of the proposed schemes. Zheng et al. [28] presented a battery pack system in a demonstrated EV with 96 cells in series and discovered the battery power fade fault during the demonstration. The preliminary analysis indicated that the internal or contact resistance increase causes the fault and calculating the Shannon entropy clearly identified the cause of the power fade fault. Rezvanizanian et al. [29] examined the mobility prediction of LiFeMnPO4 batteries for an emission-free electric vehicle. Through the comparison with an adaptive recurrent neural network (ARNN) with regression, the former performs with better accuracy in two different road types and driving modes. All of these mentioned studies have modeled under online detection and prediction on the SOH of battery system. However, the literature has rarely explored temperature fault diagnosis and prognosis issues of battery systems directly for the real-time running vehicles. The conventional threshold methods lack the ability of identifying the time and location that the abnormity occurs if the abnormal data remains within the permitted limits together with the safety data. The existing BMS technology generally cannot achieve an early warning effect of battery thermal runaway.

This paper focuses on a prognosis method for the thermal runaway of battery systems caused by a temperature fault during vehicular operations. For addressing these mentioned issues, the entropy method was employed. Furthermore, the abnormity coefficient was set up using the Z-score method to evaluate the fault severity. Accordingly, homologous management strategies were proposed to handle detected temperature fault problems and make real-time assessments of the fault levels. A vast quantity of real-time voltage monitoring data was derived from the NSMC-EV in Beijing to validate the proposed method. The results show that the proposed method can accurately forecast both the time and location of the temperature fault within battery packs.

The remainder of this paper is structured as follows: Section 2 gives a brief introduction of the proposed prognosis method. Section 3 describes the big data platform for data acquisition. Section 4 presents the detailed prognosis analysis and discussions about temperature faults for battery systems. Finally, the key conclusions is summarized in Section 5.

2. Diagnosis and Prognosis Method

Information entropy has been widely employed to judge the degree of system disorder in thermodynamics, information science, and other fields, which was firstly introduced by Laude Elwood Shannon in 1948 [30]. It generally judges the degree of system disorder in a wide range of scientific fields and is still an important method nowadays [31]. Due to the capability of measuring the information content, combined with the case of information processing, it is a useful and popular method for information entropy. The typical calculation process of the Shannon entropy is shown as follows:

$$H(X) = -\sum_{i=1}^{n} p(x_i) \log p(x_i) \tag{1}$$

where $H(X)$ is the Shannon entropy, $p(x_i)$ is the data probability density in the *i*th region, and n is the number of regions.

The Z-score denotes the standard score, which has the function of risk prediction in the fields of statistics and finance. For instance, Nanayakkara [32] developed a financial distress prediction model for Sri Lankan companies using the Z-score model. Chadha and Aloy et al. [33,34] used Altman's Z-score model to evaluate the financial performance and avoided the high cost that is associated with distress in predicting bankruptcy. However, the Z-score method has not demonstrated the ability and potential of risk prediction of mechanical or electrical faults, especially electric vehicles. In this paper, the Z-score method is applied to quantitatively evaluate the temperature fault within battery packs, which can perform real-time detection and prognosis of abnormal temperature by setting the abnormal coefficient. The voltages and temperatures of different cells are different due to the inconsistency of the battery pack. The formula of the Z-score is expressed as:

$$Z = \frac{x - \mu}{\delta} \tag{2}$$

where x is a specific score, μ is the average score and σ is the standard deviation.

In order to confirm a reasonable real-time detection and evaluation standard, the abnormity coefficient based on the Z-score is implemented as follows:

$$A = \frac{|E - E_{ave}|}{\sigma_E} \tag{3}$$

where E denotes the Shannon entropy, E_{ave} denotes the average Shannon entropies, and σ_E denotes the standard deviation of entropy.

It is worth mentioning that there are multiple iterations of the past data in the entropy calculation. However, monitoring and diagnosis are required in real-time to predict the state of the battery and connection failure, thus, the Shannon entropy calculation needs to be appropriately modified to accommodate the online implementation requirement of EVs. The diagnosis and prognosis algorithm flowchart based on the different extreme value selections for the Shannon entropy is shown in Figure 1.

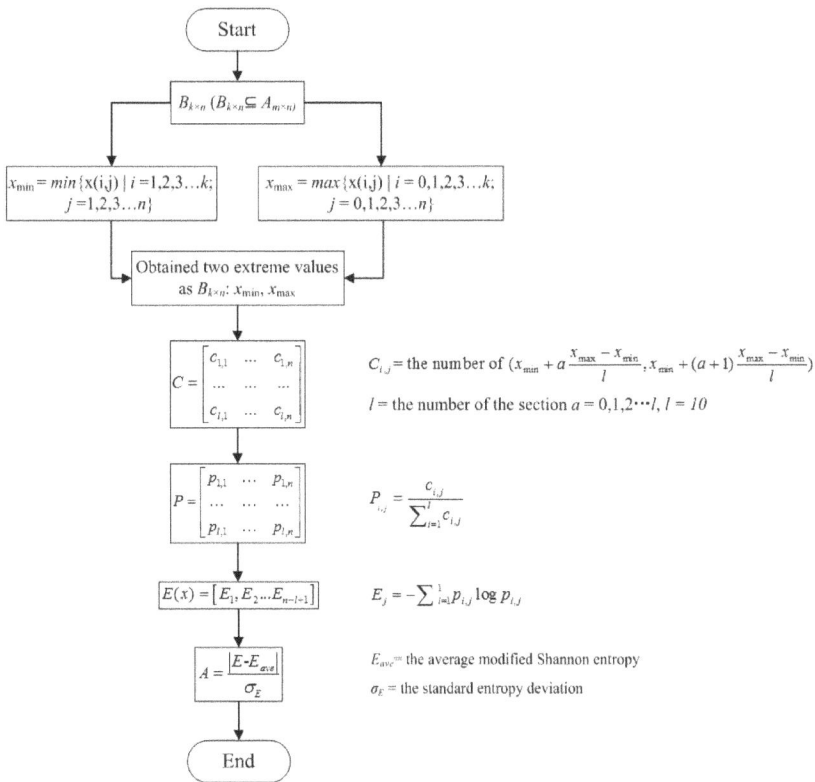

Figure 1. The diagnosis and prognosis algorithm flowchart.

3. Data Acquisition Platform

The temperature and voltage data was derived from the NSMC-EV [35], which has the functions of monitoring and collecting the real-time running data of EVs, such as the voltage and temperature of the battery systems, conducting in-depth analysis and research through big data techniques. The monitoring and management process of the NSMC-EV is shown in Figure 2. The data acquisition frequency from the monitored vehicles ranges from 0.03 Hz to 1 Hz. In addition, the failure statistics of the vehicle running state are categorized into six levels according to failure types, where the first level is the most dangerous. When anomalous information, such as the temperature reaching the limit threshold, a corresponding fault alarm will be immediately dispatched to the relevant vehicle according to the established response protocols. Eventually, the statistical statements about the vehicle-running characteristic and fault statistics will be detected in the forms of daily, weekly, and annual reports.

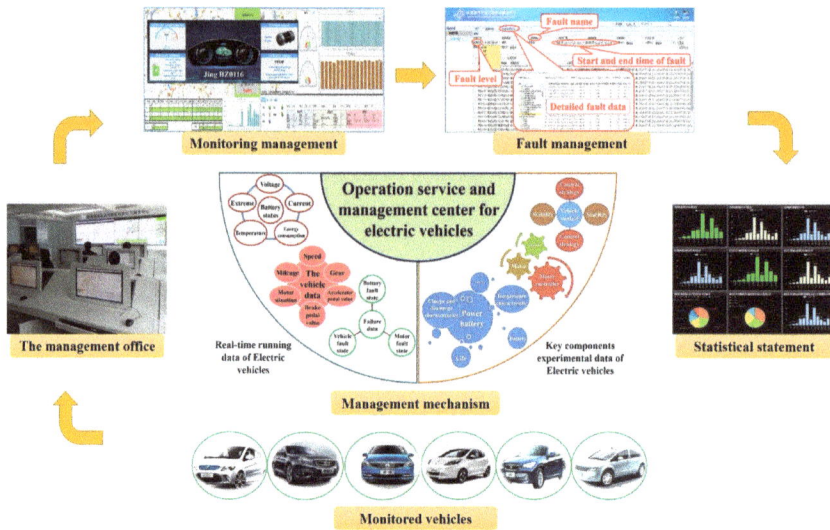

Figure 2. The monitoring and management process of the NSMC-EV.

Through the big data platform, running information and the key component states of the monitored vehicles can be obtained using the vehicle-to-platform communication. The main monitoring objects and purposes of NSMC-EV are shown in Table 1, which illustrates that there is a potential thermal runaway risk once the battery temperature reaches beyond the maximum threshold. Meanwhile it requires human intervention for identifying potential problems to safeguard vehicle operation and maintain the battery cycle life. The logical topological management structure of NSMC-EV is sketched in Figure 3, which is a multi-level structure of "acquisition/access-storage-analysis-application", implementing the fusion and centralized supervision multi-source information, one-stop query and service, as well as data-supporting the whole series of models. Until now, this center has provided around-the-clock monitoring service for over 7000 units of EVs mainly consisting of public vehicles, such as taxis, buses, and sanitation vehicles, etc.

Table 1. The main monitoring objects and purposes of NSMC-EV.

Order Number	Monitoring Object	Monitoring Purpose
1	Battery voltage	To confirm whether there is a value beyond the range.
2	Cell voltage	The low voltage will lead to insufficient capacity, and the high voltage will cause high temperature, gas precipitation, water losses, and grid corrosion of the battery.
3	Battery temperature	To identify potential problems and optimize the vehicle operation and cycle life of the battery. Once beyond the maximum value means that there is a potential thermal runaway and it requires human intervention.

Table 1. *Cont.*

Order Number	Monitoring Object	Monitoring Purpose
4	Ambient temperature	Too high an ambient temperature will shorten battery life and too low an ambient temperature will lead to battery capacity decline.
5	Temperature difference	Large temperature difference is because of the inconsistency of the battery, which will cause endurance deterioration.
6	Charge and discharge current	Provide the health state information of the battery to users, which can be used to indicate the operating state and the integrity of the battery connection.

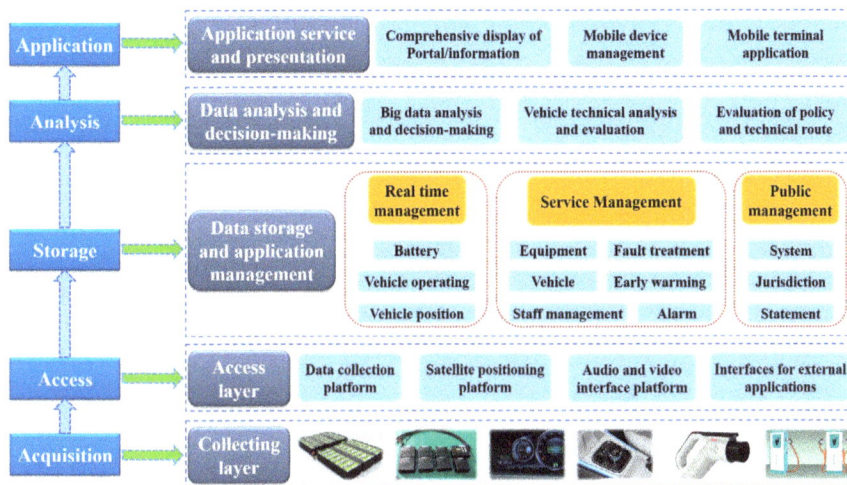

Figure 3. The logical topological management structure of the NSMC-EV.

4. The Thermal Fault Prognosis Analysis and Discussion

4.1. Thermal Management Schematic

A well-designed thermal management system possesses the function of regulating EV and HEV battery pack temperatures evenly, keeping them within the desired operating range. Proper thermal design of every module has a positive impact on overall pack thermal management with the corresponding thermal behavior. In general, a battery thermal management system (BTMS) with few battery modules, using air as the heat transfer medium, is less complicated, which is more effective than using liquid for cooling/heating. Nevertheless, a battery thermal management system with a large number of battery modules faces the opposite issues. General schematics of BTMS using air and liquid are shown in Figure 4a,b, respectively [36]. Either natural or forced air convection can be used for air BTM. Figure 4a illustrates three air BTM methods including passive air cooling, passive air cooling/heating and active air cooling/heating. As opposed to air, liquid has higher thermal conductivity and heat capacity. Liquid BTM is regarded as a better solution, which can be divided into passive or active methods, shown as Figure 4b. The thermal management system may be passive (i.e., only the ambient environment is used) or active (i.e., a built-in source provides heating and/or cooling at cold or hot temperatures). The thermal management control strategy is settled through the electronic control unit. A thermal management system probably uses air for heating/cooling ventilation or liquid as the cooling/heating insulation layer. In addition, phase change materials are another choice for cooling/heating as thermal storage. However, the combination of these three methods are the most common scheme in current BTMS.

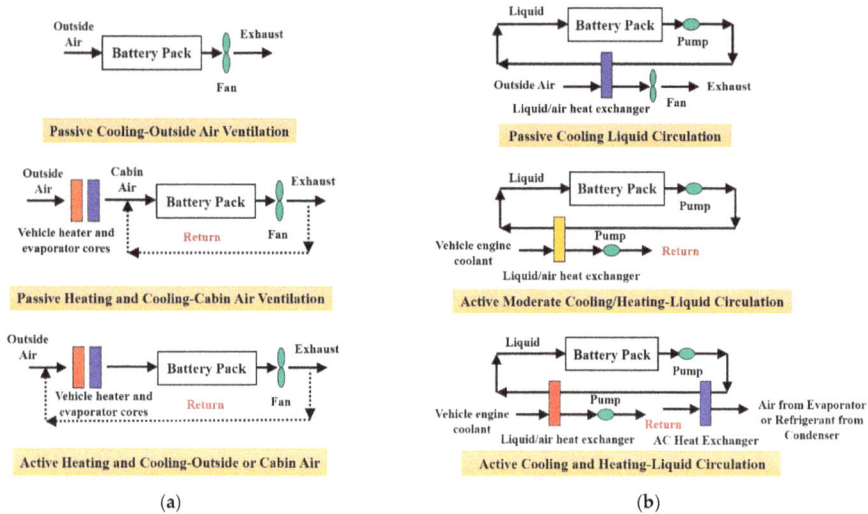

Figure 4. General schematic of BTMS using air and liquid. (**a**) Three air BTM methods; (**b**) Three liquid BTM methods.

Generally, for parallel HEVs, an air thermal management system is suggested, whereas for EVs and series HEVs, liquid-based systems are more suitable for optimum thermal performance. NiMH batteries require a more elaborate thermal management system than lithium-ion and valve-regulated lead acid (VRLA) batteries. Lithium-ion batteries need a well-behaved thermal management system due to the concerns of safety and low-temperature performance. Furthermore, the location of the battery pack has a strong impact on the type of BTMS and whether the pack is air-cooled, liquid cooled, or another method is used.

In addition to considering the temperature of a battery pack, uneven temperature distribution should also be taken into account. Temperature variation from module to module could lead to different charging/discharging behaviors for each module. This, in turn, leads to electrically-unbalanced modules or packs and reduced pack performance. Higher temperatures degrade batteries more quickly, while low temperatures reduce power and energy capabilities, resulting in cost, reliability, safety, range, or drivability implications. Therefore, battery thermal management is all-important for EVs to keep the cells in the desired temperature range, minimize cell-to-cell temperature variations, prevent the battery from going above or below acceptable limits, and maximize the useful energy from the cells and the pack with little energy for operation.

A perfect BTMS not only heats and cools the battery system as soon as possible, but also controls the system's thermal safety to prevent thermal runaway. The typical types of temperature faults in NSMC-EV are over-temperature and excessive temperature difference (TD), which are usually caused by abnormal temperature variation. Detecting when and where the abnormal temperature occurs will play an extremely important role in safe battery management. The normal operating temperature range of lithium-ion batteries is −20 to 60 °C, which is generally controlled at 15–60 °C for the safe operation of the vehicles. The maximum permissible TD is 5 °C, which means the limitation of TD within 5 °C. There are a certain amount of temperature probes in different locations of the battery pack for different vehicles, the monitoring platform of NSMC-EV will send an over-temperature alarm when any temperature probe exceeds 45 °C and an excessive TD alarm when TD > 5 °C.

4.2. The Fault Prognosis of Over-Temperature

In order to verify the feasibility and reliability of the proposed prognosis method for temperature anomaly, the cell data of Vehicle 1 (vehicle plate: Jing Q6S772, Fukuda pure electric sanitation truck, a style of 5023ZLJEV 2T dump truck, with a top speed of 45 km/h. The type of battery is a lithium-ion phosphate battery with 120 cells in series, the monomer voltage is 3.3 V and total voltage is 396 V) on March 6th, 2017 was retrieved from NSMC-EV and the work period of the monitored vehicle was 09:48:39–16:07:52 (more than 6 h), which experienced an over-temperature alarm of T > 45 °C at 11:07:20. There are 16 temperature probes in the different locations of the battery pack and the data acquisition frequency of 0.05 Hz. The temperature and SOC curves of Vehicle 1 are shown in Figure 5, which demonstrates that the temperature of Probe 1 and Probe 9 had different fluctuations form the other probes. In addition, Probe 1 experiences an over-temperature fault with the vehicle running. However, although the abnormity appeared early, it cannot be identified before the alarm occurs by the conventional temperature sensor because it is still in the normal temperature range of $T < 45$ °C.

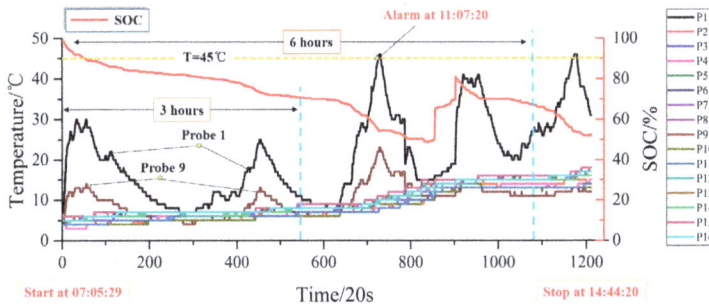

Figure 5. The temperature and SOC curves of Vehicle 1.

As for the presented entropy method in Section 2, the length of the computation window K has significant influence on the accuracy of entropy. If K is too small, the temperature fluctuations cannot be fully revealed. On the contrary, the iterations would become too few to pick out the abnormal temperature fluctuations. Furthermore, because of the graduality and stability of temperature, the temperature fluctuations are relatively small and the position of the abnormal temperature is difficult to detect in a short period of time, so $K = 100$ was selected as the length of the computation window in this study through the trial-and-error method.

With the vehicle operation and the rise of the battery temperature, the temperature of all probes will gradually stabilize. It is difficult to detect the abnormal temperature fluctuations after temperature stability or failure, so the monitoring data should be processed from the vehicle starting every day. Figure 6a shows the abnormal coefficients of Vehicle 1 in the first 6 h. Probe 1 and Probe 9, especially Probe 1, have obviously larger abnormal coefficients than the others do. This fluctuation of abnormal coefficients is consistent with the temperature fluctuation shown in Figure 5, which verifies that the proposed method can accurately identify the time and location of the abnormal temperature. In order to verify the prognosis performance of the proposed method, the first 3 h were chosen as the calculation unit, during when the over-temperature has not been triggered. The abnormal coefficient in the first 3 h of Vehicle 1 is shown in Figure 6b, which shows that both Probe 1 and Probe 9 with abnormal temperature can be detected. Therefore, the proposed method can accurately predict the over-temperature fault.

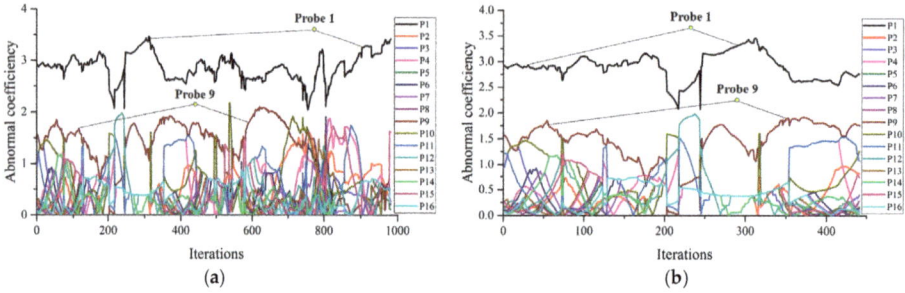

Figure 6. The abnormal coefficient in the (**a**) first 6 h and (**b**) first 3 h of Vehicle 1.

As shown in Figure 6, the anomaly coefficient curves have crosses and accidental extremes, which are not conducive to quantifying the level of the abnormal coefficient. In order to make the abnormal coefficient more readable, and to facilitate a horizontal comparison and evaluation between different temperature probes, a boxplot was employed to express the abnormal coefficient to forecast the temperature faults in this section, which is represented as A_b. Boxplots can reflect the center and spread scope of the data distribution. By drawing the boxplots of multiple sets of data on the same coordinates, the distribution difference is clearly displayed. The structure diagram of the boxplot is shown in Figure 7. The boxplot requires the statistical concept of quartiles, which means the position numbers of three segmentation points. Q1 denotes the lower quartile, which is equal to the number of 25% of all values. Q2 is the median, which is equal to the number of 50% of all values. Q3 is the upper quartile, which is equal to the number of 75% of all values. The abnormal coefficient A_b is the median of the boxplot in this paper.

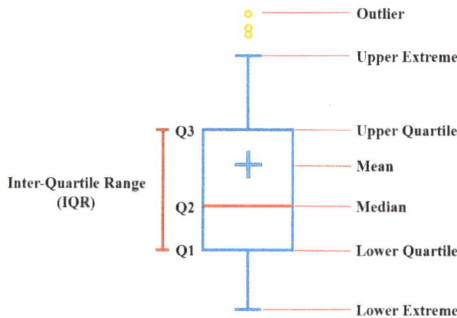

Figure 7. The structure diagram of the boxplot.

The abnormal coefficient boxplot in the first 6 h and the first 3 h of Vehicle 1 are shown in Figure 8a,b, respectively. The results reveal that both Probe 1 and Probe 9 can be easily detected and the A_b of Probe 1 is much greater than that of Probe 9 and the others. By defining certain detection thresholds as $A_b = 1$ and $A_b = 1.2$, the over-temperature fault alarm can be avoided if the abnormal temperature is detected in advance by this method. Actually, for the purpose of accurate over-temperature fault prognosis, much more monitoring data were derived from NSMC-EV. The evaluation strategy of the abnormal temperature was obtained by the trial-and-error method through a large number of analytical results, which is feasible, reliable, and can accurately forecast both the time and location of over-temperature faults. Thus, this method can effectively prevent the over-temperature fault by detecting the abnormal temperature in real-time.

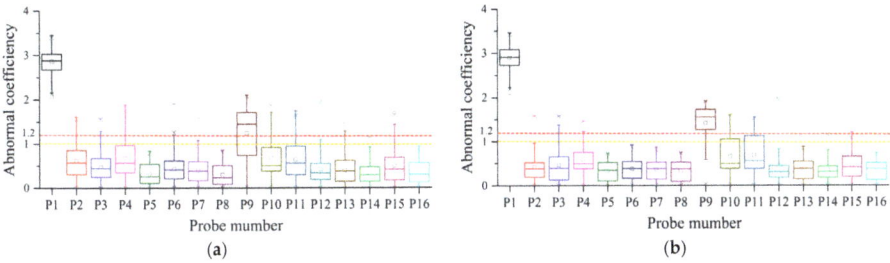

Figure 8. The abnormal coefficient boxplot of the (**a**) first 6 h and (**b**) the first 3 h of Vehicle 1.

4.3. The Fault Prognosis of Temperature Difference

The other typical thermal fault in NSMC-EV is excessive temperature difference (TD). The cell data of Vehicle 2 (vehicle plate: Jing B1Y163, CA E30 electric taxi) on November 2nd, 2016 was retrieved from NSMC-EV and the work period of the monitored vehicle was 07:55:57–23:59:54 (more than 16 h), which experienced an excessive TD fault alarm with a TD > 5 °C at 18:14:55, after the tested vehicle traveled for more than 9 h. There are 16 temperature probes in the different locations of the battery pack and the data acquisition frequency is 0.1 Hz. The temperature curves of Vehicle 2 are shown in Figure 9. This revealed that the temperature of Probe 11 has an abnormal fluctuation with the vehicle running, which directly leads to the generation of the TD fault. However, this abnormity cannot be detected by the conventional temperature sensor because it is still in the normal temperature range of 0–30 °C.

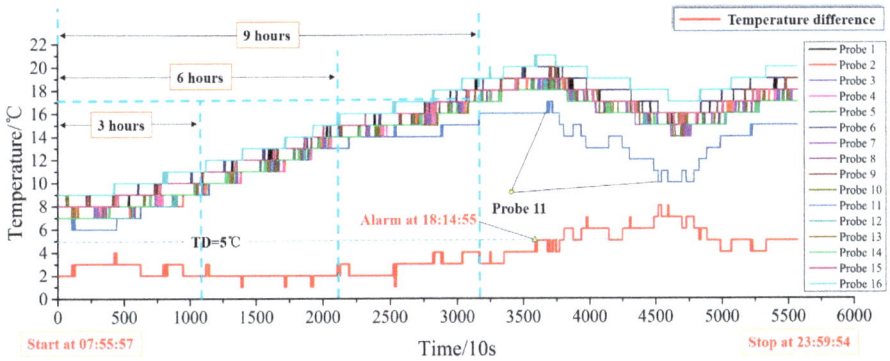

Figure 9. The temperature curves of Vehicle 2.

The SOC, speed, and TD curves of Vehicle 2 on November 2nd, 2016 are shown in Figure 10. This demonstrates that this car charged twice and parked several times at 14:58:03 and 21:47:46. In addition, the TD curves rise slowly with the increase of speed and vehicular running.

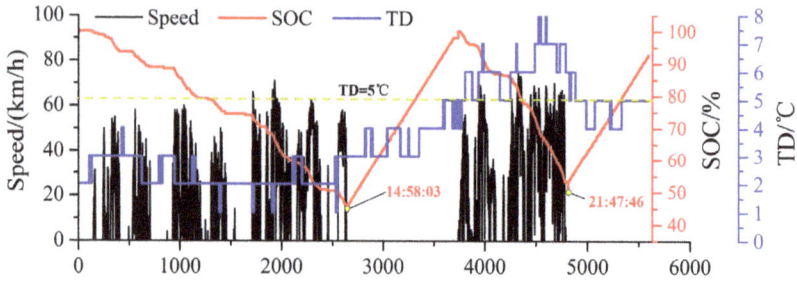

Figure 10. SOC, speed and TD curves of Vehicle 2.

The abnormal coefficient and boxplot of Vehicle 2 in the first 3 h on November 2nd, 2016 are shown in Figure 11a,b, respectively. It is observed from Figure 11a that some probes have anomalous extremum points but no probe has obviously larger abnormity coefficients than the others. Figure 11b displays that the median position of all probes that also be confirmed to $A_b < 1$, which is consistent with the temperature curves in Figure 9. Thus, all of the probes have a safe temperature status and no abnormal temperature can be detected in the first 3 h.

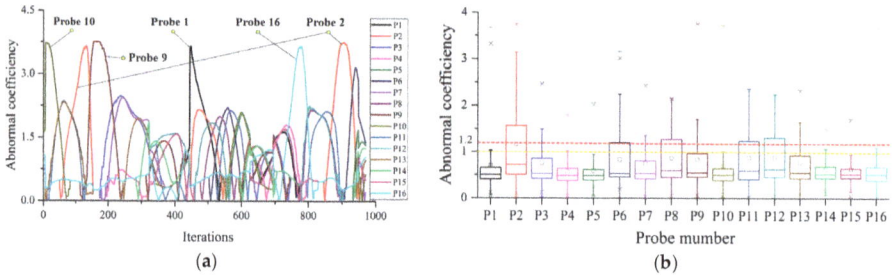

Figure 11. The abnormal coefficient and boxplot at the first 3 h of Vehicle 2. (a) The abnormal coefficient curves; (b) Boxplot of the abnormal coefficient.

Due to the design flaws of the battery box or the thermal runaway of batteries, the tendency of the temperature change of different temperature probes will have certain differences. With the vehicle operation and the rising of the battery temperature, the temperature will be gradually stabilize. It is difficult to detect the abnormal temperature fluctuations after the temperature become stable, or there is a failure, so the first 3 h from the starting point are taken as the initial calculation window, if the abnormal temperature probe cannot be detected, then continues to calculate for the next 3 h. The abnormal coefficient and boxplot of Vehicle 2 at the first 6 h and the first 9 h on November 2nd, 2016 are shown in Figures 12 and 13, respectively. Figure 12a indicates that Probe 11 has an abnormal temperature fluctuation, but is difficult to detect due to the interference of Probe 2, Probe 6 and Probe 16. Figure 12b shows that the median position of Probe 11 is greater than those of the others and the abnormal coefficient $A_b > 1$, which is consistent with the temperature curves in Figure 9. Thus, abnormal temperature of Probe 11 can be detected in the first 6 h. From Figure 13a, Probe 11 has a distinct abnormal fluctuation and is easier to detect. Figure 13b demonstrates that the median position of Probe 11 is higher compared to those of other probes and the abnormal coefficient $A_b > 1$. The results show excellent consistency with the previous temperature curves in Figure 9. The excessive TD fault of Vehicle 2 occurred after it traveled more than 9 h. Therefore, the proposed prognosis method can detect the abnormal probe in real-time and identify the fault location in advance.

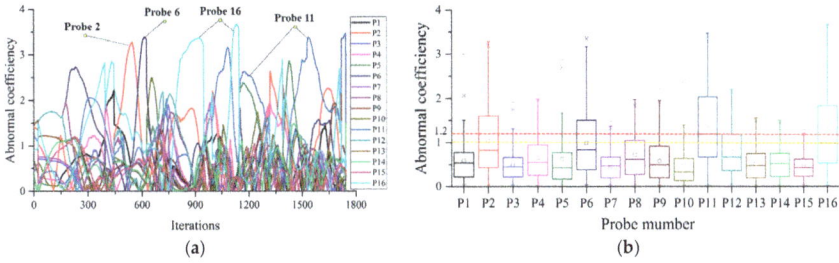

Figure 12. The abnormal coefficient and boxplot at the first 6 h of Vehicle 2. (**a**) The abnormal coefficient curves; (**b**) Boxplot of the abnormal coefficient.

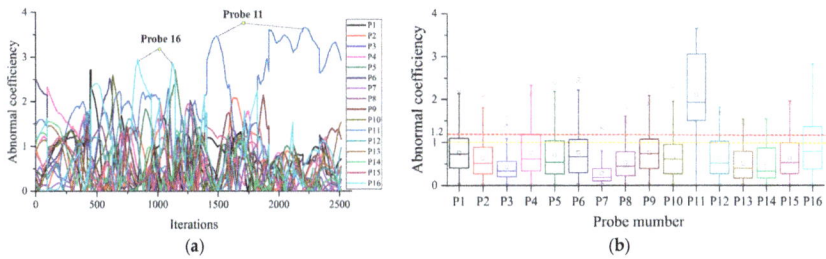

Figure 13. The abnormal coefficient at the first 9 h of Vehicle 2. (**a**) The abnormal coefficient curves; (**b**) Boxplot of the abnormal coefficient.

In order to verify the stability of this method, the cell data of Vehicle 2 on November 1st, 2016 was derived from NSMC-EV and the period of the monitoring data was 10:51:05–23:36:38. An alarm of excessive temperature difference of TD > 5 °C at 17:12:15 occurred in Vehicle 2 after the tested vehicle traveled for more than 9 h. The temperature curves of Vehicle 2 are shown in Figure 14, which illustrates that the temperature of Probe 11 has different fluctuations with the vehicle running. However, the abnormal temperature cannot be identified as long as it is still in the safe temperature range.

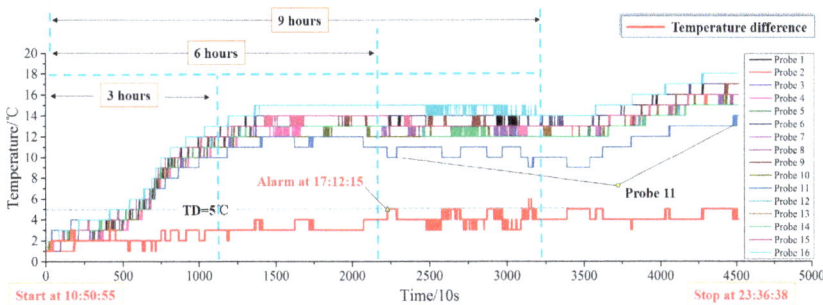

Figure 14. The temperature curves of Vehicle 2.

The abnormal coefficient and boxplot at the first 3 h of Vehicle 2 are shown in Figure 15. Figure 15a indicates that Probe 11 has an abnormal temperature fluctuation and can be detected out. Figure 15b demonstrate that Probe 11 can be easily detected and the limitation of the abnormal coefficient of Probe 11 is $A_b > 1$. However, the excessive TD fault can be avoided if the abnormal temperature is detected in advance. Actually, for accurate excessive TD fault prognosis, much more monitoring data were retrieved from NSMC-EV and analyzed, which reveals the proposed method is feasible, reliable, and stable to accurately predict the time and location of excessive TD faults within a battery pack. Thus, this method can effectively prevent the excessive TD fault by detecting the abnormal temperature in real-time.

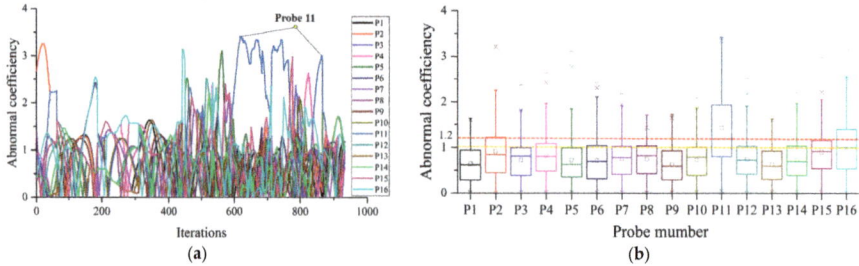

Figure 15. The abnormal coefficient and boxplot at the first 3 h of Vehicle 2. (**a**) The abnormal coefficient curves; (**b**) Boxplot of the abnormal coefficient.

4.4. The Security Management Strategy and Discussion

Through the above analysis, the over-temperature fault and excessive TD fault can be predicted using the proposed method and it has well-behaved reliability and stability. By implementing a certain detection threshold as $A_b = 1$ and $A_b = 1.2$, the cell with abnormal temperature can be detected before the thermal faults occur, which has vital significance for the future prognosis and safety management of the battery fault, especially for the prevention of thermal runaway. The prognosis strategy of the thermal fault can be obtained through analyzing much more monitoring data retrieved from NSMC-EV using the trial-and-error method. The prognosis strategy flowchart of the thermal fault is shown in Figure 16.

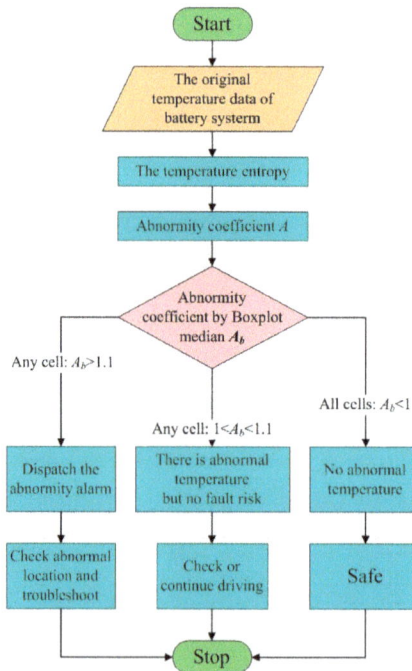

Figure 16. The prognosis strategy flowchart of a thermal fault.

NSMC-EV currently provides around-the-clock monitoring services, mainly for public vehicles apart from private cars, such as taxis, buses, and sanitation vehicles, which always have a relatively small number of cells. Nevertheless, according to the analysis and discussion of different sets of monitoring data, by setting a suitable value of the calculation window K, this technique is still valid even if the EV has a larger number of cells (i.e., Tesla, with 6000+ cells). Therefore, it has a strong timeliness and will have greater application prospects if some private cars with more cells are monitored and managed by NSMC-EV in the future, which will also provide a foundation for the establishment of safety precaution mechanisms for battery thermal runaway.

5. Conclusions

This paper presents a real-time thermal fault diagnosis and prognosis method based on the NSMC-EV in Beijing. A vast quantity of real-time voltage monitoring data was collected from this big data platform to verify the effectiveness of the presented prognosis method. The Shannon entropy was applied to analyze the monitoring data. The analysis results showed that the proposed method could detect probes with abnormal temperature, which can also effectively predict the occurrence time and location. These were achieved with a relatively small calculation effort, which makes it implementable in a real safety BMS. The feasibility, reliability, and stability of the prognosis capability were also discussed and verified by analyzing extensive monitoring data. Furthermore, the prognosis and safety management strategy for thermal faults of battery systems were also developed by applying the Z-score method, and the abnormity coefficients were implemented to make real-time evaluation on the faulty levels. The presented method is flexible in all disorder systems with abnormal fluctuations regardless the data types and application fields, so it possesses widespread application potential in not only electric vehicles, but also other areas with complex abnormally fluctuating environments.

Acknowledgments: The project is supported by the State Key Program of National Natural Science Foundation of China (No. U1564206) and the Special Innovation Methods Program of Science and Technology Ministry of China (grant No. 2015IM030100).

Author Contributions: Jichao Hong provided algorithms, analyzed the data and wrote the paper; and Zhenpo Wang and Peng Liu conceived the structure and research direction of the paper.

Conflicts of Interest: The authors declare no conflict of interest.

References

1. Tarascon, J.; Armand, M. Issues and challenges facing rechargeable lithium batteries. *Nature* **2001**, *414*, 359–367. [CrossRef] [PubMed]
2. Wang, Q.S.; Ping, P.; Zhao, X.J.; Chu, G.Q.; Sun, J.H.; Chen, C.H. Thermal runaway caused fire and explosion of lithium ion battery. *J. Power Sources* **2012**, *208*, 210–224. [CrossRef]
3. Tugce, Y.; Shawn, L.; Venkatasubramanian, V.J.; Jeremy, J.M. Plug-in hybrid electric vehicle LiFePO$_4$ battery life implications of thermal management, driving conditions and regional climate. *J. Power Sources* **2017**, *338*, 49–64.
4. Daowd, M.; Antoine, M.; Omar, N.; Lataire, P.; Van, D.P.; Van, M.J. Battery Management System-Balancing Modularization Based on a Single Switched Capacitor and Bi-Directional DC/DC Converter with the Auxiliary Battery. *Energies* **2014**, *7*, 2897–2937. [CrossRef]
5. Panchal, S.; Mathewson, S.; Fraser, R.; Culham, R.; Fowler, M. Thermal Management of Lithium-Ion Pouch Cell with Indirect Liquid Cooling using Dual Cold Plates Approach. *SAE Int. J. Altern. Powert.* **2015**, *4*, 293–307. [CrossRef]
6. Panchal, S.; Dincer, I.; Agelin-Chaab, M.; Fraser, R.; Fowler, M. Experimental and theoretical investigations of heat generation rates for a water cooled LiFePO$_4$ battery. *Int. J. Heat Mass Transf.* **2016**, *101*, 1093–1102. [CrossRef]
7. Lan, C.; Xu, J.; Qiao, Y.; Ma, Y. Thermal management for high power lithium-ion battery by minichannel aluminum tubes. *Appl. Therm. Eng.* **2016**, *101*, 284–292. [CrossRef]

8. Qian, Z.; Li, Y.; Rao, Z. Thermal performance of lithium-ion battery thermal management system by using mini-channel cooling. *Energy Convers. Manag.* **2016**, *126*, 622–631. [CrossRef]

9. Mastali, M.; Samadani, E.; Farhad, S.; Fraser, R.; Fowler, M. Three-dimensional Multi-Particle Electrochemical Model of LiFePO$_4$ Cells based on a Resistor Network Methodology. *Electrochim. Acta* **2016**, *190*, 574–587. [CrossRef]

10. Sun, F.; Xiong, R.; He, H. A systematic state-of-charge estimation framework for multi-cell battery pack in electric vehicles using bias correction technique. *Appl. Energy* **2016**, *162*, 1399–1409. [CrossRef]

11. Blanco, C.; Sanchez, L.; Gonzalez, M.; Anton, J.C.; Garcia, V.; Viera, J.C. An Equivalent Circuit Model with Variable Effective Capacity for Batteries. *IEEE Trans. Veh. Technol.* **2014**, *63*, 3592–3599. [CrossRef]

12. Xiong, R.; Sun, F.; Gong, X.; Gao, C. A data-driven based adaptive state of charge estimator of lithium-ion polymer battery used in electric vehicles. *Appl. Energy* **2014**, *113*, 1421–1433. [CrossRef]

13. Xiong, R.; Tian, J.; Mu, H.; Wang, C. A systematic model-based degradation behavior recognition and health monitor method of lithium-ion batteries. *Appl. Energy* **2017**, *5*, 124. [CrossRef]

14. Sun, F.; Xiong, R. A novel dual-scale cell state-of-charge estimation approach for series-connected battery pack used in electric vehicles. *J. Power Sources* **2015**, *274*, 582–594. [CrossRef]

15. Xiong, R.; Sun, F.; Chen, Z.; He, H. A data-driven multi-scale extended Kalman filtering based parameter and state estimation approach of lithium-ion polymer battery in electric vehicles. *Appl. Energy* **2014**, *113*, 463–476. [CrossRef]

16. Lin, X.; Perez, H.E.; Siegel, J.B.; Stefanopoulou, A.G.; Li, Y.; Anderson, R.D.; Ding, Y.; Castanier, M.P. Online parameterization of lumped thermal dynamics in cylindrical lithium ion batteries for core temperature estimation and health monitoring. *IEEE Trans. Control Syst. Technol.* **2013**, *21*, 1745–1755.

17. Forgez, C.; Do, D.V.; Friedrich, G.; Morcrette, M.; Delacourt, C. Thermal modeling of a cylindrical LiFePO$_4$/graphite lithium-ion battery. *J. Power Sources* **2010**, *195*, 2961–2968. [CrossRef]

18. Xiong, R.; Yu, Q.; Wang, L.; Lin, C. A novel method to obtain the open circuit voltage for the state of charge of lithium ion batteries in electric vehicles by using H infinity filter. *Appl. Energy* **2017**, *5*, 136. [CrossRef]

19. Seo, M.; Goh, T.; Park, M.; Koo, G.; Kim, S.W. Detection of Internal Short Circuit in Lithium Ion Battery Using Model-Based Switching Model Method. *Energies* **2017**, *10*, 76. [CrossRef]

20. Zhang, H.; Pei, L.; Sun, J.L.; Song, K.; Lu, R.G.; Zhao, Y.P.; Zhu, C.B.; Wang, T.S. Online Diagnosis for the Capacity Fade Fault of a Parallel-Connected Lithium Ion Battery Group. *Energies* **2016**, *9*, 387. [CrossRef]

21. Chen, C.; Xiong, R.; Shen, W. A lithium-ion battery-in-the-loop approach to test and validate multi-scale dual H infinity filters for state of charge and capacity estimation. *IEEE Trans. Power Electr.* **2017**, *PP*, 1. [CrossRef]

22. Liu, Z.; He, H. Model-based Sensor Fault Diagnosis of a Lithium-ion Battery in Electric Vehicles. *Energies* **2015**, *8*, 6509–6527. [CrossRef]

23. Bai, H.; Miao, S.; Zhang, P.; Bai, Z. Reliability Evaluation of a Distribution Network with Microgrid Based on a Combined Power Generation System. *Energies* **2015**, *8*, 1216–1241. [CrossRef]

24. Dey, S.; Biron, Z.A.; Tatipamula, S.; Das, N.; Mohon, S.; Ayalew, B.; Pisu, P. Model-based real-time thermal fault diagnosis of Lithium-ion batteries. *Control Eng. Pract.* **2016**, *56*, 37–48. [CrossRef]

25. Wu, C.; Zhu, C.; Ge, Y.; Zhao, Y. A diagnosis approach for typical faults of lithium-ion battery based on extended kalman filter. *Int. J. Electrochem. Sci.* **2016**, *11*, 5289–5301. [CrossRef]

26. Sidhu, A.; Izadian, A.; Anwar, S. Adaptive Nonlinear Model-Based Fault Diagnosis of Li-Ion Batteries. *IEEE Trans. Ind. Electron.* **2016**, *62*, 1002–1011. [CrossRef]

27. Chen, Z.; Lin, F.; Wang, C.; Wang, L.; Xu, M. Active Diagnosability of Discrete Event Systems and its Application to Battery Fault Diagnosis. *IEEE Trans. Control Syst. Technol.* **2014**, *22*, 1892–1898. [CrossRef]

28. Zheng, Y.; Han, X.; Lu, L.; Li, J.; Ouyang, M. Lithium ion battery pack power fade fault identification based on Shannon entropy in electric vehicles. *J. Power Sources* **2013**, *223*, 136–146. [CrossRef]

29. Rezvanizanian, S.M.; Huang, Y.; Chuan, J.; Lee, J. A Mobility Performance Assessment on Plug-in EV Battery. *Int. J. Progn. Health Manag.* **2012**, *3*, 102.

30. Shannon, C.; Weaver, W. *The Mathematical Theory of Communication*; University of Illinois Press: Champaign, IL, USA, 1949.

31. Montani, F.; Deleglise, E.; Rosso, O. Efficiency characterization of a large neuronal network: A causal information approach. *Physics A* **2014**, *401*, 58–70. [CrossRef]

32. Nanayakkara, K.G.M.; Azeezz, A.A. Predicting Corporate Financial Distress in Sri Lanka: An Extension to Z-Score Model. *Int. J. Bus. Soc. Res.* **2015**, *5*, 41–52.

33. Chadha, P. Exploring the Financial Performance of the Listed Companies in Kuwait Stock Exchange Using Altman's Z-Score Model. *Int. J. Econ. Manag. Sci.* **2016**, *5*, 341. [CrossRef]
34. Aloy, N.J.; Pratheepan, T. The Application of Altman's Z-Score Model in Predicting Bankruptcy: Evidence from the Trading Sector in Sri Lanka. *Int. J. Bus. Manag.* **2015**, *10*, 269–275.
35. Wang, Z.; Hong, J.; Liu, P.; Zhang, L. Voltage fault diagnosis and prognosis of battery systems based on entropy and Z-score for electric vehicles. *Appl. Energy* **2017**, *196*, 289–302. [CrossRef]
36. Wang, Q.; Jiang, B.; Li, B.; Yan, Y. A critical review of thermal management models and solutions of lithium-ion batteries for the development of pure electric vehicles. *Renew. Sustain. Energy Rev.* **2016**, *64*, 106–128. [CrossRef]

![energies logo] *energies*

MDPI

Article

Optimization of Battery Capacity Decay for Semi-Active Hybrid Energy Storage System Equipped on Electric City Bus

Xiaogang Wu [1,2,*] and Tianze Wang [1]

[1]   College of Electrical and Electronics Engineering, Harbin University of Science and Technology,
       Harbin 150000, China; wangtz8255@163.com
[2]   State Key Laboratory of Automotive Safety and Energy, Tsinghua University, Beijing 100084, China
*   Correspondence: xgwu@hrbust.edu.cn

Academic Editor: Rui Xiong
Received: 9 May 2017; Accepted: 6 June 2017; Published: 9 June 2017

Abstract: In view of severe changes in temperature during different seasons in cold areas of northern China, the decay of battery capacity of electric vehicles poses a problem. This paper uses an electric bus power system with semi-active hybrid energy storage system (HESS) as the research object and proposes a convex power distribution strategy to optimize the battery current that represents degradation of battery capacity based on the analysis of semi-empirical LiFePO$_4$ battery life decline model. Simulation results show that, at a room temperature of 25 °C, during a daily trip organized by the Harbin City Driving Cycle including four cycle lines and four charging phases, the percentage of battery degradation was 9.6×10^{-3}%. According to the average temperature of different months in Harbin, the percentage of battery degradation of the power distribution strategy proposed in this paper is 3.15% in one year; the electric bus can operate for 6.4 years until its capacity reduces to 80% of its initial value, and it can operate for 0.51 year more than the rule-based power distribution strategy.

Keywords: electric bus; hybrid energy storage system; energy management; convex optimization; LiFePO$_4$ battery degradation

1. Introduction

As the sole power source in a traditional electric vehicle, a battery needs to satisfy the power and energy demands of a bus under different operating conditions. When the battery is repeatedly over-charged and over-discharged in the long-term operating, the battery degradation will be accelerated. Furthermore, when the battery is operated at low temperature, its capacity degradation is more significant. The hybrid energy storage system (HESS) is composed of a battery and super capacity (SC); the battery provides the required energy and the SC satisfies the instantaneous power requirements, can effectively inhibit the battery charge and discharge current changes, and optimizes the working conditions of the energy system [1].

Currently, experts and scholars in the field of electric vehicle hybrid energy storage research are focused on the modeling and performance of the system components, system parameters matching, power distribution, etc. In terms of system component modeling and experimentation, Luo et al. [2] derived and verified a driving cycle life prediction model for LiFePO$_4$ battery based on the experimental verification of the existing capacity decay model for LiFePO$_4$ under a constant current charge/discharge condition. Abeywardana et al. [3] proposed a new type of inverter combined with boost circuits used in HESS, which eliminates the high current injected into the drive motor as compared to conventional controllers that eliminate the equivalent series resistance of the inverter. Henson et al. [4] conducted a comparative study of the battery/SC with different depths of discharge

(DODs) for minimizing the cost of the HESS. Xiong et al. [5,6] used different algorithms to estimate the relationship of the voltage to the state-of-charge (SOC) and capacity of lithium ion battery, and they also validated the accuracy of the method through hardware-in-the-loop experiments. In terms of system parameter matching and power distribution, Mid-Eum et al. [7] used the convex optimization method to optimize the power loss and battery power fluctuations considering the real-time dynamic load to propose a method for calculating the SC reference voltage. Song et al. [8,9] proposed a new semi-active topology; the operation cost of the HESS, including the battery degradation cost and electricity cost, is minimized by using the dynamic programming (DP) approach. Further, they studied four topologies and proposed a rule-based power distribution strategy with four kinds of topologies based on the optimization results. Hu et al. [10] conducted energy efficiency analysis and component selection of the plug-in hybrid power system using convex optimization.

In summary, there have been studies related to the parameter matching and system control of HESS facing battery degradation. However, to the best of our knowledge, there are no published papers that combine the climatic conditions in northern China and the corresponding urban driving cycle operating conditions to optimize the functioning of the HESS. In order to attain the full potential of the HESS to enhance the battery life of electric buses under local conditions, in this study, we considered the electric bus operating in Harbin, China as an example, and proposed a method to optimize charge/discharge current of battery through convex optimization considering the average monthly temperature change in one year.

This paper is organized as follows. In Section 2, we analyze the configuration and working modes of the HESS. In Section 3, we introduce the models of LiFePO$_4$ batteries, SCs, and vehicle. Section 4 presents a convex optimization power distribution strategy based on the semi-empirical model of battery degradation. In Section 5, the simulation results and operating years were analyzed and the results are compared with those of the rule-based strategy.

2. Analysis of Configuration and Working Modes of Hybrid Energy Storage System

As the main energy source of electric vehicles, energy-based batteries have the disadvantages of low power density and high capacity degradation [11]. In order to satisfy the peak power demand, the power density of batteries should be sufficiently high; further, considering a battery that is the only power source of a traditional electric car, in principle, the only way to increase the power density of the batteries is to increase the number of batteries. Thus, it will result in high cost and high battery degradation. However, the SCs have characteristics of high power density and low capacity degradation. A combination of the battery and SC satisfies power and energy requirements, as well as the different performance requirements of the vehicles.

According to the different connections between the battery, SC, and DC/DC converter, the HESS can be classified into three major types, namely, fully active, passive, and semi-active, as shown in Figure 1. In the fully active HESSs, both the battery and SC are connected to the DC bus via a DC/DC converter; two DC/DC converters can simultaneously control the output power of the battery and the SC. Further, it has good control margins. However, a fully active HESS has low system efficiency and a complicated system structure owing to the existence of the two converters; it also increases the system cost owing to the additional cost of the DC/DC converter. Therefore, the fully active topology can achieve a good control effect, but at the expense of system efficiency, complexity, and cost [12]. In the passive topology, the battery and SC are directly connected to the DC bus and the system structure is simple. Owing to the absence of a converter, the system efficiency of the passive topology is the highest, whereas it is uncontrollable of the energy flowing [13]. In the semi-active topology, either the battery or the SC is connected to the DC/DC converter through a unique converter to control the distribution of the output power of the two energy sources. Since the DC bus is connected to one of the battery and SC directly, a fast DC/DC converter is required to maintain DC voltage when the load is changed. However, compared to the fully active and passive topologies, the semi-active topology solved the problems of low efficiency, high cost, uncontrollability, etc. [14].

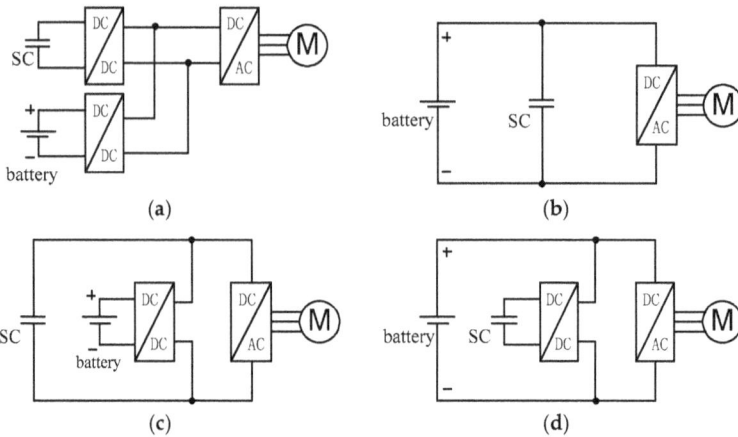

Figure 1. The topology of the electric vehicle HESS: (**a**) fully active topology; (**b**) passive parallel topology; (**c**) semi-active topology 1; and (**d**) semi-active topology 2.

The topology adopted in this study is shown in Figure 1d. Semi-active topology 2 employs a DC/DC converter to decouple the SC from the battery/DC bus. Furthermore, the DC bus voltage is equal to the battery voltage as they are directly connected. Compared to the other three topologies, the use of SC is more flexible [15], and its working mode is shown in Figure 2. In the driving mode, both the battery and SC provide power to the motor, and the SC satisfies the instantaneous high power requirements. In the braking mode, the energy charging for the SC first through the converter, then the braking energy charging for the battery when the SC is full. For the power conversion between SC and DC bus, a fast three-leg bidirectional DC/DC converter is used. It can be operated in the interleaved manner and has the merit of being commercially available [16,17]. The degradation of the SC is very small and its working life can accommodate millions of charge/discharge cycles. The power demand from the vehicle will be volatile during rapid acceleration and braking; hence, the SC plays the role of power and energy buffer when it is connected between the battery and driving motor through the DC/DC converter.

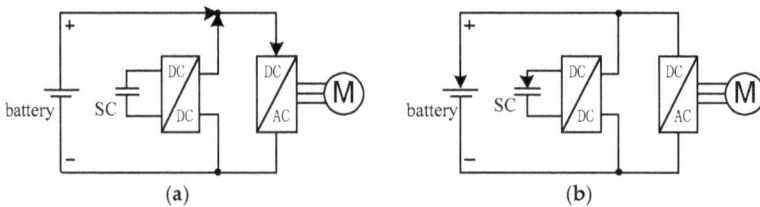

Figure 2. The different modes of operation of semi-active topology: (**a**) power flow based on driving mode; and (**b**) power flow based on braking mode.

3. Power System Modeling Based on HESS

3.1. Battery Model

Compared to the Nickel Metal Hydride (Ni-MH) power battery, lead-acid power battery, and the other driving batteries utilized in electric vehicles, the LiFePO$_4$ battery has the characteristic of good battery service life and high energy density. However, its low temperature performance is not outstanding [18]. The parameters of the LiFePO$_4$ cell used in this study are shown in Table 1.

Table 1. Basic parameters of the LiFePO$_4$ battery cell.

Item	Value
V_{bat_norm}, nominal voltage (V)	3.2
Q_{bat}, capacity (A$_h$)	180
m_{bat_cell}, cell mass (kg)	5.6
$I_{bat\_max,min}$, max dis/charge current (A)	±540

The *Rint* model shown in Figure 3 was adopted to represent the battery behavior, where U_{bat} is the battery terminal voltage and R_{bat} is the resistance of the battery. Since the voltage and SOC are strongly correlated [19] in Equation (1), and in order to adopt the subsequent power distribution strategy, the one-time curve fitting is used to find the functional relationship between the battery open circuit voltage V_{bat} and its SOC, as shown in Figure 4.

$$V_{bat} = (V_{bat1} - V_{bat0})SOC_{bat} + V_{bat0} \tag{1}$$

where V_{bat1} is the open circuit voltage corresponding to SOC_{bat} = 100% and V_{bat0} is the open circuit voltage when SOC_{bat} = 0. The storage energy E_{bat} of the battery pack can be calculated as

$$E_{bat} = \frac{1}{2}n_{bat}Q_{bat}(V_{bat}^2 - V_{bat0}^2) \tag{2}$$

where n_{bat} is the number of battery cells and Q_{bat} is the cell capacity. The battery pack output power P_{bat} can be calculated by Equation (3).

$$P_{bat} = -\frac{dE_{bat}}{dt} \tag{3}$$

The power consumption on battery internal resistance P_{bat_loss} can be calculated by the following equation, where I_{bat} is the current flowing through the battery cell. The charge resistance and discharge resistance of the cell are measured under different SOCs at a room temperature of 25 °C, as shown in Figure 5.

$$P_{bat\_loss} = n_{bat}I_{bat}^2R_{bat} \tag{4}$$

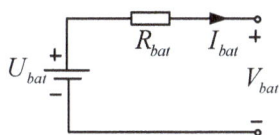

Figure 3. *Rint* model of the LiFePO$_4$ battery.

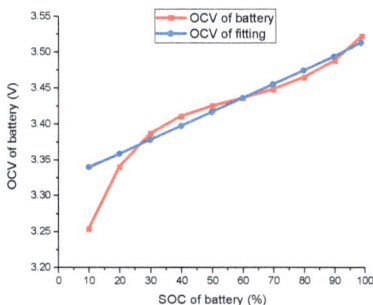

Figure 4. The relationship between V_{bat} and SOC.

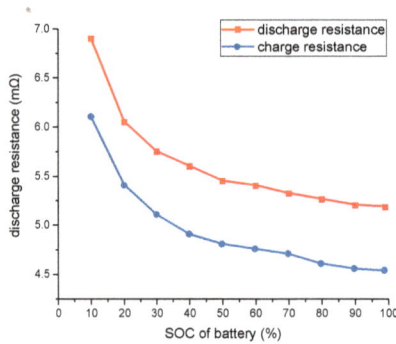

Figure 5. Charge and discharge resistances of the LiFePO$_4$ cell at room temperature of 25 °C.

3.2. Super Capacitor Model

The main purpose of using the SC is to protect the battery more effectively; it has the characteristics of high charge and discharge efficiency, long service life, and better low temperature performance. The SC equivalent circuit model is shown in Figure 6, where R$_{cap}$ is the equivalent series resistance, I$_{cap}$ is the current flowing through the SC cell, C$_{cap}$ is the capacitance of the SC cell, and V$_{cap}$ is the Open Circuit Voltage (OCV) of the SC cell. In this study, we use The Maxwell Technologies® company's super capacity and the parameters of the SC cell are listed in Table 2. Since the SC can achieve millions of charge/discharge cycles, this article ignores the capacity degradation of the SC in the entire process [20]. The open circuit voltage method is used to express the relationship between its open circuit voltage V$_{cap}$ and SOC$_{cap}$.

$$V_{cap} = V_{cap1} SOC_{cap} \tag{5}$$

where V$_{cap1}$ is the open circuit voltage corresponding to SOC$_{cap}$ = 100%. As shown in Equation (6), E$_{cap}$ is the energy released when the SC is discharged from the fully-charged state to SOC$_{cap}$.

$$E_{cap} = \frac{1}{2} n_{cap} C_{cap} V_{cap1}{}^2 (1 - SOC_{cap}) \tag{6}$$

where n$_{cap}$ is the number of SC cells. The output power of the SCs P$_{cap}$ is the first derivative of its release time, as shown in Equation (7).

$$P_{cap} = -\frac{dE_{cap}}{dt} \tag{7}$$

The power consumption on SCs internal resistance P$_{cap\_loss}$ can be calculated by Equation (8), where I$_{cap}$ is the current flowing through the SC cell.

$$P_{cap\_loss} = n_{cap} I_{cap}{}^2 R_{cap} \tag{8}$$

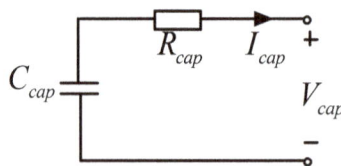

Figure 6. Equivalent circuit modelof the SC.

Table 2. Basic parameters of the SC cell.

Item	Value
V_{cap_norm}, nominal voltage (V)	2.7
C_{cap}, capacity (F)	2×10^3
R_{cap}, resistance (Ω)	3.5×10^{-4}
m_{cap_cell}, cell mass (kg)	0.36
$I_{cap\_max,min}$, max dis/charge current (A)	$\pm 1.6 \times 10^3$

3.3. Battery Degradation Model

The high cost of lithium-ion battery is one of the main factors restricting the development of electric vehicles. The conversion cost of electric vehicles can be reduced by extending the lifespan of lithium-ion battery. In recent years, researchers have made significant efforts to calculate and predict the degradation of the battery [21–23]. In Ref. [21], the authors conducted a series of charge/discharge experiments through constant current-constant voltage (CC-CV) and used a scanning electron microscope (SEM) to characterize the structure of cathode, anode, and separator in Li-ion batteries. The results have shown that the capacity fading of batteries can be attributed primarily to the loss of active Li^+ and the losses of cathode and anode active materials. In Ref. [22], a large number of charge and discharge experiments were carried out on $LiFePO_4$ batteries, and the semi-empirical formula of battery decay percentage and ambient temperature, charge/discharge rate, and cycling time were obtained. In Ref. [23], the effect of parameters such as the end of charge voltage, DOD, film resistance, exchange current density, and over voltage of the parasitic reaction on the capacity fading and battery performance were studied. However, in summary, it is very difficult to calibrate and parameterize the degree of battery degradation in an electric vehicle during actual operation. Therefore, we considered many factors that affect battery degradation and adopted the semi-empirical model used in Ref. [24]. The semi-empirical formula is as shown in Equation (9).

$$Q_{loss} = B \cdot e^{-\left(\frac{E+a \cdot n}{RT}\right)}(A_h)^x \tag{9}$$

where Q_{loss} is the percentage of battery degradation, E is the activation energy, R is the gas constant, T is the absolute temperature, A_h is the A_h-throughput, and B, a, and x are constants. The percentage of discrete battery degradation at different temperatures can be calculated using Equation (10).

$$Q_{loss\_k+1} - Q_{loss\_k} = 9.78 \times 10^{-4} e^{-\left(\frac{15162-1516 \cdot n}{0.849R(|285.75-T|+265)}\right)} \cdot \Delta A_h \cdot Q_{loss\_k}^{-0.1779} \tag{10}$$

where Q_{loss_k} and Q_{loss_k+1} are the percentages of battery capacity decay degradation for the steps k and k + 1, respectively. ΔA_h is the A_h-throughput from t_k to t_{k+1}, and it satisfies Equation (11), where Δt is the sampling time.

$$\Delta A_h = \frac{1}{3600} \cdot |I_{bat}| \cdot \Delta t \tag{11}$$

3.4. Vehicle Model

The vehicle power system model can be used to obtain the power demand of the vehicle at different times during the operation. The power demand of a vehicle should be the output power of the driving wheel for an electric bus with semi-active HESS. The output power of the driving wheel is the product of the demand torque and the required angular velocity, as shown in Equation (12).

$$P_{dem} = T_{dem} \omega_{dem} \eta_T^{-k} \tag{12}$$

where η_T is transmission system efficiency; and k is the power factor: k = 1 when the bus is in the driving state and k = −1 when the bus is in the braking state. T_{dem} and ω_{dem} are the demand torque and the demand angular velocity, respectively, which can be calculated as shown in Equation (13). The basic parameters involved in Equation (13) are listed in Table 3. In Table 3, a_ω is angular acceleration, and the total mass m_{bus} is equal to the sum of the body quality, passenger quality, batteries, and SCs mass (15% additional mass).

$$
\begin{cases}
T_{dem} = \left[(m_{bus}gc_r \cos(\theta) + m_{bus}a + m_{bus}g \sin(\theta)) + (\frac{Ja_\omega}{R_w} + \frac{1}{2}\rho Ac_d v^2) \right] \cdot \frac{R_w}{g_{final}} \\
\omega_{dem} = \frac{v \cdot g_{final}}{R_w} \\
\text{Among}: \\
m_{bus} = m_{veh} + m_p + 1.15(n_{bat}m_{bat\_cell} + n_{cap}m_{cap\_cell}) \\
a = \frac{dv}{dt}; a_\omega = \frac{a}{R_w}
\end{cases} \tag{13}
$$

In the optimization process, the minimal mileage L of more than 50 km should be considered, obtained at a constant cruising speed v_0 (50 km/h) on a flat road [25]. Accordingly, we can deduce the following inequality constraints.

$$
n_{bat} \geq \left(\frac{1}{2}\rho Ac_d v_0^2 + m_{bus}gc_r \right) \frac{L}{V_{bat\_norm}Q_{bat}\eta_T} \tag{14}
$$

We assume that the maximum required power in the drive mode and brake mode is provided by the SCs (SCs output power in driving mode, and SCs absorb power in braking mode). The following inequality constraints should be satisfied when selecting the number of SCs.

$$
n_{cap} \geq \frac{\max\{|P_{dem}|\}}{|I_{cap\_max}|V_{cap\_norm}} \tag{15}
$$

Table 3. Basic parameters of the vehicle.

Item	Value
m_{veh}, body quality	1.3×10^4
m_p, passenger quality (F)	3×10^3
c_r, rolling resistance coefficient	0.007
ρ, air density (kg/m^3)	1.18
J, total inertia (kgm^2)	143.41
A, frontal area (m^2)	7.83
c_d, aerodynamic drag coefficient	0.75
R_w, wheel radius (m)	0.51
g_{final}, final gear ratio	6.2
η_{DC}, DC/DC converter efficiency	0.9
η_T, powertrain efficiency	0.9

4. Convex Optimal Control Strategy Based on the Battery Degradation

Through the driving cycle, we can calculate the power demand of the bus at every step. In the driving mode, the required power P_{dem} should be the output power of the drive motor on its output shaft. The electrical power output on the DC bus is equal to the motor output power plus the motor power loss, which is obtained by the fitting. The electrical power output on the DC bus at this time is the total power to be satisfied by the batteries and the SCs (considering the efficiency of DC/DC converter).

Assuming that the SCs are in the same energy storage state at the beginning and end of the driving cycle, the degradation percentage Q_{loss_sum} of the battery is calculated using the following equation.

$$
Q_{loss\_sum} = \sum_{k=1}^{N} Q_{loss}(k) \tag{16}
$$

In order to reduce the computation time in the convex optimization, as shown in Figure 7, the relationship between the battery current I_{bat} and the battery degradation percentage Q_{loss} is determined at room temperature 25 °C based on Equation (9). Irrespective of the braking or driving modes, Q_{loss} always increases as the charge or discharge current I_{bat} increases. Hence, the optimization of the entire driving cycle of battery degradation can be equivalent to optimizing the entire driving cycle of the battery current in the absolute value. There is the following relationship:

$$\text{Minimize}\left\{\sum_{k=1}^{N} Q_{loss}(k)\right\} \Leftrightarrow \text{Minimize}\left\{\sum_{k=1}^{N} |I_{bat}(k)|\right\} \tag{17}$$

Therefore, as shown in Equation (18), the equivalent optimization target for the purpose of optimizing the battery degradation could be obtained. According to the constraints of the optimization target in the convex optimization [26], Equation (18) satisfies the requirement that the convex optimization must be a convex function or an affine function for the objective function.

$$J = \text{Minimize}\left\{\sum_{k=1}^{N} |I_{bat}(k)|\right\} \tag{18}$$

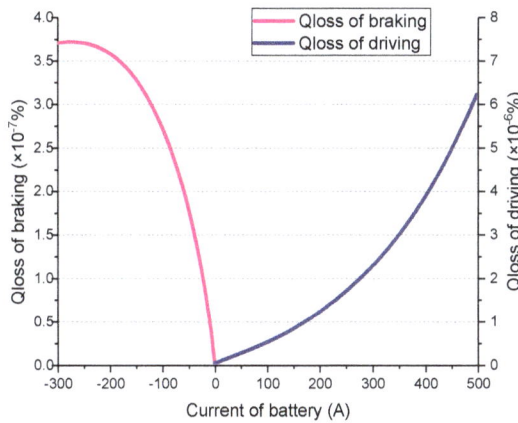

Figure 7. The relationship between battery charge/discharge current I_{bat} and battery degradation percentage Q_{loss}.

The implementation process of the convex optimization strategy used in this study is shown in Figure 8. In order to satisfy the constraints on the number of battery cells and SC cells in Equations (14) and (15), the number of selected battery and SC cells is $n_{bat} = 120$ and $n_{cap} = 240$, respectively. The battery pack and SC pack in the HESS are all grouped by n series and one parallel connection, and the unbalance between the cells is ignored [27]. As the energy source of the vehicle, the LiFePO$_4$ battery pack is not only for the drive motor to provide energy, but also for the super capacitor when SOC$_{cap}$ is low. The SC pack is between the LiFePO$_4$ battery pack and the drive motor, acting as an energy and power buffer, and absorbing the braking energy from the drive motor during braking.

According to the output voltage changes of the battery and the SC cell, we can constrain the output energy range of the battery pack and capacitor group,

$$\begin{cases} E_{bat}(k) \in \frac{n_{bat}Q_{bat}}{2}\left([V_{bat\_min}^2, V_{bat\_max}^2] - V_{bat0}^2\right) \\ E_{cap}(k) \in \frac{n_{cap}C_{cap}}{2}\left([V_{cap\_min}^2, V_{cap\_max}^2] - V_{cap0}^2\right) \end{cases} \tag{19}$$

where V_{bat_min}, V_{bat_max}, V_{cap_min}, and V_{cap_max} are the maximum and minimum voltages corresponding to the battery and the SOC of the SC, given by Equation (20).

$$\begin{cases} V_{bat\_min,max} = SOC_{bat\_min,max}(V_{bat1} - V_{bat0}) + V_{bat0} \\ V_{cap\_min,max} = SOC_{SC\_min,max}V_{SC1} \end{cases} \tag{20}$$

The range of the output power of the battery and SC pack can be limited according to the maximum charge/discharge current of the battery and SC cell.

$$\begin{cases} P_{bat} \in [I_{bat\_min}, I_{bat\_max}]n_{bat}V_{bat} \\ P_{cap} \in [I_{cap\_min}, I_{cap\_max}]n_{cap}V_{cap} \end{cases} \tag{21}$$

The power consumed on the battery pack and the SC pack internal resistance can be calculated from Equation (22).

$$\begin{cases} P_{bat\_loss} = n_{bat}I_{bat}{}^2R_{bat} \\ P_{cap\_loss} = n_{cap}I_{cap}{}^2R_{cap} \end{cases} \tag{22}$$

where I_{bat} and I_{cap} are the currents flowing through the battery and the SC cell, respectively. Both I_{bat} and I_{cap} have the following constraints.

$$\begin{cases} I_{bat} \in [I_{bat\_min}, I_{bat\_max}] \\ I_{cap} \in [I_{cap\_min}, I_{cap\_max}] \end{cases} \tag{23}$$

For the battery pack and the SC pack, it is necessary to satisfy the total power demand in different cases, by satisfying the following equation constraints.

$$\begin{cases} P_{batopen} + P_{capopen}\eta_{DC} = P_{emloss} + P_{dem}/\eta_T & P_{dem} \geq 0 \\ P_{batopen} + P_{capopen}/\eta_{DC} = P_{emloss} + P_{dem}\eta_T & P_{dem} < 0 \end{cases} \tag{24}$$

where $P_{batopen}$ and $P_{capopen}$ are the output powers of the battery pack and SC pack, respectively. P_{emloss} is the motor power loss and can be interpolated by the motor power loss curve.

In this study, we use Equation (18) as an optimization target, and the battery pack energy E_b, SC pack energy E_{cap}, battery pack port output power $P_{batopen}$, and SC pack port output power $P_{capopen}$ as the convex optimization variables. The battery pack power P_{bat}, SC pack power P_{cap}, battery pack power loss P_{bat_loss}, and SC pack power loss P_{cap_loss} are used as the equation constraints of the convex optimization. The overall optimization function is given in Table 4.

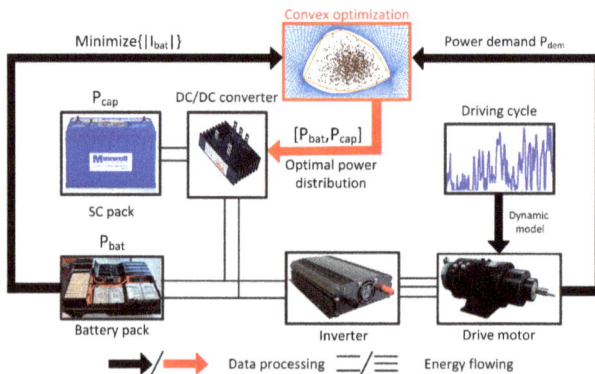

Figure 8. The implementation process of the convex optimization strategy.

Table 4. Convex optimization function of hybrid energy storage system

Variables	$E_{bat}(N+1), E_{cap}(N+1), P_{batopen}(N), P_{capopen}(N)$
Minimize	$\sum\limits_{k=1}^{N} |I_{bat}(k)|$
Subject to	$\begin{cases} P_{batopen}(k) + P_{capopen}(k)\eta_{DC} = P_{emloss}(k) + P_{dem}(k)/\eta_T P_{dem}(k) \geq 0 \\ P_{batopen}(k) + P_{capopen}(k)/\eta_{DC} = P_{emloss}(k) + P_{dem}(k)\eta_T P_{dem}(k) \geq 0 \end{cases}$ $P_{bat\_loss}(k) = n_{bat}I_{bat0}{}^{(k)2}R_{bat}; P_{cap\_loss} = n_{cap}I_{cap0}(k)^2 R_{cap}$ $P_{bat}(k) = -\Delta E_{bat}(k); P_{cap}(k) = -\Delta E_{cap}(k)$ $E_{bat}(k) \in ([V_{bat\_min}{}^2, V_{bat\_max}{}^2] - V_{bat0}{}^2)n_{bat}Q_{bat}/2$ $E_{cap}(k) \in ([V_{cap\_min}{}^2, V_{cap\_max}{}^2] - V_{cap0}{}^2)n_{cap}C_{cap}/2$ $P_{bat}(k) \in [I_{bat\_min}, I_{bat\_max}]n_{bat}V_{bat}(k)$ $P_{cap}(k) \in [I_{cap\_min}, I_{cap\_max}]n_{cap}V_{cap}(k)$ $I_{bat} \in [I_{bat\_min}, I_{bat\_max}]; I_{cap} \in [I_{cap\_min}, I_{cap\_max}]$
	$\forall k \in \{0, \ldots, N\}$

According to the description of the convex optimization problem in Ref. [26], the constraint conditions in Table 4 are convex or affine functions; the entire convex optimization problem satisfies the convex optimization requirement, and the convex optimization implementation flow is shown in Figure 9.

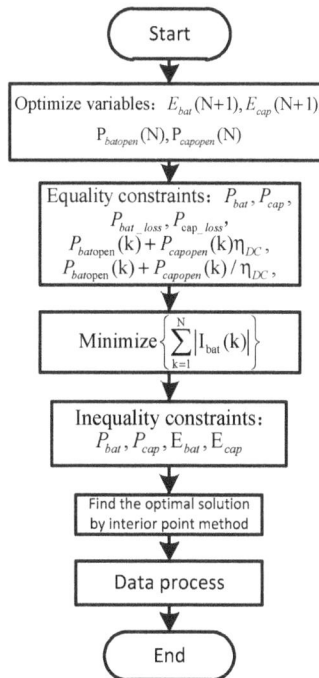

Figure 9. Convex optimization implementation flow.

5. Simulation and Results Analysis

In order to verify the effectiveness of the HESS optimization proposed in this study, we compared its performance with that of the rule-based power distribution strategy under the Harbin city driving cycle [28]. The Harbin city driving cycle and the power demand is shown in Figure 10. The convex optimization result is shown in Figure 11. It can be observed that the battery only provides a small range of demand for power fluctuations, whereas the SC satisfies instantaneous high power requirements. When the required power is greater than a certain threshold, the SC and the battery pack together provide the power required. However, owing to the presence of DC/DC converter efficiency and powertrain efficiency, the sum of the power of the battery pack and the SC pack is much greater than the power demand. Furthermore, the same as the result of two efficiency effects, in the braking mode, the SC cannot recover all the braking energy. In addition, as shown in Figure 11b, the battery only provides a small demand for power and the SC satisfies instantaneous high power requirements; this can be reflected in the relationship between power demand and SC power. When the power demand fluctuates in the range of 0 kW to 20 kW, the SC does not output power. This part of the power is borne by the battery and when the power demand fluctuates in the range of 20 kW to the highest power demand, the SC satisfies the power demand. Owing to the efficiency of DC/DC converter and the efficiency of the powertrain, the slope of the fitted line k is slightly greater than 1 when the demand power is greater than 20 kW. When the power demand fluctuates in the negative range, the SC absorbs the braking energy, and the slope of the fitting line k is less than 1 owing to the DC/DC converter efficiency and the efficiency of the powertrain. The entire convex optimization process consumes 1.33×10^7 J, i.e., approximately 3.70 kWh. Since the optimization target is absolute value of the battery current, while the terminal voltage drop of the battery pack is very small in one driving cycle. Therefore, the optimization goal also has a significant role in optimizing the battery energy consumption.

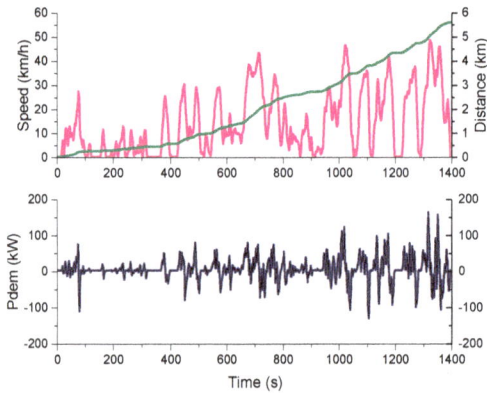

Figure 10. Harbin city driving cycle and the power demand.

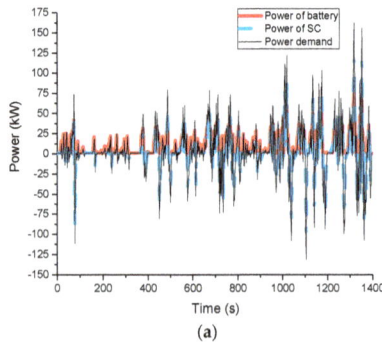

(a)

Figure 11. *Cont.*

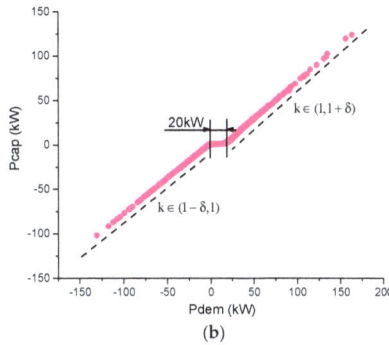

Figure 11. Convex optimization results: (**a**) power distribution results based on convex optimization; and (**b**) the relationship between power demand and SC power.

Figure 12 shows the battery current and battery degradation percent at a room temperature of 25 °C based on convex optimization. In the latter part of the driving cycle, the acceleration/deceleration of the vehicle is violent; hence, the battery current undergoes greater fluctuations. However, the maximum value of the battery current does not exceed 200 A. The battery runs within 1C discharge magnification rate altogether. In addition, according to the relationship between the battery charge/discharge current I_{bat} and the battery degradation percentage Q_{loss} in Figure 7, it can be observed that the battery current has approximately an exponential relationship with the battery degradation percentage Q_{loss} when the battery current is greater than zero. Therefore, Q_{loss} obtained in Figure 12 is more intense than the battery current fluctuation. At room temperature of 25 °C, the battery degradation percent in one driving cycle is $2.16 \times 10^{-4}\%$.

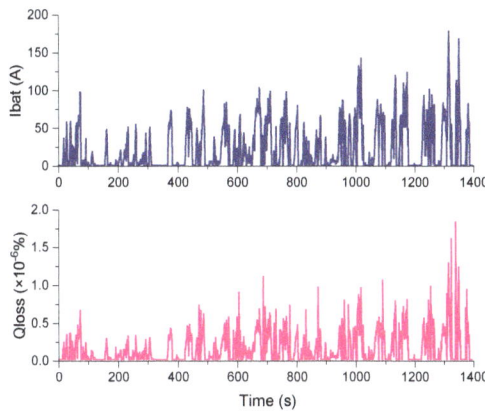

Figure 12. Battery current and battery degradation percentage at room temperature of 25 °C based on convex optimization.

In order to verify the effectiveness of the proposed optimization method in this paper for the electric bus equipped with semi-active HESS, it is compared with the rule-based strategy in Harbin driving cycle. The rule-based strategy is shown in Figure 13 [9,29]. When the power demand $P_{dem} \geq 0$, the battery only provides the threshold power P_{thr}, and when $P_{dem} > P_{thr}$, the battery will provide the part of power demand exceed according to the SOC_{cap}. When the power demand $P_{dem} < 0$, it is also need to charge for SC according to the charge of state of the SC, when the SC pack is fully charged and then to the battery pack for energy braking. According to the characteristics of the SC power and the power demand relationship obtained from the convex optimization, the threshold power P_{thr} in the rule-based strategy is set to 20 kW, and the same power system model and initial parameters are used to calculate the power distribution based on the rules strategy, as shown in Figure 14.

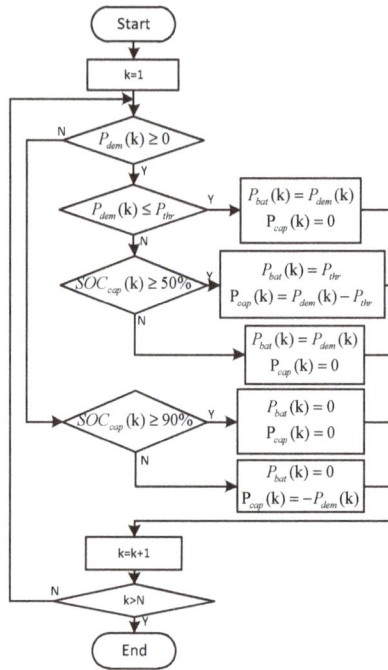

Figure 13. The implementation process of rule-based strategy.

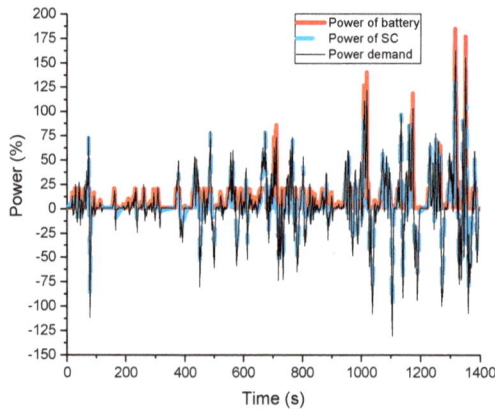

Figure 14. Power distribution results based on the rule-based strategy.

Figure 15 shows the battery current and battery degradation percent at 25 °C based on the rule strategy. In the latter part of the driving cycle, it can be seen that the SC obtained the energy from braking cannot satisfy the power demand; this part of the power demand difference can only be borne by the battery. Therefore, the battery experiences a large output power fluctuation. The battery current and battery decay percentages have several significant peaks in the later stages based on the rule strategy. However, in contrast, the SC satisfies the instantaneous high power ripple and the battery outputs a smaller power fluctuation in the convex optimization. According to the energy management strategy evaluation method in Ref. [30], the standard deviation of the battery current in two strategies has been calculated, as shown in Equation (25), where the I_{b_avg} is the average of battery current, s is the standard deviation of battery current.

$$s = \sqrt{\frac{\sum\limits_{k=1}^{N} (I_b(k) - I_{b\_avg})^2}{N}} \tag{25}$$

The standard deviation of the battery current based on the convex optimization strategy and rule-based strategy is 31.03 and 43.99, respectively, the battery current fluctuation based on the convex optimization strategy is smaller than rule-based strategy. Figure 16 shows the battery capacity degradation curve based on convex optimization and rule-based strategy. At approximately t = 1050 s, t = 1320 s, and t = 1360 s, the battery degradation percent based on the rule strategy has three significant rising intervals, which correspond to three battery current pulses based on the rules strategy; further, it can be reflected from the battery current and battery degradation percentage result of the rule-based strategy shown in Figure 15. In contrast, the battery power fluctuation is smoother when based on the convex optimization strategy, thus protecting the battery better. The rule-based strategy consumes 1.35×10^7 J in one driving cycle, i.e., approximately 3.75 kWh, and the percentage of battery degradation is $2.31 \times 10^{-4}\%$.

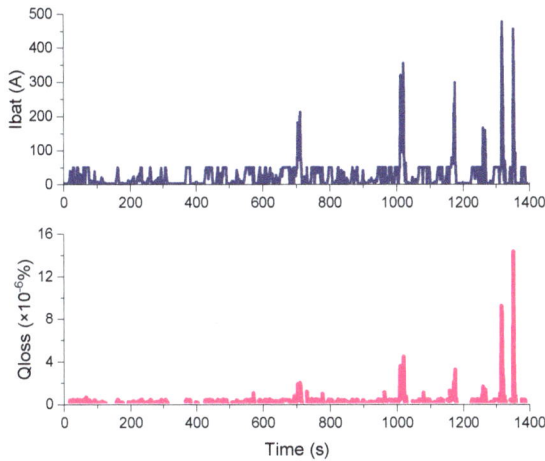

Figure 15. Battery current and battery degradation percentage at room temperature of 25 °C based on the rule strategy.

In order to further reflect the real situation of a bus running on Harbin city roads in a day, and quantitatively analyze the potential of convex optimization relative to the rule-based strategy to improve battery life, we use nine Harbin city driving cycles (total 50.4 km) to simulate a cycle line: a bus from the bus terminal, followed by a cycle line, and subsequently back to the bus terminal. When the bus arrives at the terminal again, the CC-CV is used to recharge the battery. When the battery pack is recharged, the SOC_{bat} returns to 0.9. Table 5 shows the indicators of the battery pack when the bus drives in one cycle line based on the two strategies.

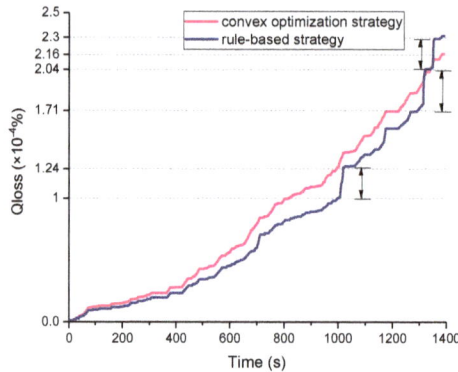

Figure 16. The percentage of battery degradation under two strategies.

Table 5. The indicators of two strategies under one cycle line.

Strategy	Total Length	Initial SOC	Final SOC	ΔSOC	Recharge Time
Based on convex	50.4 km	0.9	0.4223	0.4777	1561 s
Based on rules	50.4 km	0.9	0.4112	0.4888	1598 s

For the LiFePO$_4$ battery pack with 120 cells in series and one in parallel, the SOC of the battery varies from SOC$_{bat}$ = 0.9 to SOC$_{bat}$ = 0.4223 through one cycle line under convex optimization, and the SOC of the battery varies from SOC$_{bat}$ = 0.9 to SOC$_{bat}$ = 0.4112 through one cycle line under rule-based strategy. In the process of the recharge phase, the recharge magnification is set to 1C. The entire recharge state continues for 1561 s in convex optimization and 1598 s in rule-based strategy. After the recharge, the battery SOC of the convex optimization strategy and the rule-based strategy are returned from SOC$_{bat}$ = 0.4223 and SOC$_{bat}$ = 0.4112 to SOC$_{bat}$ = 0.9, respectively. According to the actual operation, the electric bus run four line cycles (approximately 200 km) in one day, and the electric bus needs to recharge the battery four times. Assuming that the battery is fully charged (SOC$_{bat}$ = 0.9) in the first trip every day, the battery SOC change curve based on the convex optimization strategy and rule-based strategy can be obtained as shown in Figure 17.

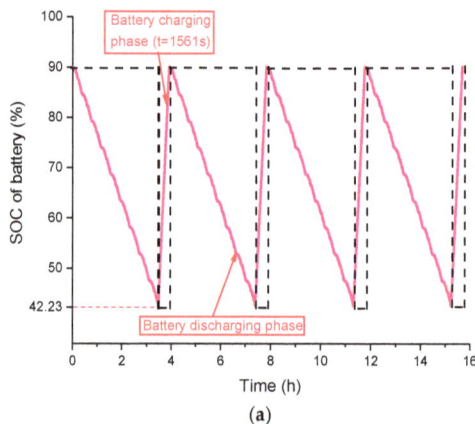

(a)

Figure 17. *Cont.*

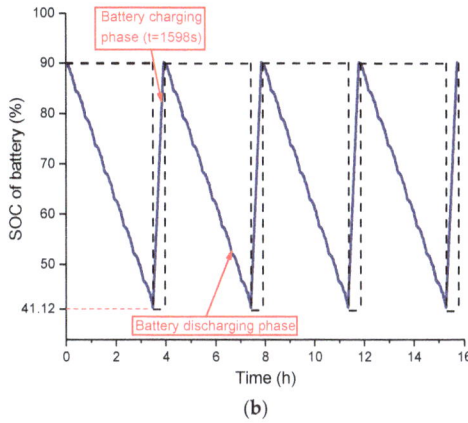

Figure 17. The battery SOC change based on different power distribution strategies in one day: (**a**) the battery SOC based on the optimization strategy described in this paper; and (**b**) the battery SOC based on the rules of the power distribution strategy.

The battery capacity degradation will occur during the entire charge/discharge process. Therefore, we can obtain the battery degradation percentage curve of the battery pack using the semi-empirical model in one day according to the battery charge/discharge situation. At room temperature of 25 °C, Q_{loss} base on the convex optimization strategy and the rules strategy are shown in Figure 18. After a day of operation (four cycle lines and four charging phases), Q_{loss} based on convex optimization is 9.6×10^{-3}%, and the battery pack can be used for approximately 2.08×10^3 days when the battery degradation is attenuated to 80% of its initial capacity, whereas Q_{loss} based on rules strategy is 10.3×10^{-3}%, and the battery pack can be used for approximately 1.94×10^3 days when the battery is attenuated to 80% of its initial capacity.

Notably, there are two main causes of the greater battery degradation under rule-based strategy: (1) The battery current I_{bat} based on the convex optimization is more stable; hence, this results in a smaller Q_{loss} than the rule-based strategy, and this is reflected in the comparison of battery degradation under the two strategies shown in Figure 16; (2) Since the rule-based strategy consumes more energy in one day, the battery discharges deeper during operation ($SOC_{bat} \in [0.4223, 0.9]$ during one cycle line based on convex optimization and $SOC_{bat} \in [0.4112, 0.9]$ during one cycle line based on rules). Similarly, in order to return the battery back to the initial SOC, the rule-based strategy must be charging deeper than the convex optimized strategy (charging time t = 1561 s based on the convex optimization and t = 1598 s based on the rules). Consequently, the rule-based strategy increases the SOC operating range of the battery used in the same road cycle as compared to the strategy based on the convex optimization, which also increases the percentage of battery capacity degradation.

The battery usage time based on the two strategies is approximately 15 h in one day, but the usage time of the rule-based strategy is slightly longer because the charging time is longer. In Figure 18, Q_{loss} slope of the battery charging phase is larger than Q_{loss} at the time of traveling on the road; hence, the battery degradation caused by charging the battery at a charging magnification of 1C is greater than that caused by running on the road.

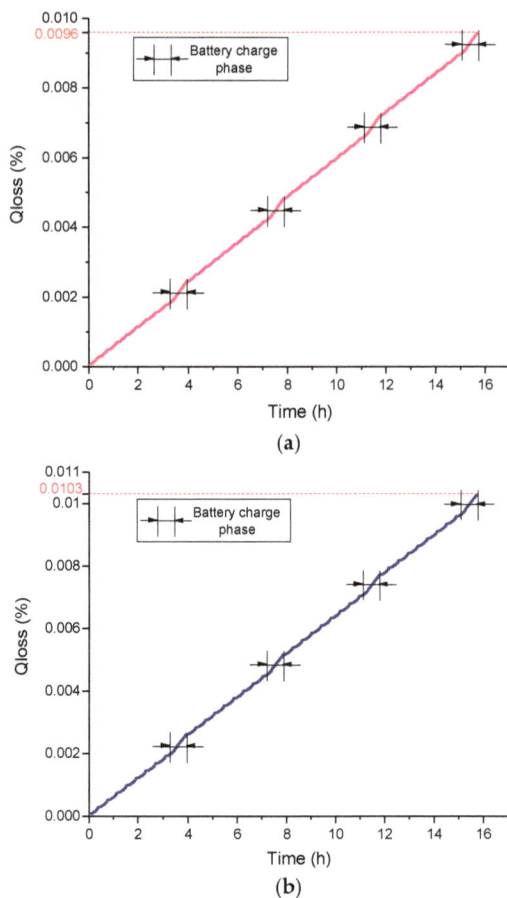

Figure 18. Q_{loss} based on the two strategies in one day: (**a**) battery degradation percentage based on convex optimization strategy; and (**b**) battery degradation percentage based on rules strategy.

In order to obtain further analysis of Q_{loss} for one year in Harbin, we assume that the Harbin city bus will run for cycle lines in one day, and the bus operates for 30 days in a month. Therefore, the bus will operate for 360 days in a year. According to the average temperature of different months in Harbin, the daily value of Q_{loss} for every month can be calculated. It is assumed that the vehicle in each month is running at the monthly average temperature value and ignores the self-heating effect of the battery [31]. The semi-empirical model is also used to calculate the battery degradation. The average monthly temperature in Harbin is given in Table 6 [32], i.e., the temperature range of $T \in [-18.3\,^{\circ}\text{C}, 23.0\,^{\circ}\text{C}]$.

Table 6. The average temperature of the city in Harbin.

Month	Jan	Feb	Mar	Apr	May	Jun
Temp (°C)	−18.3	−13.6	−3.4	7.1	14.7	20.4
Month	Jul	Aug	Sep	Oct	Nov	Dec
Temp (°C)	23.0	21.1	14.5	5.6	−5.3	−14.8

According to the different average temperatures in different months, we can calculate the electric bus battery degradation percentage in one day for different months, as shown in Figure 19.

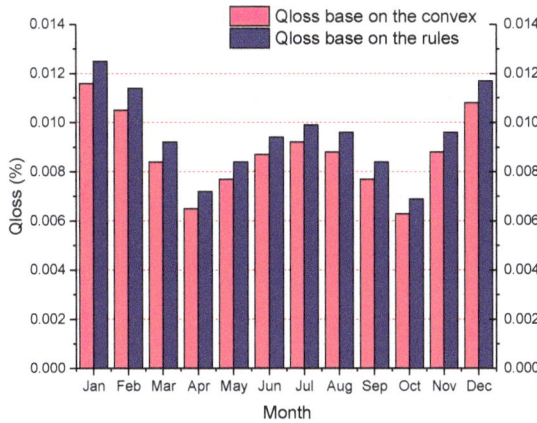

Figure 19. Percentage of battery degradation Q_{loss} in one day for different months.

As evident in Figure 19, Q_{loss} in different months based on the rules strategy is larger than that based on the convex optimization strategy. It can also be observed that, when the temperature is within $\pm 10\,^{\circ}C$ during spring and autumn, Q_{loss} is minimal, whereas Q_{loss} is large in summer and winter with higher or lower temperatures. Especially in the winter months of December, January, and February, the two strategies under daily operation of the battery degradation reached $\geq 0.01\%$. This directly reflects the problem of large degradation of LiFePO$_4$ battery at low temperatures. The total battery degradation percentage of the year is shown in Table 7.

Table 7. The total battery degradation percentage in different month of the year.

Month	Jan	Feb	Mar	Apr	May	Jun
Q_{loss} base on the convex/rules (%)	0.348/0.375	0.315/0.342	0.252/0.276	0.195/0.216	0.231/0.252	0.261/0.282
Month	Jul	Aug	Sep	Oct	Nov	Dec
Q_{loss} base on the convex/rules (%)	0.276/0.297	0.264/0.288	0.231/0.252	0.189/0.207	0.264/0.288	0.324/0.351

In Table 7, Q_{loss} in one year based on the convex optimization strategy is 3.15%, and Q_{loss} based on the rules strategy is 3.43%. A quantitative analysis of battery life extension is addressed in this paper, assuming that the battery cannot be used when its capacity reduces to 80% of the initial value. Subsequently, the year of usage of the battery can be calculated using Equation (26).

$$Y_{ope} = 20 / \sum_{m=1}^{12} Q_{loss\_month}(m) \qquad (26)$$

where Y_{ope} is the year of usage of battery and Q_{loss_month} is the percentage of Q_{loss} per month. It can be calculated that the bus equipped with HESS can operate for 6.35 years based on the convex optimization strategy. However, the bus equipped with HESS can operate for 5.84 years based on the rules strategy. The superiority of the convex optimization strategy is reflected, further illustrating the effectiveness of the use of convex optimization.

6. Conclusions

This paper used the electric bus power system with semi-active HESS as the research object, in the Harbin city driving cycle and proposed a convex optimization power distribution strategy target to optimize the battery current that represents battery degradation. According to the average temperature of different months of the year in the northern city of Harbin, the percentage of battery degradation of the electric bus with the HESS is analyzed and calculated. Simulation results show that using the convex optimization strategy proposed in this paper, at a room temperature of 25 °C, in a daily trip composed of the Harbin City Driving Cycle—including four cycle lines and four charging stages—the percentage of the battery degradation is 9.6×10^{-3}%, whereas the battery degradation is 10.3×10^{-3}% based on the rules under the same conditions. Assuming the daily mileage of an electric bus is approximately 200 km, and it will operate for 360 days in a year, the percentage of battery degradation is 3.15% in one year in Harbin. Before the battery capacity reduces to 80% of the initial value, the electric bus can run for 6.35 years based on the strategy proposed in this paper. However, the battery degrades 3.43% per year in Harbin using the rule-based strategy, and the bus can run for 5.84 years. Thus, the convex-based optimization strategy can operate for 0.51 year more than the rule-based strategy.

Acknowledgments: This work was supported by the State Key Laboratory of Automotive Safety and Energy under Project No. KF16062 and Science Funds for the Young Innovative Talents of HUST, No. 201503.

Author Contributions: Xiaogang Wu and Tianze Wang conceived and designed the simulations; Xiaogang Wu analyzed the data; Tianze Wang wrote the paper.

Conflicts of Interest: The authors declare no conflict of interest.

References

1. Conte, M.; Genovese, A.; Ortenzi, F.; Vellucci, F. Hybrid battery-supercapacitor storage for an electric forklift: A life-cycle cost assessment. *J. Appl. Electrochem.* **2014**, *44*, 523–532. [CrossRef]
2. Yutao, L.; Feng, W.; Hao, Y.; Xiutian, L. A Study on the Driving-cycle-based Life Model for LiFePO4 battery. *Automot. Eng.* **2015**, *37*, 881–885.
3. Abeywardana, D.B.W.; Hredzak, B.; Agelidis, V.G.; Demetriades, G.D. Supercapacitor Sizing Method for Energy-Controlled Filter-Based Hybrid Energy Storage Systems. *IEEE Trans. Power Electron.* **2017**, *32*, 1626–1637. [CrossRef]
4. Henson, W. Optimal battery/ultracapacitor storage combination. *J. Power Sources* **2008**, *179*, 417–423. [CrossRef]
5. Xiong, R.; Sun, F.; Chen, Z.; He, H. A data-driven multi-scale extended Kalman filtering based parameter and state estimation approach of lithium-ion olymer battery in electric vehicles. *Appl. Energy* **2014**, *113*, 463–476. [CrossRef]
6. Zhang, Y.; Xiong, R.; He, H.; Shen, W. Lithium-Ion Battery Pack State of Charge and State of Energy Estimation Algorithms Using a Hardware-in-the-Loop Validation. *IEEE Trans. Power Electron.* **2017**, *32*, 4421–4431. [CrossRef]
7. Choi, M.; Kim, S.; Seo, S. Energy Management Optimization in a Battery/Supercapacitor Hybrid Energy Storage System. *IEEE Trans. Smart Grid* **2012**, *3*, 463–472. [CrossRef]
8. Song, Z.; Li, J.; Han, X.; Xu, L.; Lu, L.; Ouyang, M.; Hofmann, H. Multi-objective optimization of a semi-active battery/supercapacitor energy storage system for electric vehicles. *Appl. Energy* **2014**, *135*, 212–224. [CrossRef]
9. Song, Z.; Hofmann, H.; Li, J.; Han, X.; Zhang, X.; Ouyang, M. A comparison study of different semi-active hybrid energy storage system topologies for electric vehicles. *J. Power Sources* **2015**, *274*, 400–411. [CrossRef]
10. Hu, X.; Murgovski, N.; Johannesson, L.; Egardt, B. Energy efficiency analysis of a series plug-in hybrid electric bus with different energy management strategies and battery sizes. *Appl. Energy* **2013**, *111*, 1001–1009. [CrossRef]
11. Lu, L.; Han, X.; Li, J.; Hua, J.; Ouyang, M. A review on the key issues for lithium-ion battery management in electric vehicles. *J. Power Sources* **2013**, *226*, 272–288. [CrossRef]

12. Amjadi, Z.; Williamson, S.S. Prototype Design and Controller Implementation for a Battery-Ultracapacitor Hybrid Electric Vehicle Energy Storage System. *IEEE Trans. Smart Grid* **2012**, *3*, 332–340. [CrossRef]
13. Ma, T.; Yang, H.; Lu, L. Development of hybrid battery-supercapacitor energy storage for remote area renewable energy systems. *Appl. Energy* **2015**, *153*, 56–62. [CrossRef]
14. Hung, Y.; Wu, C. An integrated optimization approach for a hybrid energy system in electric vehicles. *Appl. Energy* **2012**, *98*, 479–490. [CrossRef]
15. Vinot, E.; Trigui, R. Optimal energy management of HEVs with hybrid storage system. *Energy Convers. Manag.* **2013**, *76*, 437–452. [CrossRef]
16. Yoo, H.; Sul, S.; Park, Y.; Jeong, J. System integration and power-flow management for a series hybrid electric vehicle using supercapacitors and batteries. *IEEE Trans. Ind. Appl.* **2008**, *44*, 108–114. [CrossRef]
17. Li, Y.; Han, Y. A Module-Integrated Distributed Battery Energy Storage and Management System. *IEEE Trans. Power Electron.* **2016**, *31*, 8260–8270. [CrossRef]
18. Zhao, N.; Li, Y.; Zhao, X.; Zhi, X.; Liang, G. Effect of particle size and purity on the low temperature electrochemical performance of LiFePO$_4$/C cathode material. *J. Alloys Compd.* **2016**, *683*, 123–132. [CrossRef]
19. Chen, C.; Xiong, R.; Shen, W. A lithium-ion battery-in-the-loop approach to test and validate multi-scale dual H infinity filters for state of charge and capacity estimation. *IEEE Trans. Power Electron.* **2017**, *PP*, 1. [CrossRef]
20. Marzougui, H.; Amari, M.; Kadri, A.; Bacha, F.; Ghouili, J. Energy management of fuel cell/battery/ultracapacitor in electrical hybrid vehicle. *Int. J. Hydrogen Energy* **2017**, *42*, 8857–8869. [CrossRef]
21. Wen-gang, L.; Bo, Z.; Xiao-dan, W.; Jun-kui, G.; Xing-jiang, L. Capacity fading of 18650 Li-ion cells with cycling. *Chin. J. Power Sources* **2012**, *3*, 306–309.
22. Wang, J.; Liu, P.; Hicks-Garner, J.; Sherman, E.; Soukiazian, S.; Verbrugge, M.; Tataria, H.; Musser, J.; Finamore, P. Cycle-life model for graphite-LiFePO$_4$ cells. *J. Power Sources* **2011**, *196*, 3942–3948. [CrossRef]
23. Ramadass, P.; Haran, B.; Gomadam, P.M.; White, R.; Popov, B.N. Development of First Principles Capacity Fade Model for Li-Ion Cells. *J. Electrochem. Soc.* **2004**, *151*, A196–A203. [CrossRef]
24. Song, Z.; Hofmann, H.; Li, J.; Hou, J.; Zhang, X.; Ouyang, M. The optimization of a hybrid energy storage system at subzero temperatures: Energy management strategy design and battery heating requirement analysis. *Appl. Energy* **2015**, *159*, 576–588. [CrossRef]
25. New Energy Vehicles Pure Electric Driving Range and Special Inspection Standards, N.E.V.P. Available online: http://news.xinhuanet.com/auto/2015-05/19/c_127817177.htm (accessed on 6 June 2017).
26. Boyd, S. *Convex Optimization*; Cambridge University Press: Cambridge, UK, 2004; pp. 7–13.
27. Sun, F.; Xiong, R.; He, H. A systematic state-of-charge estimation framework for multi-cell battery pack in electric vehicles using bias correction technique. *Appl. Energy* **2016**, *162*, 1399–1409. [CrossRef]
28. Hu, C.; Wu, X.; Li, X.; Wang, X. Construction of Harbin City Driving Cycle. *J. Harbin Univ. Sci. Technol.* **2014**, *19*, 85–89.
29. Song, Z.; Hofmann, H.; Li, J.; Han, X.; Ouyang, M. Optimization for a hybrid energy storage system in electric vehicles using dynamic programing approach. *Appl. Energy* **2015**, *139*, 151–162. [CrossRef]
30. Choi, M.; Lee, J.; Seo, S. Real-Time Optimization for Power Management Systems of a Battery/Supercapacitor Hybrid Energy Storage System in Electric Vehicles. *IEEE Trans. Veh. Technol.* **2014**, *63*, 3600–3611. [CrossRef]
31. Forgez, C.; Vinh Do, D.; Friedrich, G.; Morcrette, M.; Delacourt, C. Thermal modeling of a cylindrical LiFePO$_4$/graphite lithium-ion battery. *J. Power Sources* **2010**, *195*, 2961–2968. [CrossRef]
32. Analysis of Climate Background in Harbin. Available online: http://www.weather.com.cn/cityintro/101050101.shtml (accessed on 6 June 2017).

energies

MDPI

Article

PTC Self-heating Experiments and Thermal Modeling of Lithium-ion Battery Pack in Electric Vehicles

Chengning Zhang [1,2], Xin Jin [1,2] and Junqiu Li [1,2,*]

[1] National Engineering Laboratory for Electric Vehicles, Beijing Institute of Technology, Beijing 100081, China; mrzhchn@bit.edu.cn (C.Z.); jx493067818@126.com (X.J.)
[2] Collaborative Innovation Center of Electric Vehicles in Beijing, Beijing Institute of Technology, Beijing 100081, China
* Correspondence: lijunqiu@bit.edu.cn; Tel.: +86-10-6891-2947

Academic Editor: Hailong Li
Received: 10 March 2017; Accepted: 18 April 2017; Published: 22 April 2017

Abstract: This paper proposes a positive temperature coefficient (PTC) self-heating method, in which EVs can be operated independently of external power source at low temperature, with a lithium-ion battery (LIB) pack discharging electricity to provide PTC material with power. Three comparative heating experiments have been carried out respectively. With charge/discharge tests implemented, results demonstrate the superiority of the self-heating method, proving that the discharge capability, especially the discharge capacity of the self-heated pack is better than that of the external power heated pack. In order to evaluate the heating effect of this method, further studies are conducted on temperature distribution uniformity in the heated pack. Firstly, a geometric model is established, and heat-generation rate of PTC materials and LIB are calculated. Then, thermal characteristics of the self-heating experiment processes are numerically simulated, validating the accuracy of our modeling and confirming that temperature distributions inside the pack after heating are kept in good uniformity. Therefore, the PTC self-heating method is verified to have a significant effect on the improvement of performance of LIB at low temperature.

Keywords: lithium-ion battery; PTC self-heating method; self-heating experiment; thermal modeling

1. Introduction

Improving the performance of the Lithium-ion batteries (LIBs) at low temperature has become an urgent problem to be solved, since some problems may exist, such as dramatic decrease of discharge rate and serious degradation of discharge capacity [1,2], which may lead to shortened driving range, deteriorated dynamic performance, restricted feasibility and applicability for electric vehicles (EVs) [3]. One of the relatively feasible methods proved to be effective is heating the LIB pack [4–8].

Generally, two types of heating methods for LIB have been adopted: the external heating and the internal heating.

As for the researches on the external heating, Wang et al. [9] have put forward a battery bottom-heating method with a design of a bench test, and their experimental results verified the bottom-heating method could effectively improve discharge capacity of LIB pack. Based on charge-/discharge performance experiments on the heated cell, Zhang et al. [10] proved a method of wide-line metal film heating, which could be efficiently applied to enhance cell performances at low temperature. With a low-temperature-heating model for the cell established and a method of electro-thermal-film heater recommended, Cun-shan et al. [11] confirmed the practical efficacy of their method via simulations and experiments. On the basis of experiments on a convection heat transfer method, in which the battery pack was heated by using heated wires to transfer heat through the air, Wang et al. [12]

found that the surface temperature of the battery pack could be increased from −15 °C to 0 °C after about 21 min of heating.

For the researches on the internal heating methods [13], Hand and Stuart [14–18] have proposed a method with which battery electrolyte be directly heated with Alternating Current(AC). However, they also warned that Direct Current (DC) should not be applied as it might generate a lot of gas inside and thus damage the battery. Zhang J. et al. [19] pointed out that within a certain scope, the higher the sine AC rose, the lower the frequency might be, and the more rapidly the battery temperature would be increased. Comparing the heat generation process of battery charge and discharge, Zhao et al. [20] proposed that LIB could be heated at low temperature by combining the large pulse discharge with the small pulse charging.

It can be deduced from the above research that the heating method with external resistance is to heat the battery via heat conduction. The battery heats up rapidly, while the size and weight of the battery pack may be increased. In the heating method with convective heat transfer, there are still some problems such as longer heating time, lower heat exchange efficiency, and poorer temperature uniformity inside the battery pack after heating. Even for a heating method that applies AC to a battery to heat it, problems still exist; for instance, the circuits used in the external heating are complex, and the impact of AC heating on battery life remains to be verified. Moreover, the heating methods discussed above are highly dependent on an external power supply, which will be an inconvenience for EVs. For example, when an external power supply is not available in a cold environment, EVs may not function normally at low temperatures.

To solve the above problems, a positive temperature coefficient (PTC) self-heating method is proposed in this paper. With this method, EVs can be operated independently of external power at low temperature, with the LIB pack being heated on its own when it discharges electricity to provide PTC material power. The main advantages of the method are: (1) Making full use of the PTC heat generation characteristics. Namely, when the temperature is low, and the PTC resistances are small, heating power will be increased. When the temperature rises to a certain extent, the PTC resistances increase sharply, and heating power will be reduced to avoid overheating the LIB pack. (2) Improving heating effects. In a low-temperature environment, as the LIB internal resistances rise, the heat generated by the resistances will also increase, which can be partially transferred directly within the LIB, thus further promoting the heating effects. Based on the PTC self-heating method, three comparative heating experiments, including an external power source heating experiment and two self-heating experiments, were carried out. With further tests on charge/discharge performance implemented, results reveal that the discharge performance of a self-heated battery is better than that of a pack heated by an external power source.

In order to evaluate the heating effectiveness of this method, further study has been conducted to examine temperature distribution uniformity of the heated LIB pack. The heat transfer of the external PTC materials and the internal heat of LIB are analyzed theoretically, and a geometric model is established. The thermal parameters of LIB are obtained, and the heat generation rate of the PTC materials and LIB are calculated with our experimental data. The thermal characteristics of the self-heating process are numerically simulated. With comparison of the simulation results and the experimental results, the accuracy of the modeling and simulation is verified. The results also demonstrate that temperature inside the LIB pack after heating is kept in good uniformity, which proves that this method has the advantage of improving heating effectiveness.

2. Experimental Method

2.1. Design of PTC Self-heating for LIB Pack

The LIB used in this paper is the 35Ah square aluminum-plastic-film $LiMn_2O_4$ cell, its basic parameters are specified in Table 1. When charged and discharged under different temperatures with different rates, its discharge capacity and 1C charge characteristics are shown in Table 2 and in Figure 1, respectively.

Table 1. Basic parameters of the LiMn$_2$O$_4$ cell.

Parameters	Unit
Mass	1.08 (kg)
Length/width/height	246/180/14.7 (mm)
Rated Voltage	3.7 (V)
Rated Capacity	35 (Ah)
Maximum Voltage	4.2 (V)
Minimum Voltage	3.0 (V)
Resistance	$\leq$1 mΩ

Table 2. Discharge capacity of a cell at different temperatures and discharge rates (Unit: Ah).

Temperature \ Rates	0.3C (10 A)	1C (35 A)	2C (70 A)
20 °C	36.1	35.2	33.8
0 °C	33.7	32.4	32.0
−20 °C	20.3	15.6	14.8
−40 °C	6.9	0.2	0.0

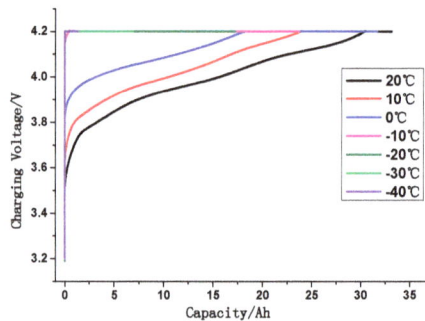

Figure 1. The 1C charge characteristics of a cell at different temperatures.

From the above results, it can be noted that with the decrease of temperature, the discharge capacity corresponding to the same discharge rate, will diminish, and the low-temperature charging performance is more obviously subject to temperature. The battery cannot be charged at a rate of 1C below −10 °C, which means it is very difficult to charge the battery in winter. Therefore, heating LIB at low temperature seems to be quite necessary.

Accordingly, a PTC self-heating method is proposed in this paper. In the design, PTC resistance bands are embedded in slotted aluminum plates, which are arranged between two sides of each cell. PTC is heated by electricity derived from LIB pack, and the heat generated by PTC materials is rapidly transferred to the cell through those aluminum plates. The extra slots on the aluminum plates will be formed as air ducts to dissipate heat at high temperature. This design achieves an integration of low-temperature heating and high-temperature cooling, with a schematic diagram shown in Figure 2. The product of LIB pack with PTC material is exhibited in Figure 3. From Figure 3, there are 24 cells in each of the two columns of the test pack, and in each column, there are 23 aluminum plates placed between two sides of every two cells, ensuring that each cell has at least one side contacting with the aluminum plates. Besides, a temperature sensor is installed at the center of each cell side and on the positive/negative column of each cell. The characteristic of PTC material is shown in Figure 4. Figure 4a depicts the resistance characteristic of PTC material. The resistances will be increased exponentially when the temperature rises to the Curie temperature (T_C).

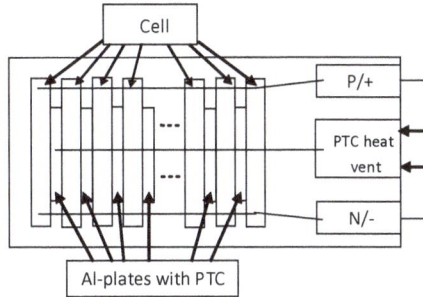

Figure 2. Schematic diagram of PTC self-heating method.

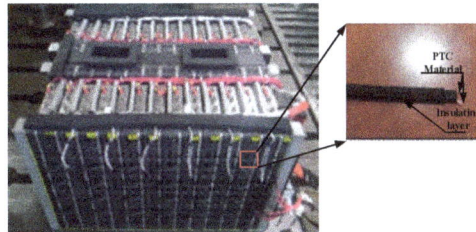

Figure 3. Product of battery pack with PTC.

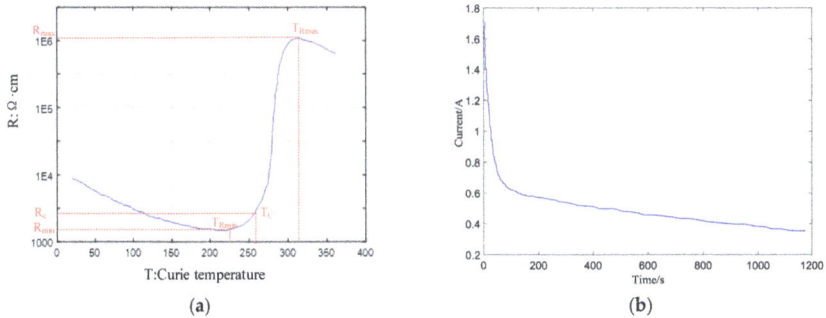

Figure 4. The characteristics of the PTC material: (**a**) The resistance characteristic of PTC; and (**b**) the current curve in the heating process at −40 °C.

Thereafter, the resistances of the PTC material remain steady and the heat generation rate will be kept constant. Figure 4b is a current change curve when the PTC resistance bands are provided with an external power (220 V AC) at −40 °C. It can be seen from the figure that the temperature of the PTC material will reach the T_C point in a very short time (about 40 s) after the power supply is switched on; the current is then basically kept at a constant state, and the PTC material is sustained at a state of constant power to heat the LIB pack.

2.2. Experiment

The experiment consists of three procedures:

(1)   External power source heating experiment: when the tested pack is at SOC = 100%, heat is generated only by PTC resistance bands, and the external electricity power (220 V AC) serves as a supplying power;

(2)   Self-heating Experiment I: when the pack is at SOC = 100%, heat is generated by PTC resistance bands and the internal resistances of the battery, and the source of supplying power is the pack itself;

(3)   Self-heating Experiment II: when the pack is at SOC = 60%, heat is generated by PTC resistance bands and the internal resistances of the battery, and the source of supplying power is the pack itself.

Detailed steps of the external power source heating experiment are as follows:

Step 1   Soak the tested pack into a −40 °C incubator for more than 5 h, to maintain the average temperature inside the battery at −40 °C;

Step 2   Connect PTC materials with 220 V AC current, then start to heat the pack;

Step 3   Suspend the first heating process when the lowest temperature collected in the pack is raised to −20 °C;

Step 4   Test the pack with hybrid pulse power characteristic (HPPC) specification by Digatron EVT500-500 (Digatron Power Electronics Company, Aachen, Germany);

Step 5   Repeat Step 2;

Step 6   Stop the second heating process when the lowest temperature collected in the pack is raised to 0 °C;

Step 7   Repeat Step 4;

Step 8   Test the pack with 1C constant-rate discharge until the discharge cutoff voltage is reached, then terminate the experiment.

As can be seen from the above steps, the experiment is composed of two heating processes. The lowest temperature of each cell in the pack rises from −40 °C to −20 °C during the first heating process; and rises from −20 °C to 0 °C during the second process. An HPPC test is carried out after each process to study the recovery of charge/discharge performance. Furthermore, a 1C constant-current (CC) discharge rate test is conducted to investigate the recovery of the LIB capacity after the second process.

The experimental steps for Self-heating Experiment I and II are virtually identical to those of the external power source heating experiment, with the only difference being that the supply power is derived from the pack itself, which can only be used to supply DC current, rather than from the external power source (220 V AC). Therefore, the current transmitted through the PTC material is DC rather than AC. The devices used in the experiments are listed in Table 3 and the photos of experiments are shown in Figure 5.

Table 3. Main devices needed in the experiments.

Serial Number	Device
1	Battery Management System (BMS)
2	Digatron EVT 500V-500A
3	Incubator

Figure 5. The scene photos of the experiments.

3. Comparison of Experiment Results

3.1. Comparison of Heating Results

The results of two heating experiments are shown in Tables 4 and 5.

Table 4. Results acquired in the three experiments after the first heating process.

Experiment Types	Time (min)	Temperature (°C)	Rate (°C/min)
External power source heating	31	−39.8 to −20.3	0.629
Self-heating I	34.2	−39.4 to −20.7	0.459
Self-heating II	43.33	−32 to −20.3	0.270

Table 5. Results acquired in the three experiments after the second heating process.

Experiment Types	Time (min)	Temperature (°C)	Rate (°C/min)
External power source heating	45	−23.2 to −0.5	0.504
Self-heating I	48	−19.3 to −2.4	0.352
Self-heating II	52	−19.7 to −2.7	0.327

As can be observed in Tables 4 and 5:

(1) During the process of external power source heating experiment, the temperature of the pack increases quickly with the highest heating rate, as the externally applied 220 V AC is high and stable and the PTC material generates heat quickly.
(2) Although the discharge capability of the pack is poor at low temperatures, the heating effect of the pack cannot be neglected when it is heated via supplying power to the PTC resistance bands. In the fully-charged state (SOC = 100%), the self-heating rate will be about 70% of the external power source heating rate.
(3) As the battery SOC grows larger, better heating effects will be achieved, and a higher heating rate can be obtained.

However, some underlying information may have not been reflected from data presented in the above tables, which only represents the surface temperature of the LIB and cannot accurately reflect the actual temperature inside the battery. Reasons are that during processes of heating, in addition to the heat produced by the external PTC resistance bands, the internal Joule heat is also generated due to battery discharge. Hence, the temperature values collected by temperature sensors (attached to cell surfaces) are comparatively lower than the actual temperature.

3.2. Variations of Total Voltage and Average Temperature of the Pack in Self-Heating Process

In the external power source heating experiment, since there is no charge/discharge process for the pack, the variations of total voltage and average temperature of the pack are discussed only in the self-heating experiment.

With the pack at 100% SOC, the variations of total voltage and average temperature of the pack in the Self-heating Experiment I are shown in Figures 6 and 7.

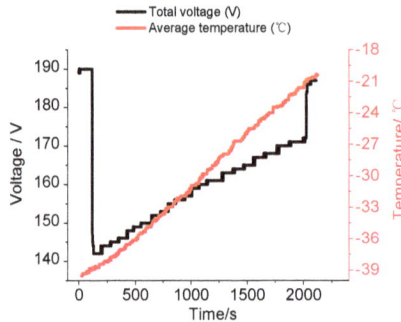

Figure 6. The first heating curves of Self-heating Experiment I.

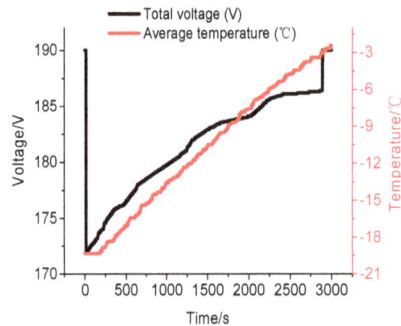

Figure 7. The second heating curves of Self-heating Experiment I.

In the self-heating processes, the total voltage of the pack rises with the increase in temperature. As can be seen from Figure 6, the initial total voltage of the pack is 190 V at −39.4 °C. At the moment when the pack is supplying power to PTC resistance bands, the total voltage dramatically drops to 142 V since the internal resistances are relatively large at that time. After 34.2 min of heating, the temperature of the pack will go up to −20.7 °C, and then the total voltage will rise to 172 V. That is because during the heating process, though the pack consumes part of its energy in discharging, its charge transfer is speeded up and the voltage platform is elevated gradually as the temperature increases. When the circuit is disconnected and the heating process terminated, the total voltage will be raised to 187 V, which illustrates that, after heating, the internal resistances being subjected to the temperature rise are significantly decreased.

Similarly, the curves of the pack heated from −19.3 °C to −2.4 °C are shown in Figure 7. The voltage platform rises from 172 V to 186 V as the temperature increases; the total voltage is restored to 189 V after heating is stopped, signifying that the temperature may become higher and the internal resistances of the pack grow smaller with a longer heating time. However, the duration of heating time has been increased to about 48 min, since the resistances of the PTC material increase rapidly with the rise in temperature, resulting in a slow heating rate. Further details about the heat generation rate curves will be given in Section 4.

When the pack is at 60% SOC, the variations in total voltage and average temperature of the pack during Self-heating Experiment II are shown in Figures 8 and 9.

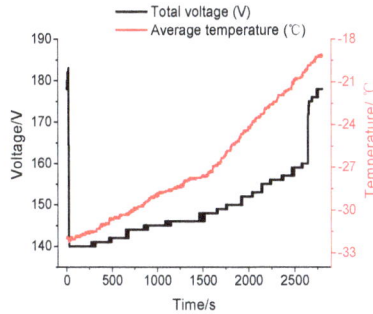

Figure 8. The first heating curves of Self-heating Experiment II.

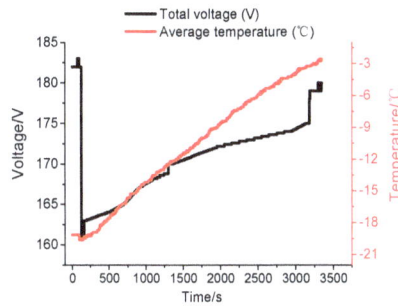

Figure 9. The second heating curves of Self-heating Experiment II.

As shown in Figure 8, the initial total voltage of the pack is indicated as 183 V and the initial average temperature is −32 °C. After supplying power to the PTC resistance bands, its total voltage drops quickly to 140 V, which means that at this moment large resistances exist inside the pack. After 43.3 min of heating, its temperature is raised from −32 °C to −20.3 °C, with the voltage platform lifted, and the total voltage rises from 140 V to 160 V. When heating is ended, the voltage is restored to 178 V, signifying a significant decrease in its internal resistances.

Again similarly, the initial total voltage of the pack in Figure 9 is indicated as 178 V and its initial average temperature is −19.7 °C. After supplying power to the PTC resistance bands, the pack's total voltage goes down to 161 V. With 52 min of heating, the temperature goes up to −2.7 °C. When heating is terminated, the total voltage is restored to 177 V.

Comparing the results of the two self-heating experiments, we find that the pack with larger SOC can provide higher voltage, so the heating time will be shortened and the temperature will be raised more quickly, and a better heating effect will be achieved. From Figures 6–9, it should also be taken into account that when the total voltage data is being collected, the curves in all figures, in view of the sensor acquisition accuracy, may have different degrees of right-angle folding, which, nevertheless, will not affect voltage variations in the heating process.

3.3. Pulse Charge–Discharge Capability of the Heated LIB Pack

HPPC tests are implemented to check the performance recovery of pulse charge–discharge capability for the pack after each heating process.

Figures 10 and 11 are the test results after each heating in the external power source heating experiment.

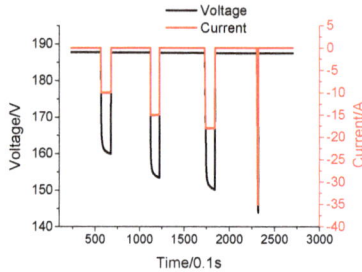

Figure 10. Pulse discharge capability of the pack after the first heating with an external power source.

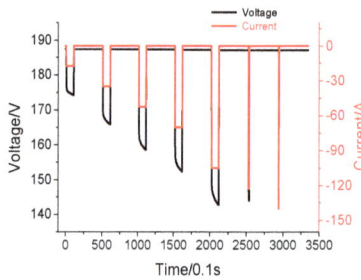

Figure 11. Pulse discharge capability of the pack after the second heating with an external power source.

After the first heating, the temperature rises to −20.3 °C, the battery can be discharged at the 0.5C rate for 10 s, but fails to be discharged at the 1C rate, and its discharge curve is shown in Figure 10. As can be seen from Figure 11, after the second heating, the temperature rises to 0.5 °C, the discharge performance of the pack is remarkably improved, and then the pack can be discharged at the 3C rate for 10 s, but not at the 3.5C rate.

Figures 12 and 13 exhibit graphs of HPPC tests after each heating in Self-heating Experiment I.

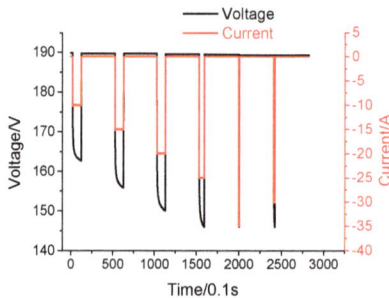

Figure 12. Pulse discharge capability of the pack after the first heating in Self-heating Experiment I.

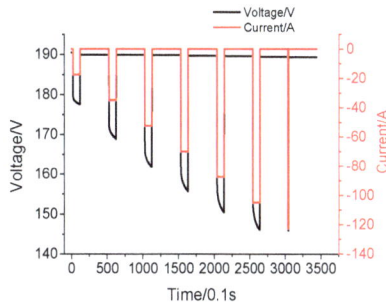

Figure 13. Pulse discharge capability of the pack after the second heating in Self-heating Experiment I.

From Figures 12 and 13, we find that after the first heating the pack can be discharged at the 0.57C rate for 10 s when heated to a temperature of −20.7 °C, yet not at the 1C rate; after the second heating, when the pack is heated to −2.4 °C, and its performance has been significantly improved, it can be discharged at the 3C rate, but not at the 3.5C rate.

Figures 14 and 15 give the test results after each heating in Self-heating Experiment II.

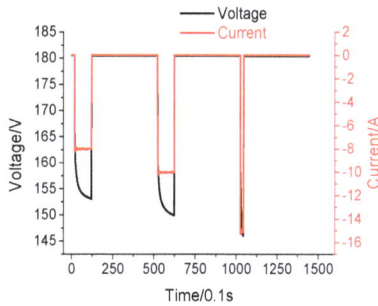

Figure 14. Pulse discharge capability of the pack after the first heating in Self-heating Experiment II.

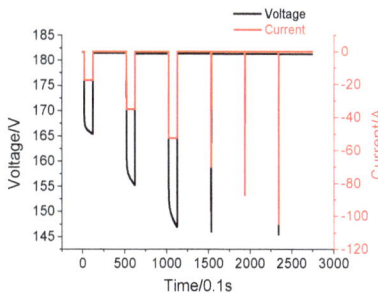

Figure 15. Pulse discharge capability of the pack after the second heating in Self-heating Experiment II.

As seen in Figure 14, when it is heated to −20.3 °C after the first heating, the pack can be discharged at the 0.29C rate for 10 s, but fails at the 0.43C rate for 10 s. As seen in Figure 15, when it is heated to −2.7 °C after the second heating, the pack can be discharged at the 1.5C rate for 10 s, but fails at the 2C rate for 10 s.

Results of charging performance recovery are obtained as follows:

(1) Only pulse dynamic tests can be used to verify the pack's charge capability when it is at 100% SOC. The results of charging performance recovery for the external power source heating experiment are consistent with those of Self-heating Experiment I. After the first heating, the pack cannot be charged at the 0.5C rate; after the second heating, the pack can be charged at the 0.5C rate for 10 s, but fails at 1C.

(2) In Self-heating Experiment II, after the first heating, the pack (at SOC = 60%) cannot be charged at the 0.29C rate; after the second heating, it can be charged at the 0.34C rate for 10 s, but fails at the 0.5C rate.

Comparing the three heating experiments, we can conclude that:

(1) The charge–discharge performance recovery of the pack after the second heating is obviously superior to that after the first heating, due to the fact that the second heating temperature tends to be relatively higher, LIB electrolyte viscosity is decreased, and the charge transfer is speeded up. It is also confirmed that the low-temperature discharge capability of $LiMn_2O_4$ LIB is much better than its charge ability.

(2) When the pack is at 100% SOC, though the pack in Self-heating Experiment I consumes part of its energy, there is no significant difference between its discharge capacity during self-heating and external power source heating. Moreover, the former is comparatively better than the latter.

(3) The charge/discharge capability of LIB in Self-heating Experiment II is relatively poor when LIB is at 60% SOC, mainly due to influential factors such as smaller SOC, lower voltage platform, slower heating rate, longer heating time, more consumption of energy, etc.

3.4. Constant-Current Discharge Capability of the Heated LIB Pack

The HPPC tests prove that when the pack is at 100% SOC in self-heating experiments, its charge/discharge performance is relatively better. In order to further study its capacity recovery, 1C CC discharge rate tests (based on the PNGV Battery Test Manual, the Freedom CAR Battery Test Manual for Electric Vehicles is jointly compiled by the authors and the battery manufacturers) are implemented on three heated packs. The test results are depicted in Figure 16.

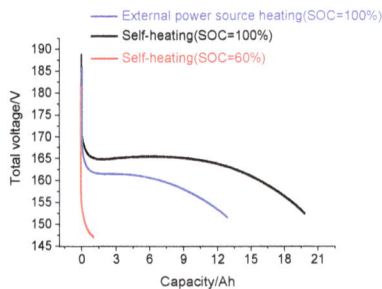

Figure 16. Comparison of 1C CC discharge tests.

As can be seen in Figure 16:

(1) Although the pack in Self-heating Experiment I consumes part of its energy, it has discharged up to 19.834 Ah energy, with the highest capacity at 1C CC among the three packs. In contrast, the discharge capacity of the pack in the external power source heating experiment is only 12.853 Ah. Thus, the internal heat is proved to be of non-negligible value. Compared with the external power source heating method, which depends solely on the PTC heating material, the self-heating method can integrate the internal heat with the external heat to effectively promote the restorability of battery discharge capacity. Therefore, even when the self-heated LIB

pack is at a state of SOC < 100% after heating, its discharge capacity is still larger than that of the external power source heated pack at 100% SOC. Thereby, the superiority of the self-heating method is validated.

(2)   When the pack is at 60% SOC and heated in the Self-heating Experiment II, the pack discharges the least capacity (<2 Ah). This is mainly because the initial SOC of the pack is comparatively small; the voltage platform is lower than that of the pack at 100% SOC and its heating rate is low, so the two heating processes will require a longer time and consume more energy, causing lower residual capacity. Another reason is that the 1C discharge rate is too large for the pack when SOC < 60% at low temperatures. Moreover, in order to protect the pack from being damaged during the experiment, its discharge cutoff voltage is set to be comparatively higher than its minimum cutoff voltage, leading to the consequence that virtually no energy is discharged.

4. Modeling for PTC self-heating

The above experimental results show when the pack at 100% SOC is heated from −40 °C to 0 °C by using the self-heating method and the external power source heating method, its pulse charging/discharging capabilities are equivalent to each other, and yet the 1C CC discharge capacity of the former pack is far greater than that of the latter pack. One possible reason for that result may be that the temperature is not distributed uniformly. Accordingly, a self-heating model is established to further study the temperature distribution uniformity of the pack after heating.

Generally, thermal models for a battery come in a variety of types, such as an electrochemical–thermal coupled model, an electro-thermal coupled model, a thermal abuse model, a 1D model, a 2D model, a 3D model, etc. [21–25]. An electrochemical–thermo-coupled model is a battery thermal model established on the basis of the thermo–chemical reaction of the battery, in which the temperature in the battery is considered to be distributed uniformly, with the distribution of current density on the battery pole pieces being ignored. Tiedemann et al. [26], Pollard and Newman [27], and Bernardi et al. [28] have performed many in-depth studies on the electrochemical–thermo-coupled model. The equation of the heat generation rate model proposed by Bernardi, which is one of the most widely used models, will be adopted in this paper.

In this paper, the discussed heat resources generated by heating a LIB pack with self-heating method are derived from two sources: internal heat and external heat. (1) The internal heat refers to the amount of heat produced by the internal resistances, when the LIB is supplying power to the PTC resistance bands; (2) the external heat means the amount of heat produced by the PTC resistance bands. Thus, thermal modeling usually begins with an analysis of the heat generation theories of the internal and external power source.

4.1. Theoretical Analysis of Heat Generation

4.1.1. Theoretical Analysis of the Internal Heat Generation

The internal heat generation of the LIB is the internal heat source of the thermal model, so the calculation accuracy of heat generation directly influences the accuracy of the thermal model. As the assumption of Bernardi heat generation model is not consistent with the actual temperature distribution in the LIB, some researchers have made improvements in the Bernardi heat generation model based on the introduction of the current density.

(1)   The Bernardi heat generation rate model.

The equation for Bernardi's heat generation rate model is as follows [28]:

$$q_B = I(E_0 - E) - IT(dE_0/dT), \tag{1}$$

where q_B denotes the heat generation rate, the unit is W; I denotes the charge/discharge current, the unit is A, and in the charging process, I has a negative value, while in the discharging process,

I has a positive value; *T* denotes the temperature, the unit is K, and the average temperature is adopted when calculating q_B; dE_0/dT denotes the change rate of open circuit voltage with the temperature; Joule heat and reaction heat are mainly considered in q_B. In Equation (1), $I(E_0 - E)$ denotes joule heat, and $IT(dE_0/dT)$ denotes reaction heat.

(2)   The advanced Bernardi heat generation rate model.

The equation for the advanced Bernardi heat generation rate model is as follows:

$$q_B = J[(E_0 - E) - T(dE_0/dT)], \tag{2}$$

where *J* denotes the charge/discharge current density of the positive/negative plate.

According to the LIB temperature variation during the charging/discharging process at different currents, the accuracies of the two models are compared and the results are shown in Figure 17.

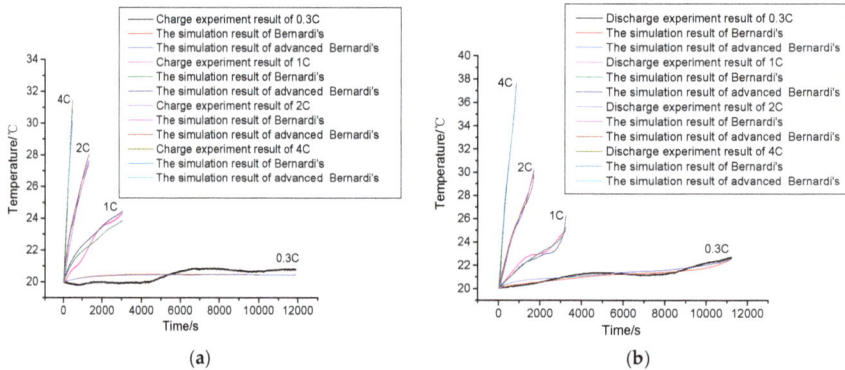

Figure 17.   Comparison of simulation results of two models and the experimental results. (**a**) Temperature variation in the charge process; and (**b**) temperature variation in the discharge process.

It can be seen from Figure 17 that the calculation accuracy of the advanced Bernardi heat generation model is not superior to that of the initial one, and the advanced model is also highly computationally demanding, which is not conducive to calculations for the heating model of the LIB pack. Therefore, this paper adopts the Bernardi heat generation model.

4.1.2. Analysis of Internal Heat Conduction

Before establishing the model, we initially set up a differential equation of heat conduction for a LIB cell, which will help us to understand at the microscopic level the heat generation, heat transfer, and temperature rise inside the cell.

Also before establishing the thermal differential equation, we put forward some assumptions: (1) The inner part of a cell composed of different materials should be simplified into an isotropic continuous medium; (2) if heat is generated inside a cell, the internal heat should be uniformly distributed.

Next, we assume that a micro-unit will be taken out from a cell; as shown in Figure 18, q_x, q_y, q_z represent the heat fluxes flowing respectively into the micro-unit from its left side, from below, and from the back, respectively; $q_{x+dx}, q_{y+dy}, q_{z+dz}$ stand for the heat fluxes flowing out of the micro-unit from the right side, from the top, and from the front, respectively. According to Fourier's law, Equation (3) can be obtained:

$$q_{x+dx} = q_x + \frac{\partial q_x}{\partial x}dx \quad q_{y+dy} = q_y + \frac{\partial q_y}{\partial y}dy \quad q_{z+dz} = q_z + \frac{\partial q_z}{\partial z}dz. \tag{3}$$

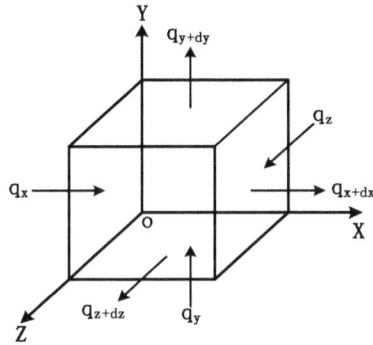

Figure 18. The micro-unit of the cell.

Here, the internal instantaneous heat power of the micro-unit is $\dot{q}dxdydz$. If there is no internal heat source in a micro-unit, then $\dot{q} = 0$; and the thermodynamic increment of a micro-unit is $\rho c \frac{\partial T}{\partial \tau} dxdydz$, where $\dot{q}$ denotes the heat power per unit volume; ρ denotes the density; c denotes the specific heat capacity; T denotes the temperature; and τ denotes the time.

According to the conservation of energy, a general expression of the heat conduction differential equation in the rectangular coordinate system can be obtained:

$$\rho c \frac{\partial T}{\partial \tau} = \lambda_x \frac{\partial^2 T}{\partial x^2} + \lambda_y \frac{\partial^2 T}{\partial y^2} + \lambda_z \frac{\partial^2 T}{\partial z^2} + \dot{q}. \tag{4}$$

4.1.3. Analysis of the External Heat Conduction

As the heat generated externally is originated primarily from the PTC material, this paper analyzes the external heat transfer, with two aspects considered—heat conduction and heat convection—and with the radiation heat transfer being ignored.

The theoretical basis of heat transfer is Fourier's law, which can be expressed as: the heat flux at any point and any time is proportional to the temperature gradient at that point. The formula is as follows:

$$\vec{q} = -\lambda gradT = -\lambda \frac{\partial T}{\partial n} \vec{n}, \tag{5}$$

where $\vec{q}$ denotes the heat flux density of heat conduction; λ denotes the coefficient of heat conductivity; and n denotes the direction of the outer normal. Since the heat is transferred from the points of high temperature to those of low temperature, the resulting value in Equation (5) is negative.

As the sides of the cells are heated directly by the aluminum plates arranged between the sides of every two cells, the other sides of the cells can only be heated by the heated air flow, which can be calculated by using the Newton cooling formula:

$$q = h(T_w - T_f), \tag{6}$$

where h denotes the heat transfer coefficient and the unit is $W/(m^2 \cdot K)$.

4.1.4. Heat Generation and Conduction Model

Combining the heat conduction differential equation and the Bernardi heat generation rate model, the heat generation and conduction model can be obtained:

$$\rho c \frac{\partial T}{\partial t} = \lambda_x \frac{\partial^2 T}{\partial x^2} + \lambda_y \frac{\partial^2 T}{\partial y^2} + \lambda_z \frac{\partial^2 T}{\partial z^2} + \beta I[(E_0 - E) - IT(dE_0/dT)], \tag{7}$$

where β is the correction coefficient of the heat generation rate.

The initial and boundary conditions are as follows:

$$\begin{cases} T(x,y,z;0) = T_0 \\ -\lambda \frac{\partial T}{\partial n}\Big|_w = q(t) \\ -\lambda \frac{\partial T}{\partial n}\Big|_w = h(T_w - T_f) \end{cases} \tag{8}$$

4.2. The Establishment of a Geometric Model

Several assumptions are made to simplify the geometry model of the pack. (1) As the thickness of each element inside the LIB is very small, the amount of calculation work will become considerable if the three-dimensional model is established with every element. Therefore, the cell composed of different materials is simplified into an isotropic continuous medium. (2) The PTC heat generation rate is assumed to be equivalent to the heat generation rate of aluminum plates; because the thermal conductivity of aluminum plates is relatively larger and the PTC resistance bands are embedded in the aluminum-plate containers, the heat produced by the PTC material can be rapidly transferred to the aluminum plates. (3) The connectors, wires, and insulation objects inside the pack are ignored, and the irregular structures of the pack are assumed to be regular cuboids. All the above assumptions can help simplify the model and shorten the calculation. (4) Based on the symmetry principle, the 48 s battery pack can be reduced to a 1/4 model composed of 12 cells. Simulations of the temperature field of the whole LIB pack can thus be simulated with boundary conditions.

Based on the above simplifications, the 1/4 geometry model falls into four aspects: (1) 12 cells; (2) 12 aluminum plates, including an aluminum plate at the center of the original 48s pack; (3) the LIB pack shell; and (4) the air flow inside the pack. Relevant parameters for modeling are listed in Table 6 and a geometric model is shown in Figure 19.

Table 6. The parameters of the 1/4 geometry model.

Components	Size (Length × Width × Height, Unit: mm³)
A cell	180 × 14.7 × 246
Aluminum plate	170 × 5 × 198
shell of the container	220 × 262.5 × 296

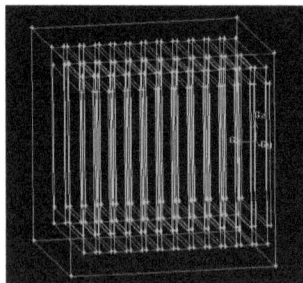

Figure 19. The 1/4 geometry model.

4.3. Acquisition of Model Parameters

4.3.1. Acquisition of the Thermo-Physical Parameters

The general thermo-physical parameters of each components inside $LiMn_2O_4$ cell are as shown in Table 7.

Table 7. The thermo-physical parameters.

Parameters Components	Density (kg/m^3)	Specific Heat Capacity J/(kg·K)	Heat Conductivity Coefficient W/(m·K)
Anode	2840	839	3.9
Aluminum foil	2710	903	238
Cathode	1671	1064	3.3
Copper foil	8933	385	398
Separator	659	1978	0.33
Shell of the cell	1636	1377	0.427

Yet, some thermo-physical parameters still need to be calculated, such as: density, specific heat capacity, heat conductivity coefficient, etc.

(1) The cell density can be calculated by using the mean method, and the general formula is:

$$\rho = \frac{M}{V},\tag{9}$$

where M is the total quality of the cell, and V the total volume of the cell.

(2) The specific heat capacity of the cell can be calculated by using the theoretical equation expressed as Equation (10):

$$C_p = \frac{1}{M}\sum_{i=1}^{n}m_iC_i = \sum_{i=1}^{n}(\rho V)_iC_i / \sum_{i=1}^{n}(\rho V)_i.\tag{10}$$

(3) The heat conductivity coefficient of the cell can be calculated by using the thermal resistance method, which leads to Equations (11) and (12):

$$\lambda_x = \frac{L_x}{(L_{xp}/\lambda_p) + (L_{xn}/\lambda_n) + (L_{xs}/\lambda_s) + (L_{xw}/\lambda_w)}\tag{11}$$

$$\lambda_y = \lambda_z = \frac{L_{xp}\lambda_p + L_{xn}\lambda_n + L_{xs}\lambda_s + L_{xw}\lambda_w}{L_x}\tag{12}$$

where L_x is the thickness of the cell; and L_{xp}, L_{xn}, L_{xs}, L_{xw} and λ_p, λ_n, λ_s, λ_w denote the lengths and the heat conductivity coefficients of cathode plates and anode plates, separator, and shell, respectively.

Based on Equations (9)–(12), the calculated results are: density 2182.7 kg/m^3; specific heat capacity 1100 J/(K·kg); and heat conductivity coefficient 0.895 W/(m·K).

4.3.2. Internal and External Heat Generation Rates

The discharge voltage and current in the heating processes are acquired with self-heating experiments. The calculation approach discussed in Section 4.1 is used to determine the internal and external heat generation rate. The heat generation rates are obtained many times to fully demonstrate the real-time changes in heating, so the simulation may be kept closer to the actual experiment. Graphs of the heating current and power during the first and second heating experiments are shown in Figures 20 and 21, respectively.

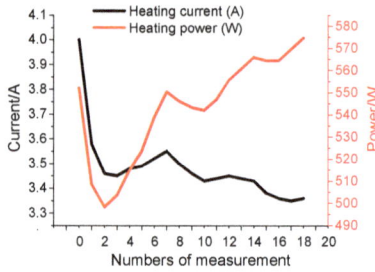

Figure 20. Heating current and heating power in the first self-heating process.

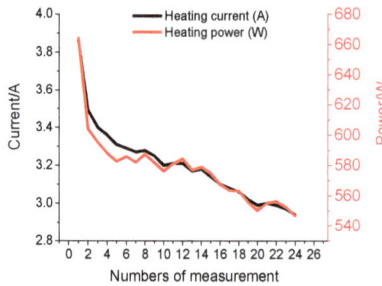

Figure 21. Heating current and heating power in the second self-heating process.

Comparing Figure 20 with Figure 21, we find:

(1) In the first self-heating process, the heating current tends to decrease, while the heating power keeps increasing. The reasons may be that the PTC material resistances increase as the temperature rises; at the same time, the pack voltage platform is also being raised, and the degree of resistance increase is greater than that of the voltage rise.

(2) In the second self-heating process, although the temperature is still rising, the battery voltage platform has been basically stabilized. At this time, as the PTC material resistances increase dramatically, the heating current will decrease and the heating power is gradually reduced.

With the calculation results of the internal and external heat generation rate, we can acquire a graph of the heat generation rate for the pack from $-40\,^{\circ}$C to $0\,^{\circ}$C. As shown in Figure 22, the mean value of the external heat generation rate is 72.386 kw/m^3, and the mean value of the internal heat generation rate is 8.337 kw/m^3, which is about 11.5% of the former.

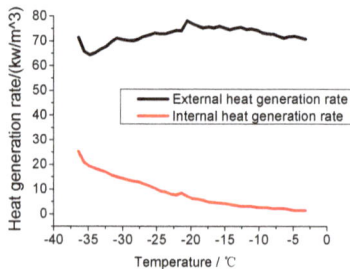

Figure 22. Variation of the internal and external heat generation rates of the whole self-heating process.

5. Results and discussion

We write the program according to the results in Figure 22 and import the geometric model in to our simulation design, then conduct a numerical simulation.

Although the heat generation rate inside the battery is rather small, it cannot be ignored because of its direct effect on the battery. In order to better verify the characteristics of the internal heat generation, two cases are also included in the simulation:

(1)   Considering the co-generation of both the internal and the external heat in the heating process, which is in line with the actual situation.
(2)   Ignoring the internal heat, with only the external heat considered.

The curves of temperature rise, acquired from experiments and simulations on the battery during two heating processes, are shown in Figures 23 and 24, respectively.

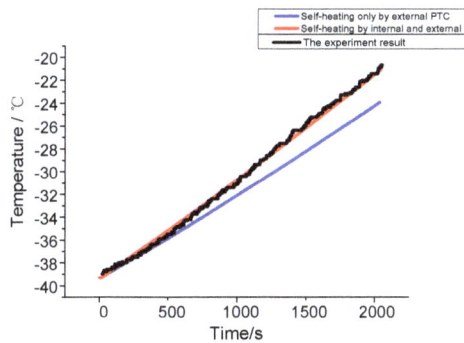

Figure 23. Comparison of simulation and experimental results in the first self-heating process.

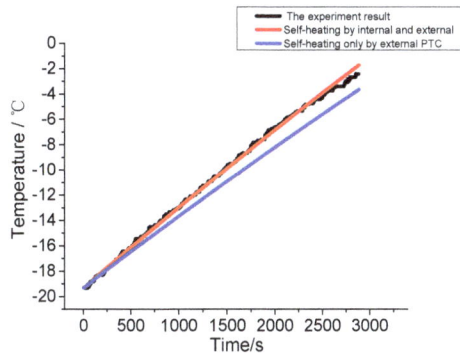

Figure 24. Comparison of simulation and experimental results in the second self-heating process.

As shown in Figures 23 and 24, simulation curves of the internal and external co-heating are in good agreement with the temperature rise curves obtained during the experiments. The average and maximum temperature difference in the first heating process are 0.201 °C and 0.938 °C, respectively. The average and maximum temperature difference in the second heating process are 0.164 °C and 0.783 °C, respectively. The above results verify the accuracy of the simulation results. The simulation curve, without regard for the internal heat, is significantly lower than the curve in the experimental process, which fully illustrates the significance of internal heat on the LIB temperature rise. Even if the heat rate is comparatively tiny, it should not be ignored.

Next, utilizing the function of temperature contour, we further analyze the temperature distribution inside the pack after heating. The results are shown in Figures 25 and 26.

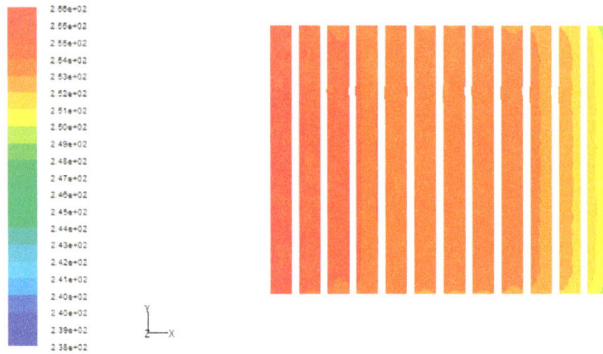

Figure 25. Temperature distribution of 12 cells on the $z = 0$ section (the center) at the end of the first self-heating process.

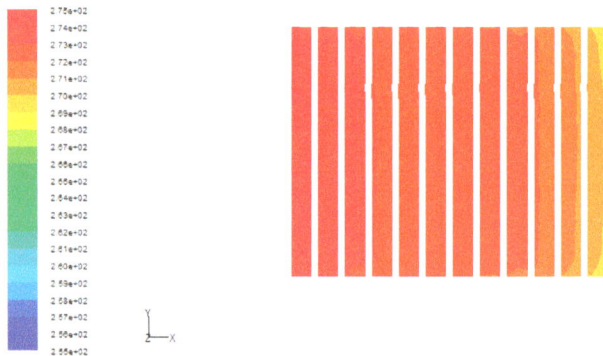

Figure 26. Temperature distribution of 12 cells on the $z = 0$ section (the center) at the end of the second self-heating process.

As can be seen from Figure 25, for the cell on the far left, namely in the middle of the original 48-s battery pack, it is raised to the highest temperature of $-18.8\,°C$ after heating; for the cell on the far right, namely close to the pack container shell, the lowest temperature only reaches $-23.47\,°C$ after heating. The maximum temperature difference inside the pack amounts to $4.67\,°C$. Similarly, as can be seen in Figure 26, the highest and lowest temperatures within the pack after heating are $0.248\,°C$ and $-3.258\,°C$, respectively, with a maximum temperature difference of $3.506\,°C$.

Although the quantity of heat generation for each cell is virtually identical, the temperature of the cell on the far left is the highest and the temperature of the cell on the far right is the lowest. The reason may be that the position of the cell on the far left is at the center of the original 48-s pack, and the heat convection is slow, as a result, it is kept at the highest temperature and vice versa, as the cell on the far right is placed next to the outer shell of the pack and the heat convection is rapid, it is maintained at the lowest temperature. The temperature difference after the second heating is smaller than that of the first heating, which may be explained by the fact that with the extension of heating time, the heat generated inside the battery is more favorable for the uniform temperature distribution.

6. Conclusions

The work presented in this paper can be summarized as follows:

(1) A PTC self-heating method is proposed, in which the supplying power is the pack itself rather than the external power source, and EVs can be operated independently of external power source in cold areas.

(2) Although the power consumption of the pack truly exists in the self-heating experiment process (the consumed energy in the whole self-heating experiment is approximately 13% of the total pack energy), the experimental results show that the charge/discharge capability and capacity recovery of the pack in Self-heating Experiment I are superior to that of the pack in the other two heating experiments. The results fully illustrate the superiority of the method and the pack with a high SOC is more helpful to the recovery of LIB performance.

(3) Although the average internal heat generation rate is only 11.5% of the external heat generation rate, the impact of temperature rise on capacity recovery cannot be ignored because the internal heat has a direct effect on the inside of LIB.

(4) The simulation results verify the accuracy of the modeling and simulation, and demonstrate that temperature distribution inside the pack after heating is kept uniform, which further proves that this method is of great value for the performance improvement of LIB and can be utilized to effectively promote the feasibility and applicability of EVs at low temperatures. However, if the initial SOC of the LIB pack is small and self-heating is unavailable, external power source heating is an alternative approach. The main reasons are that the energy and capacity of the pack with the external power source heating cannot be reduced. In our future work, we will work at a strategy that can incorporate the advantages of both internal and external heating to reduce the restrictions on initial SOC.

Acknowledgments: The authors would like to thank the Collaborative Innovation Center of Electric Vehicles in Beijing Institute of Technology for the support of this research project.

Author Contributions: Chengning Zhang and Junqiu Li contributed to the conception of the study; Xin Jin did the experiments and established the model; Junqiu Li and Xin Jin performed the data analyses and wrote the manuscript; Chengning Zhang helped perform the analysis with constructive discussions.

Conflicts of Interest: The authors declare no conflict of interest.

References

1. Ng, S.S.; Xing, Y.; Tsui, K.L. A naive bayes model for robust remaining useful life prediction of lithium-ion battery. *Appl. Energy* **2014**, *118*, 114–123. [CrossRef]
2. Wang, C.; Xiong, R.; He, H.; Ding, X.; Shen, W. Efficiency analysis of a bidirectional DC/DC converter in a hybrid energy storage system for plug-in hybrid electric vehicles. *Appl. Energy* **2016**, *183*, 612–622. [CrossRef]
3. Zhang, C.; Zhao, T.; Xu, Q.; An, L.; Zhao, G. Effects of operating temperature on the performance of vanadium redox flow batteries. *Appl. Energy* **2015**, *155*, 349–353. [CrossRef]
4. Zhang, Y.; Xiong, R.; He, H.; Shen, W. A lithium-ion battery pack state of charge and state of energy estimation algorithms using a hardware-in-the-loop validation. *IEEE Trans. Power Electron.* **2016**, *32*, 4421–4431. [CrossRef]
5. Song, Z.; Hofmann, H.; Li, J.; Hou, J.; Zhang, X.; Ouyang, M. The optimization of a hybrid energy storage system at subzero temperatures: Energy management strategy design and battery heating requirement analysis. *Appl. Energy* **2015**, *159*, 576–588. [CrossRef]
6. Qiao, Y.; Tu, J.; Wang, X.; Gu, C. The low and high temperature electrochemical performances of Li 3 V 2 (PO 4) 3/C cathode material for Li-ion batteries. *J. Power Sources* **2012**, *199*, 287–292. [CrossRef]
7. Zhang, S.; Xu, K.; Jow, T. Electrochemical impedance study on the low temperature of Li-ion batteries. *Electrochim. Acta* **2004**, *49*, 1057–1061. [CrossRef]
8. Fan, J.; Tan, S. Studies on charging lithium-ion cells at low temperatures. *J. Electrochem. Soc.* **2006**, *153*, A1081–A1092. [CrossRef]

9. Wang, Z.P.; Lu, C.; Liu, P. Reviews of Studying on Thermal Management System in EV/HEV Battery Pack. *Adv. Mat. Res.* **2011**, *291–294*, 1674–1678. [CrossRef]
10. Zhang, C.-N.; Lei, Z.-G.; Dong, Y.-G. Method for heating low-temperature lithium battery in electric vehicle. *Trans. Beijing Inst. Technol.* **2012**, *32*, 921–925.
11. Liu, C.S.; Zhang, H.W. Research on heating method at low temperature of electric vehicle battery. *Chin. J. Power Sources* **2015**, *39*, 1645–1647. (In Chinese).
12. Wang, F.-C.; Zhang, J.-Z.; Wang, L.-F. Design of electric air-heated box for batteries in electric vehicles. *Chin. J. Power Sources* **2013**, *7*, 1184–1187.
13. Wang, T.; Tseng, K.; Zhao, J.; Wei, Z. Thermal investigation of lithium-ion battery module with different cell arrangement structures and forced air-cooling strategies. *Appl. Energy* **2014**, *134*, 229–238. [CrossRef]
14. Zhang, S.; Xu, K.; Jow, T. The low temperature performance of li-ion batteries. *J. Power Sources* **2003**, *115*, 137–140. [CrossRef]
15. Hande, A.; Stuart, T. Ac Heating for EV/HEV Batteries. In Proceedings of the 2002 Power Electronics in Transportation, Auburn Hills, MI, USA, 24–25 October 2002; pp. 119–124.
16. Hande, A. A High Frequency Inverter for Cold Temperature Battery Heating. In Proceedings of the 2004 IEEE Workshop on Computers in Power Electronics, Piscataway, NJ, USA, 15–18 August 2004; pp. 215–222.
17. Stuart, T.; Hande, A. Hev battery heating using ac currents. *J. Power Sources* **2004**, *129*, 368–378. [CrossRef]
18. Hande, A.; Stuart, T. Effects of high frequency ac currents on cold temperature battery performance. In Proceedings of the 9th IEEE International Power Electronics Congress (IICPE 2004), Mumbai, India, 17–22 October 2004.
19. Zhang, J.; Ge, H.; Li, Z.; Ding, Z. Internal heating of lithium-ion batteries using alternating current based on the heat generation model in frequency domain. *J. Power Sources* **2015**, *273*, 1030–1037. [CrossRef]
20. Zhao, X.W.; Zhang, G.Y.; Yang, L.; Qiang, J.X.; Chen, Z.Q. A new charging mode of li-ion batteries with life PO4/C composites under low temperature. *J. Therm. Anal. Calorim.* **2011**, *104*, 561–567. [CrossRef]
21. Jiang, J.; Ruan, H.; Sun, B.; Zhang, W.; Gao, W.; Wang, L.Y.; Zhang, L. A reduced low-temperature electro-thermal coupled model for lithium-ion batteries. *Appl. Energy* **2016**, *177*, 804–816. [CrossRef]
22. Sato, N. Thermal behavior analysis of lithium-ion batteries for electric and hybrid vehicles. *J. Power Sources* **2001**, *99*, 70–77. [CrossRef]
23. Lin, C.; Mu, H.; Xiong, R.; Shen, W. A novel multi-model probability battery state of charge estimation approach for electric vehicles using h-infinity algorithm. *Appl. Energy* **2016**, *166*, 76–83. [CrossRef]
24. Sun, F.; Xiong, R.; He, H. A systematic state-of-charge estimation framework for multi-cell battery pack in electric vehicles using bias correction technique. *Appl. Energy* **2016**, *162*, 1399–1409. [CrossRef]
25. Sun, F.; Xiong, R. A novel dual-scale cell state-of-charge estimation approach for series-connected battery pack used in electric vehicles. *J. Power Sources* **2015**, *274*, 582–594. [CrossRef]
26. Tiedemann, W.H.; Newman, J.; Gross, S. Battery design and optimization. *J. Electrochem. Soc.* **1979**, *79*, 23.
27. Pollard, R.; Newman, J. Mathematical modeling of the lithium-aluminum, iron sulfide battery I. Galvanostatic discharge behavior. *J. Electrochem. Soc.* **1981**, *128*, 491–502. [CrossRef]
28. Bernardi, D.; Pawlikowski, E.; Newman, J. A general energy balance for battery systems. *J. Electrochem. Soc.* **1985**, *132*, 5–12. [CrossRef]

energies

MDPI

Article

Impact of Silicon Carbide Devices on the Powertrain Systems in Electric Vehicles

Xiaofeng Ding *, Jiawei Cheng and Feida Chen

School of Automation Science and Electrical Engineering, Beihang University, Beijing 100191, China; chengjiawei0218@126.com (J.C.); dage820@buaa.edu.cn (F.C.)
* Correspondence: dingxiaofeng@buaa.edu.cn; Tel.: +86-10-8233-9498

Academic Editor: Rui Xiong
Received: 7 February 2017; Accepted: 11 April 2017; Published: 14 April 2017

Abstract: The DC/DC converters and DC/AC inverters based on silicon carbide (SiC) devices as battery interfaces, motor drives, etc., in electric vehicles (EVs) benefit from their low resistances, fast switching speed, high temperature tolerance, etc. Such advantages could improve the power density and efficiency of the converter and inverter systems in EVs. Furthermore, the total powertrain system in EVs is also affected by the converter and inverter system based on SiC, especially the capacity of the battery and the overall system efficiency. Therefore, this paper investigates the impact of SiC on the powertrain systems in EVs. First, the characteristics of SiC are evaluated by a double pulse test (DPT). Then, the power losses of the DC/DC converter, DC/AC inverter, and motor are measured. The measured results are assigned into a powertrain model built in the Advanced Vehicle Simulator (ADVISOR) software in order to explore a direct correlation between the SiC and the performance of the powertrain system in EVs, which are then compared with the conventional powertrain system based on silicon (Si). The test and simulation results demonstrate that the efficiency of the overall powertrain is significantly improved and the capacity of the battery can be remarkably reduced if the Si is replaced by SiC in the powertrain system.

Keywords: DC/DC converters; DC/AC inverters; silicon carbide (SiC); electric vehicles (EV); powertrain system; battery

1. Introduction

Because of the global energy crisis and environmental pollution, the past decade has witnessed the rapid development of new energy vehicle technologies, such as Electric Vehicles (EVs), Hybrid Electric Vehicles (HEVs), Plug-in Hybrid Electric Vehicles (PHEVs), etc. As a result, more and more companies that produce transportation vehicles are developing new technologies for EVs/HEVs/PHEVs [1–11]. The main challenge of EVs/HEVs/PHEVs development remains the limited cruising range due to the small battery capacity and the long charging times according to the available battery charging technologies, especially for pure electric vehicles [3–7]. Hence, it is important to maximize the efficiency of each component in the powertrain system of EVs [8]. The high efficiency of the system will distinctly reduce the burden of the battery and extend the cruising range.

A typical powertrain system in an EV and the corresponding losses are shown in Figure 1. The main contributions of the total losses are copper and iron losses in a permanent magnet synchronous machine (PMSM), and switching and conduction losses of switching devices in both the DC/DC converter and inverter, respectively.

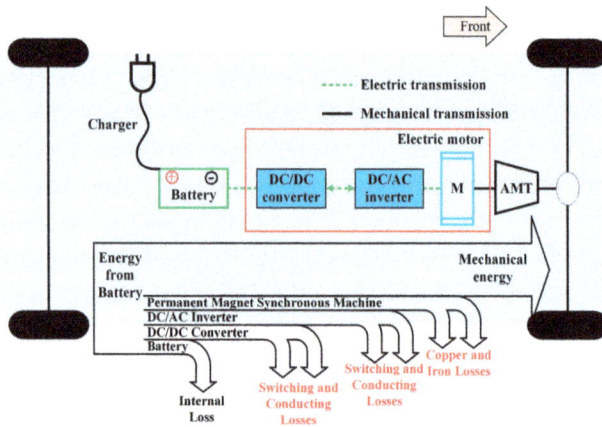

Figure 1. The components of the powertrain system in an electric vehicle (EV) and the corresponding losses.

Currently, the conventional insulated-gate bipolar transistor (IGBT) or Si metal-oxide-semiconductor field-effect transistor (MOSFET) technologies based on silicon (Si) material dominate the semiconductor fields in the application of power converters and inverters. However, the Si IGBT/MOSFET is now reaching the material's theoretical limits. Higher efficiency, higher power density, and higher temperature application are the urgent requirements in traction converters and inverters of EVs. Recently, wide band-gap (WBD) silicon carbide (SiC) MOSFET has exhibited great potential to replace Silicon (Si) as the dominant transistor technology according to its extreme advantages, such as faster switching speed, lower voltage-drop, higher operating temperature, etc. [1,2,12–17]. Such outstanding characteristics of SiC are due to the advantages of its material and structure, such as higher electron velocity, higher energy gap, higher thermal conductivity, etc., which are shown in Figure 2. As a result, a motor drive system, which consists of a converter plus an inverter, based on SiC devices can provide higher efficiency and higher power density in comparison with their Si counterparts [9,10,18–21].

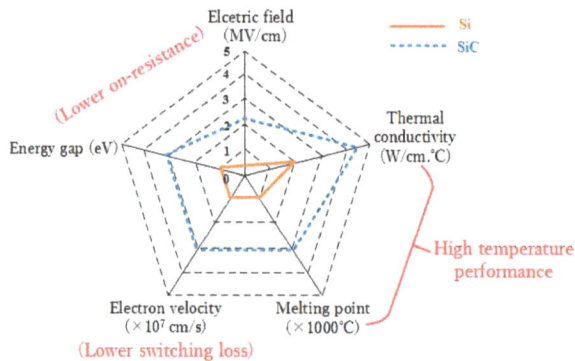

Figure 2. The components of the powertrain system in an EV and the corresponding losses.

Much research involving SiC has been widely conducted by many researchers [1,2,9–23]. Most of the work reflects an enormous effort investigating the switching and conduction losses of SiC [1,9,12,23]. For example, Ref. [22,23] developed a loss model of SiC, and both the conduction loss and switching loss of SiC are less than that of Si. The efficiency of the inverter based on SiC is 99.1% while the efficiency of the Si-inverter is 97.1%. The above work mostly focuses on the losses of the SiC-inverter or converter. However, the impact of SiC on the overall-powertrain system has not been explored systematically yet.

Therefore, this paper comprehensively investigates the impact of SiC on the powertrain systems in EVs. First, the characteristics of SiC are measured by a double pulse test (DPT). Then, the power losses of the DC/DC converter, DC/AC inverter, and motor are tested. The experimental results are assigned into a powertrain model built in the Advanced Vehicle Simulator (ADVISOR) software in order to explore a direct correlation between the SiC and the EV's battery capacity as well as the overall powertrain system efficiency. The SiC results are compared with the conventional powertrain system based on Si.

The remainder of this paper is organized as follows. In Section 2, the characteristics of SiC and Si are evaluated. In Section 3, the efficiencies of the DC/DC converter, DC/AC inverter, and motor are measured. In Section 4, simulations of the powertrain are implemented in ADVISOR. Conclusions are drawn in the final section.

2. SiC Device Characterization

The characteristics of a SiC power device are investigated experimentally in this Section, and compared with the Si counterparts. Both dynamic and static characteristics are tested. The test bench adopts a double pulse test (DPT), which is the widely accepted test method. Figure 3 shows the DPT test bench. A high accuracy current probe and a high accuracy voltage probe are used, namely a current probe TCPA300 plus TCP303 and a voltage probe P5100A.

The switching transition waveforms of SiC and Si are described in Figures 4 and 5, respectively. Both the turn-on and turn-off speed of SiC are faster than the Si counterparts. The quantified results are summarized in Table 1. The turn-on time of SiC is 85 ns while the turn-on time of Si is 143 ns at 25 °C. Meanwhile, the turn-on time of SiC is 79 ns while the turn-on time of Si is 145 ns at 175 °C. The turn-on times of both SiC and Si are around a constant value as the temperature changes. However, the turn-off time of Si is 752 ns at 175 °C, which is almost twice the value of 328 ns when Si operates at 25 °C. The turn-off time of SiC is not sensitive to the change of temperature, as well as its turn-on time. Such characteristics indicate that the switching loss of SiC is not increasing as the temperature rises. However the switching loss of Si incredibly rises as the temperature increases.

The variation tendencies of the voltage-drops of SiC and Si are different with their switching losses, as shown in Table 2. The voltage-drop of SiC becomes higher as the temperature increases, while the Si counterpart reduces as the temperature rises. However, the voltage-drop of Si is almost 20 times the value of SiC at 25 °C, which demonstrates that the conduction loss of SiC is much smaller than that of Si. Therefore, the test results of both the dynamic and static characteristics of SiC show that the power losses of SiC are much smaller than the Si counterparts.

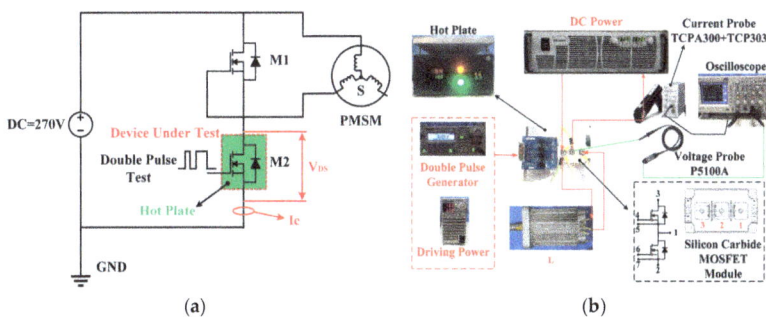

(a)   (b)

Figure 3. A double pulse test (DPT) bench. (**a**) Simplified DPT circuit; (**b**) Actual components in the test bench.

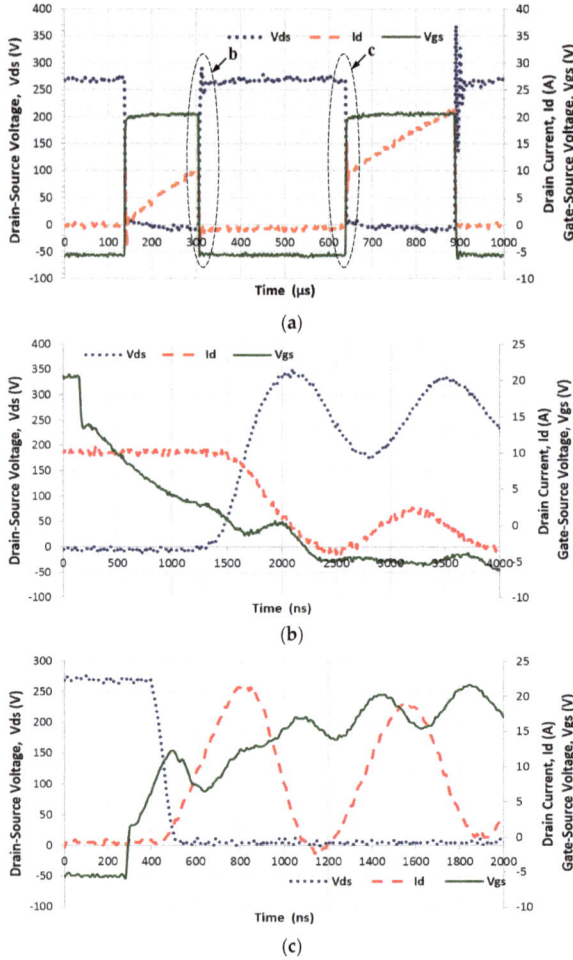

Figure 4. Switching transition waveform of SiC. (**a**) Total switching transition; (**b**) Turn-off transition; (**c**) Turn-on transition.

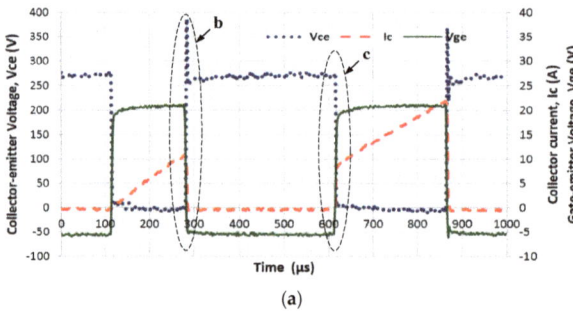

Figure 5. *Cont.*

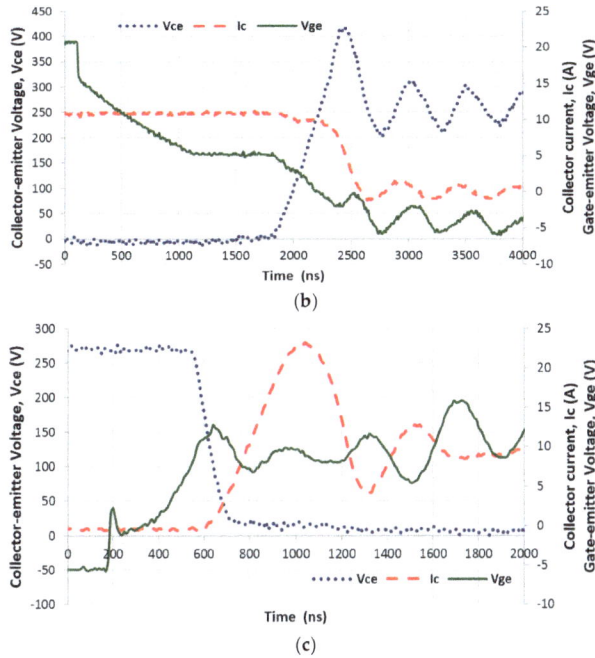

Figure 5. Switching transition waveform of Si. (**a**) Total switching transition; (**b**) Turn-off transition; (**c**) Turn-on transition.

Table 1. Characteristic comparison between SiC and Si.

	SiC (Cree CAS300M12BM2)		Si Insulated-Gate Bipolar Transistor (IGBT) (Infineon FF400R12KE3)	
	25 °C	175 °C	25 °C	175 °C
DC voltage	270 V	270 V	270 V	270 V
Turn-on time	85 ns	79 ns	143 ns	145 ns
Turn-off time	153 ns	148 ns	328 ns	752 ns
On-state resistance	4.8 mΩ	7.84 mΩ	/	/
Collector-emitter saturation voltage	/	/	832.5 mV (9.5 A)	583.0 mV (9.4 A)
Output capacitance	12.7 nF	13.4 nF	32.7 nF	35.3 nF

Table 2. Voltage-drop comparison between SiC and Si under different temperatures.

Temperature	SiC (Cree CAS300M12BM2)		Si (Infineon FF400R12KE3)	
	V_{ds}	I_d	V_{ce}	I_d
25 °C	44.2 mV	9.2 A	832.5 mV	9.5 A
100 °C	60.1 mV	9.2 A	686.9 mV	9.3 A
175 °C	73.7 mV	9.4 A	583.0 mV	9.4 A

3. Efficiency of the Powertrain System

The efficiency of the powertrain system is investigated in this Section. An experimental setup for a powertrain system is shown in Figure 6. There are three main components in the powertrain, namely the DC/DC converter, inverter, and PMSM. The main losses are switching and conduction losses of switching devices in both the DC/DC converter and the inverter, and iron loss and copper loss in the PMSM.

Both the voltages and currents of the DC/DC converter and DC/AC inverter are measured, respectively, by a Power Analyzer. Then, the efficiencies of both the converter and the inverter are obtained. Meanwhile, the output torque and speed of the motor are measured by a dynamometer, which are employed in the calculation of the efficiency of the PMSM.

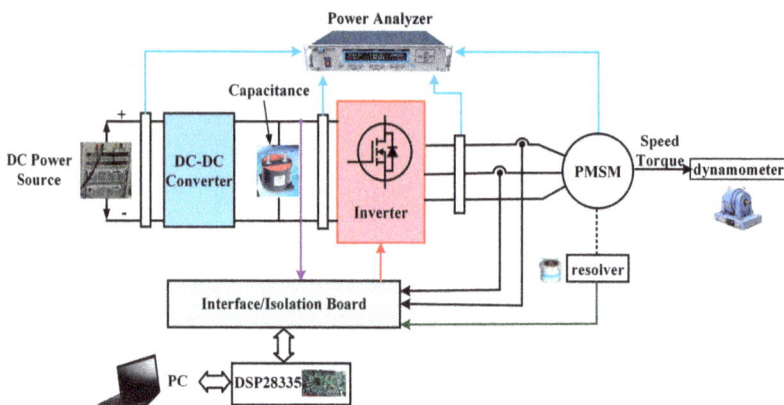

Figure 6. The structure of the experimental setup.

3.1. Efficiency of a Buck-Boost DC/DC Converter

The fundamental topology of a buck-boost DC/DC converter is shown in Figure 7. The buck-boost converter consists of two SiC MOSFETs (or Si IGBTs), two parallel diodes, an inductance, and two capacitors. The switching loss and conduction loss of SiC are the main contributions for the power losses of the DC/DC converter. The switching losses can be expressed as a function of the integration voltage and current during commuted intervals,

$$E_{switching} = E_{T\_on} + E_{T\_off} = \int_{t_1}^{t_2} V_{ds} \cdot I_d dt + \int_{t_3}^{t_4} V_{ds} \cdot I_d dt \tag{1}$$

where E_{T_on} and E_{T_off} are the turn-on and turn-off losses, respectively. t_1, t_2, t_3, and t_4 represent the start and end of turn-on and turn-off, respectively. The switching losses of Si can be calculated the same as those of SiC.

The conduction loss of SiC can be calculated directly by the current and voltage,

$$P_{conduction} = I_d^2 \cdot R_{ds(on)} = V_{ds} \cdot I_d \tag{2}$$

Additionally, the power losses of the converter also contain the loss of inductance resistance and the capacitors' equivalent series resistances (ESR). Hence, the total losses, namely the efficiency of the converter, can be measured and calculated as follows,

$$\eta_{converter} = (U_2 \cdot i_2)/(U_1 \cdot i_1) \tag{3}$$

where U_1 and U_2 are the input and output voltages of the converter, respectively. i_1 and i_2 are the input and output currents of the converter, respectively.

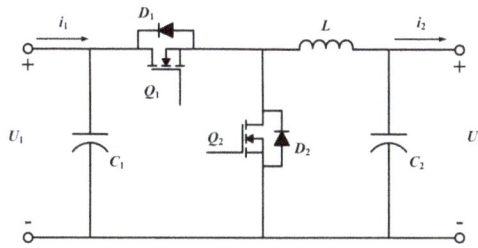

Figure 7. Topology of the DC/DC converter.

Figure 8 shows the efficiency comparison between the SiC- and Si-converters. The blue curve and black curve are measured by the experiments while the red curve and yellow curve are fitted based on the measured results. The peak efficiency of the SiC-inverter is nearly 93% at the output power of 30 kW, which is approximate 1% higher than that of the Si-converter.

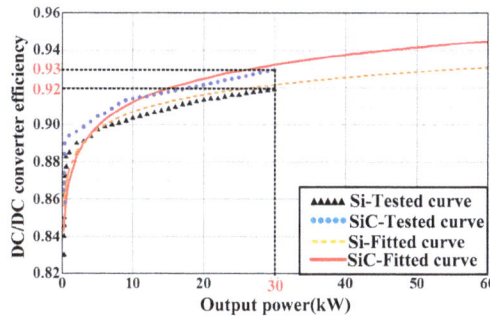

Figure 8. Efficiency comparison between the Si- and SiC-converters.

3.2. Efficiency of Inverter-PMSM

Figure 9 shows the fundamental topology of the DC/AC inverter and PMSM. The inverter contains six SiC MOSFETs (or Si IGBTs), six parallel diodes, one capacitor, etc. The switching loss and conduction loss of SiC are the main contributions for the power losses of the inverter, which is the same as that of the DC/DC converter. The loss model of the inverter can be found in Ref. [24–26]. This Section presents the measurement method and the test results.

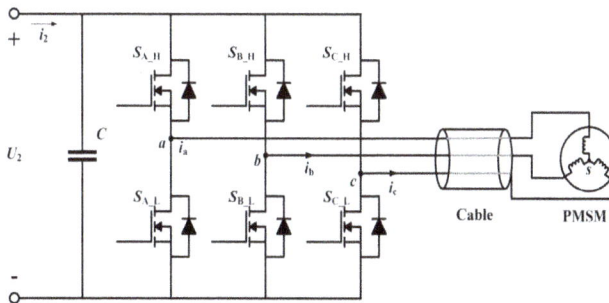

Figure 9. Topology of the DC/AC inverter-permanent magnet synchronous motor (PMSM).

The input power of the inverter is equal to the output power of the converter, as shown in Equation (4). The output power of the inverter is equal to the input power of the motor. The measurement method is shown in Figure 10. i_a, i_b, i_c represent the currents of Phase A, Phase B, and Phase C, respectively. U_{ab}, U_{bc}, U_{ac} are the three line-line voltages. Hence, the expression of the output power of the inverter is shown in Equation (5). The efficiency of the inverter can be calculated according to Equation (6).

$$P_{input} = U_2 \cdot i_2 \tag{4}$$

$$P_{output} = U_{ac} \cdot i_a + U_{bc} \cdot i_b \tag{5}$$

$$\eta_{inverter} = (U_{ac} \cdot i_a + U_{bc} \cdot i_b)/(U_2 \cdot i_2) \tag{6}$$

For the PMSM, the output power is mechanical energy, which is measured by a dynamometer. Hence, the efficiency of the motor is expressed as

$$\eta_{motor} = T \cdot n / [9.55(U_{ac} \cdot i_a + U_{bc} \cdot i_b)] \tag{7}$$

where T is the output mechanical torque of the motor and n is the mechanical speed of the motor.

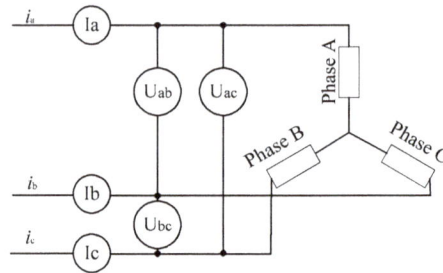

Figure 10. Measurement points of the DC/AC inverter.

The waveforms of Phase A and the $\alpha\beta$ currents are shown in Figure 11. It is clearly seen that the currents of the Si-drive system include more harmonic components, which will induce more power losses in the motor. The phase current is dependent on the output phase voltage of the inverter. The distortions of the phase voltage are contributed by the voltage-drop, turn-on time and turn-off time of the switching devices, and the dead time of the phase leg [27–29]. The voltage-drop, turn-on time, and turn-off time of SiC MOSFETs are smaller than their Si IGBTs counterparts, which were measured and are shown in Tables 1 and 2 of Section 2 in this paper. Meanwhile, the shorter dead time could be set in the SiC-drive due to the faster turn-on and turn-off speed. Therefore, the distortions of the phase voltage of the Si-drive are more than the SiC-drive counterparts, resulting in more harmonic components in the current of the Si-drive system.

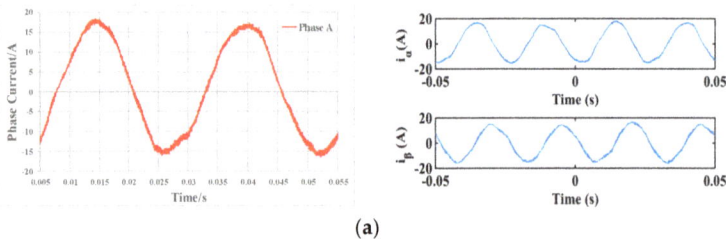

(a)

Figure 11. *Cont.*

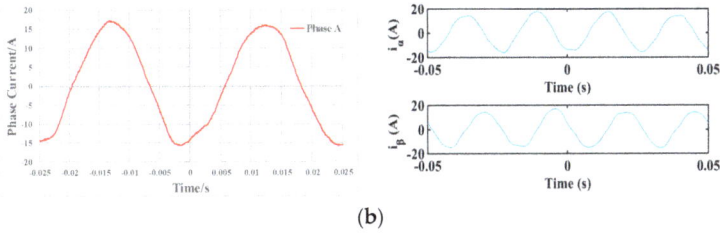

(b)

Figure 11. Phase current and $\alpha\beta$ current waveforms under 800 rpm and 8 N·m of the motor. (**a**) Current waveforms of the Si-drive system; (**b**) Current waveforms of the SiC-drive system.

There are two main losses in the motor, namely core losses in the iron core and copper losses in the winding. Both the core losses and copper losses include two parts; one part is induced by the fundamental current, and the other part is induced by the harmonic currents [24,30–32]. The core losses in the motor are composed of the eddy current loss P_e and hysteresis loss P_h. Both types of losses are caused by variation of the flux density in the core. Bertotti's model [30] is shown as,

$$P_c = P_e + P_h = k_e \sum_{n=1}^{\infty} (n\omega_1)^2 B_{p,n}^2 + k_h \sum_{n=1}^{\infty} (n\omega_1) B_{p,n}^x \tag{8}$$

where k_e and k_h are the eddy loss coefficient and the hysteresis loss coefficient, respectively, ω_1 is the fundamental angular frequency of the applied voltage, and n is the order of the harmonic. The peak flux density of the nth-order harmonic $B_{pcurrent,n}$ due to the nth-order current is predicted by [31],

$$B_{pcurrent,n} = \mu_0 \frac{2W}{\pi\delta} \sum_{n=1}^{\infty} I_n \sum_v \frac{1}{v} K_{sov} K_{dpv} F_v(r) \tag{9}$$

The copper losses also include the fundamental component loss and the losses related with the harmonic currents [32]. The harmonics induce eddy currents in the conductors, which cause a non-uniform distribution of the current density within the cross-sectional area of each conductor. Such non-uniform distribution of the current density in a conductor according to its own current is called the skin effect, while that according to the currents in adjacent conductors is called the proximity effect. The expression of the copper losses can be written as follows [32],

$$P_{cu} = I_{rms}^2 R_{dc} + \sum_{h=3}^{\infty} I_n^2 R_{n,ac} \tag{10}$$

where I_{rms} is the rms current, R_{dc} is the dc resistance, and I_n is the rms current of the nth current. $R_{n,ac}$ is the value of the nth harmonic resistance, which is determined by its dc value R_{dc} multiplying the ac skin and proximity gain,

$$R_{n,ac} = R_{dc}(K_{n,se} + K_{n,pe}) \tag{11}$$

where $K_{n,se}$ is the nth resistance gain caused by the skin effect, and $K_{n,pe}$ is the nth resistance gain caused by the proximity effect. Therefore, the phase current of the Si-drive system includes more harmonic components, resulting in more power losses in the motor.

Meanwhile, the amplitude of the phase current in the Si-drive system is higher than the SiC-system counterpart when the output powers of the two systems are the same. Hence, the efficiencies of both the inverter and the motor in the SiC-drive system are higher than the Si-system counterpart as shown in Figure 12. As a result, the efficiency of the overall inverter-motor system based on SiC is higher than that of the Si-system, especially with light loads.

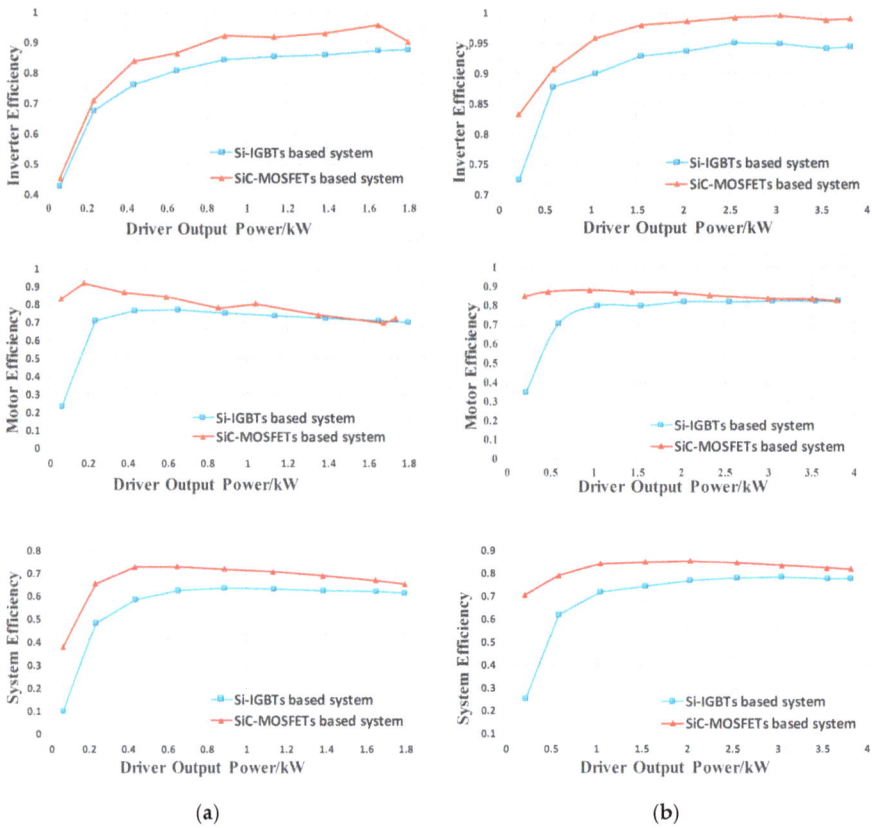

Figure 12. Efficiency comparison of Si and SiC based inverter-motor systems under different output powers. (**a**) 400 rpm; (**b**) 1000 rpm.

Due to the limited power rating of the dynamometer in our lab, the efficiency of the overall inverter-motor system during the full power range is expanded from the limited tested results shown in Figure 13, which will be adopted in the next Section. When the speed was smaller than 800 rpm and the torque was below 350 N·m, the efficiency of the overall inverter-motor system was measured by experiments. Then the range of the efficiency was expanded when the speed was higher than 800 rpm and the torque was bigger than 350 N·m through fitting formulas. There are several fitting algorithms to fit formulas based on MATLAB, such as Gaussian, Interpolant, Polynomial, Wei bull, etc. Among these fitting methods, the formulas fitted by the Polynomial had the best fitting degree, and the optimal fitting degree was 0.9682. Hence, the polynomial was adopted and the corresponding fitting formulas are shown as Equations (12) and (13), respectively.

$$\eta = 0.01941n^3 - 0.01153n^2T - 0.001971nT^2 + 0.0205T^3 - 0.0405n^2$$
$$+0.02037nT - 0.02738T^2 + 0.006666n - 0.0154T + 0.9134 \tag{12}$$

$$\eta = 0.02362n^3 - 0.01249n^2T - 0.0010351nT^2 + 0.02103T^3 - 0.05183n^2$$
$$+0.02005nT - 0.02814T^2 - 0.007325n - 0.0152T + 0.953 \tag{13}$$

(a)

(b)

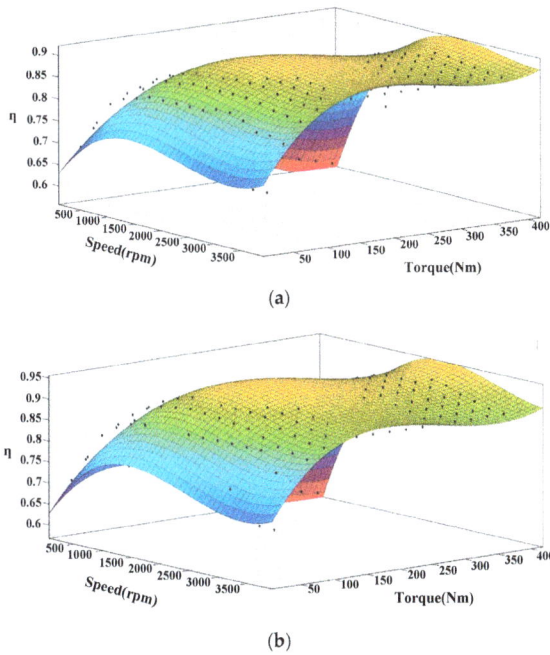

Figure 13. Efficiency comparison of Si and SiC based inverter-motor systems under full power range. (a) Si-system; (b) SiC-system.

4. Simulations in ADVISOR

The ADVISOR software provides a simulation environment based on MATLAB for EVs. The powertrain architecture of an EV in ADVISOR is shown in Figure 14. The efficiencies of the motor, inverter, and converter could be assigned to the motor and control module, respectively. Then, the comprehensive comparison between the SiC based powertrain and the Si based powertrain is implemented through the simulations. The U.S. Environmental Protection Agency Urban Dynamometer Driving Schedule (UDDS) cycle was adopted in these simulations. UDDS represents city driving conditions for a light duty vehicle as shown in Figure 15. Two typical topologies are investigated, namely Topology A: battery-inverter-motor, and Topology B: battery-converter-inverter-motor.

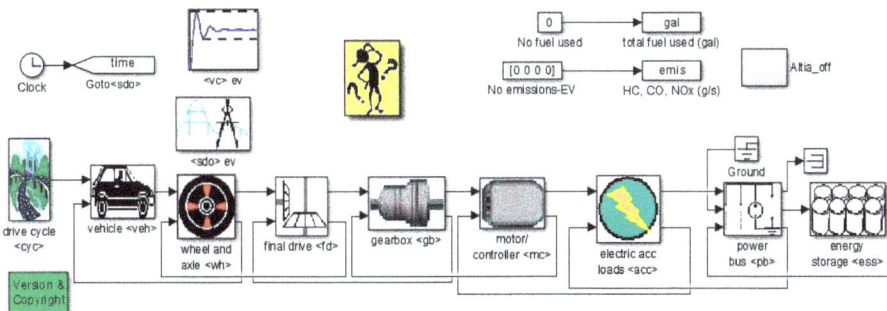

Figure 14. Powertrain architecture of the Electric Vehicle in ADVISOR.

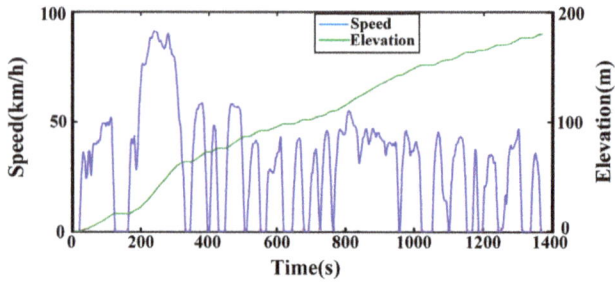

Figure 15. Urban Dynamometer Driving Schedule (UDDS) cycle.

4.1. Topology A: Battery-Inverter-Motor

The topology A is shown in Figure 16, which consists of battery, DC/AC inverter and PMSM. The efficiency contours and actual operating points for the Si- and SiC-inverter-motor systems in ADVISOR are shown in Figure 17, which are drawn based on the experimental results. The efficiency of the SiC-inverter-motor system is higher than the Si-system counterpart during the full power range. Hence, the power loss of the SiC-inverter-motor system is much smaller than that of the Si-system during the total UDDS cycle as shown in Figure 18. Meanwhile, the SOC of the battery in the SiC-system is also higher than that in the Si-system due to the higher efficiency of the SiC-inverter-motor system shown in Figure 19. The quantified results are given in Tables 3 and 4.

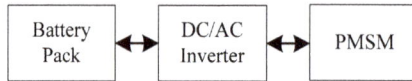

Figure 16. Topology A: Battery-inverter-PMSM.

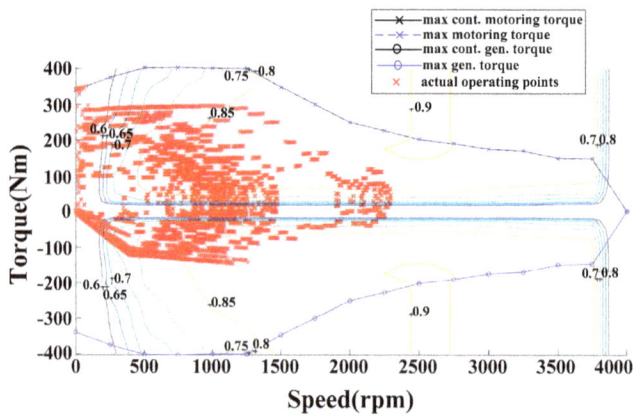

(a)

Figure 17. *Cont.*

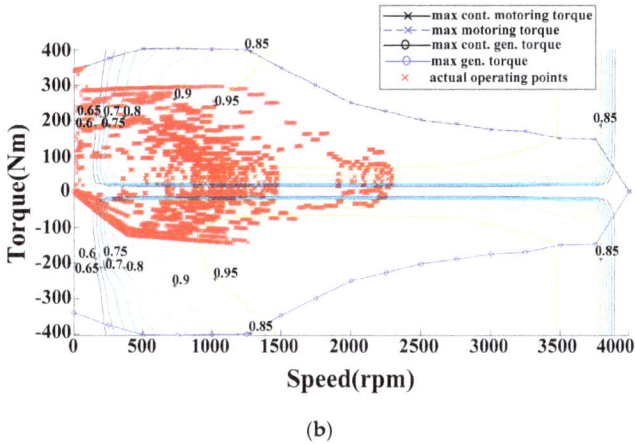

(b)

Figure 17. The efficiency contours and actual operating points of the Si- and SiC-inverter-motor system at different speeds and different torques. (**a**) Si-inverter-motor system; (**b**) SiC-inverter-motor system.

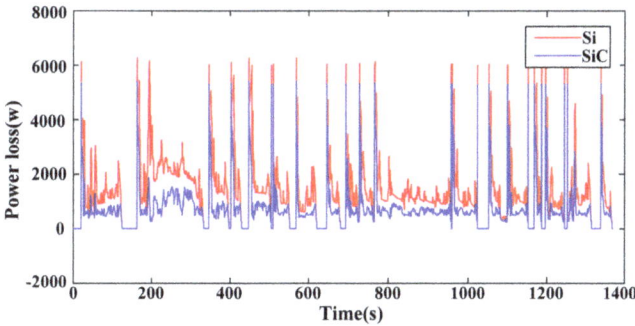

Figure 18. Power loss comparison of the SiC- and Si-inverter-motor systems.

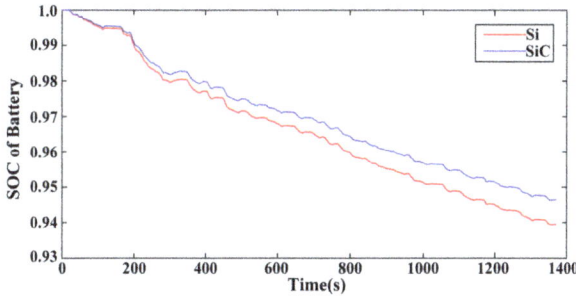

Figure 19. Battery SOC comparison of the SiC- and Si-inverter-motor systems.

Different elevations are considered, namely 0, 1.5%, 3.0%, and 4.5%. All the power consumptions in the SiC-system are smaller than those in the Si-system, as shown in Tables 3 and 4. When the elevation is 0, the improvements of equivalent fuel, output energy of the battery, braking energy recuperated, and efficiency of the overall system are 12.50%, 8.85%, 14.98%, and 13.00%, respectively.

Such incredible enhancements are due to the higher efficiency and more sinusoidal waveform phase current of the SiC-inverter-motor system.

Additionally, the higher elevation increases the power consumption due to more output torque required. For example, the output energy of the battery in the SiC system is 15,307 KJ when the elevation is 4.5%, while the output energy of the battery is 9783 KJ when the elevation is 1.5%.

Table 3. Comparison of SiC- and Si-systems at zero elevation for Topology A.

Description	SiC (Cree CAS300M12BM2)	Si Insulated-Gate Bipolar Transistor (IGBT) (Infineon FF400R12KE3)	Improve (%)
Elevation	0	0	-
Equivalent fuel (L/100 km)	2.1	2.4	12.50
Drive distance (km)	12	12	-
Output energy of battery (kJ)	7285	7992	8.85
Efficiency of inverter-motor system (%)	87.5	79	10.76
Braking energy recuperated (kJ)	990	861	14.98
Efficiency of overall system (%)	36.5	32.3	13.00

Table 4. Comparison of SiC- and Si-systems at 1.5%, 3.0%, and 4.5% elevations for Topology A.

Description	SiC (Cree CAS300M12BM2)			Si IGBT (Infineon FF400R12KE3)		
Elevation	1.5	3	4.5	1.5	3	4.5
Equivalent fuel (L/100 km)	2.9	3.7	4.5	3.2	4.1	5
Drive distance (km)	12	12	12	12	12	12
Output energy of battery (kJ)	9783	12,480	15,307	10,736	13,719	16,847
Efficiency of inverter-motor system (%)	88.4	89.9	91.06	81.37	82.75	83.56
Braking energy recuperated (kJ)	742	559	442	631	484	365
Efficiency of overall system (%)	25.6	19.4	15.5	22.9	17.7	14

4.2. Topology B: Battery-Converter-Inverter-Motor

In this section, the DC/DC converter is taken into account. The topology is shown in Figure 20. There are the battery, DC/DC converter, DC/AC inverter, and PMSM in the powertrain system. Actually, many EV powertrain systems adopt this topology, which can easily adjust the DC voltage in the system. The efficiency of the SiC-converter-inverter-motor system is higher than the Si-system counterpart during the full power range, which is the same as topology-B. Hence, the power loss of the SiC-converter-inverter-motor system is much smaller than that of the Si-system during the total UDDS cycle, as shown in Figure 21. Additionally, the battery SOC for the SiC-system is higher than that of the Si-system, as shown in Figure 22. The quantified results are summarized in Tables 5 and 6.

For topology B, four scenarios are also considered, namely 0, 1.5%, 3.0%, and 4.5% elevations. The output energy of the battery in the SiC-system is smaller than those of the Si-system for every case shown in Tables 5 and 6. Its improvement is 14.31%, which is more than the improvement of 8.85% for topology A. The improvements of the other parameters of topology B, such as equivalent fuel, efficiency of the converter-inverter-motor, braking energy recuperated, and efficiency of the overall system, are all higher than their counterparts of topology A, which is due to taking the SiC DC/DC converter into account in the powertrain system. Its efficiency is higher than that of the Si DC/DC converter. Table 6 also shows that the higher the elevation, the more power consumption is needed.

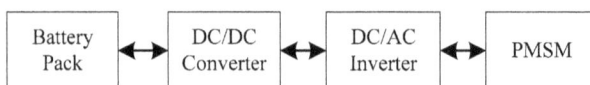

Figure 20. Topology B: Battery-converter-inverter-PMSM.

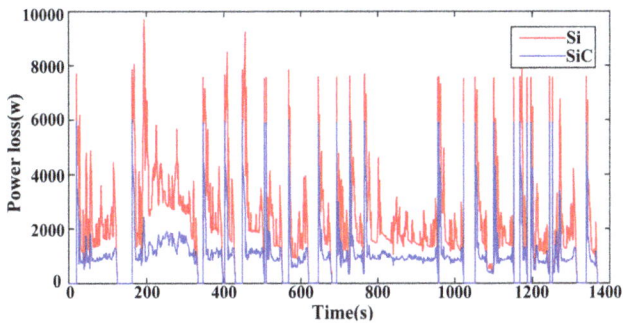

Figure 21. Power loss comparison of the SiC- and Si-converter-inverter-motor systems.

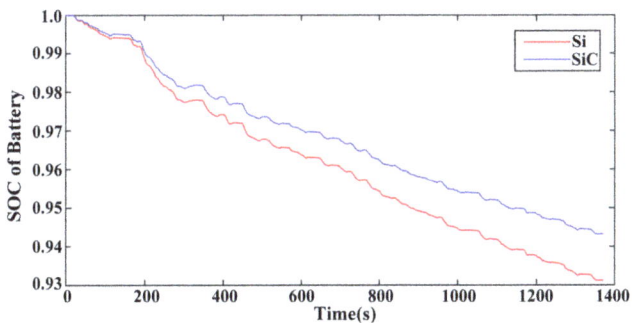

Figure 22. Battery SOC comparison of the SiC- and Si-converter-inverter-motor systems.

Table 5. Comparison of the SiC- and Si-systems at zero elevation for Topology B.

Description	SiC (Cree CAS300M12BM2)	Si IGBT (Infineon FF400R12KE3)	Improve (%)
Elevation	0	0	-
Equivalent fuel (L/100 km)	2.2	2.6	15.38
Drive distance (km)	12	12	-
Output energy of battery (kJ)	7599	8821	14.31
Efficiency of converter-inverter-motor system (%)	83.59	70.94	17.83
Braking energy recuperated (kJ)	915	696	31.47
Efficiency of overall system (%)	34.4	28.4	21.14

Table 6. Comparison of the SiC- and Si-systems at 1.5%, 3.0%, and 4.5% elevations for Topology B.

Description	SiC (Cree CAS300M12BM2)			Si IGBT (Infineon FF400R12KE3)		
Elevation	1.5	3	4.5	1.5	3	4.5
Equivalent fuel (L/100 km)	3	3.8	4.7	3.5	4.5	5.5
Drive distance (km)	12	12	12	12	12	12
Output energy of battery (kJ)	10,148	12,897	15,765	11,778	15,009	18,382
Efficiency of converter-inverter-motor system (%)	86.51	88.41	89.62	73.7	75.25	76.26
Braking energy recuperated (kJ)	681	511	383	514	386	287
Efficiency of overall system (%)	24.4	18.7	15	20.5	15.8	12.7

5. Conclusions

This paper comprehensively investigated the impact of SiC power devices on the powertrain of EVs. The characteristics of SiC were measured and demonstrated outstanding advantages, such

Energies **2017**, *10*, 533

as faster switching speed, lower voltage drop, and more stability than the Si counterparts. Hence, both the DC/DC converter and DC/AC inverter based on SiC exhibited higher efficiency than that of the Si systems. Furthermore, there are low harmonic components in the phase currents of the SiC-system. Therefore, the high efficiency of the motor benefits from the more sinusoidal waveform of the phase current.

Two typical topologies of drive systems were analyzed by ADVISOR. Both topologies A and B based on SiC represented remarkable enhancements of the performances, such as smaller equivalent fuel consumption, smaller output energy of the battery, more efficiency and braking energy recuperated, etc. The smaller output energy of the battery, higher efficiency, and more braking energy recuperated means that EVs could adopt a smaller battery pack, reduce their weight, increase cruise distance, etc. Therefore, this work is meaningful for improving the performances of EVs.

Acknowledgments: This work was supported in part by the National Natural Science Foundation of China under Project 51407004 and in part by the Aeronautical Science Foundation of China 20162851016.

Author Contributions: Xiaofeng Ding proposed the main idea for this paper and wrote this paper. Jiawei Cheng did the simulation and drew the related figures. Feida Chen evaluated the characteristics of the devices by the experiments. All authors carried out the theoretical analysis and contributed to writing the paper.

Conflicts of Interest: The authors declare no conflict of interest.

References

1. Shamsi, P.; McDonough, M.; Fahimi, B. Wide-Bandgap Semiconductor Technology: Its impact on the electrification of the transportation industry. *IEEE Electr. Mag.* **2013**, *1*, 59–63. [CrossRef]
2. Hamada, K.; Nagao, M.; Ajioka, M.; Kawai, F. SiC—Emerging Power Device Technology for Next-Generation Electrically Powered Environmentally Friendly Vehicles. *IEEE Trans. Electr. Devices* **2015**, *62*, 278–285. [CrossRef]
3. Xiong, R.; Sun, F.; Chen, Z. A data-driven multi-scale extended Kalman filtering based parameter and state estimation approach of lithium-ion polymer battery in electric vehicles. *Appl. Energy* **2014**, *113*, 463–476. [CrossRef]
4. Ruan, J.; Walker, P.; Zhang, N. A comparative study energy consumption and costs of battery electric vehicle transmissions. *Appl. Energy* **2016**, *165*, 119–134. [CrossRef]
5. Zhang, Y.; Xiong, R.; He, H.; Shen, W. A lithium-ion battery pack state of charge and state of energy estimation algorithms using a hardware-in-the-loop validation. *IEEE Trans. Power Electr.* **2017**, *32*, 4421–4431. [CrossRef]
6. Jaguemont, J.; Boulon, L.; Dubé, Y. A comprehensive review of lithium-ion batteries used in hybrid and electric vehicles at cold temperatures. *Appl. Energy* **2016**, *164*, 99–114. [CrossRef]
7. Sun, F.; Xiong, R.; He, H. A systematic state-of-charge estimation framework for multi-cell battery pack in electric vehicles using bias correction technique. *Appl. Energy* **2016**, *162*, 1399–1409. [CrossRef]
8. Pohlenz, D.; Böcker, J. Efficiency improvement of an IPMSM using Maximum Efficiency operating strategy. In Proceedings of the 2010 14th International Power Electronics and Motion Control Conference (EPE/PEMC), Ohrid, Macedonia, 6–8 September 2010; pp. 15–19.
9. Shang, F.; Arribas, A.P.; Krishnamurthy, M. A comprehensive evaluation of SiC devices in traction applications. In Proceedings of the 2014 IEEE Transportation Electrification Conference and Expo (ITEC), Dearborn, MI, USA, 15–18 June 2014; pp. 1–5.
10. Zhang, H.; Tolbert, L.M.; Ozpineci, B. Impact of SiC Devices on Hybrid Electric and Plug-In Hybrid Electric Vehicles. *IEEE Trans. Ind. Appl.* **2011**, *47*, 912–921. [CrossRef]
11. Jahdi, S.; Alatise, O.; Fisher, C.; Ran, L.; Mawby, P. An Evaluation of Silicon Carbide Unipolar Technologies for Electric Vehicle Drive-Trains. *IEEE J. Emerg. Sel. Top. Power Electr.* **2014**, *2*, 517–528. [CrossRef]
12. Kassakian, J.G.; Jahns, T.M. Evolving and Emerging Applications of Power Electronics in Systems. *IEEE J. Emerg. Sel. Top. Power Electr.* **2013**, *1*, 47–58. [CrossRef]
13. Wang, F.; Zhang, Z.; Ericsen, T.; Raju, R.; Burgos, R.; Boroyevich, D. Advances in Power Conversion and Drives for Shipboard Systems. *Proc. IEEE* **2015**, *103*, 2285–2311. [CrossRef]
14. Tanimoto, S.; Matsui, K. High Junction Temperature and Low Parasitic Inductance Power Module Technology for Compact Power Conversion Systems. *IEEE Trans. Electr. Devices* **2015**, *62*, 258–269. [CrossRef]

15. Xu, F.; Guo, B.; Xu, Z.; Tolbert, L.M.; Wang, F.; Blalock, B.J. Paralleled Three-Phase Current-Source Rectifiers for High-Efficiency Power Supply Applications. *IEEE Trans. Ind. Appl.* **2015**, *51*, 2388–2397. [CrossRef]
16. Josifović, I.; Popović-Gerber, J.; Ferreira, J.A. Improving SiC JFET Switching Behavior under Influence of Circuit Parasitics. *IEEE Trans. Power Electr.* **2012**, *27*, 3843–3854. [CrossRef]
17. Sun, K.; Wu, H.; Lu, J.; Xing, Y.; Huang, L. Improved Modeling of Medium Voltage SiC MOSFET within Wide Temperature Range. *IEEE Trans. Power Electr.* **2014**, *29*, 2229–2237. [CrossRef]
18. Brown, C.D.; Sarlioglu, B. Reducing Switching Losses in BLDC Motor Drives by Reducing Body Diode Conduction of MOSFETs. *IEEE Trans. Ind. Appl.* **2015**, *51*, 1864–1871. [CrossRef]
19. Swamy, M.M.; Kang, J.; Shirabe, K. Power Loss, System Efficiency, and Leakage Current Comparison between Si IGBT VFD and SiC FET VFD with Various Filtering Options. *IEEE Trans. Ind. Appl.* **2015**, *51*, 3858–3866. [CrossRef]
20. Shirabe, K.; Swamy, M.; Kang, J.K.; Hisatsune, M.; Das, M.; Callanan, R.; Lin, H. Design of 400 V class inverter drive using SiC 6-in-1 power module. In Proceedings of the 2013 IEEE Energy Conversion Congress and Exposition (ECCE), Denver, CO, USA, 15–19 September 2013; pp. 2363–2370.
21. Merkert, A.; Krone, T.; Mertens, A. Characterization and Scalable Modeling of Power Semiconductors for Optimized Design of Traction Inverters with Si- and SiC-Devices. *IEEE Trans. Power Electr.* **2014**, *29*, 2238–2245. [CrossRef]
22. Zhao, T.; Wang, J.; Huang, A.Q.; Agarwal, A. Comparisons of SiC MOSFET and Si IGBT Based Motor Drive Systems. In Proceedings of the Conference Record of the 2007 IEEE Industry Applications Conference, 42nd IAS Annual Meeting, New Orleans, LA, USA, 23–27 September 2007; pp. 331–335.
23. Reed, J.K.; McFarland, J.; Tangudu, J.; Vinot, E.; Trigui, R.; Venkataramanan, G.; Gupta, S.; Jahns, T. Modeling power semiconductor losses in HEV powertrains using Si and SiC devices. In Proceedings of the 2010 IEEE Vehicle Power and Propulsion Conference (VPPC), Lille, France, 1–3 September 2010; pp. 1–6.
24. Ding, X.F.; Liu, G.L.; Du, M.; Guo, H.; Duan, C.W.; Qian, H. Efficiency Improvement of Overall PMSM-Inverter System Based on Artificial Bee Colony Algorithm Under Full Power Range. *IEEE Trans. Mag.* **2016**, *52*, 1–4. [CrossRef]
25. Hassan, W.; Wang, B. Efficiency optimization of PMSM based drive system. In Proceedings of the 2012 7th International Power Electronics and Motion Control Conference (IPEMC), Harbin, China, 2–5 June 2012; Volume 2.
26. Mestha, L.K.; Evans, P.D. Analysis of on-state losses in PWM inverters. *Electr. Power Appl. IEE Proc. B* **1989**, *136*, 189–195. [CrossRef]
27. Choi, J.W.; Sul, S.K. Inverter output voltage synthesisusing novel dead time compensation. *IEEE Trans. Power Electr.* **1996**, *11*, 221–227. [CrossRef]
28. Zhang, Z.; Xu, L. Dead-Time Compensation of InvertersConsidering Snubber and Parasitic Capacitance. *IEEE Trans. Power Electr.* **2014**, *29*, 3179–3187. [CrossRef]
29. Bedetti, N.; Calligaro, S.; Petrella, R. Self-Commissioning of Inverter Dead-Time Compensation by Multiple Linear Regression Based on a Physical Model. *IEEE Trans. Ind. Appl.* **2015**, *51*, 3954–3964. [CrossRef]
30. Boglietti, A.; Cavagnino, A.; Lazzari, M.; Pastorelli, M. Predicting iron losses in soft magnetic materials with arbitrary voltage supply: An engineering approach. *IEEE Trans. Magn.* **2003**, *39*, 981–989. [CrossRef]
31. Zhu, Z.Q.; Howe, D. Instantaneous magnetic field distribution in permanent magnet brushless DC motors. IV. Magnetic field on load. *IEEE Trans. Magn.* **1993**, *29*, 152–158. [CrossRef]
32. Key, T.S.; Lai, J.S. Costs and benefits of harmonic current reduction for switch-mode power supplies in a commercial office building. *IEEE Trans. Ind. Appl.* **1996**, *32*, 1017–1025. [CrossRef]

Article

Data-Driven Predictive Torque Coordination Control during Mode Transition Process of Hybrid Electric Vehicles

Jing Sun [1], Guojing Xing [2] and Chenghui Zhang [2],*

[1] School of Information and Electronic Engineering, Shandong Technology and Business University, Yantai 264005, China; sunjing@sdu.edu.cn
[2] School of Control Science and Engineering, Shandong University, Jinan 250061, China; xgjsdu@sdu.edu.cn
*   Correspondence: zchui@sdu.edu.cn; Tel.: +86-531-8839-5717

Academic Editor: Joe (Xuan) Zhou
Received: 16 November 2016; Accepted: 22 March 2017; Published: 1 April 2017

Abstract: Torque coordination control significantly affects the mode transition quality during the mode transition dynamic process of hybrid electric vehicles (HEV). Most of the existing torque coordination control methods are based on the mechanism model, whose control effect heavily depends on the modeling accuracy of the HEV powertrain. However, the powertrain structure is so complex, that it is difficult to establish its precise mechanism model. In this paper, a torque coordination control strategy using the data-driven predictive control (DDPC) technique is proposed to overcome the shortcomings of mechanism model-based control methods for a clutch-enabled HEV. The proposed control strategy is only based on the measured input-output data in the HEV powertrain, and no mechanism model is needed. The conflicting control requirements of comfortability and economy are included in the cost function. The actual physical constraints of actuators are also explicitly taken into account in the solving process of the data-driven predictive controller. The co-simulation results in Cruise and Simulink validate the effectiveness of the proposed control strategy and demonstrate that the DDPC method can achieve less vehicle jerk, faster mode transition and smaller clutch frictional losses compared with the traditional model predictive control (MPC) method.

Keywords: mode transition; torque coordination; data-driven predictive control (DDPC); hybrid electric vehicle (HEV)

1. Introduction

The multi-energy powertrain system is the most distinctive feature that makes hybrid electric vehicle (HEV) more energy efficient than the traditional vehicle, and its key technology directly determines the economy, reliability, safety and comfortability in HEV. The control of the HEV multi-energy powertrain system can be classified into two kinds of core problems. The first is the energy distribution and efficiency optimization of multi-energy sources, and the second is the dynamic torque coordination between the multi-power sources. The former aims to improve the fuel economy and reduce emissions at arbitrary driving cycles. It belongs to the research category of energy management strategy and has been widely concerned [1–3]. By contrast, the latter, which is critical to the ride comfortability, switching rapidity and durability in HEV, is relatively less studied, but it directly determines people's purchase intention and influences the industrialization process of HEV. In fact, in order to improve the fuel economy, frequent transitions among basic operation modes are required, such as the motor-only mode, the engine-only mode, the compound driving mode and the regenerative braking mode. However, mode transitions are often accompanied by the target torque mutation of the engine, the clutch and the motor. The vehicle impact, jitter and clutch excessive wear will appear with unfavorable coordination; thus, the comfortability and clutch durability will be influenced in the vehicle driving. In fact, torque coordination control has become not only the key problem of the HEV multi-energy powertrain system, but also a tough tradeoff commonly concerned by the business circle and the academia.

It is well known that the HEV powertrain structure is very complicated. To facilitate the torque coordination control problem, early research works neglected the clutch dynamic characteristic, which is a crucial factor to the mode transition performance. The torque compensation control strategy using the fast response ability of the motor was an early torque coordination control method [4,5]. The engine torque was estimated online, and the motor's fast response characteristic was used to compensate the output torque lag of the engine to decrease the total driving torque fluctuation of the mode transition process, which could thus improve the comfortability. The advantages of this method are that it is simple and easy to realize. A model matching control (MMC) method was proposed in [6]. The control idea was similar to the torque compensation control strategy. Compared with the torque compensation method, the motor torque demand was not simply equal to the difference between the total demand torque and the estimated engine torque, but was obtained by a model matching controller using the actual total torque and acceleration pedal as its inputs.

Taking the clutch dynamic characteristic into account, many research works handled the torque coordination problem from the perspective of state equations under different running modes. A fuzzy adaptive sliding mode approach was applied to the mode transition control in [7], and the switched hybrid theory was applied to the control of a parallel hybrid electric vehicle drivetrain [8]. The model predictive control (MPC) method was also used to manage HEV mode transitions in [9,10], in which two model predictive controllers were needed to accomplish the comfortable transition and synchronize the two drivetrain parts, which were divided by the clutch. Considering the discontinuity of the clutch, a model reference control (MRC) law was proposed to coordinate the engine torque, the clutch torque and the motor torque during the mode transition dynamic process from motor-only mode to compound driving mode [11]. The MPC method was combined with the MRC in [12], and good mode transition performance has been achieved in the MATLAB/Simulink environment. Only one model predictive controller was needed, which greatly reduced the calculation amount of the control strategy and was more suitable for the real-time control application. However, when the MPC method is further applied in the vehicle simulation platform, the mode transition performance is unsatisfactory, especially the vehicle jerk outdistances the recommended value, which is 10 m/s^3 in Germany and 17.64 m/s^3 in China [13]. This motivates us to find a more practical solution to solve the torque coordination control problem.

It is worth pointing out that most of the above HEV torque coordination control methods are based on the mechanism model, whose control effect heavily depends on the modeling accuracy of the controlled process. The HEV powertrain is a multi-input and multi-output, nonlinear and strong coupling system, so it is difficult to establish its precise mechanism model. Even if the global exact mathematical model of HEV powertrain system can be obtained, the model is surely high order nonlinear, which makes it difficult to use the mechanism model-based control methods. The mechanism modeling is simplified to some extent for the convenience of controller design in the existing research, which is bound to bring about many unmodeled dynamics and a poor control effect. Although a seven-order mathematical model is used in [9,10], the quadratic terms proportional to vehicle speed are neglected in the resistance torque model in order to facilitate the control design; thus, the dynamic characteristics of the HEV powertrain system cannot be precisely described.

In recent years, the data-driven theory has provided a new path for the modeling and controlling of complex systems [14,15]. There are abundant online and offline measurement data in the HEV operation process, which provide the possibility for the realization of the data-driven modeling and control for the HEV mode transition dynamic process. As one of the efficient data-driven control methods, the data-driven predictive control (DDPC) algorithm was proposed in recent years, which combines the modeling superiority of the data-driven and the multiple-constraint handling capability of the predictive control [16]. It was firstly proposed in [17]. Its basic idea is to use the data-driven method to establish the data model of the controlled object and use the predictive control method to design the controller. DDPC has been used for a multidimensional blast furnace system [18], wind power generation [19], the distributed solar collector field [20], biped robots [21], a waste water system [22], a vapor compression refrigeration cycle system [23] and other industrial applications. Furthermore, a small quantity of application research works relative to DDPC have been used for fast and dynamic situations, such as automated manual transmission (AMT) vehicle starting [24], dual-clutch transmission [25] and high-speed train operation [26]. Since the actual engine torque, the clutch torque and the motor torque and their change rates are all limited by actuator saturation, the torque coordination control of HEV powertrain is a multiple-constraint optimal control problem. Therefore, it might be a good choice to solve the HEV's torque coordination control problem by the DDPC method.

To the best of our knowledge, the proposed torque coordination control strategy based on the DDPC method is a novel contribution for the HEV dynamic process control during the mode transition process. Taking the representative mode transition process from motor-only mode to compound driving mode as an example, a data-driven predictive controller is designed for the torque coordination control problem of HEV, which overcomes the shortcomings of the traditional mechanism model-based control methods. This controller is directly obtained only based on the input-output data of HEV powertrain system, and no accurate mechanism model is required. A multi-objective function is constructed, which deals with the tradeoff between the transition rapidity and the riding comfortability. The time-domain hard constraints of the actuating units are also explicitly taken into account during the optimization control solution. The Cruise simulation results demonstrate that the designed data-driven predictive controller works very well during the mode transition process. A model predictive controller is also designed for comparison. Cruise simulation results show that the DDPC method achieves better mode transition quality compared with the MPC method, i.e., shorter transition duration, better transition comfortability (smaller vehicle jerk) and lower clutch abrasion (less frictional losses). This further validates the superiority of the DDPC method.

The remainder of this paper is organized as follows. In Section 2, the HEV model is built in Cruise simulation software. In Section 3, the detailed implementation of the proposed torque coordination control strategy based on DDPC method is introduced for the mode transition process from the motor-only mode to the compound mode. In Section 4, Cruise simulation results validate the effectiveness of the proposed DDPC method and its advantages over MPC method. Finally, the concluding remarks are given in Section 5.

2. HEV Model in Cruise and Problem Formulation

In order to obtain the input-output data, which can fully reflect HEV dynamics, a complete clutch-enabled HEV model is established in the professional automotive simulation software Cruise, which is used by Volkswagen, General Motors, Qoros and other automobile companies; the block diagram of the powertrain configuration is shown in Figure 1. The detailed vehicle parameters are shown in Table 1. This HEV model can well describe the HEV's transient dynamic characteristics, e.g., the delay characteristics of engine output torque, the torsional vibration characteristics of the clutch and the driving shaft, the tire slip characteristics, etc. The engine power is delivered to the drive-line using the friction torque of the clutch. The rule-based energy management strategy is adopted in the supervisory controller.

To improve the fuel economy and reduce emissions, multiple operation modes are required, including stop mode, motor-only mode, engine-only mode, compound driving mode, regenerative brake mode, charging mode, and so on. Therefore, various mode transition dynamic processes may appear in HEV operation. The mode transition processes related to the clutch slipping phase are relatively complicated and deserve intensive study. Among these processes, the mode transition from motor-only mode to compound driving mode is the most typical dynamic process, which will be studied in this paper.

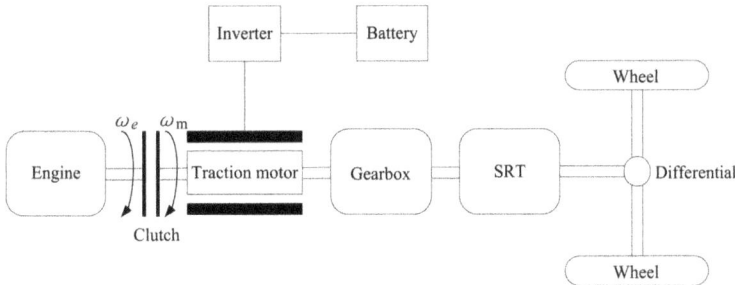

Figure 1. The block diagram of the powertrain configuration.

Table 1. Key vehicle parameters in the cruise simulation platform.

Component	Parameter	Value and Unit
Vehicle	Vehicle mass	1430 kg
	Air density	1.29 kg/m^3
	Vehicle frontal area	2.15 m^2
	Aerodynamic drag coefficient	0.3
Engine	Moment of inertia	0.19 kg·m^2
	Maximum speed	5800 r/min
	Maximum torque	130 Nm
Clutch	Moment of inertia	0.001 kg·m^2
	Maximum transferable torque	200 Nm
Motor	Moment of inertia	0.15 kg·m^2
	Maximum speed	8000 r/min
	Maximum torque	65 Nm
Battery	Maximum charge	50 Ah
Tyre	Moment of inertia	1.1 kg·m^2
	Dynamic rolling radius	0.308 m
Gear Box	Moment of inertia	0.005 kg·m^2
	Gear ratio	3.62, 2.22, 1.51, 1.08, 0.85
SRT	Moment of inertia	0.018 kg·m^2
	Transmission ratio	5.5
Differential	Torque split factor	1.0
	Moment of inertia	0.02 kg·m^2

For the convenience of research, we use the vehicle speed of 27 km/h as the mode transition threshold. When the vehicle speed is lower than 27 km/h, the HEV is propelled only by the electric motor, and the two sides of the clutch are separated. When the vehicle speed exceeds 27 km/h, the two sides of the clutch remain separated; the vehicle is still propelled by the electric motor separately; and the engine is started. When the clutch slip speed is less than a given threshold ε, the mode transition dynamic process from motor-only mode to compound driving mode begins. The clutch enters into the slipping phase, during which time the coordination control of the engine, clutch and motor will directly determine the transition quality. When the clutch slip speed is zero, the HEV works in compound driving mode; the motor propels the HEV along with the engine; and the clutch is locked.

The indices to evaluate the torque coordination quality during the mode transition dynamic process mainly are comprised of riding comfortability, rapidity and economy.

a. Riding comfortability refers to whether the impact due to the acceleration change is acceptable to the passengers during the mode transition dynamic process. It can be expressed by the vehicle jerk, namely the change rate of acceleration.

$$j = \frac{da_v}{dt}$$

where j represents the vehicle jerk whose unit is m/s^3 and a_v represents the vehicle acceleration whose unit is m/s^2. The smaller the jerk, the better the riding comfortability.

b. Rapidity means whether the response speed to the mode transition command can satisfy the driver's requirement. It can be represented by the mode transition duration. The shorter the transition duration, the better the rapidity.

c. Economy refers to the torque coordination strategy's influence on the component service life in the powertrain system. For the mode transition process related to clutch slipping phase, frictional losses can be used to represent the clutch abrasive wear resulting from the control strategy.

$$W_{sl} = \int_{t_0}^{t_f} T_c \,|\Delta\omega|\, dt$$

where W_{sl} represents the clutch frictional losses, T_c is the clutch torque and t_0 and t_f are the initial time and terminal time of clutch slipping phase, respectively. $\Delta\omega = \omega_m - \omega_e$ is the clutch slip speed; ω_m is the motor speed; ω_e is the engine speed; and $|\cdot|$ is the symbol for absolute value. It can be seen that the clutch slip speed, the clutch torque and the clutch slipping duration mainly contribute to frictional losses. The less the frictional losses, the longer the clutch working life and the better the mode transition economy.

3. Data-Driven Predictive Controller

In this section, we propose a data-driven predictive controller, which is directly based on the input-output data of the Cruise simulation platform in order to deal with the torque coordination control problem during the mode transition from motor-only mode to compound driving mode. Firstly, we deduce the subspace predictor from the state space equation in brief and get the computational formula of subspace matrices. Then, we represent the parts that can be finished offline, i.e., open-loop data sample, the identification of subspace matrices and subspace predictor verification. Finally, we translate the torque coordination control problem considering physical constraints into a quadratic programming problem; thus, we can accomplish the data-driven predictive torque coordination control online.

3.1. Derivation of the Subspace Predictor Equation

At the k-th sampling instant, the discrete state equations of the mode transition dynamic process from motor-only mode to compound driving mode are as follows:

$$\mathbf{x}(k+1) = A\mathbf{x}(k) + B\mathbf{u}(k) \qquad (1.a)$$

$$\mathbf{y}(k) = C\mathbf{x}(k) \qquad (1.b)$$

$$\mathbf{y}^b(k) = C^b\mathbf{x}(k) \qquad (1.c)$$

where $\mathbf{u}(k) \in R^l$ is the input variable, $\mathbf{y}(k) \in R^m$ is the output variable, $\mathbf{y}^b(k) \in R^{m_b}$ is the constrained output variable, $\mathbf{x}(k) \in R^n$ is the state variable and $A \in R^{n \times n}$, $B \in R^{n \times l}$, $C \in R^{m \times n}$, $C^b \in R^{m_b \times n}$ are the state, input, output and constrained output gain matrix, respectively.

Notably, the state space Equation (1) is just used to reveal the relationship between the state space equation and the subspace predictor. The state space matrices are not required for the design of the proposed data-driven predictive controller. We are only interested in identifying the subspace matrices.

Equation (1) is a three-input three-output system for the studied mode transition dynamic process. Its input is selected as $\mathbf{u}(k) = [T_e(k) \quad T_c(k) \quad T_m(k)]^T$, where T_e is the engine torque, T_c is the clutch torque, which is proportional to the normal pressure between two sides of the clutch and can be controlled by the displacement of clutch release bearing, and T_m is the motor torque. $\mathbf{y}(k) = \Delta\omega(k)$ is selected as the system output. The engine speed ω_e and motor speed ω_m constitute the constrained output $\mathbf{y}^b(k) = [\omega_e(k) \quad \omega_m(k)]^T$; thus, $l = 3, m = 1, m_b = 2$.

The open-loop data collection of the input, the output and the constrained output $\mathbf{u}(k)$, $\mathbf{y}(k)$ and $\mathbf{y}^b(k)$ for $k \in \{1, 2, 3, \ldots, 2i+j-1\}$ are collected through the Cruise simulation platform, whose details are shown in Section 3.2.

Next, the data block Hankel matrices $\mathbf{U}_p$, $\mathbf{U}_f$, $\mathbf{Y}_p$, $\mathbf{Y}_f$, $\mathbf{Y}_p^b$ and $\mathbf{Y}_f^b$ are constructed to identify the subspace matrices, which are used to design the data-driven predictive torque coordination controller [16].

The data block Hankel matrices $\mathbf{U}_p$ and $\mathbf{U}_f$ for $\mathbf{u}(k)$ are denoted by:

$$\mathbf{U}_p = \begin{bmatrix} \mathbf{u}(1) & \mathbf{u}(2) & \cdots & \mathbf{u}(j) \\ \mathbf{u}(2) & \mathbf{u}(3) & \cdots & \mathbf{u}(j+1) \\ \mathbf{u}(3) & \mathbf{u}(4) & \cdots & \mathbf{u}(j+2) \\ \vdots & \vdots & \ddots & \vdots \\ \mathbf{u}(i) & \mathbf{u}(i+1) & \cdots & \mathbf{u}(i+j-1) \end{bmatrix}, \tag{2}$$

$$\mathbf{U}_f = \begin{bmatrix} \mathbf{u}(i+1) & \mathbf{u}(i+2) & \cdots & \mathbf{u}(i+j) \\ \mathbf{u}(i+2) & \mathbf{u}(i+3) & \cdots & \mathbf{u}(i+j+1) \\ \mathbf{u}(i+3) & \mathbf{u}(i+4) & \cdots & \mathbf{u}(i+j+2) \\ \vdots & \vdots & \ddots & \vdots \\ \mathbf{u}(2i) & \mathbf{u}(2i+1) & \cdots & \mathbf{u}(2i+j-1) \end{bmatrix}. \tag{3}$$

The data block Hankel matrices $\mathbf{Y}_p$ and $\mathbf{Y}_f$ for $\mathbf{y}(k)$ are constructed as follows:

$$\mathbf{Y}_p = \begin{bmatrix} \mathbf{y}(1) & \mathbf{y}(2) & \cdots & \mathbf{y}(j) \\ \mathbf{y}(2) & \mathbf{y}(3) & \cdots & \mathbf{y}(j+1) \\ \mathbf{y}(3) & \mathbf{y}(4) & \cdots & \mathbf{y}(j+2) \\ \vdots & \vdots & \ddots & \vdots \\ \mathbf{y}(i) & \mathbf{y}(i+1) & \cdots & \mathbf{y}(i+j-1) \end{bmatrix}, \tag{4}$$

$$\mathbf{Y}_f = \begin{bmatrix} \mathbf{y}(i+1) & \mathbf{y}(i+2) & \cdots & \mathbf{y}(i+j) \\ \mathbf{y}(i+2) & \mathbf{y}(i+3) & \cdots & \mathbf{y}(i+j+1) \\ \mathbf{y}(i+3) & \mathbf{y}(i+4) & \cdots & \mathbf{y}(i+j+2) \\ \vdots & \vdots & \ddots & \vdots \\ \mathbf{y}(2i) & \mathbf{y}(2i+1) & \cdots & \mathbf{y}(2i+j-1) \end{bmatrix}. \tag{5}$$

where p and f denote the past and the future, respectively. The matrices above have i-block rows and j-block columns. The constrained output Hankel matrices $\mathbf{Y}_p^b$ and $\mathbf{Y}_f^b$ for $\mathbf{Y}^b(k)$ can be constructed in the same way. The past and future state sequences are defined as follows [16]:

$$\mathbf{X}_p = \begin{bmatrix} \mathbf{x}(1) & \mathbf{x}(2) & \mathbf{x}(3) & \cdots & \mathbf{x}(j) \end{bmatrix}, \tag{6}$$

$$\mathbf{X}_f = \begin{bmatrix} \mathbf{x}(i+1) & \mathbf{x}(i+2) & \mathbf{x}(i+3) & \cdots & \mathbf{x}(i+j) \end{bmatrix}. \tag{7}$$

where $i \le j$.

By recursive substitution of (1.a) and (1.b), the matrix output equations used in subspace identification are obtained as below:

$$\mathbf{Y}_p = \Gamma_i \mathbf{X}_p + H_i \mathbf{U}_p, \tag{8}$$

$$\mathbf{Y}_f = \Gamma_i \mathbf{X}_f + H_i \mathbf{U}_f, \tag{9}$$

where:

$$\Gamma_i = \begin{bmatrix} C \\ CA \\ CA^2 \\ \vdots \\ CA^{i-1} \end{bmatrix}, \tag{10}$$

$$H_i = \begin{bmatrix} 0 & 0 & \cdots & 0 & 0 \\ CB & 0 & \cdots & 0 & 0 \\ CAB & CB & \cdots & 0 & 0 \\ \vdots & \vdots & \ddots & \vdots & 0 \\ CA^{i-2}B & CA^{i-3}B & \cdots & CB & 0 \end{bmatrix}. \tag{11}$$

The following matrix equation can be reduced from the iteration of (1.a):

$$\mathbf{X}_f = A^i \mathbf{X}_p + \Delta_i^d \mathbf{U}_p, \tag{12}$$

where $\Delta_i^d = \begin{bmatrix} A^{i-1}B & A^{i-2}B & \cdots & B \end{bmatrix}$.

The following formula is obtained by solving the matrix Equation (8):

$$\mathbf{X}_p = \Gamma_i^\dagger (\mathbf{Y}_p - H_i \mathbf{U}_p), \tag{13}$$

where † represents the Moore-Penrose pseudo-inverse.

When (13) is substituted into (12), the following formula can be obtained:

$$\mathbf{X}_f = A^i \Gamma_i^\dagger \mathbf{Y}_p + (\Delta_i^d - A^i \Gamma_i^\dagger H_i) \mathbf{U}_p. \tag{14}$$

The following formula is derived when (14) is substituted into (9):

$$\mathbf{Y}_f = \Gamma_i A^i \Gamma_i^\dagger \mathbf{Y}_p + \Gamma_i (\Delta_i^d - A^i \Gamma_i^\dagger H_i) \mathbf{U}_p + H_i \mathbf{U}_f. \tag{15}$$

With enough measurement data, (15) can be written as the following optimal predictive output:

$$\hat{\mathbf{Y}}_f = L_w \mathbf{W}_p + L_u \mathbf{U}_f, \tag{16}$$

where:

$$\mathbf{W}_p = \begin{bmatrix} \mathbf{Y}_p \\ \mathbf{U}_p \end{bmatrix}. \tag{17}$$

Equation (16) is the subspace predictor equation. It shows that the future output can be predicted based on the past input-output data and the future input. $\mathbf{W}_p$ is the past input-output data matrix. $\mathbf{U}_f$ is the future input data matrix. L_w and L_u are the coefficient matrix of $\mathbf{W}_p$ and $\mathbf{U}_f$, respectively.

The least-square prediction $\hat{\mathbf{Y}}_f$ can be found by solving the least square problem:

$$\min_{L_w, L_u} \left\| \mathbf{Y}_f - \begin{pmatrix} L_w & L_u \end{pmatrix} \begin{pmatrix} \mathbf{W}_p \\ \mathbf{U}_f \end{pmatrix} \right\|^2.$$

$\hat{\mathbf{Y}}_f$ can be found by the orthogonal projection of the row space of $\mathbf{Y}_f$ into the row space spanned by $\mathbf{W}_p$ and $\mathbf{U}_f$:

$$\hat{\mathbf{Y}}_f = \mathbf{Y}_f \Big/ \begin{bmatrix} \mathbf{W}_p \\ \mathbf{U}_f \end{bmatrix}.$$

$$\begin{bmatrix} L_w & L_u \end{bmatrix} = \mathbf{Y}_f \begin{bmatrix} \mathbf{W}_p \\ \mathbf{U}_f \end{bmatrix}^{\dagger}$$

$$= \mathbf{Y}_f \begin{bmatrix} \mathbf{W}_p^T & \mathbf{U}_f^T \end{bmatrix} \left(\begin{bmatrix} \mathbf{W}_p \\ \mathbf{U}_f \end{bmatrix} \begin{bmatrix} \mathbf{W}_p^T & \mathbf{U}_f^T \end{bmatrix} \right)^{-1}. \tag{18}$$

The terms Γ_i^b, H_i^b, $\hat{\mathbf{Y}}_f^b$, $\mathbf{W}_p^b$, L_w^b, and L_u^b of constrained output $y^b(k)$ can be obtained in the same way as (10), (11) and (16)–(18).

3.2. Identification of Subspace Matrices

The prediction accuracy of the subspace matrix identification method is sensitive to different open-loop data. During the open-loop data collection phase, to get an accurate subspace predictor equation, we should design the input data to fully excite the system dynamics relevant to the control goal. That is to say, the designed input data should be as diverse as possible.

As shown in Figure 2a–c, the input data T_e, T_c and T_m, which can fully excite the system characteristics of the HEV powertrain, are constructed. The system output $\Delta\omega$ and constrained outputs ω_e, ω_m can be obtained after the input data act on the built HEV dynamics model, as shown in Figure 2d–f. The sample time T_s is chosen to be 0.01 s; the number of rows i in the input-output data matrix is chosen to be 20; and the number of columns j in the input-output data matrix is chosen to be 180. Thus, the open-loop data at 219 sampling instants are collected during the studied mode transition dynamic process.

Based on the sampling data as shown in Figure 2, the data block Hankel matrices $\mathbf{U}_p$, $\mathbf{U}_f$, $\mathbf{Y}_p$, $\mathbf{Y}_f$, $\mathbf{Y}_p^b$, $\mathbf{Y}_f^b$ of input $\mathbf{u}(k)$, output $\mathbf{y}(k)$ and constrained output $\mathbf{y}^b(k)$ can be constructed. Then, the subspace matrices L_w, L_u, L_w^b and L_u^b can be solved using the corresponding formulas.

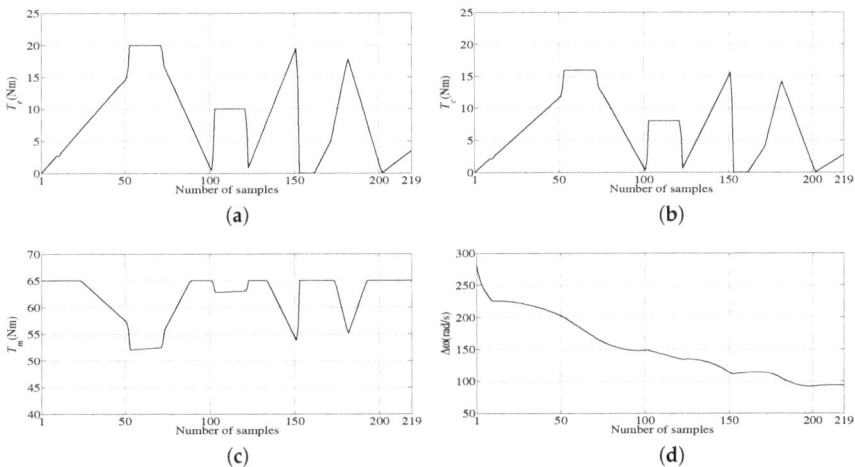

Figure 2. *Cont.*

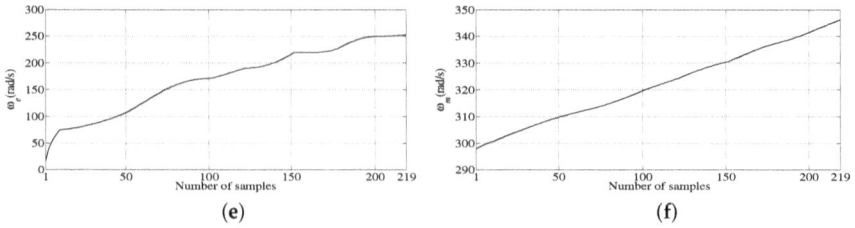

(e)

(f)

Figure 2. Input and output data of 219 sample points for identification: (**a**) Engine torque T_e; (**b**) Clutch torque T_c; (**c**) Motor torque T_m; (**d**) The clutch slip speed $\Delta\omega$; (**e**) Engine speed ω_e; (**f**) Motor speed ω_m.

3.3. Verification of the Subspace Predictor Equation

The input-output sampling data from the 220th–599th sampling instant are used to verify the effectiveness of the identified subspace predictor equation (i.e., whether the identified subspace predictor equation can well reflect the HEV's dynamic characteristics). The constructed fully-excited input data T_e, T_c, T_m and the specific identification prediction effect are shown in Figure 3.

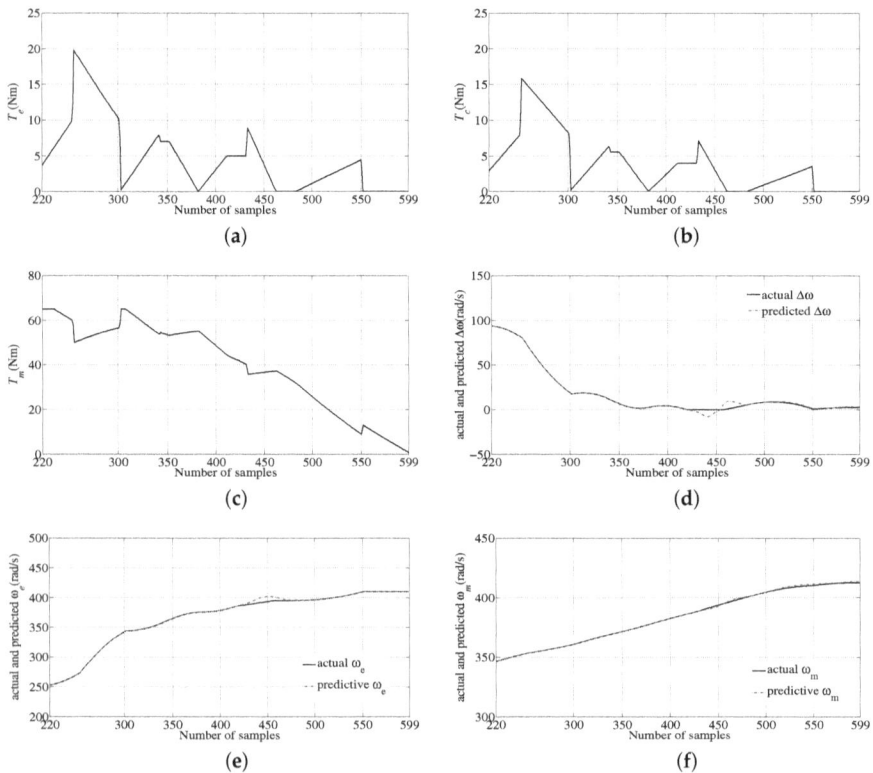

(a)

(b)

(c)

(d)

(e)

(f)

Figure 3. Input and output data for verification and identification results: (**a**) Engine torque T_e; (**b**) Clutch torque T_c; (**c**) Motor torque T_m; (**d**) The identified results of the clutch slip speed $\Delta\omega$; (**e**) The identified results of constrained engine speed ω_e; (**f**) The identified results of constrained motor speed ω_m.

As shown in Figure 3d–f, the predictive output can fit the actual output very well when $\Delta \omega$ is not equal to zero. Since the data used to identify the predictive model do not include the case that $\Delta \omega$ is equal to zero, a small predictive error appears when $\Delta \omega$ is equal to zero. The subject investigated in this paper is the clutch slipping phase during which $\Delta \omega$ is not equal to zero, so the predictive error appearing in Figure 3 will not influence the design of the proposed data-driven predictive controller.

3.4. Description of the Optimization Problem

The control input sequence increment $\Delta u_f(k)$ to be optimized at time k can be expressed as follows:

$$\Delta u_f(k) = \begin{bmatrix} \Delta u(k) \\ \Delta u(k+1) \\ \vdots \\ \Delta u(k+N_u-1) \end{bmatrix},$$

where:

$$\Delta u(k+q) = \begin{bmatrix} \Delta T_e(k+q) \\ \Delta T_c(k+q) \\ \Delta T_m(k+q) \end{bmatrix}, q = 0, 1, \ldots, N_u - 1.$$

The predictive control output sequence $\hat{y}_f(k+1)$ is defined as follows:

$$\hat{y}_f(k+1) = \begin{bmatrix} \hat{y}(k+1) \\ \hat{y}(k+2) \\ \vdots \\ \hat{y}(k+N_p) \end{bmatrix}.$$

N_p and N_u are used to represent the prediction horizon and the control horizon, respectively.

The cost function is usually defined as a quadratic form in order to make the predictive output as close as possible to a given reference output. A penalty term is added to the cost function to punish the violent changes of control variables. Therefore, considering the input and output constraints, the optimal control problem during the mode transition dynamic process from motor-only mode to compound driving mode can be described as below:

$$\min_{\Delta u_f(k)} J(y(k), \Delta u_f(k), N_p, N_u)$$

$$J = \left\| \Gamma_y(\hat{y}_f(k+1) - R_e(k+1)) \right\|^2 + \left\| \Gamma_u \Delta u_f(k) \right\|^2 \tag{19.a}$$

$$s.t. \ u_{min}(k+q) \leq u(k+q) \leq u_{max}(k+q), q = 0, \ldots, N_u - 1, \tag{19.b}$$

$$\Delta u_{min}(k+q) \leq \Delta u(k+q) \leq \Delta u_{max}(k+q), q = 0, \ldots, N_u - 1, \tag{19.c}$$

$$y_{bmin}(k+s) \leq y_b(k+s) \leq y_{bmax}(k+s), s = 1, \ldots, N_p, \tag{19.d}$$

where:

$$\Gamma_y = \begin{bmatrix} \tau_{y,1} & 0 & \cdots & 0 \\ 0 & \tau_{y,2} & \cdots & 0 \\ \vdots & \vdots & \ddots & \vdots \\ 0 & 0 & \cdots & \tau_{y,N_p} \end{bmatrix}, \Gamma_u = \begin{bmatrix} \tau_{u,1} & 0 & \cdots & 0 \\ 0 & \tau_{u,2} & \cdots & 0 \\ \vdots & \vdots & \ddots & \vdots \\ 0 & 0 & \cdots & \tau_{u,N_u} \end{bmatrix},$$

$$\mathbf{R}_e(k+1) = \begin{bmatrix} r(k+1) \\ r(k+2) \\ \vdots \\ r(k+N_p) \end{bmatrix} = \begin{bmatrix} \alpha \mathbf{y}(k) \\ \alpha^2 \mathbf{y}(k) \\ \vdots \\ \alpha^{N_p} \mathbf{y}(k) \end{bmatrix}.$$

In this cost function, the control objective of the first part is to force the clutch slip speed to converge to zero, which realizes the fast mode transition and reduces the wear and tear to the clutch. The control objective of the second part is to limit the torque changing rate of the engine, the clutch and the motor to ensure the mode transition comfortability. Obviously, these two control objectives are contradictory. The weighting factors Γ_y and Γ_u are given to trade off the two conflicting objectives. The bigger the weighting factor Γ_y, the faster the mode transition process. The larger the weighting factor Γ_u, the less the vehicle jerk.

$\mathbf{R}_e(k+1)$ is the reference sequence of $\Delta \omega$ within the prediction horizon, and $\alpha \in (0,1)$ is an adjustable parameter. The smaller α is, the faster the clutch slip speed reaches zero. Thus, the clutch can finish the engagement more quickly.

3.5. Predictive Output Equation

In order to deal with the optimal control problem (19.a), the predictive output equation will be deduced in this section based on the data-driven method and the predictive control method.

To guarantee regulation with zero steady-state error for the reference input, the subspace matrix incremental input-output expressions are as follows:

$$\Delta \hat{\mathbf{Y}}_f = L_w \Delta \mathbf{W}_p + L_u \Delta \mathbf{U}_f, \tag{20}$$

and:

$$\Delta \hat{y}_f(k) = L_w(1:mN_p,:) \begin{bmatrix} \Delta y_p \\ \Delta u_p \end{bmatrix} + L_u(1:mN_p, 1:lN_u) \Delta u_f(k). \tag{21}$$

where:

$$\Delta \mathbf{W}_p = \begin{bmatrix} \Delta \mathbf{Y}_p \\ \Delta \mathbf{U}_p \end{bmatrix}, \Delta y_p = \begin{bmatrix} \Delta \mathbf{y}(k-i+1) & \Delta \mathbf{y}(k-i+2) & \cdots & \Delta \mathbf{y}(k) \end{bmatrix}^T,$$

$$\Delta u_p = \begin{bmatrix} \Delta \mathbf{u}(k-i) & \Delta \mathbf{u}(k-i+1) & \cdots & \Delta \mathbf{u}(k-1) \end{bmatrix}^T.$$

Therefore, the optimal prediction of the future outputs can be derived as the following form:

$$\hat{y}_f(k+1|k) = y(k) + L_w^\Delta(1:mN_p,:) \begin{bmatrix} \Delta y_p \\ \Delta u_p \end{bmatrix} + S_{N_p,N_u} \Delta u_f(k)$$

$$= F + S_{N_p,N_u} \Delta u_f(k), \tag{22}$$

where S_{N_p,N_u} is the $mN_p \times lN_u$ dynamic matrix containing the step response coefficients and formed from L_u:

$$S_{N_p,N_u} = L_u(1:mN_p, 1:lN_u) \begin{bmatrix} I_{l \times l} & 0 & \cdots & 0 \\ I_{l \times l} & I_{l \times l} & \cdots & 0 \\ \vdots & \vdots & \ddots & \vdots \\ I_{l \times l} & I_{l \times l} & \cdots & I_{l \times l} \end{bmatrix} \tag{23}$$

$$y(k) = \begin{bmatrix} \mathbf{y}(k) & \mathbf{y}(k) & \cdots & \mathbf{y}(k) \end{bmatrix}^T \tag{24}$$

L_w^Δ is constructed from L_w, and **F** is the free response for the case of measured disturbances:

$$L_w^\Delta(k,:) = \sum_{i=1}^{k} L_w(i,:), 1 \le k \le mN_p \tag{25}$$

$$\mathbf{F} = \mathbf{y}(k) + L_w^\Delta(1:mN_p,:) \begin{bmatrix} \Delta y_p \\ \Delta u_p \end{bmatrix} \tag{26}$$

S_{N_p,N_u}^b, $y^b(k)$, $L_w^{b\Delta}$, $\mathbf{F}^b$ about $\hat{y}_f^b(k+1|k)$ can be calculated in the same way as (23)–(26).

3.6. Data-Driven Predictive Controller with Constraints

The DDPC problem (19.a) is usually converted into the following quadratic programming problem:

$$\min_{\Delta u_f(k)} \frac{1}{2} \Delta u_f(k)^T H \Delta u_f(k) + G(k+1|k)^T \Delta u_f(k), \tag{27}$$

$$s.t. \ C_u \Delta u_f(k) \le b.$$

First, the expression of H and G will be briefly inferred. When (21) is substituted into (19.a), the following formula can be obtained:

$$\begin{aligned} J &= \Delta u_f(k)^T (S_{N_p,N_u})^T \Gamma_y^T \Gamma_y S_{N_p,N_u} \Delta u_f(k) + \Delta u_f(k)^T \Gamma_u^T \Gamma_u \Delta u_f(k) \\ &\quad - 2E(k+1)^T \Gamma_y^T \Gamma_y S_{N_p,N_u} \Delta u_f(k) + E(k+1)^T \Gamma_y^T \Gamma_y E(k+1) \\ &= J' + E(k+1)^T \Gamma_y^T \Gamma_y E(k+1), \end{aligned} \tag{28}$$

where $E(k+1) = R_e(k+1) - \mathbf{F}$.

The last term in J is irrelevant to $\Delta u_f(k)$, so the cost function will be minimized if J' obtains the minimum.

$$J' = \frac{1}{2} \Delta u_f(k)^T H \Delta u_f(k) + G(k+1|k)^T \Delta u_f(k), \tag{29}$$

where:

$$H = 2((S_{N_p,N_u})^T \Gamma_y^T \Gamma_y S_{N_p,N_u} + \Gamma_u^T \Gamma_u),$$
$$G(k+1|k) = -2(S_{N_p,N_u})^T \Gamma_y^T \Gamma_y E(k+1).$$

Therefore, the optimal problem (19.a) is converted into the quadratic programming problem (27).

Then, the constraint condition matrix C_u and b of the quadratic programming problem (27) will be inferred from (19.b)–(19.d).

Let:

$$u_f(k) = u_f = \begin{bmatrix} \mathbf{u}(k) \\ \mathbf{u}(k+1) \\ \vdots \\ \mathbf{u}(k+N_u-1) \end{bmatrix}, u_{fmin}(k) = u_{fmin} = \begin{bmatrix} \mathbf{u}_{min}(k) \\ \mathbf{u}_{min}(k+1) \\ \vdots \\ \mathbf{u}_{min}(k+N_u-1) \end{bmatrix},$$

$$u_{fmax}(k) = u_{fmax} = \begin{bmatrix} \mathbf{u}_{max}(k) \\ \mathbf{u}_{max}(k+1) \\ \vdots \\ \mathbf{u}_{max}(k+N_u-1) \end{bmatrix}, \Delta u_f(k) = \Delta u_f = \begin{bmatrix} \Delta \mathbf{u}(k) \\ \Delta \mathbf{u}(k+1) \\ \vdots \\ \Delta \mathbf{u}(k+N_u-1) \end{bmatrix},$$

$$\Delta u_{fmin}(k) = \Delta u_{fmin} = \begin{bmatrix} \Delta \mathbf{u}_{min}(k) \\ \Delta \mathbf{u}_{min}(k+1) \\ \vdots \\ \Delta \mathbf{u}_{min}(k+N_u-1) \end{bmatrix}, \Delta u_{fmax}(k) = \Delta u_{fmax} = \begin{bmatrix} \Delta \mathbf{u}_{max}(k) \\ \Delta \mathbf{u}_{max}(k+1) \\ \vdots \\ \Delta \mathbf{u}_{max}(k+N_u-1) \end{bmatrix},$$

$$y^b = \hat{y}_f^b(k+1|k) = \begin{bmatrix} \mathbf{y}^b(k+1) \\ \mathbf{y}^b(k+2) \\ \vdots \\ \mathbf{y}^b(k+N_p) \end{bmatrix}, \tag{30}$$

$$y_{min}^b = \begin{bmatrix} \mathbf{y}_{min}^b(k+1) \\ \mathbf{y}_{min}^b(k+2) \\ \vdots \\ \mathbf{y}_{min}^b(k+N_p) \end{bmatrix}, y_{max}^b = \begin{bmatrix} \mathbf{y}_{max}^b(k+1) \\ \mathbf{y}_{max}^b(k+2) \\ \vdots \\ \mathbf{y}_{max}^b(k+N_p) \end{bmatrix},$$

the constraint conditions (19.b)–(19.d) change to the following form:

$$u_{fmin} \le u_f \le u_{fmax}, \tag{31.a}$$

$$\Delta u_{fmin} \le \Delta u_f \le \Delta u_{fmax}, \tag{31.b}$$

$$y_{min}^b \le y^b \le y_{max}^b. \tag{31.c}$$

For the constraint condition (31.a),

$$u_f = C_1 u(k-1) + C_2 \Delta u_f,$$

where:

$$C_1 = \begin{bmatrix} I_{3\times3} \\ I_{3\times3} \\ I_{3\times3} \\ I_{3\times3} \\ I_{3\times3} \end{bmatrix}, C_2 = \begin{bmatrix} I_{3\times3} & 0 & 0 & 0 & 0 \\ I_{3\times3} & I_{3\times3} & 0 & 0 & 0 \\ I_{3\times3} & I_{3\times3} & I_{3\times3} & 0 & 0 \\ I_{3\times3} & I_{3\times3} & I_{3\times3} & I_{3\times3} & 0 \\ I_{3\times3} & I_{3\times3} & I_{3\times3} & I_{3\times3} & I_{3\times3} \end{bmatrix}.$$

Therefore, (31.a) can be changed to the following form:

$$\begin{bmatrix} C_2 \\ -C_2 \end{bmatrix} \Delta u_f \le \begin{bmatrix} u_{fmax} - C_1 u(k-1) \\ -u_{fmin} + C_1 u(k-1) \end{bmatrix}.$$

Constraint Condition (31.b) can be changed to the following form:

$$\begin{bmatrix} I_{3N_u \times 3N_u} \\ -I_{3N_u \times 3N_u} \end{bmatrix} \Delta u_f \le \begin{bmatrix} \Delta u_{fmax} \\ -\Delta u_{fmin} \end{bmatrix}.$$

Constraint Condition (31.c) can be changed to the following form from (22) and (30):

$$\begin{bmatrix} S^b_{N_p,N_u} \\ -S^b_{N_p,N_u} \end{bmatrix} \Delta u_f(k) \leq \begin{bmatrix} y^b_{max} - \mathbf{F}^b \\ -y^b_{min} + \mathbf{F}^b \end{bmatrix}.$$

Overall, the constraint condition matrices can be derived as the following form:

$$C_u = \begin{bmatrix} C_2 \\ -C_2 \\ I_{3N_u \times 3N_u} \\ -I_{3N_u \times 3N_u} \\ S^b_{N_p,N_u} \\ -S^b_{N_p,N_u} \end{bmatrix}, b = \begin{bmatrix} u_{fmax} - C_1 u(k-1) \\ -u_{fmin} + C_1 u(k-1) \\ \Delta u_{fmax} \\ -\Delta u_{fmin} \\ y^b_{max} - \mathbf{F}^b \\ -y^b_{min} + \mathbf{F}^b \end{bmatrix}.$$

According to the receding horizon principle of predictive control, only the first element of $\Delta u_f(k)$ is implemented, and the calculation is repeated at each time instant. Hence, at the k-th sampling instant, the control law is given as follows:

$$\Delta u(k) = \begin{bmatrix} I_{3\times3} & 0_{3\times3} & 0_{3\times3} & 0_{3\times3} & 0_{3\times3} \end{bmatrix} \Delta u_f(k).$$

4. Results and Discussion

The co-simulation of Cruise and MATLAB/Simulink R2010a is conducted under the urban driving cycle (UDC), which is shown in Figure 4a. The simulation time step is equal to 0.01 s. The maximum torque and maximum speed of the engine, clutch and motor are shown in Table 1. N_p is chosen to be 20, and N_u is chosen to be five. $\varepsilon = 30\text{rad/s}$; $\omega_{emin} = 720$ r/min; $\omega_{mmin} = 0$ r/min; $\alpha = 0.8$; $\tau_{y,1} = \tau_{y,2} = \cdots = \tau_{y,20} = 0.8$; and $\tau_{u,1} = \tau_{u,2} = \cdots = \tau_{u,5} = \begin{bmatrix} 4 & 0 & 0 \\ 0 & 1 & 0 \\ 0 & 0 & 2 \end{bmatrix}$. To improve the transition comfortability, the variation ranges of control increments are as follows: -50 Nm/s$\leq \Delta T_e \leq 50$ Nm/s, -50 Nm/s$\leq \Delta T_c \leq 50$ Nm/s and -100 Nm/s $\leq \Delta T_m \leq 100$ Nm/s.

4.1. Data-Driven Predictive Control Method

Figure 4b–f show the Cruise simulation results of the proposed data-driven predictive torque coordination control strategy considering the constraints. At 60.44 s, the vehicle speed reaches 27 km/h; the mode transition from motor-only mode to parallel hybrid operation begins; the engine is started with a constant torque 60 Nm; and the two sides of the clutch are still separated. At 61.22 s, the engine speed rises very close to the motor speed (the clutch slip speed reaches 30 rad/s), and the clutch enters into the slipping phase. At about 61.49 s, the clutch slip speed reaches zero, then the clutch torque increases with a predetermined pattern to lock up reliably according to its characteristics. It can be seen that good transition quality can be achieved. The amplitude of vehicle jerk is 9.96 m/s^3, which avoids the discomfort of the passengers during the mode transition process. The mode transition duration is about 1.06 s, which can well meet the driver's transition demand. The slipping duration of the clutch is only 0.27 s, and the total friction work is only about 8.59 J. This helps to prolong the clutch life.

The control effect of different switch triggers ε on the vehicle jerk, mode transition duration, clutch slipping duration and clutch frictional losses is shown in Table 2. Additionally, the vehicle jerk of the DDPC method under different ε is shown in Figure 5. It can be seen that a small switch trigger implies less usage of clutch torque. The smaller the switch trigger, the smaller the vehicle jerk and the less the clutch frictional losses. However, the smaller the switch trigger, the longer the mode transition duration. Taking sensor inaccuracy and actuator delay into account, it is hard to achieve small jerk with a too small ε [12], so ε is selected to be 30 rad/s.

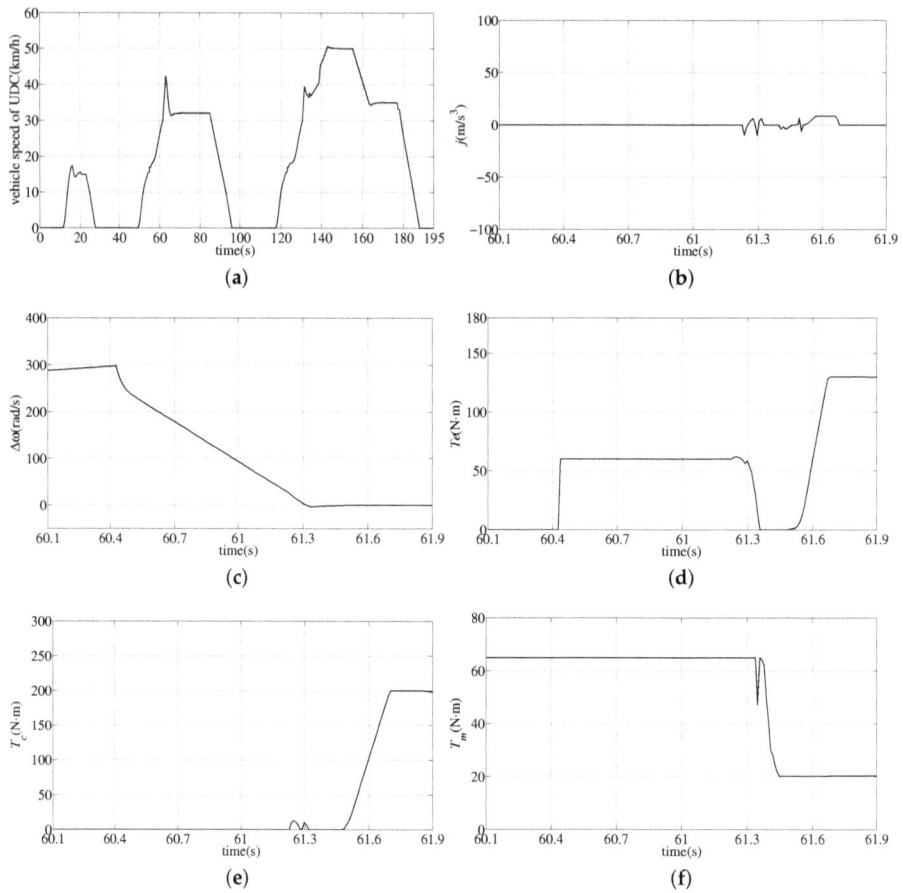

Figure 4. Cruise simulation results of the data-driven predictive controller considering the constraints: (a) Urban driving cycle (UDC); (b) Jerk; (c) The clutch slip speed $\Delta\omega$; (d) The engine torque T_e; (e) The clutch torque T_c; (f) The motor torque T_m.

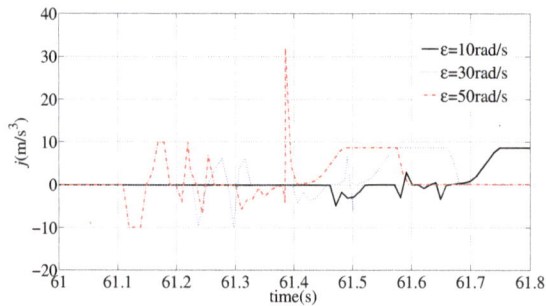

Figure 5. Vehicle jerk of data-driven predictive controller considering the constraints under different ε.

Table 2. Cruise simulation results of data-driven predictive controller under different speed difference thresholds.

ε	Jerk Range	Transition Duration	Clutch Slipping Duration	Clutch Frictional Losses
10 rad/s	-4.8–2.96 m/s^3	1.18 s	0.32 s	0.002 J
30 rad/s	-9.96–6.76 m/s^3	1.06 s	0.29 s	8.59 J
50 rad/s	-9.96–32.1 m/s^3	1.01 s	0.27 s	42.92 J

4.2. Comparison with the Model Predictive Control Method

In this subsection, the torque coordination control effect of the DDPC method is compared with that of the MPC method to further illustrate the advantages of the proposed mode transition strategy.

To get the model predictive torque coordination controller, the built HEV vehicle model as shown in Figure 1 can be simplified to the structure in Figure 6.

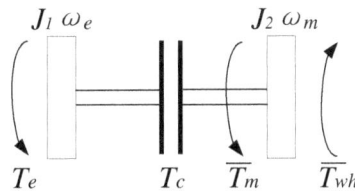

Figure 6. Simplified model of the HEV drive-line.

The equivalent inertia moment of the clutch input side $J_1 = J_e$, where J_e is the inertia moment of the engine.

The equivalent inertia moment of the clutch output side can be expressed as:

$$J_2 = J_c + J_m + J_{GB} + \frac{J_{SRT}}{i_{GB}^2} + \frac{mR^2}{i_{GB}^2 i_{SRT}^2} + \frac{J_D}{i_{GB}^2 i_{SRT}^2},$$

where J_c, J_m, J_{GB}, J_{SRT} and J_D are the inertia moments of the clutch, motor, gearbox, single ratio transmission (SRT) and differential, respectively.

The vehicle resistance torque can be expressed as:

$$T_{wh} = [mg\sin\alpha + f_r mg\cos\alpha + \frac{\rho}{2}c_D A_v(\omega_{wh}R)^2]R.$$

For the selected UDC, the road inclination angle is equal to zero; thus:

$$T_{wh} = [f_r mg + \frac{\rho}{2}c_D A_v(\omega_{wh}R)^2]R.$$

In the existing torque coordination control literature using the MPC method [10,11], for the convenience of the model predictive controller design, the contributions of the quadratic term of ω_{wh}^2 to T_{wh} are all considered small since the development of this powertrain model is built for the clutch engagement at low vehicle speeds (below 27 km/h). Thus, without significant loss of accuracy, the resistance torque T_{wh} is simplified as $T_{wh} = f_r mg$.

The equivalent vehicle resistance torque can be expressed as:

$$\overline{T}_{wh} = \frac{T_{wh}}{i_{GB}i_{SRT}} = \frac{T_{wh}}{i}.$$

The state variable, control vector, output and constrained output are defined as:

$$\mathbf{x} = \begin{bmatrix} w_e \\ \Delta w \end{bmatrix}, \mathbf{u} = \begin{bmatrix} T_e \\ T_c \\ T_m \end{bmatrix}, \mathbf{y}_c = \Delta w, \mathbf{y}_b = w_e.$$

Therefore, the dynamic equation in the mode transition of the HEV drive-line can be written as:

$$\dot{\mathbf{x}} = A_c\mathbf{x} + B_{cu}\mathbf{u} + B_{cd}d$$
$$\mathbf{y}_c = C_c\mathbf{x} \tag{32}$$
$$\mathbf{y}_b = C_b\mathbf{x}$$

where:

$$A_c = \begin{bmatrix} -\frac{b_e}{J_1} & 0 \\ \frac{b_e}{J_1} - \frac{b_m}{J_2} & -\frac{b_m}{J_2} \end{bmatrix}, B_{cu} = \begin{bmatrix} \frac{1}{J_1} & -\frac{1}{J_1} & 0 \\ -\frac{1}{J_1} & \frac{1}{J_1}+\frac{1}{J_2} & \frac{1}{J_2} \end{bmatrix}, B_{cd} = \begin{bmatrix} 0 \\ -\frac{1}{J_{2i}} \end{bmatrix},$$

$$d = f_r mg, C_c = \begin{bmatrix} 0 & 1 \end{bmatrix}, C_b = \begin{bmatrix} 1 & 0 \end{bmatrix}.$$

Then, applying zero-order hold discretization with a sampling period T_s to (32) gives the discrete state-space model:

$$\mathbf{x}(k+1) = A\mathbf{x}(k) + B_u\mathbf{u}(k) + B_d d(k)$$
$$\mathbf{y}_c(k) = C_c\mathbf{x}(k) \tag{33}$$
$$\mathbf{y}_b(k) = C_b\mathbf{x}(k)$$

where $A = e^{A_c T_s}$, $B_u = (\int_0^{T_s} e^{A_c\tau}d\tau)B_{cu}$.

The increment form of (33) is as follows:

$$\Delta\mathbf{x}(k+1) = A\Delta\mathbf{x}(k) + B_u\Delta\mathbf{u}(k)$$
$$\mathbf{y}_c(k) = C_c\Delta\mathbf{x}(k) + \mathbf{y}_c(k-1) \tag{34}$$
$$\mathbf{y}_b(k) = C_b\Delta\mathbf{x}(k) + \mathbf{y}_b(k-1)$$

The predictive output at the k-th instant is defined as:

$$\mathbf{y}_c(k+1|k) = \begin{bmatrix} \mathbf{y}_c(k+1|k) \\ \mathbf{y}_c(k+2|k) \\ \vdots \\ \mathbf{y}_c(k+20|k) \end{bmatrix} = \begin{bmatrix} \Delta w(k+1|k) \\ \Delta w(k+2|k) \\ \vdots \\ \Delta w(k+20|k) \end{bmatrix}.$$

The optimal control input sequence is defined as:

$$\Delta \mathbf{u}_f(k) = \begin{bmatrix} \Delta\mathbf{u}(k) \\ \Delta\mathbf{u}(k+1|k) \\ \Delta\mathbf{u}(k+2|k) \\ \Delta\mathbf{u}(k+3|k) \\ \Delta\mathbf{u}(k+4|k) \end{bmatrix},$$

where:

$$\Delta\mathbf{u}(k+i|k) = \begin{bmatrix} \Delta T_e(k+i) \\ \Delta T_c(k+i) \\ \Delta T_m(k+i) \end{bmatrix}, i = 0, 1, 2, 3, 4.$$

The iteration form of (34) is:

$$y_c(k+1|k) = S_{c,x}\Delta x(k) + I_c y_c(k) + S_{c,u}\Delta u_f(k),$$

where:

$$S_{c,x} = \begin{bmatrix} C_c A & 0 & \cdots & 0 \\ C_c A & C_c A & \cdots & 0 \\ \vdots & \vdots & \ddots & \vdots \\ C_c A & C_c A & \cdots & C_c A \end{bmatrix}, \Delta x(k) = \begin{bmatrix} \Delta x(k) \\ \Delta x(k+1) \\ \vdots \\ \Delta x(k+20) \end{bmatrix}, I_c = I_{20\times1},$$

$$S_{c,u} = \begin{bmatrix} C_c B_u & 0 & 0 & 0 & 0 \\ C_c B_u & C_c B_u & 0 & 0 & 0 \\ C_c B_u & C_c B_u & C_c B_u & 0 & 0 \\ C_c B_u & C_c B_u & C_c B_u & C_c B_u & 0 \\ C_c B_u & C_c B_u & C_c B_u & C_c B_u & C_c B_u \\ \vdots & \vdots & \vdots & \vdots & \vdots \\ C_c B_u & C_c B_u & C_c B_u & C_c B_u & C_c B_u \end{bmatrix}.$$

The cost function of the model predictive controller is selected the same as that of the data-driven predictive controller, which is shown in (19.a). The constrained model predictive control law can also be solved as a quadratic programming problem, where $H = 2(S_{c,u}^T \Gamma_y^T \Gamma_y S_{c,u} + \Gamma_u^T \Gamma_u)$, $G = -2S_{c,u}^T \Gamma_y^T \Gamma_y (R_e(k+1) - S_{c,x}\Delta x(k) - I_c y_c(k))$.

The results of the proposed DDPC method and the traditional MPC method are shown in Figure 7. To visually demonstrate the advantages of DDPC, the detailed data that can indicate the transition quality are shown in Table 3. As shown in the table, compared with the MPC method, the DDPC method can achieve smaller jerk, shorter mode transition duration, shorter clutch slipping duration and less clutch wear and tear under the same simulation conditions.

Table 3. Cruise simulation results compare between the DDPC method and the MPC method.

Indices	DDPC	MPC
Jerk Fluctuation Range	−9.96–6.76 m/s³	−82.47–87.92 m/s³
Mode Transition Duration	1.05 s	1.24 s
Clutch Slipping Duration	0.27 s	0.5 s
Clutch Frictional Losses	8.59 J	19.3 J

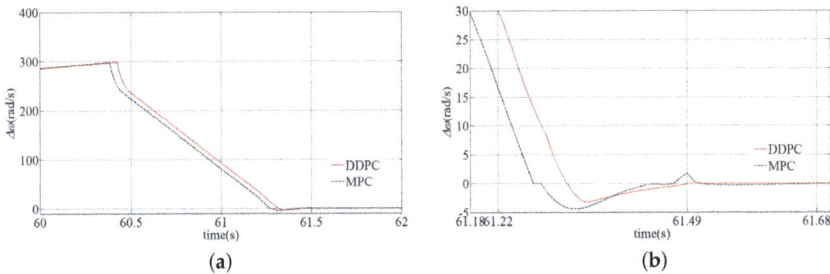

(a)

(b)

Figure 7. *Cont.*

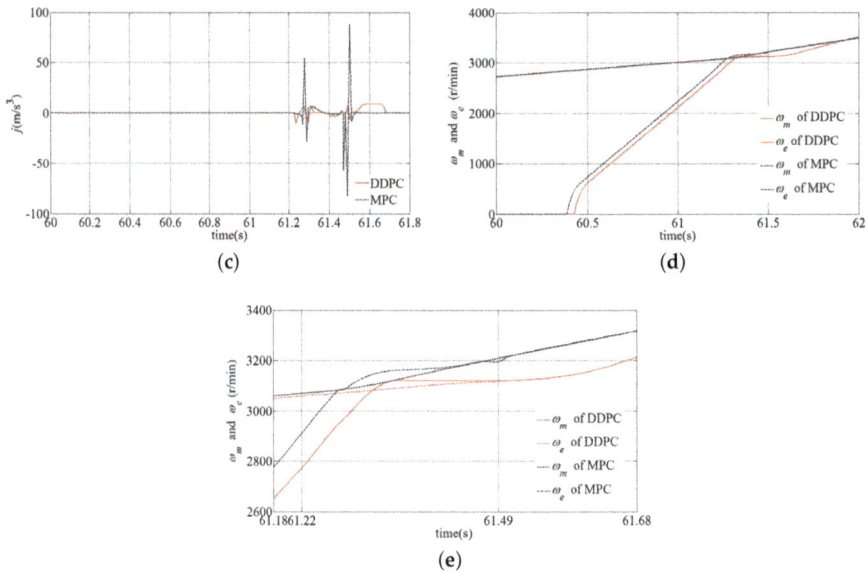

Figure 7. Comparison between the DDPC method and the MPC method for the torque coordination control problem during the mode transition dynamic process from the motor-only mode to compound driving mode: (**a**) The clutch slip speed $\Delta\omega$; (**b**) Partially enlarged view of (**a**); (**c**) Jerk; (**d**) The motor speed ω_m and the engine speed ω_e; (**e**) Partially enlarged view of (**d**).

5. Conclusions

A new torque coordination control strategy based on the DDPC method has been proposed to solve the torque coordination problem during the HEV mode transition dynamic process to improve the mode transition quality. The conflicting control objectives of the mode transition, which are small jerk and short transition duration, have been simultaneously considered in the optimal objective function by tracking a properly-selected output reference sequence and limiting the change rates of the actuators. Cruise simulation results have validated the effectiveness of the proposed DDPC method. The results compared with the MPC have further shown that the DDPC can achieve higher mode transition quality, which contributes to the improvement of the riding comfortability and the economy of HEV greatly.

In this paper, the mode transition dynamic process from the motor-only mode to the compound driving mode has been taken as an example of the proposed DDPC method. As time goes on, the system characteristics and the component parameters may change with the long-term aging and the diverse driving conditions, so the predictive model should be updated using the real vehicle operation data, which are newly collected online so that the performance of the proposed strategy can be guaranteed steadily in the long term. The performance monitoring method of the data-driven predictive controller will be our future work.

Acknowledgments: This work was supported by the National Natural Science Foundation of China (Grant Numbers 61403236, 51277116, 61304130, 61304033 and 61573218) and the Doctoral Scientific Research Foundation of Shandong Technology and Business University (Grant Number BS201511).

Author Contributions: Jing Sun carried out the main research tasks and wrote the full manuscript. Guojing Xing provided important suggestions on the writing of the paper. Chenghui Zhang polished the paper.

Conflicts of Interest: The authors declare no conflicts of interest.

References

1. Cairano, S.D.; Bernardini, D.; Bemporad, A.; Kolmanovsky, I.V. Stochastic MPC with learning for driver-predictive vehicle control and its application to HEV energy management. *IEEE Trans. Control Syst. Technol.* **2014**, *3*, 1018–1031.
2. Larsson, V.; Johannesson, L.; Egardt, B. Analytic solutions to the dynamic programming sub-problem in hybrid vehicle energy management. *IEEE Trans. Veh. Technol.* **2015**, *4*, 1458–1467.
3. Chen, Z.Y.; Xiong, R.; Wang, K.Y.; Jiao, B. Optimal energy management strategy of a plug-in hybrid electric vehicle based on a particle swarm optimization algorithm. *Energies.* **2015**, *8*, 3661–3678.
4. Tong, Y. Real-time simulation and research on control algorithm of parallel hybrid electric vehicle. *Chin. J. Mech. Eng.* **2003**, *10*, 156–161.
5. Davis, R.I.; Lorenz, R.D. Engine torque ripple cancellation with an integrated starter alternator in a hybrid electric vehicle: Implementation and control. *IEEE Trans. Ind. Appl.* **2003**, *6*, 1765–1774.
6. Li, M.H.; Luo, Y.G.; Yang, D.G.; Li, K.Q.; Lian, X.M. A dynamic coordinated control method for parallel hybrid electric vehicle based on model matching control. *Automot. Eng.* **2007**, *3*, 203–207.
7. Chiang, C.J.; Chen, Y.C.; Lin, C.Y. Fuzzy sliding mode control for smooth mode changes of a parallel hybrid electric vehicle. In Proceedings of the 2014 IEEE 11th International Conference on Control and Automation, Taichung, Taiwan, 18–20 June 2014; pp. 1072–1077.
8. Koprubasi, K.; Westervelt, E.R.; Rizzoni, G. Toward the systematic design of controllers for smooth hybrid electric vehicle mode changes. In Proceedings of the 2007 American Control Conference, New York, NY, USA, 11–13 July 2007; pp. 2985–2990.
9. Minh, V.T.; Rashid, A.A. Modeling and model predictive control for hybrid electric vehicles. *Int. J. Autom. Technol.* **2012**, *3*, 477–485.
10. Beck, R.; Saenger, S.; Richert, F.; Bollig, A.; Scholt, T.; Noreikat, K.E.; Abel, D. Model predictive control of a parallel hybrid vehicle drivetrain. In Proceedings of the 2005 44th IEEE Conference on Decision and Control, and the European Control Conference, Seville, Spain, 12–15 December 2005; pp. 2670–2675.
11. Chen, L. Torque coordination control during mode transition for a series-parallel hybrid electric vehicle. *IEEE Trans. Veh. Technol.* **2012**, *7*, 2936–2949.
12. Sun, J.; Xing, G.J.; Liu, X.D.; Fu, X.L.; Zhang, C.H. A novel torque coordination control strategy of a single-shaft parallel hybrid electric vehicle based on model predictive control. *Math. Probl. Eng.* **2015**, *1*, 1–12.
13. Guo, L.; Ge, A.; Zhang, T.; Yue, Y. AMT shift process control. *Trans. Chin. Soc. Agric. Mach.* **2003**, *2*, 1–3.
14. Xiong, R.; Sun, F.C.; Chen, Z; He, H.W. A data-driven multi-scale extended Kalman filtering based parameter and state estimation approach of lithium-ion polymer battery in electric vehicles. *Appl. Energy* **2014**, *1*, 463–476.
15. Sun, F.C.; Xiong, R.; He, H.W. A systematic state-of-charge estimation framework for multi-cell battery pack in electric vehicles using bias correction technique. *Appl. Energy* **2016**, *1*, 1399–1409.
16. Kadali, R.; Huang, B.; Rossiter, A. A data driven subspace approach to predictive controller design. *Control Eng. Pract.* **2003**, *3*, 261–278.
17. Favoreel, W.; Moor, B.D. SPC: Subspace predictive control. In Proceedings of the 1999 14th World Congress, Beijing, China, 5–9 July 1999; pp. 1–11.
18. Gao, C.H.; Jian, L.; Liu, X.Y.; Chen, J.M.; Sun, Y.X. Data-driven modeling based on volterra series for multidimensional blast furnace system. *IEEE Trans. Neural Netw.* **2011**, *12*, 2272–2283.
19. Kusiak, A.; Song, Z.; Zheng, H.Y. Anticipatory control of wind turbines with data-driven predictive models. *IEEE Trans. Energy Convers.* **2009**, *3*, 766–774.
20. Gil, P.; Henriques, J.; Cardoso, A.; Carvalho, P.; Dourado, A. Affine neural network-based predictive control applied to a distributed solar collector field. *IEEE Trans. Control Syst. Technol.* **2014**, *2*, 585–596.
21. Ge, S.S.; Li, Z.J.; Yang, H.Y. Data driven adaptive predictive control for holonomic constrained under-actuated biped robots. *IEEE Trans. Control Syst. Technol.* **2012**, *3*, 787–795.
22. Wahab, N.A.; Katebi, R.; Balderud, J.; Rahmat, M.F. Data-driven adaptive model-based predictive control with application in wastewater systems. *IET Control Theory Appl.* **2011**, *6*, 803–812.
23. Yin, X.H.; Li, S.Y.; Wu, J.; Li, N.; Cai, W.J.; Li, K. Data-driven based predictive controller design for vapor compression refrigeration cycle systems. In Proceedings of the 2013 9th Asian Control Conference, Istanbul, Turkey, 23–26 June 2013; pp. 1–6.

24. Lu, X.H.; Chen, H.; Wang, P.; Gao, B.Z. Design of a data-driven predictive controller for start-up process of AMT vehicles. *IEEE Trans. Neural Netw.* **2011**, *12*, 2201–2212.

25. Lu, X.H.; Chen, H.; Gao, B.Z.; Zhang, Z.W.; Jin, W.W. Data-driven predictive gearshift control for dual-clutch transmissions and FPGA implementation. *IEEE Trans. Ind. Electron.* **2015**, *1*, 599–610.

26. Zhong, L.S.; Yan, Z.; Yang, H.; Qi, Y.P.; Zhang, K.P.; Fan, X.P. Predictive control of high-speed train based on data driven subspace approach. *J. China Railw. Soc.* **2013**, *4*, 77–83.

energies

MDPI

Article

Optimal Energy Management Strategy for a Plug-in Hybrid Electric Vehicle Based on Road Grade Information

Yonggang Liu [1,2,*], Jie Li [1], Ming Ye [2], Datong Qin [1], Yi Zhang [3] and Zhenzhen Lei [1]

[1] State Key Laboratory of Mechanical Transmissions & School of Automotive Engineering, Chongqing University, Chongqing 400044, China; lijiecqu2015@163.com (J.L.); dtqin@cqu.edu.cn (D.Q.); zhenlei@umich.edu (Z.L.)

[2] Key Laboratory of Advanced Manufacture Technology for Automobile Parts, Ministry of Education, Chongqing University of Technology, Chongqing 400054, China; cqyeming@cqut.edu.cn

[3] Department of Mechanical Engineering, University of Michigan-Dearborn, Dearborn, MI 48128, USA; anding@umich.edu

* Correspondence: andylyg@umich.edu; Tel.: +86-23-65106249

Academic Editor: Hailong Li
Received: 9 January 2017; Accepted: 18 March 2017; Published: 23 March 2017

Abstract: Energy management strategies (EMSs) are critical for the improvement of fuel economy of plug-in hybrid electric vehicles (PHEVs). However, conventional EMSs hardly consider the influence of uphill terrain on the fuel economy and battery life, leaving vehicles with insufficient battery power for continuous uphill terrains. Hence, in this study, an optimal control strategy for a PHEV based on the road grade information is proposed. The target state of charge (*SOC*) is estimated based on the road grade information as well as the predicted driving cycle on uphill road obtained from the GPS/GIS system. Furthermore, the trajectory of the *SOC* is preplanned to ensure sufficient electricity for the uphill terrain in the charge depleting (CD) and charge sustaining (CS) modes. The genetic algorithm is applied to optimize the parameters of the control strategy to maintain the *SOC* of battery in the CD mode. The pre-charge mode is designed to charge the battery in the CS mode from a reasonable distance before the uphill terrain. Finally, the simulation model of the powertrain system for the PHEV is established using MATLAB/Simulink platform. The results show that the proposed control strategy based on road-grade information helps successfully achieve better fuel economy and longer battery life.

Keywords: plug-in hybrid electric vehicles; energy management strategy; road grade; state of charge

1. Introduction

Plug-in hybrid electric vehicles (PHEVs) achieve a longer all-electric range (AER) with a higher battery capacity compared to conventional hybrid electric vehicles (HEVs). Hence, the PHEV has improved fuel economy, as it replaces more fossil fuel with cheaper grid electricity [1]. The fuel economy and power battery are two important PHEV research fields.

Energy management strategies (EMSs) and driving conditions (such as road conditions, traffic conditions, and weather conditions) strongly influence the fuel economy of PHEVs. Hence, the EMSs of conventional HEVs, which aim at optimizing the power split rate between the engine and motor, such as the rules-based control strategy [2], the equivalent-consumption minimization strategy [3], and the EMS based on the driving-pattern recognition [4,5], have been extensively analyzed in previous studies. Furthermore, the vehicles can obtain considerable information of driving conditions, such as traffic lights, traffic congestion, and road grade, with the development of intelligent vehicle technologies. In previous studies, the terrain and trip distance were used to optimize the EMS of an HEV [6,7].

For the power battery of a PHEV, relevant studies focus on the battery state estimation and charging schedule designs. Several estimation approaches have been employed to estimate the accurate *SOC* of batteries, such as estimation approaches based on data-driven multi-scale extended Kalman filtering [8], battery *SOC* estimators using a bias correction technique [9] and an adaptive H infinity filter method [10]. Furthermore, optimization algorithms are employed for the charging schedule design; the objectives include load following/stabilization [11], battery health [12], etc.

Different from the EMSs of conventional HEVs, the EMSs for PHEVs are divided into two categories: the charge depleting–charge sustaining (CD–CS) control strategy and blended control strategy. For the CD–CS strategy, PHEVs can operate in either the CD mode or the CS mode. In the CD mode, the motor is the primary power source, consuming considerably cheaper and cleaner grid electricity. The PHEV switches to the CS mode to avoid excessive battery discharge when the state of charge (*SOC*) of the battery reaches the minimum boundary. In the CS mode, the PHEV is driven by both the engine and the motor to maintain the *SOC* near a specific value like conventional HEVs [13]. Figure 1 shows the trajectory of the *SOC* for the CD–CS control strategy. The advantage of the CD–CS strategy is that it has a simple control rule to achieve real-time control. However, the optimal fuel economy cannot be achieved when the range is not within the AER, which is validated [14]. A blended control strategy is developed based on the assumption that the entire driving cycle is known. The optimization algorithms, such as dynamic programming [15] and Pontryagin's Minimum Principle [16], are utilized to achieve global optimization in the blended strategy. Prof. Xiong in Ref. [15] present a procedure for the design of a near-optimal power management strategy for the hybrid battery and ultracapacitor energy storage system in a plug-in hybrid electric vehicle considering battery durability and longevity performance, and this approach shows excellent performance against uncertain diving cycles and battery packs. In the blended control strategy, electricity and fossil fuel are blended for the consumption at the beginning of the range, and the minimum boundary of the *SOC* of the battery is reached until the end of range. However, the future driving cycle must be provided accurately. In recent studies, the future driving cycle was predicted using the Markov chain model [17] and the neural network mode [18] using historical traffic data. In addition, the future driving cycle can be predicted using the traffic information obtained from intelligent transportation systems [19,20]. However, an accurate prediction of the driving cycle usually requires considerable traffic data and involves complex calculations. For the blended strategy, the tradeoff between the accuracy of driving-cycle prediction and computational complexity needs further study.

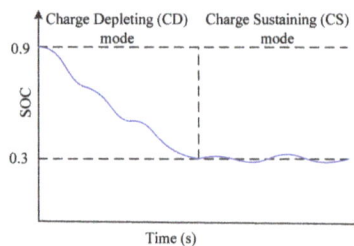

Figure 1. The trajectory of the *SOC* in the CD–CS control strategy.

Furthermore, for the real-time control strategies of PHEVs, a good vehicle performance, in terms of fuel economy, power, and battery life, cannot be ensured, if the driving conditions are not considered. The slope is proven to be an important factor that influences the fuel consumption of the HEV [21,22]. PHEVs usually operate in the hybrid-driving mode while driving uphill, and the engine and motor coordinately work together to ensure that the engine operates in the high-efficiency region. However, the *SOC* of the battery tends to reach the minimum boundary. Moreover, increasing the depth of discharge of the *SOC* will degrade the battery life [23]. Hence, the PHEV is driven only by the engine to avoid degrading the battery life when the *SOC* is lower than the minimum boundary. On the other hand, when the vehicle is driven only by the engine, both the power of the vehicle as well as the fuel economy are affected, because the engine cannot be ensured to operate in the high-efficiency region. Currently, to solve this problem, there are two major solutions: (1) maintaining the minimum boundary of the *SOC* at a higher level in mountainous terrain, such as 50%, to avoid excessive battery discharge during the hybrid driving phase in the CS mode [24]. However, this solution requires more battery capacity to satisfy the requirement of the AER, which will increase the cost of the vehicle; (2) temporarily extending the minimum boundary of the *SOC* to meet the requirement of excessive electricity during the hybrid driving phase in the CS mode [25]. However, to restore the power of the battery, it needs to be continuously charged by the engine after driving through the uphill terrain, which increases the fuel consumption. Moreover, the increased depth of discharge of the *SOC* will reduce the battery life. Hence, it is necessary to develop an optimal EMS that can improve the fuel economy and the battery life considering the condition of the road.

This study focuses on two problems for the basic control strategy while driving continuously uphill. Firstly, in the CD mode, the PHEV may switch to the CS mode in uphill terrain, as the *SOC* reaches the minimum boundary. Therefore, the engine becomes the primary power source to drive the vehicle, which increases the fuel consumption. Secondly, in the CS mode, considerably low *SOC* will reduce the power or cause excessive battery discharge while driving uphill.

In this study, an optimal EMS for the PHEV based on the road-grade information is proposed to improve both the fuel economy and battery life. The most important issues are analyzed, including the prediction of the electricity consumption for the uphill terrain and the planning of trajectory of the *SOC*. First, the electricity consumption for the uphill terrain is predicted based on the road grade and the average velocity of the traffic obtained from the GPS/GIS system. Moreover, the trajectory of the *SOC* is preplanned based on the target *SOC* for the uphill terrain in the CD and CS modes. The proposed EMS was simulated on the MATLAB/Simulink (R2014a, MathWorks, Natick, MA, USA) platform under the comprehensive driving cycle. The effectiveness of the proposed control strategy was validated by comparing with the results of the basic control strategy.

The rest of the paper is organized as follows: Section 2 presents the structure and the basic control strategy of the powertrain system. Section 3 explains the prediction of the road grade and the electricity consumption for the uphill terrain. Section 4 proposes the energy management of the PHEV considering the condition of the road. Section 5 provides the simulation results and analyses of the proposed approach. Section 6 gives the conclusions of this study.

2. Basic Control Strategy

2.1. The Structure and Parameters of the Powertrain System

Figure 2 shows the structure of the PHEV analyzed in this study. The power sources are the engine and the motor, which is an integrated starter generator (ISG). The engine and ISG are coaxially arranged to achieve dynamic coupling, and the master clutch is placed between the engine and the ISG. Furthermore, a dual-clutch transmission (DCT) is used to meet the different requirements of the driving conditions in terms of speed and torque. The powertrain system of the PHEV can operate in one of six working modes, including the pre-charge mode, the driving and charging mode, the engine-driving mode, the electric-driving mode, the hybrid-driving mode, and the regenerative-braking mode. Moreover, the working modes are switched by changing the status of the master clutch and dual clutches. Table 1 gives the basic parameters of the PHEV.

As the length of the paper is limited, the basic control strategy of the PHEV based on the driving-pattern recognition is introduced directly and used as the reference for the proposed control strategy [26]. The main idea of the basic control strategy is stated as follows: first, the basic control strategy is divided into the CD and CS modes based on the *SOC* of the battery. Six benchmark driving cycles, namely, ECE, NYCC, UDDS, LA92, HWFET, and US06HWY, are considered as typical driving cycles to represent the different road types, namely urban congestion, suburban, and highway conditions. The driving-pattern recognition is realized using the clustering-analysis method [27]. Moreover, the key control parameters are optimized via the genetic algorithm (GA) under the different typical driving cycles, and the corresponding optimized results are saved in the database. Based on the results of the driving-pattern recognition in real time, the PHEV is controlled by the corresponding optimal parameters, which are obtained from database.

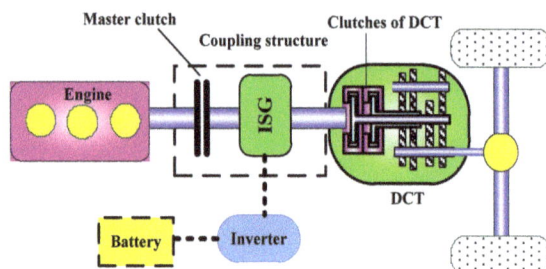

Figure 2. The powertrain configuration of the PHEV.

Table 1. Basic parameters of the PHEV.

Description	Parameters	Value
Basic parameters of the vehicle	vehicle mass/kg	1325
	frontal area/m^2	2.275
	drag coefficient	0.3146
	rolling radius/m	0.308
	rolling resistance coefficient	0.00995
ISG	peak power/kW	30
	maximum torque/N·m	115
Engine	peak power/kW	80
	maximum torque/N·m	140
NI-H power battery	capacity/A·h	38.5
	rated voltage/V	288
	initial *SOC*	0.85
	minimum *SOC*	0.3
DCT	speed ratio	3.917/2.429/1.436 1.021/0.848/0.667
	efficiency	0.95

2.2. Control Strategy in CD Mode

The CD mode of the PHEV comprises the electric-driving and hybrid-driving modes. In the CD mode, the PHEV is mainly driven by the motor to make full use of the battery. If the required power of the vehicle is less than the maximum power of the motor, only the motor will drive the vehicle; otherwise, both the engine and the motor will drive the vehicle together. In our previous research, the efficiency optimization models for the powertrain system under different working modes were developed to obtain the mode-switching schedules [26]. Figure 3 shows the mode-switching schedule of the CD mode.

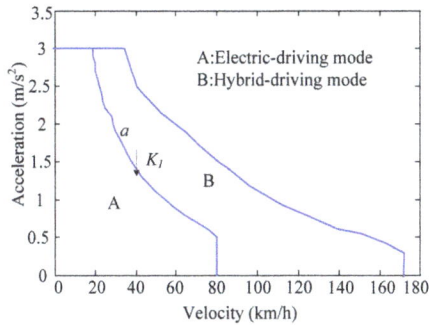

Figure 3. The mode-switching schedule in the CD mode.

To achieve better fuel economy, the mode-switching schedule needs to be adaptively modified based on the driving-pattern recognition. Hence, the mode-switching schedule is adjusted by a control parameter K_1, which is limited to the range $[0, 1]$ to avoid exceeding the maximum power of the motor (borderline a). Hence, the borderline a will adaptively fluctuate by multiplying the control parameter K_1. Furthermore, the control parameter K_1 is regarded as a design variable to be optimized offline under different typical driving cycles, and the optimized results are saved in the database.

2.3. Control Strategy in CS Mode

The CS mode of the PHEV comprises the electric-driving mode, driving and charging mode, engine-driving mode, and hybrid-driving mode. In the CS mode, the vehicle is mainly driven by the engine to maintain the battery SOC. Through the optimal torque distribution between the two power sources, the engine operates in the high-efficiency region to improve fuel economy. Figure 4 shows the mode-switching schedule of the CS mode. Similar to the CD mode, the control parameters K_2, K_3, and K_4 are used to modify the mode-switching schedule based on the result of the driving-pattern recognition. The three control parameters are limited to the range $[0, 1]$. Furthermore, the three control parameters are optimized offline via GA. The controller will select the corresponding optimized parameters from the database based on the result of the driving-pattern recognition.

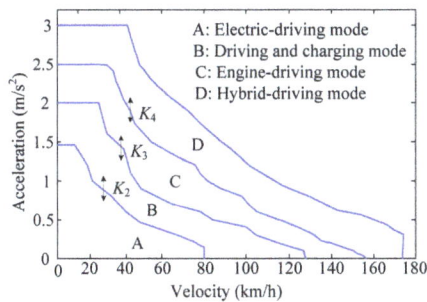

Figure 4. The mode-switching schedule in the CS mode.

3. The Prediction of the Electricity Consumption for the Uphill Terrain

The electricity consumption for the uphill terrain is predicted to determine the target *SOC* of the battery. The road grade and the velocity profile are crucial factors affecting the energy consumption of the HEV, which was validated in a previous research [21]. To determine the target *SOC* for the uphill terrain, the road grade and the driving cycle need to be predicted.

3.1. Road Grade Information with GPS/GIS System

The PHEV is assumed to be equipped with a GPS/GIS system, which is a reasonable assumption. If the origin and destination of the vehicle are entered in the digital map, the GPS/GIS system can provide the elevation and the length of the road [28]. The location and the distance of the PHEV with respect to the uphill road can be obtained from the GPS, thereby facilitating the prediction of the road grade. Figure 5 illustrates the estimation of the road grade using the length of the uphill road and the difference in altitude between the starting and ending points.

The road grade is estimated using the following expression:

$$i = \tan\left(\arcsin\frac{E_2 - E_1}{L}\right) \times 100\%. \tag{1}$$

where E_2 is the elevation of the road for the sampling point of the road ahead, E_1 is the elevation of the road for the current location of the PHEV, and L is the traveling distance between the two sampling points.

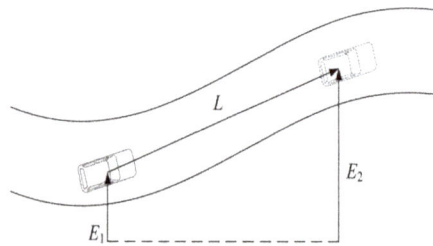

Figure 5. Road grade estimation using GPS/GIS.

3.2. The Prediction of the Driving Cycle along the Slope

Traffic information, such as traffic flow, velocity limits, and average velocity of the traffic, can be obtained from GPS/GIS. The average velocity of the traffic is the average velocity of all the vehicles on a specific road segment. The average velocity of the traffic on the uphill road is used to predict the driving cycle. First, the uphill road is divided into several segments to sample the average vehicle velocity, and the average velocities of the different segments are used to construct the reference velocity profile in the spatial domain, as shown in Figure 6. The length of the segment for the sampling of the average velocity plays a key role in the accuracy of the prediction. The length of the segment is set as 160 m [29].

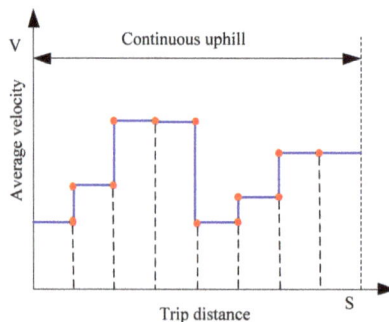

Figure 6. An example of the velocity profile in the spatial domain.

The velocity profile in the time domain, which is defined as the reference driving cycle, is obtained by domain transformation. The domain transformation is expressed as follows:

$$\begin{cases} \Delta t(k \Rightarrow k+1) = \frac{\Delta d}{V_s(k)} \\ t(k+1) = t(k) + \Delta t(k \Rightarrow k+1) \\ k = 1, 2, 3..., \frac{S}{\Delta d} \\ V_t(k) = V_s(k) \end{cases} \tag{2}$$

where $\Delta t(k \Rightarrow k+1)$ is the time period for the average-velocity sampling of the segment k in the time domain, $t(k)$ is the initial time for the sampling of the segment k in the time domain, Δd is the length of the segment for the average-velocity sampling in the spatial domain, $V_s(k)$ is the average velocity of the vehicle of the segment k in the spatial domain, $V_t(k)$ is the average velocity of the vehicle of the segment k in the time domain, and S is the length of the continuous uphill road.

However, the reference driving cycle is not reasonable as the real-driving conditions, as the velocity changes abruptly. To achieve a reasonable driving cycle that reflects the actual driving cycle to the highest possible degree, a second order Butterworth filter with a cut-off frequency of 0.4 Hz is utilized to filter the reference velocity profile. Figure 7 shows the filtering procedure. The filtered velocity profile is selected as the predicted driving cycle for the uphill road. As it is difficult to obtain the traffic information from a simulation environment, this study assumes that the average velocity of the PHEV driving through a specific segment is the same as the average velocity of the traffic obtained from the GPS/GIS system. Hence, the prediction of the driving cycle on the uphill road can be achieved, as shown in Figure 8. The average velocity of the traffic is set as the reference driving cycle, which is obtained from the GPS/GIS. The predicted driving cycles can then be constructed using the Butterworth filter.

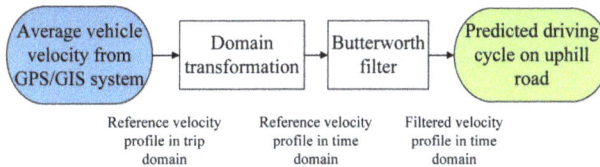

Figure 7. Filtering procedure for the velocity profile.

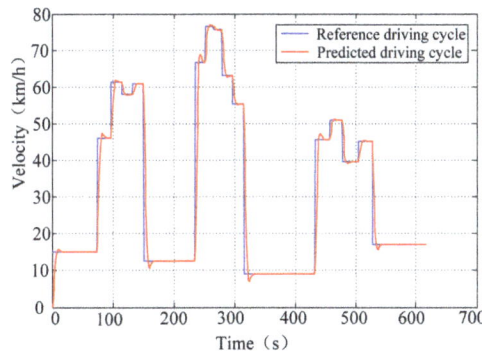

Figure 8. An example of the prediction for the driving cycle.

3.3. Calculation of Target SOC for the Uphill Terrain

The variation in the *SOC* during the uphill terrain is calculated based on the road grade and the predicted driving cycle. The traffic information obtained from the GPS/GIS system can be updated every 300 s [30]. Hence, during the interval of the update of the traffic information, the variation in the *SOC* can be simulated under the predicted driving cycle in time. The simulation model of the powertrain system of the PHEV has been extensively studied, and is not discussed further in this paper. Figure 9 shows the structure of the powertrain-system model.

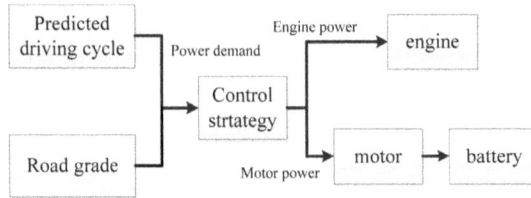

Figure 9. Structure of PHEV powertrain system model.

Based on the variation in the *SOC* during the uphill terrain, the target *SOC* for the uphill terrain is determined as follows:

$$SOC_t = SOC_0 + \Delta SOC \tag{3}$$

where SOC_t is the target *SOC* for the uphill terrain, SOC_0 is the minimum boundary of the *SOC*, and ΔSOC is the variation in the *SOC* during the uphill terrain.

4. Energy Management Strategy Based on Road Grade Information

In the paper, in order to avoid excessive battery discharge and reduce the fuel consumption of the PHEV during a continuous uphill terrain trip, an energy management strategy based on road grade information has been proposed for PHEV, whose control diagram is shown in Figure 10. The proposed control strategy is developed based on the basic control strategy.

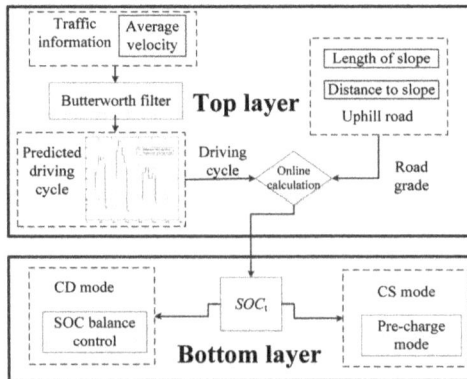

Figure 10. Control diagram for the proposed control strategy.

In the top layer, driving cycles on the uphill road can be predicted with the average vehicle velocity of the traffic obtained from GPS/GIS, and the target *SOC* for the uphill terrain can be determined by online calculation.

In the bottom layer, the *SOC* trajectory is preplanned in the CD and CS modes. The *SOC* trajectory planning strategy can make sure sufficient electricity to drive motor during the uphill terrain. Therefore, the engine can operate in the high-efficiency region with the optimal torque distribution between the engine and motor, which effectively improves the overall efficiency of whole system. The battery *SOC* is maintained at the target value SOC_t through the battery SOC balance control method in the CD mode, and the battery SOC is charged to the target value of SOC_t through the pre-charge method in the CS mode. The details of the proposed energy management strategy will be introduced in the following sections.

4.1. Control Strategy for CD Mode

4.1.1. SOC Trajectory Planning Strategy

Due to the sufficient electricity of the power battery in CD mode, it's necessary to make a judgment whether the current SOC would meet the demand for uphill. The driving mileage per SOC in CD mode denoted by a (km·%$^{-1}$) was determined offline under different typical driving cycles. The driving mileage per battery SOC, a, which is shown in Table 2, was used to estimate the variation of SOC in the CD mode.

Table 2. Driving mileage per SOC (km·%$^{-1}$) in CD mode.

Driving Cycle	ECE	NYCC	NEDC	UDDS	LA92	HWFET	US06HWY
a	1.401	1.222	0.786	0.783	0.592	0.599	0.525

The diagram of SOC trajectory planning in the CD mode is shown in Figure 11, and the control flow is described as follows:

(1) Road grade estimation: The destination of travel, the road path, and the elevation as well as length of the road are determined from the vehicle navigation system (GPS/GIS) before traveling. Then the road grade can be calculated by Equation (1).
(2) Driving cycle prediction: The predicted driving cycle on the uphill road is constructed through average velocity of the traffic on the uphill road obtained from the GPS/GIS.
(3) Target SOC calculation: The variation in SOC during the uphill terrain is calculated, and the target value SOC_t for uphill is then determined.
(4) Driving mileage lookup: The type of real-time driving cycle is obtained based on driving pattern recognition, and the driving mileage, a, per SOC is referred from Table 2 based on the real-time driving cycle of PHEV.
(5) SOC control decision: The initial battery SOC for encountering the uphill terrain is calculated before the vehicle took the uphill road. The initial battery SOC for the uphill terrain, which is defined as SOC_g, can be calculated using Equation (4). If SOC_g is greater than SOC_t, the vehicle is still on the electric-driving mode to consume electric energy; otherwise the SOC trajectory is planned by the SOC balance control method:

$$SOC_g = SOC(t) - \frac{L(t)}{a} \qquad (4)$$

where SOC (t) is the current SOC of power battery, L (t) is the distance between the vehicle current position and uphill road, and a is the driving mileage per SOC in CD mode. Because the result of driving pattern recognition changes with time, the SOC_g can be dynamically calculated according to real-time driving cycle.

(6) SOC balance control: The PHEV controller switches to SOC balance control to ensure that the target value SOC_t is maintained. As a result, there is sufficient battery power to encounter the uphill terrain when the PHEV took uphill road. Note that the vehicle is controlled by basic strategy after entering the uphill road.

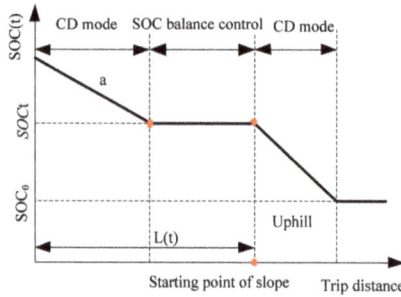

Figure 11. Diagram of the battery *SOC* trajectory in CD mode.

4.1.2. SOC Balance Control

In the *SOC* balanced control mode, engine and motor work together to maintain a roughly constant battery SOC. It is similar to the basic control strategy in the CS mode. However, the basic control strategy in the CS mode cannot ensure sufficient battery power for uphill terrains, since the battery *SOC* may change too much during certain driving conditions. Therefore, the *SOC* balance control method has been proposed, which takes the fuel economy as well as the *SOC* variation into consideration. The key control parameters of this control strategy are optimized by GA. The optimization target is to obtain optimal parameters to restrict the variation in battery SOC, which can also ensure the fuel economy of PHEV.

(1)   Optimization objective function

In order to restrict the battery *SOC* variation, *SOC* variation is converted to the corresponding fuel consumption. The equivalent fuel consumption is integrated into the optimization objective function. The *SOC* correction method has been used for objective function in this paper as follows:

$$\Delta fuel = \frac{\Delta SOC \cdot Q_{cap} \cdot \overline{U_{bat}} \cdot \overline{\eta_{eng\_chg}}}{1000 \cdot \rho} \tag{5}$$

where $\Delta fuel$ is the equivalent fuel consumption (L), ΔSOC is the variation of battery *SOC* between the initial and final values, Q_{cap} is the capacity of battery (Ah), $\overline{U_{bat}}$ is the average value of battery bus voltage during drive cycles (V), $\overline{\eta_{eng\_chg}}$ is the average value of the engine's power efficiency (g/kW·h), and ρ is the density of gasoline (g/L).

To optimize the fuel economy and prevent excessive variation in the *SOC*, the fuel consumption and equivalent fuel consumption are integrated in the optimization objective function. Therefore, the optimal control parameters can make sure both the fuel consumption and the *SOC* variation are the minimum. The fitness function is given by:

$$Min\ f(x) = \int Fuel_{use(t)}dt + w_p \cdot |\Delta fuel_p| \tag{6}$$
$$s.t.\quad x_i^l \leq x_i \leq x_i^k \quad i = 1, 2, 3, ..., n$$

where n is the number of optimization variables; x_i^l and x_i^k are the upper and lower boundaries of the optimization variables, respectively; w_p is the weight coefficient of $\Delta fuel$, which is used for adjusting the limitation of *SOC* variation. The weight coefficient is set as 1.2 using an enumerative technique based on experience and simulation.

(2)   Optimized parameters

The mode-switching schedule of the *SOC* balance control method is the same as that of the basic control strategy in the CS mode, shown in Figure 4. To optimize the fuel economy of the PHEV, the control parameters K_2, K_3, and K_4 are set as the design variables to modify the mode-switching schedule. Furthermore, the electric-driving mode and the driving and charging mode are the most critical modes that influence the variation in the *SOC*. Hence, two other control parameters SOC_{ele} and SOC_{charge} are created to modify the mode-switching schedule of the electric-driving mode and driving and charging mode, which are given in Equations (7) and (8):

$$SOC_{ele} = SOC_{min} + K_e \tag{7}$$

$$SOC_{charge} = SOC_{min} + K_c \tag{8}$$

where SOC_{ele} is the upper boundary of the *SOC* in the *SOC* balance control. The controller switches to the electric-driving mode to consume surplus electric energy when the *SOC* reaches the upper boundary. K_e is the control parameter that modifies SOC_{ele}. SOC_{charge} is the threshold for switching to the driving and charging mode. To avoid unnecessary fuel consumption, the controller switches to the driving and charging mode only when the *SOC* is lower than this threshold. K_c is the control parameter that modifies SOC_{charge}. SOC_{min} is the lower boundary of the *SOC* in the *SOC* balance control, which is considered SOC_t based on the calculation of the target *SOC* for the uphill terrain, given in Section 3.

The five control parameters K_e, K_2, K_3, K_4, and K_c are set as the design variables for the optimization. Table 3 lists the range and initial values of the design variables. The range is selected to ensure that the engine operates in the high-efficiency region and that different working modes will not interfere with each other.

Table 3. Range of the design variables.

Design Variable	Initial Value	Range
K_e	0	0~0.1
K_2	1	0.6~1
K_3	1	0.6~1
K_4	1	0.6~1
K_c	0	0~0.1

(3)   Optimization results

The five design variables are optimized using the GA under the different typical driving cycles. For the optimization of the parameters, the maximum number of generations is set as 80, and the population is initialized by 100 random individuals. Moreover, the elitist amount is set as 10, and the mutation rate is set as 0.4. Table 4 gives the results of the optimization. Figure 12 shows the history of the optimization process for the typical driving cycle LA92.

Table 4. Optimization results of design variables.

Driving Cycle	K_e	K_2	K_3	K_4	K_c
ECE	0.078	0.774	0.808	1	0.012
NYCC	0.005	0.846	0.993	0.889	0.000
NEDC	0.000	0.718	0.643	0.610	0.058
UDDS	0.063	0.948	0.972	0.845	0.054
LA92	0.060	0.998	0.931	0.886	0.001
HWFET	0.001	0.654	0.663	0.628	0.048
US06HWY	0.077	0.794	0.991	0.962	0.003

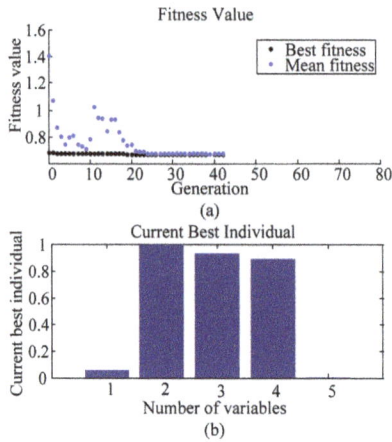

Figure 12. Optimization result under the driving cycle LA92. (**a**) Best fitness of optimization under the driving cycle LA92; (**b**) Best individual of optimization under the driving cycle LA92.

Figure 12a shows the development of the GA. The points terminate after 43 generations, and the fitness value, i.e., the fuel-consumption is 0.674. According to the optimization results, the variation in *SOC* is maintained within 0.006, 0.002, 0.013, 0.002, 0.003, 0.008 and 0.019, respectively, under the six typical driving cycles. These values satisfy the requirements of the *SOC* balance control in terms of the consistency of SOC. Figure 13 shows the trajectory of the *SOC* for the parameter optimization, taking the driving cycle LA92 as an example. The battery *SOC* is effectively maintained at the initial value.

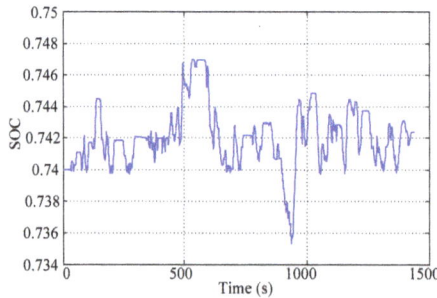

Figure 13. The trajectory of the *SOC* under the driving cycle LA92.

4.2. Control Strategy for the CS Mode

In the CS mode of the PHEV, the pre-charge mode is designed considering the road grade information based on the basic control strategy. The controller switches to the pre-charge mode before the uphill road based on the prediction of the target *SOC* for the uphill road, given in Section 3. The pre-charge mode comprises the driving and charging mode, engine-driving mode, and hybrid-driving mode; however, while the electric-driving mode is temporarily terminated to increase the *SOC* rapidly. As Figure 4 given in Section 2 indicates, the powertrain system of the PHEV will operate in the driving and charging mode when the system behavior is within sections A and B. In the driving and charging mode, the engine operates at the optimal operating points to improve fuel economy, and the remaining power is then used to charge the battery. The vehicle operates only in the electric-driving mode when the velocity of the vehicle is below the launch-speed limit of the engine. The engine launch speed is set as 30 km/h. First, the driving mileage in the pre-charge mode is determined. The driving mileage per battery-*SOC* variation in the pre-charge mode, defined as b (km·%$^{-1}$), is determined under different typical driving cycles simulated offline, given in Table 5.

Table 5. Driving mileage per *SOC* (km·%$^{-1}$) in the pre-charge mode.

Driving Cycle	ECE	NYCC	NEDC	UDDS	LA92	HWFET	US06HWY
b	0.157	0.330	0.443	0.641	0.783	1.636	8.026

Figure 14 shows the diagram of the *SOC* trajectory planning in the CS mode. The control flow is described as follows:

(1) Road-grade estimation: The destinations of travel, the path of the road, and the elevation as well as the length of the road are determined from the vehicle navigation system (GPS/GIS) before starting. Then, the road grade can be calculated using Equation (1).
(2) Driving cycle prediction: The predicted driving cycle on the uphill road is developed using the average velocity of the traffic obtained from the GPS/GIS system.
(3) Target-*SOC* calculation: The variation in the *SOC* during the uphill terrain is calculated, and the target-battery SOC_t for the uphill road is then determined.
(4) Driving mileage lookup: The type of real-time driving cycle is obtained through the driving-pattern recognition, and the driving mileage b per *SOC* is referred from Table 5 based on the real-time driving cycle of the PHEV.
(5) Pre-charge mode: The particular location between the vehicle and the uphill road is determined. When the PHEV arrives at the particular location, the controller switches to the pre-charge mode at this point. Hence, the *SOC* increases to the target value when the vehicle enters onto the uphill road. Note that the vehicle is controlled by basic strategy after entering the uphill road. The starting time of the pre-charge mode is given by:

$$(SOC_t - SOC_0) \cdot b = L'(t) \tag{9}$$

where $L'(t)$ is the distance between the vehicle and the uphill road when the controller switches to the pre-charge mode, SOC_t is the target battery *SOC* for the uphill road, and SOC_0 is the minimum boundary of the *SOC*.

By employing Equations (3) and (9), the distance to the slope at which the controller switches to the pre-charge mode is calculated using Equation (10):

$$L'(t) = \Delta SOC \cdot b \tag{10}$$

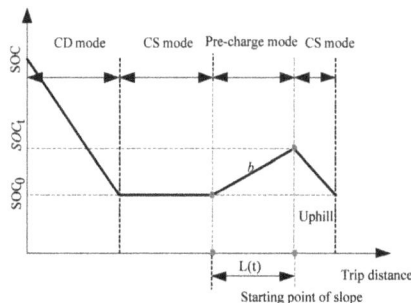

Figure 14. Diagram of the battery *SOC* trajectory in CD mode.

5. Simulation and Analysis

To verify the effectiveness of the proposed control strategy, a backward simulation model has been established using the MATLAB/Simulink platform. The proposed pre-charge mode may lead to double energy conversion losses. In order to evaluate the performance of the proposed strategy, the efficiency models of the key components have been integrated into the simulation model. The efficiency models consist of ISG motor, battery and inverter. The efficiencies of inverter and DCT are modeled as the fixed values 0.92 and 0.95, respectively. The ISG motor working efficiency, which is shown in Figure 15, is obtained from the experiment data. The efficiency model of the power battery is a function of charging/discharging power and *SOC* as shown in Figure 16. The proposed strategy has been simulated in the CD and CS modes, and the basic strategy is selected as a benchmark strategy for comparison. Table 1 lists the parameters for the simulation.

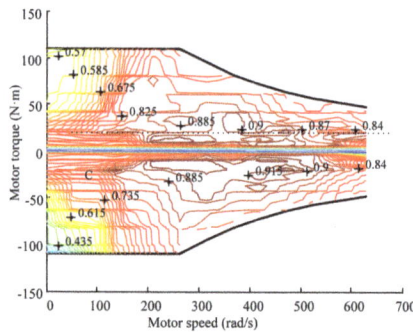

Figure 15. ISG motor efficiency map.

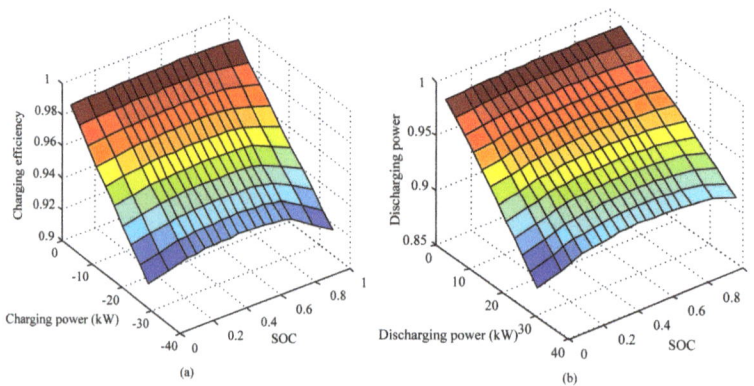

Figure 16. Power battery efficiency model. (**a**) Charging efficiency model of battery; (**b**) Discharging efficiency model of battery.

5.1. Simulation in the CD Mode

Five types of driving cycles (New York bus, 1015, Manhattan, WVUSUB, and HL07) were successively combined into a comprehensive cycle to simulate and evaluate the performance of the proposed control strategy. The information of the road grade and average velocity of the traffic are assumed to be known. The continuous uphill road has a road grade of 4%, an overall length of 3.3 km, and appears at 4724 s, as shown in Figure 17. Table 6 and Figure 18 give the comparison results between the control strategy based on the road grade information and the basic control strategy.

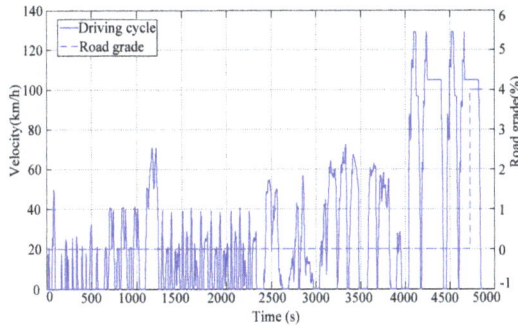

Figure 17. Test driving cycle for the CD mode.

Table 6. Comparing results for the CD mode.

Result	Basic Strategy	Proposed Strategy	Savings (%)
Fuel consumption (L)	0.640	0.596	6.878
Electricity consumption (kw·h)	5.591	5.693	−1.789
Initial *SOC* for uphill	0.322	0.391	-
Final *SOC* for uphill	0.303	0.304	-
Cost of energy consumption (RMB)	7.897	7.605	3.688

Different from the conventional HEV, a PHEV is mainly driven by the motor in the CD mode, and the power comes from two components: the electricity from the power grid and the fossil fuel. Hence, the cost of energy consumption, which is defined as Q_c, is adopted to evaluate the energy economy of the PHEV in the CD mode [31]. The cost of energy consumption is expressed using the following equation:

$$Q_c = J_f Q_f + J_e Q_e. \tag{11}$$

where J_f is the fuel price (RMB/L), which is set as 7.8 RMB/L; J_e is the electricity price (RMB/kW·h) which is set as 0.52 RMB/kW·h; Q_f is the fuel consumption (L); Q_e is the electricity consumption (kW·h).

As shown in Figure 18b, the proposed strategy switches to the *SOC* balance control at 4565 s, and the target *SOC* is 0.402. The *SOC* balance control is maintained for 156 s, and the *SOC* of the proposed strategy effectively maintains at *SOC*$_t$, as the initial *SOC* for the uphill terrain only slightly changes to 0.391. However, the initial *SOC* for the uphill terrain in the basic strategy drops to 0.322, as the road grade is not considered. As shown in Figure 18a, the fuel consumption of the proposed strategy is higher than that of the basic strategy during the *SOC* balance control, as the engine works in tandem with the motor to conserve the battery power. When the PHEV enters the uphill road, the proposed strategy maintains the vehicle operate in the CD mode. The electric motor becomes the primary power source for the proposed control strategy; hence, the fuel consumption reduces. However, for the basic strategy, the PHEV switches to the CS mode at 4756 s because it reaches the minimum boundary of the *SOC*. Moreover, the engine becomes the primary power source, thereby increasing the fuel consumption of the basic control strategy.

Compared to the basic strategy, the fuel consumption and the cost of energy consumption of the proposed strategy improve by 6.878% and 3.688%, respectively. Hence, the proposed control strategy effectively improves the fuel economy in the CD mode.

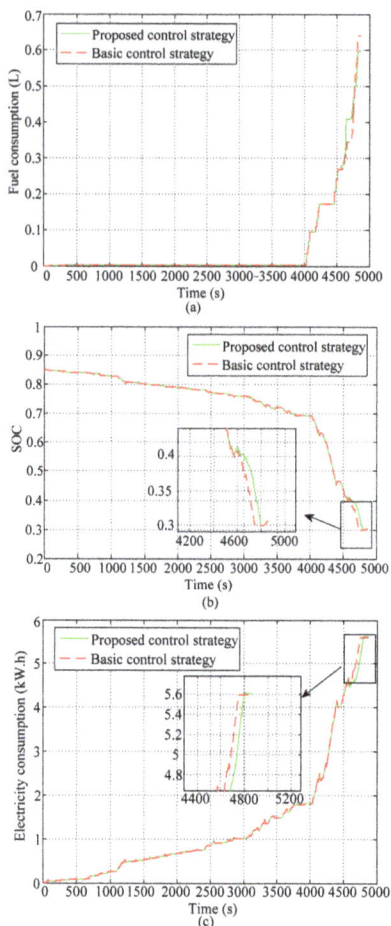

Figure 18. Simulation results in the CD mode. (**a**) Fuel consumption under the test driving cycle; (**b**) *SOC* trajectories under the test driving cycle; (**c**) Electricity consumption under the test driving cycle.

5.2. Simulation in the CS Mode

The five selected types of driving cycles (New York bus, 1015, Manhattan, WVUSUB, and HL07) were successively combined into a longer comprehensive cycle for the simulation process to ensure that the range of variation in the *SOC* includes the CS mode. The information of the road grade and average velocity of the traffic are assumed to be known. The continuous uphill road has a road grade of 8%, an overall length of 5.1 km, and appears at 9350 s, as shown in Figure 19. When the *SOC* is lower than the minimum boundary and the required power of vehicle is more than that generated by the engine at the optimal operating point, the basic strategy has two different cases: (1) the vehicle is driven only by the engine to avoid excessive battery discharge [13]; (2) the vehicle is driven by both the engine and the motor together while the range of the *SOC* in the CS mode is temporarily extended to ensure fuel economy and power [25]. The two cases of basic strategy have both been simulated for comparison with the proposed strategy. Table 7 and Figure 20 give the simulation results of the proposed control strategy and the basic control strategy.

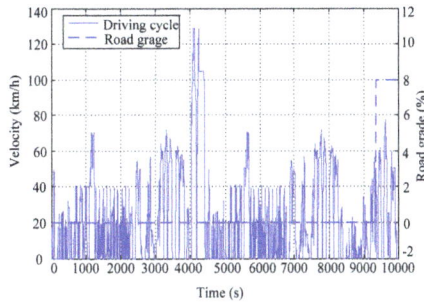

Figure 19. Test driving cycle for CS mode.

Table 7. Comparing results for CS mode.

Result	Proposed Strategy	Basic Strategy		Savings (%)	
		Case 1	Case 2 (Corrected)	Case 1	Case 2 (Corrected)
Fuel consumption (L)	1.453	1.462	1.494	0.616%	2.744%
Fuel consumption for uphill (L)	0.582	0.770	0.585	24.431%	0.547%
Electricity consumption (kw·h)	4.135	4.135	4.135	0	0
Initial *SOC* for uphill	0.355	0.308	0.308	-	-
Final *SOC* for uphill	0.301	0.303	0.251	-	-

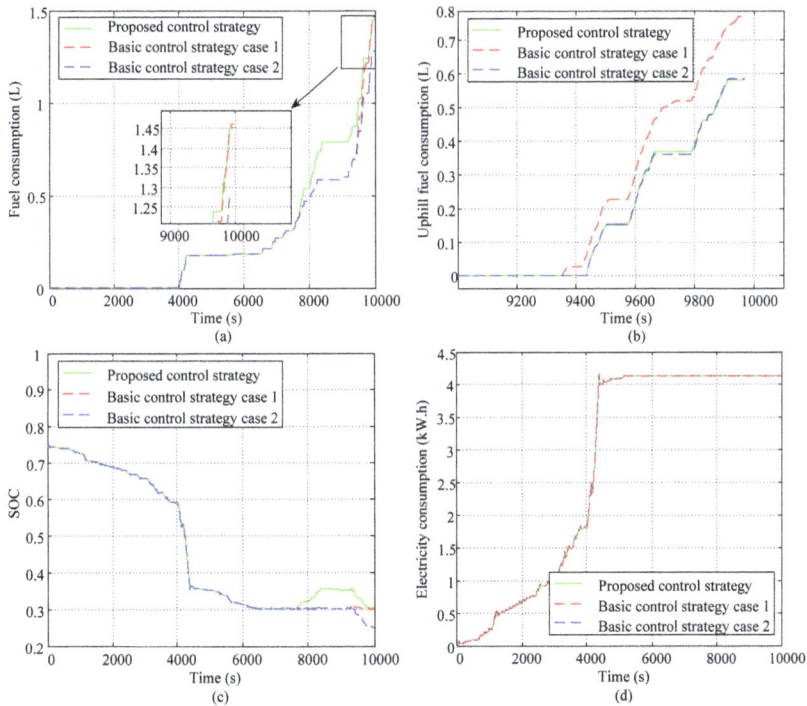

Figure 20. Simulation results in CS mode. (**a**) Fuel consumption under the test driving cycle; (**b**) Fuel consumption during the uphill terrain; (**c**) *SOC* trajectories under the test driving cycle; (**d**) Electricity consumption under the test driving cycle.

Note that the electricity consumption is only estimated in the CD mode. Generally, the electricity consumption is estimated only using the electricity from the power grid. However, in the CS mode, the consumed electricity mainly comes from the fossil fuel in the driving and charging mode. As slopes are not encountered during the CD mode, with the same control rules in the CD mode, the electricity consumptions of the two strategies are both 4.135 kW·h, as shown in Figure 20d.

In the first case, for the basic strategy, the vehicle is driven only by the engine when the *SOC* is lower than the minimum boundary. The proposed strategy effectively improves the fuel economy. As shown in Figure 20c, the controller switches to the pre-charge mode at 7640 s, and the *SOC* increases to 0.355 when the vehicle enters the slope. After the uphill terrain, the final *SOC* of the proposed control strategy is 0.301, which is close to the minimum boundary without excessively discharging the battery. As shown in Figure 20b, the proposed strategy significantly reduces the fuel consumption on the uphill road owing to the optimal torque distribution between the engine and motor. Figure 21 illustrates that, during the uphill terrain, the proposed strategy can ensure that the engine operates at relatively optimal operating points; furthermore, the vehicle can be driven in the electric-driving mode at low speed and torque, which effectively improves the efficiency and reduces the emission of the engine. Compared to the basic control strategy, the proposed control strategy improves the fuel consumptions by 0.616% and 24.431% in the whole trip and during the uphill terrain, respectively.

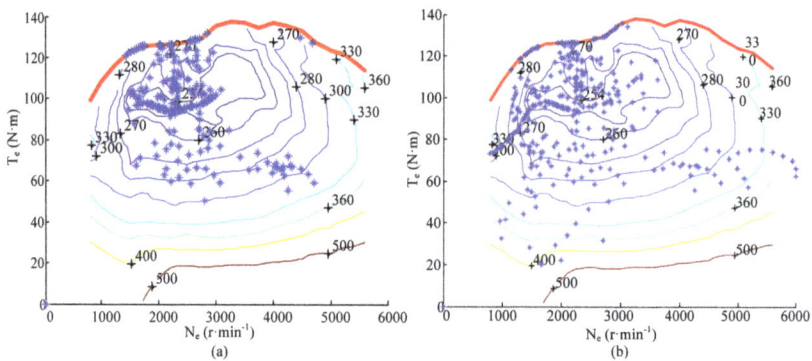

Figure 21. Engine operating points during uphill terrain. (**a**) Engine working point for the proposed control strategy; (**b**) Engine working point for the basic control strategy.

In the second case, for the basic strategy, the motor works in tandem with the engine during the uphill terrain. The proposed control strategy effectively improves the fuel economy and battery life of the PHEV. As shown in Figure 20c, for the proposed strategy, the electricity of the battery is sufficient because the controller switches to the pre-charge mode before encountering the uphill road; whereas in the basic strategy, the *SOC* drops to 0.251 after the uphill terrain. Hence, the depth of discharge in the basic strategy is lower than the minimum boundary by 16.43%, thereby excessively discharging the battery. There is an evident difference between the ending *SOCs* of the two strategies. Hence, the *SOC* correction method based on the SAE J1711 standard is necessary to calculate the equivalent fuel consumption that compensates for the difference in the *SOC* [32]. By using this method, the fuel economy of the proposed strategy improves by 2.744%. Furthermore, the battery discharges excessively in the basic strategy, which reduces the battery life.

6. Conclusions

An optimal EMS for the PHEV based on the road grade information is proposed in this study. The effectiveness of the proposed strategy is validated via the simulation, which improves both the fuel economy and battery life of the PHEV. The following work was conducted in this study:

(1) An algorithm that predicts the electricity consumption during the uphill terrain was developed based on the information obtained from the GPS/GIS system. The road-grade information is obtained from the GPS system, and the driving cycles on the uphill road are predicted through the average velocity of the traffic obtained from the GIS system. Furthermore, the target *SOC* for the uphill road is calculated based on the road grade information and predicted driving cycles.

(2) In the CD mode, the trajectory of the *SOC* is preplanned using the *SOC* balance control method based on the target *SOC*. The key control parameters are optimized using the GA to balance the *SOC* and improve the fuel economy. Compared to the basic control strategy, the simulation results show that the proposed strategy improves the fuel consumption and cost of energy consumption by 6.878% and 3.688%, respectively.

(3) In the CS mode, the trajectory of the *SOC* is preplanned using the pre-charge mode. The power battery is charged to the target *SOC* before entering the uphill road. Moreover, the proposed strategy improves the fuel consumption in the CS mode compared to that of the basic control strategy, and the excessive battery discharge is avoided during the continuous uphill terrain.

Acknowledgments: The work presented in this paper is funded by the China Postdoctoral Science Foundation (No. 2016M602925XB), and the Fundamental Research Funds for the Central Universities (No. 106112016CDJXY330001), the Key Laboratory of Advanced Manufacture Technology for Automobile Parts, Ministry of Education (No. 2016KLMT06) and the Chongqing Key Technology Innovation Project of Key Industries(No. CSTC2015ZDCY-ZTZX60013).

Author Contributions: Yonggang Liu wrote the paper and provided algorithms; Jie Li and Zhenzhen Lei completed the simulation for case studies; Ming Ye analyzed the simulation results; Datong Qin and Yi Zhang conceived the structure and research direction of the paper.

Conflicts of Interest: The authors declare no conflict of interest.

References

1. Gong, Q.; Li, Y.; Peng, Z.-R. Optimal power management of plug-in HEV with intelligent transportation system. In Proceedings of the 2007 IEEE/ASME International Conference on Advanced Intelligent Mechatronics, Seattle, WA, USA, 11–15 November 2007.
2. Hofman, T.; Steinbuch, M.; Van Druten, R.; Serrarens, A. Rule-based energy management strategies for hybrid vehicles. *Int. J. Electr. Hybrid Veh.* **2007**, *1*, 71–94. [CrossRef]
3. Sciarretta, A.; Back, M.; Guzzella, L. Optimal control of parallel hybrid electric vehicles. *IEEE Trans. Control Syst. Technol.* **2004**, *12*, 352–363. [CrossRef]
4. Jeon, S.-I.; Jo, S.-T.; Park, Y.-I.; Lee, J.-M. Multi-mode driving control of a parallel hybrid electric vehicle using driving pattern recognition. *J. Dyn. Syst. Meas. Control* **2002**, *124*, 141–149. [CrossRef]
5. Zhang, S.; Xiong, R. Adaptive energy management of a plug-in hybrid electric vehicle based on driving pattern recognition and dynamic programming. *Appl. Energy* **2015**, *155*, 68–78. [CrossRef]
6. Zhang, C.; Vahidi, A. Route preview in energy management of plug-in hybrid vehicles. *IEEE Trans. Control Syst. Technol.* **2012**, *20*, 546–553. [CrossRef]
7. Zhang, C.; Vahidi, A.; Pisu, P.; Li, X.; Tennant, K. Role of terrain preview in energy management of hybrid electric vehicles. *IEEE Trans. Veh. Technol.* **2010**, *59*, 1139–1147. [CrossRef]
8. Xiong, R.; Sun, F.; Chen, Z.; He, H. A data-driven multi-scale extended Kalman filtering based parameter and state estimation approach of lithium-ion olymer battery in electric vehicles. *Appl. Energy* **2014**, *113*, 463–476. [CrossRef]
9. Sun, F.; Xiong, R.; He, H. A systematic state-of-charge estimation framework for multi-cell battery pack in electric vehicles using bias correction technique. *Appl. Energy* **2016**, *162*, 1399–1409. [CrossRef]
10. Zhang, Y.; Xiong, R.; He, H.; Shen, W. A lithium-ion battery pack state of charge and state of energy estimation algorithms using a hardware-in-the-loop validation. *IEEE Trans. Power Electr.* **2016**, *32*, 4421–4431. [CrossRef]
11. Callaway, D.S.; Hiskens, I.A. Achieving controllability of electric loads. *Proc. IEEE* **2011**, *99*, 184–199. [CrossRef]
12. Ma, Z.; Zou, S.; Liu, X. A distributed charging coordination for large-scale plug-in electric vehicles considering battery degradation cost. *IEEE Trans. Control Syst. Technol.* **2015**, *23*, 2044–2052. [CrossRef]

13. Banvait, H.; Anwar, S.; Chen, Y. A rule-based energy management strategy for plug-in hybrid electric vehicle (PHEV). In Proceedings of the 2009 American Control Conference, St. Louis, MO, USA, 10–12 June 2009.

14. Karbowski, D.; Pagerit, S.; Kwon, J.; Rousseau, A.; von Pechmann, K.-F.F. *"Fair" Comparison of Powertrain Configurations for Plug-In Hybrid Operation Using Global Optimization*; SAE Technical Paper 0148-7191; SAE: Warrendale, PA, USA, 2009.

15. Zhang, S.; Xiong, R.; Cao, J. Battery durability and longevity based power management for plug-in hybrid electric vehicle with hybrid energy storage system. *Appl. Energy* **2016**, *179*, 316–328. [CrossRef]

16. Zhang, S.; Xiong, R.; Zhang, C. Pontryagin's minimum principle-based power management of a dual-motor-driven electric bus. *Appl. Energy* **2015**, *159*, 370–380. [CrossRef]

17. Sun, C.; Hu, X.; Moura, S.J.; Sun, F. Velocity predictors for predictive energy management in hybrid electric vehicles. *IEEE Trans. Control Syst. Technol.* **2015**, *23*, 1197–1204.

18. Di Cairano, S.; Bernardini, D.; Bemporad, A.; Kolmanovsky, I.V. Stochastic MPC with learning for driver-predictive vehicle control and its application to HEV energy management. *IEEE Trans. Control Syst. Technol.* **2014**, *22*, 1018–1031. [CrossRef]

19. Zheng, C.; Xu, G.; Xu, K.; Pan, Z.; Liang, Q. An energy management approach of hybrid vehicles using traffic preview information for energy saving. *Energy Convers. Manag.* **2015**, *105*, 462–470. [CrossRef]

20. Valera, J.; Heriz, B.; Lux, G.; Caus, J.; Bader, B. Driving cycle and road grade on-board predictions for the optimal energy management in EV-PHEVs. In Proceedings of the 2013 World Electric Vehicle Symposium and Exhibition (EVS27), Barcelona, Spain, 17–20 November 2013.

21. Younes, Z.; Boudet, L.; Suard, F.; Gérard, M.; Rioux, R. Analysis of the main factors influencing the energy consumption of electric vehicles. In Proceedings of the 2013 IEEE International Electric Machines & Drives Conference (IEMDC), Chicago, IL, USA, 12–15 May 2013.

22. Khayyer, P.; Wollaeger, J.; Onori, S.; Marano, V.; Özgüner, Ü.; Rizzoni, G. Analysis of impact factors for plug-in hybrid electric vehicles energy management. In Proceedings of the 2012 15th International IEEE Conference on Intelligent Transportation Systems, Anchorage, AK, USA, 16–19 September 2012.

23. Serrao, L.; Onori, S.; Sciarretta, A.; Guezennec, Y.; Rizzoni, G. Optimal energy management of hybrid electric vehicles including battery aging. In Proceedings of the 2011 American Control Conference, San Francisco, CA, USA, 29 June–1 July 2011.

24. Bockstette, J.; Habermann, K.; Ogrzewalla, J.; Pischinger, M.; Seibert, D. Performance Plus Range: Combined Battery Concept for Plug-In Hybrid Vehicles. *SAE Int. J. Altern. Powertrains* **2013**, *2*, 156–171. [CrossRef]

25. Kamichi, K.; Yamamoto, M.; Fushiki, S.; Yoda, T.; Kurachi, S.; Kojima, K. *Development of Plug-in Hybrid System for Midsize Car*; SAE Technical Paper 0148–7191; SAE: Warrendale, PA, USA, 2012.

26. Qin, D.; Zhao, X.; Su, L.; Yang, G. Variable parameters energy management strategy for plug-in hybrid electric vehicle. *China J. Highway Transp.* **2015**, *28*, 112–118. (In Chinese).

27. Gu, B.; Rizzoni, G. An adaptive algorithm for hybrid electric vehicle energy management based on driving pattern recognition. In Proceedings of the ASME 2006 International Mechanical Engineering Congress and Exposition, Chicago, IL, USA, 5–10 November 2006.

28. Rajagopalan, A.; Washington, G. *Intelligent Control of Hybrid Electric Vehicles Using GPS Information*; SAE Technical Paper 0148–7191; SAE: Warrendale, PA, USA, 2002.

29. Sun, C.; Moura, S.J.; Hu, X.; Hedrick, J.K.; Sun, F. Dynamic traffic feedback data enabled energy management in plug-in hybrid electric vehicles. *IEEE Trans. Control Syst. Technol.* **2015**, *23*, 1075–1086.

30. Chen, C.; Petty, K.; Skabardonis, A.; Varaiya, P.; Jia, Z. Freeway performance measurement system: Mining loop detector data. *Transp. Res. Rec. J. Transp. Res. Board* **2001**, *1748*, 96–102. [CrossRef]

31. Moura, S.J.; Callaway, D.S.; Fathy, H.K.; Stein, J.L. Tradeoffs between battery energy capacity and stochastic optimal power management in plug-in hybrid electric vehicles. *J. Power Sources* **2010**, *195*, 2979–2988. [CrossRef]

32. Hou, C.; Ouyang, M.; Xu, L.; Wang, H. Approximate Pontryagin's minimum principle applied to the energy management of plug-in hybrid electric vehicles. *Appl. Energy* **2014**, *115*, 174–189. [CrossRef]

Article

Online Reliable Peak Charge/Discharge Power Estimation of Series-Connected Lithium-Ion Battery Packs

Bo Jiang [1,2], **Haifeng Dai** [1,2,*], **Xuezhe Wei** [1,2], **Letao Zhu** [1,2] **and Zechang Sun** [1,2]

[1] National Fuel Cell Vehicle & Powertrain System Research & Engineering Center, No. 4800, Caoan Road, Shanghai 201804, China; jiangbo@tonzhan.com (B.J.); weixzh@tongji.edu.cn (X.W.); leao1217@163.com (L.Z.); sunzechang@tongji.edu.cn (Z.S.)
[2] School of Automotive Studies, Tongji University, No. 4800, Caoan Road, Shanghai 201804, China
[*] Correspondence: tongjidai@tongji.edu.cn; Tel.: +86-21-695-83847

Academic Editor: Rui Xiong
Received: 22 January 2017; Accepted: 7 March 2017; Published: 19 March 2017

Abstract: The accurate peak power estimation of a battery pack is essential to the power-train control of electric vehicles (EVs). It helps to evaluate the maximum charge and discharge capability of the battery system, and thus to optimally control the power-train system to meet the requirement of acceleration, gradient climbing and regenerative braking while achieving a high energy efficiency. A novel online peak power estimation method for series-connected lithium-ion battery packs is proposed, which considers the influence of cell difference on the peak power of the battery packs. A new parameter identification algorithm based on adaptive ratio vectors is designed to online identify the parameters of each individual cell in a series-connected battery pack. The ratio vectors reflecting cell difference are deduced strictly based on the analysis of battery characteristics. Based on the online parameter identification, the peak power estimation considering cell difference is further developed. Some validation experiments in different battery aging conditions and with different current profiles have been implemented to verify the proposed method. The results indicate that the ratio vector-based identification algorithm can achieve the same accuracy as the repetitive RLS (recursive least squares) based identification while evidently reducing the computation cost, and the proposed peak power estimation method is more effective and reliable for series-connected battery packs due to the consideration of cell difference.

Keywords: power estimation; parameter identification; ratio vector; cell difference; recursive least squares

1. Introduction

With the global issues of energy shortage and environmental degradation, the lithium-ion battery, because of its high energy and power density and long service lifetime, has become one of the most readily available and low-cost energy storage components in electric vehicles (EVs). However, the lithium-ion battery cells are sensitive to over- and under-voltage, over- and under-temperature, and some extreme working conditions may even lead to safety issues. Thus, to guarantee the safety of the vehicle and to enhance the performance of the batteries, a battery system is generally equipped with a battery management system (BMS). The most important task of a BMS is to provide an accurate and real-time estimation of the internal states of the battery, such as state of Charge (*SOC*), state of health (*SOH*) and peak power etc. [1].

The estimation of the peak power in the vehicular application is generally used to evaluate the maximum charge and discharge capability of the battery system, and thus help to optimally control the power-train system to meet the requirement of acceleration, gradient climbing and regenerative braking while achieving a high energy efficiency [2].

The peak power can be determined by three methods. The first one is the hybrid pulse power characterization (HPPC) proposed by Partnership for New Generation Vehicles (PNGVs) [3]; the second method considers the *SOC* limit of the battery, i.e., the peak power is calculated based on the limitation of the permitted maximum and minimum *SOC* [4]; and the third method is the voltage-limited method [4–12]. No matter what method is used, in a real BMS, the peak power estimation can be implemented with two types of techniques: techniques based on a characteristic map and on dynamic battery models [13].

Within the techniques based on the characteristic maps [14,15], the static interdependence existing among the peak power, battery states (e.g., *SOC* and *SOH*), working conditions (e.g., voltage and temperature) and pulse duration is applied. The dependencies are stored in the non-volatile memory of the BMS in the form of look-up tables. When the battery system is working, the BMS determines the peak power of the battery according to the present battery states, working conditions and the requirement of the duration of power delivery. Normally, the characteristic map can be obtained in advance by various test procedures, for example, the HPPC test procedure [3]. The main advantages of this technique are its simplicity and straight forward implementation. However, this technique suffers from the drawbacks that only static battery characteristics are considered, massive experiments should be implemented to obtain the characteristic map, and a significant amount of non-volatile memory is required which increases the cost of the BMS.

Another technique is the model-based estimation [4–12]. So far, there have been many researches of model-based power estimation, and the main difference among the existing researches lies in the type of models they used. If the model can track the battery dynamics well, then the peak power can also be estimated accurately. One more important aspect that should be taken into consideration is that the model should be adaptive to different aging states and temperatures. Thus, model adaption techniques are often applied [5,7,8], in which the model parameters are identified online to improve estimation accuracy. Generally, the model-based power estimation is more promising, and the above mentioned researches have been validated and proven to be effective for the power estimation of a single battery cell.

One common drawback of the above mentioned techniques is that the characteristic difference among the cells of a battery pack has not been considered. As we know, in EV applications, due to the requirement of voltage and power, the battery system is generally composed of tens to hundreds of cells connected in series. Because of the restrictions of production technology and tolerances, material defects and contaminations, small differences among cells may exist. Furthermore, in real applications, the working conditions, e.g., temperature distributions, are also different among cells. This non-homogeneity among cells leads to the peak power of the battery system being limited by the weakest cell. Although this is a big challenge of battery management, to the best of our knowledge, there are few researches on this problem so far.

An ideal solution of this problem is to estimate the peak power for each individual cell online, i.e., to design an estimator which works well for estimating cell peak power, and to replicate that estimator N times to estimate the peak power for all the N series-connected cells in the battery systems. With the accurate power estimation of all the cells, we can determine the peak power for the battery system by considering the limitation of the weakest cell. This method provides the best estimation of the peak power for the battery system, however, it incurs a high computation cost, thus is not suitable to implement online within a low-cost microcontroller-based BMS. In reference [5], Waag et al. proposed a power estimation technique with consideration of the difference between the characteristics of individual cells in a battery pack. The method has been validated with a software-in-the-loop test. The adaption is implemented through a simplified relationship between cell voltage and pack averaged voltage. The method is enlightening, however, it assumes the dynamic of each cell to be equal, which may not necessarily hold true in real battery systems; moreover, the difference of polarization effect of each individual cell is not strictly considered.

In this paper, as the main scientific contribution, a peak power estimation technique which comprehensively considers the cell difference in a pack is proposed. A novel model parameter identification algorithm considering cell difference is firstly put forward. A mean battery cell model is fabricated to describe the average characteristics of the battery pack. The average parameter of the battery pack is estimated with the mean battery model, then several ratio vectors describing the characteristic difference between the battery pack and each battery cell are used to yield the parameter estimation for each cell. The ratio vectors are deduced strictly based on a comprehensive analysis of the pack's and cells' characteristics. Based on the parameter identification, the peak power estimation of the battery pack is further developed. Since the parameters for each cell are obtained, the limitation, imposed by the weakest cell, on peak power is also taken into consideration in power estimation. Some validation experiments are implemented in which the power estimations with 1 s, 10 s and 30 s durations are obtained. The results indicate that with the online parameter identification considering cell difference, the proposed power estimation method is adaptive to different aging states, working currents and cell inconsistency.

2. Power Estimation for One Single Battery Cell

Peak power means, based on the present conditions, the maximum power that can be maintained continuously for a specific time period, e.g., 1 s or 10 s, without violating the preset operational limits on the cells. From this definition, the peak power is limited by the safe operation area of the battery, which is normally defined by temperature, voltage, current and *SOC* etc. Since the forecast period of power prediction is less than tens of seconds, the influence of temperature and *SOC* changes can be neglected because they do not change rapidly. Thus, in this paper, the limitations by voltage and current are mainly considered in power estimation.

2.1. Lumped Parameter Battery Model

Normally, in the model-based peak power estimation, the lumped-parameter battery model is needed, and the equivalent circuit model shown in Figure 1 is widely used because of its simplicity and acceptable accuracy. The model can be described in a mathematical way with Equation (1), where U_{OC} represents the open circuit voltage (*OCV*) which relates with *SOC* directly. R_O is the ohmic resistance, and R_{TH} and C_{TH} are the impedance parameters normally corresponding to charge transfer and double layer capacitor effects. U_B and U_{TH} are the terminal voltage and the voltage on C_{TH} respectively, I_B is the working current. All parameters (R_O, R_{TH} and C_{TH}) change with different temperatures, *SOC* and *SOH* of the battery.

$$\begin{cases} U_B = U_{oc} - U_{TH} - I_B R_O \\ \dot{U}_{TH} = -\dfrac{U_{Th}}{R_{TH}C_{TH}} + \dfrac{I_B}{C_{TH}} \end{cases} \tag{1}$$

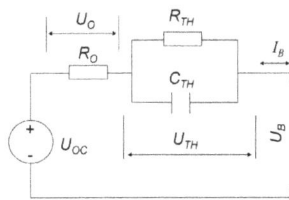

Figure 1. The equivalent circuit model.

2.2. Online Model Parameter Identification

The accuracy of the battery model influences the estimate of the peak power. As we mentioned above, all the parameters in the model vary along with the actual working conditions of the battery. Moreover, the parameters cannot be obtained from direct measurements using sensors. Thus, the accurate identification of the model parameters in various conditions is the key to guarantee the accuracy of the power estimate.

For the online model parameter identification of li-ion batteries, many researches can be found. Basically, the parameter identification techniques are the recursive least squares (RLS) type or adaptive filtering (AF) type of methods [16–25], e.g., Xiong et al. [23] proposed a data-driven estimation approach which can simultaneously obtain the model parameter and the internal state of the battery. These techniques have been widely proven to be effective in online parameter identification. Because of the recursive computation process, these methods are easy to implement in real time. In such techniques, the parameters of the battery model are identified with only real-time measurements of current and voltage needed. In this paper, we use the RLS-based parameter identification technique.

If we define the dynamic voltage response caused by the cell impedance as:

$$U_d = U_B - U_{OC}(SOC) \tag{2}$$

then, the difference equation of the dynamic voltage shown in Figure 1 can be expressed with:

$$U_{d,k+1} = aU_{d,k} + bI_{B,k+1} + cI_{B,k} \tag{3}$$

$$\begin{cases} a = e^{-\frac{\Delta t}{\tau}} \\ b = -R_O \\ c = e^{-\frac{\Delta t}{\tau}} R_O - (1 - e^{-\frac{\Delta t}{\tau}}) R_{TH} \\ \tau = R_{TH} C_{TH} \end{cases} \tag{4}$$

where Δt is the sampling period of current and voltage, and k is the sampling point. Define:

$$\begin{cases} z_k = U_{d,k} \\ h_k = [U_{d,k-1}, I_{B,k}, I_{B,k-1}] \\ \hat{\theta}_{LS,k} = [a, b, c] \end{cases} \tag{5}$$

then the RLS-based model parameter identification can be implemented recursively with the following equation:

$$\begin{cases} \hat{\theta}_{LS,k+1} = \hat{\theta}_{LS,k} + L_{k+1}(z_{k+1} - h_{k+1}^T \hat{\theta}_{LS,k}) \\ L_{k+1} = P_k h_{k+1}(1 + h_{k+1}^T P_k h_{k+1}) \\ P_{k+1} = P_k - L_{k+1} h_{k+1}^T P_k \end{cases} \tag{6}$$

Note that, in this parameter identification method, the battery *SOC* should be obtained in advance. The *SOC* estimation has been well studied in a lot of previous researches, for example, Xiong et al. [26] proposed an online battery *SOC* estimation method. This method was applied based on the hardware-in-loop (HIL) setup, where the novel adaptive H infinity filter was proposed to realize the real-time estimation of the battery *SOC*. The experiment results indicated the high estimation accuracy and strong robustness of the method to the model uncertainty and measurement noise. Thus, the *SOC* estimation is not further studied in this paper.

Another factor which should be taken into account is the influence of current and its direction on the parameters. Normally, the parameters of the equivalent circuit model change not only with different temperature, *SOC* and *SOH*, but also with the current and current direction. In this paper, however, the focus is the consideration of the influence of cell difference. Thus, the influence of the current and its direction on the parameters has not been well considered. To take the influence of current and its direction on the parameters into account, the identification algorithm can be divided into to two parts; each part deals with the parameter identification in a specific current direction. In this case, the RLS-based parameter identification algorithms introduced in the paper will be divided into two parts in a similar way. We believe that, with this design, the accuracy of the identification will be improved. This will be considered in our future works.

2.3. Power Estimation

Normally, when the battery cells are used in an EV, the battery supplier will provide a current limitation of the batteries. In any case, the current of the battery cell should never exceed this limitation value. Besides this limitation, the peak power of the battery in real applications is also limited by voltage. Thus, in this paper, the limitations by both voltage and current are considered in the power estimation.

(1) Limitation by voltage

With a pulse current lasting for a period of $m \times \Delta t$, the terminal voltage of the battery will be [6]:

$$U_{m,m+k} = U_{oc}(SOC_k) - I_{m+k}[\frac{\Delta t}{C}\frac{dU_{oc}(SOC)}{d(SOC)}\Big|_{SOC=SOC_k} + R_O + (1-e^{-\frac{\Delta t}{\tau}})R_{TH}\sum_{i=1}^{m-1}(e^{-\frac{\Delta t}{\tau}})^{m-1-i}] - (e^{-\frac{\Delta t}{\tau}})^m U_{TH,k} \quad (7)$$

thus, the maximum charge and discharge current considering the voltage limitation are:

$$
\begin{cases}
I_{min,vol,k+m} = \dfrac{U_{oc}(SOC_k) - U_{TH,k}(e^{-\frac{\Delta t}{\tau}})^m - U_{max}}{\frac{\Delta t}{C}\frac{dU_{oc}(SOC)}{d(SOC)}\Big|_{SOC=SOC_k} + R_O + R_{TH}(1-e^{-\frac{\Delta t}{\tau}})\sum_{i=1}^{m-1}(e^{-\frac{\Delta t}{\tau}})^{m-1-i}} \\[4mm]
I_{max,vol,k+m} = \dfrac{U_{oc}(SOC_k) - U_{TH,k}(e^{-\frac{\Delta t}{\tau}})^m - U_{min}}{\frac{\Delta t}{C}\frac{dU_{oc}(SOC)}{d(SOC)}\Big|_{SOC=SOC_k} + R_O + R_{TH}(1-e^{-\frac{\Delta t}{\tau}})\sum_{i=1}^{m-1}(e^{-\frac{\Delta t}{\tau}})^{m-1-i}}
\end{cases}
\quad (8)
$$

where U_{max} and U_{min} are the allowed maximum and minimum voltages of the battery, I_{min} is the maximum charge current and I_{max} is the maximum discharge current, k is the sampling point, and Δt is the sampling period.

(2) Power estimation

Considering all the limitations by current and voltage, the maximum charge and discharge current of the battery lasting for the period of $m \times \Delta t$ from sampling step k can be determined by:

$$
\begin{cases}
I_{min} = \max(I_{min,batt}, I_{min,vol,k+m}) \\
I_{max} = \min(I_{max,batt}, I_{max,vol,k+m})
\end{cases}
\quad (9)
$$

where $I_{min,batt}$ and $I_{max,batt}$ are the permitted maximum charge and discharge currents suggested by the battery suppliers. Then, the peak power estimation of the battery can be obtained by:

$$
\begin{cases}
P_{min} = U_{m,k+m}I_{min} = \{U_{oc}(SOC_k) - I_{min}[\frac{\Delta t}{C}\frac{dU_{oc}(SOC)}{d(SOC)}\Big|_{SOC=SOC_k} + R_O + R_{TH}(1-e^{-\frac{\Delta t}{\tau}})\sum_{i=1}^{m-1}(e^{-\frac{\Delta t}{\tau}})^{m-1-i}] - (e^{-\frac{\Delta t}{\tau}})^m U_{TH,k}\}I_{min} \\[3mm]
P_{max} = U_{m,k+m}I_{max} = \{U_{oc}(SOC_k) - I_{max}[\frac{\Delta t}{C}\frac{dU_{oc}(SOC)}{d(SOC)}\Big|_{SOC=SOC_k} + R_O + R_{TH}(1-e^{-\frac{\Delta t}{\tau}})\sum_{i=1}^{m-1}(e^{-\frac{\Delta t}{\tau}})^{m-1-i}] - (e^{-\frac{\Delta t}{\tau}})^m U_{TH,k}\}I_{max}
\end{cases}
\quad (10)
$$

According to the elaborations above, the online adaptive peak power estimation of a single battery cell can be illustrated in Figure 2. The process illustrated in Figure 2 has now been widely used and proven to be effective in the power estimation of a single battery cell.

Energies **2017**, *10*, 390

Initialization

$$\hat{\theta}_{LS,0} \quad P_0$$

Online model parameter identification

Signal measurement and preprocessing

RLS base parameter identification

$$\begin{cases} \hat{\theta}_{LS,k+1} = \hat{\theta}_{LS,k} + L_{k+1}(z_{k+1} - h_{k+1}^T \hat{\theta}_{LS,k}) \\ L_{k+1} = P_k h_{k+1}(1 + h_{k+1}^T P_k h_{k+1}) \\ P_{k+1} = P_k - L_{k+1} h_{k+1}^T P_k \end{cases} \quad \text{Eq. (6)}$$

$$U_{d,k+1} = aU_{d,k} + bI_{B,k+1} + cI_{B,k} \quad \text{Eq. (3)}$$

current

SOC

Dynamic voltage

OCV

Terminal voltage

R_O, R_{TH}, C_{TH}

$$\begin{cases} I_{min,vol,k+m} = \dfrac{U_{oc}(SOC_k) - U_{TH,k}(e^{-\frac{\Delta t}{\tau}})^m - U_{max}}{\dfrac{\Delta t}{C} \dfrac{dU_{oc}(SOC)}{d(SOC)}\bigg|_{SOC=SOC_r} + R_O + R_{TH}(1 - e^{-\frac{\Delta t}{\tau}})\sum_{i=1}^{m-1}(e^{-\frac{\Delta t}{\tau}})^{m-1-i}} \\[6pt] I_{max,vol,k+m} = \dfrac{U_{oc}(SOC_k) - U_{TH,k}(e^{-\frac{\Delta t}{\tau}})^m - U_{min}}{\dfrac{\Delta t}{C} \dfrac{dU_{oc}(SOC)}{d(SOC)}\bigg|_{SOC=SOC_r} + R_O + R_{TH}(1 - e^{-\frac{\Delta t}{\tau}})\sum_{i=1}^{m-1}(e^{-\frac{\Delta t}{\tau}})^{m-1-i}} \end{cases} \quad \text{Eq. (8)}$$

$I_{min,batt}$
$I_{max,batt}$

$$\begin{cases} I_{min} = \max(I_{min,batt}, I_{min,vol,k+m}) \\ I_{max} = \min(I_{max,batt}, I_{max,vol,k+m}) \end{cases} \quad \text{Eq. (9)}$$

$$\begin{cases} P_{min} = U_{L,k+m}I_{min} \\ P_{max} = U_{L,k+m}I_{max} \end{cases} \quad \text{Eq. (10)}$$

End — Power estimation

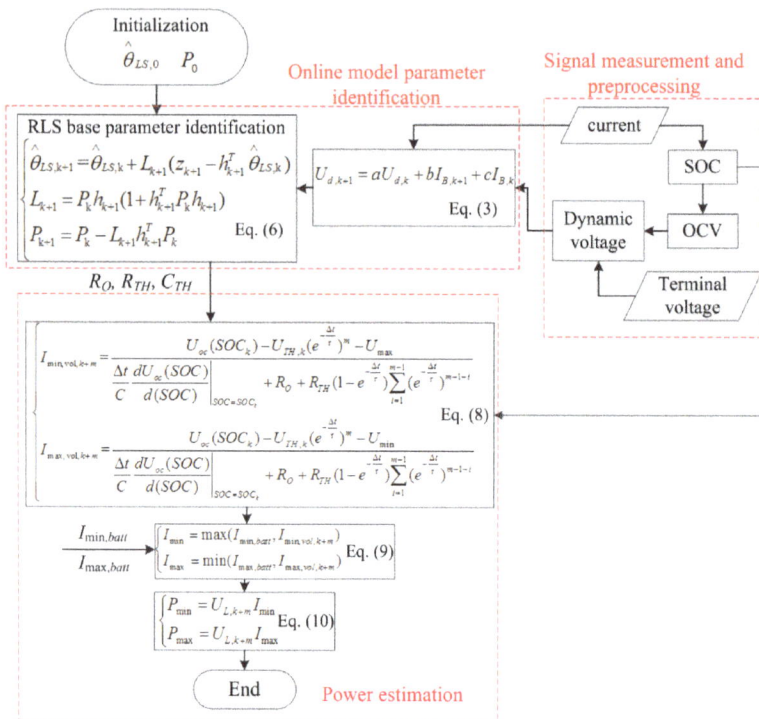

Figure 2. Peak power estimation of a single battery cell. *SOC*: state of charge; and *OCV*: open circuit voltage.

3. Power Prediction of Series-Connected Battery Packs

For a battery pack consisting of tens to hundreds of cells connected in series, it is the performance of each individual cell which limits the peak power. In a battery pack, the peak power is actually limited by the weakest cell, which is the cell that first reaches the predefined voltage or current limit during charging or discharging. Normally, the weakest cell limiting power delivery is the cell with the largest impedance. The *SOC* of each cell also influences the power capability; in any case, in a full-featured BMS, the problem caused by *SOC* imbalance can be alleviated by cell balancing [27] or cell *SOC* estimation [28]. Thus, in this paper, the influence of *SOC* imbalance to power capability is neglected, and the main focus of this study is the influence of cell impedance.

To determine the weakest cell, a straight forward method is to online identify the parameters of all individual cells. This method requires a huge computation cost, and is not suitable to be implemented on a low-cost microcontroller-based BMS. In reference [29], Roscher et al. proposed a reliable state estimation of multi-cell Li-ion battery systems, where a dimensionless vector reflecting the ratio of all the cells' total impedances is used to determine the impedance parameters of all individual cells, and Waag et al. [5] used a similar method to deal with the problem caused by cell difference. The works of Roscher and Waag are enlightening, however, they assume the dynamic of each cell to be equal, which may not necessarily hold true in real battery systems; moreover, the difference of polarization effect of each individual cell is not strictly considered. We propose a peak power estimation technique which comprehensively considers the cell difference in a pack.

3.1. Improved Parameter Identification for Series Connected Battery Systems

(1) Basic idea

For simplicity, a small battery pack consisting of two series-connected cells is taken as an example to develop the improved parameter identification for series-connected battery systems. The impedance parameters of the two cells are R_{O1}, R_{HT1}, C_{TH1}, and R_{O2}, R_{TH2}, C_{TH2}. A battery cell, called "mean cell", with the mean characteristics of the two individual cells is constructed; the impedance parameters of the mean cell are R_{Om}, R_{THm}, C_{THm}. With the fabricated mean cell, the battery pack can be considered to be composed of two same mean cells, and the system is shown in Figure 3.

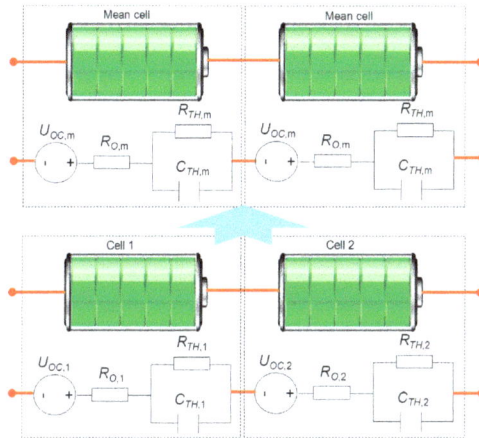

Figure 3. A simplified battery pack consisting of two series-connected individual cells.

We define three ratio vectors, which reflect the differences among the battery characteristics:

$$\begin{cases} A = [\ \frac{R_{O1}}{R_{Om}}, \quad \frac{R_{O2}}{R_{Om}}\ ] \\ B = [\ \frac{\exp(-\Delta t/\tau_1)}{\exp(-\Delta t/\tau_m)}, \quad \frac{\exp(-\Delta t/\tau_2)}{\exp(-\Delta t/\tau_m)}\ ] \\ C = [\ \frac{R_{TH1}}{R_{THm}}, \quad \frac{R_{TH2}}{R_{THm}}\ ] \end{cases} \quad (11)$$

For the battery pack, during charge and discharge, the mean parameters can be easily identified online with the RLS-based algorithm as introduced above when we consider that the pack is composed of two same mean battery cells. Then, the parameters of each individual cell can be obtained by combining the ratio vectors and the online identified mean parameters by the following equation. We can see that the determination of the ratio vectors is critical in the proposed method.

$$\begin{cases} R_{O1} = A[1]R_{Om} \\ R_{O2} = A[2]R_{Om} \\ \exp(-\Delta t/\tau_1) = B[1]\exp(-\Delta t/\tau_m) \\ \exp(-\Delta t/\tau_2) = B[2]\exp(-\Delta t/\tau_m) \\ R_{TH1} = C[1]R_{THm} \\ R_{TH2} = C[2]R_{THm} \end{cases} \quad (12)$$

(2) Determination of ratio vector A

When the battery pack works under a current I, we have:

$$R_{O1}/R_{O2} = U_{O1}/U_{O2} \quad (13)$$

where U_{O1} and U_{O2} are the voltages of R_{O1} and R_{O2} under current I (shown in Figure 1). Then, theoretically, vector A can be determined with:

$$A = [\ \tfrac{R_{O1}}{R_{Om}}, \quad \tfrac{R_{O2}}{R_{Om}}\ ] = [\ \tfrac{U_{O1}}{U_{Om}}, \quad \tfrac{U_{O2}}{U_{Om}}\ ] \tag{14}$$

In real applications, it is difficult to obtain U_{O1}, U_{O2} and U_{Om} online because they cannot be directly measured, thus it is difficult to get the value of A with Equation (14). However, according to the model shown in Figure 1, we can find that, if the current changes suddenly, R_O will cause a sudden voltage change. Based on this analysis, we can determine A with the sudden voltage changes of each individual cell and the mean cell, as shown in Equation (15), where ΔU_{s1}, ΔU_{s2} and ΔU_{sm} are the sudden voltage changes of the cells respectively.

$$A = [\ \tfrac{R_{O1}}{R_{Om}}, \quad \tfrac{R_{O2}}{R_{Om}}\ ] = [\ \tfrac{\Delta U_{s1}}{\Delta U_{sm}}, \quad \tfrac{\Delta U_{s2}}{\Delta U_{sm}}\ ] \tag{15}$$

We here illustrate the voltage responses of the two cells under a current cycle, as shown in Figure 4a. The RLS-based identification results of R_O for the cells in this case are shown in Figure 4b. We can find, from Figure 4, that the sudden voltage change ratio of the cells is very close to the R_O ratio of the cells (in this case, the ratio is close to 1.05), which proves that Equation (15) can be used to determine vector A.

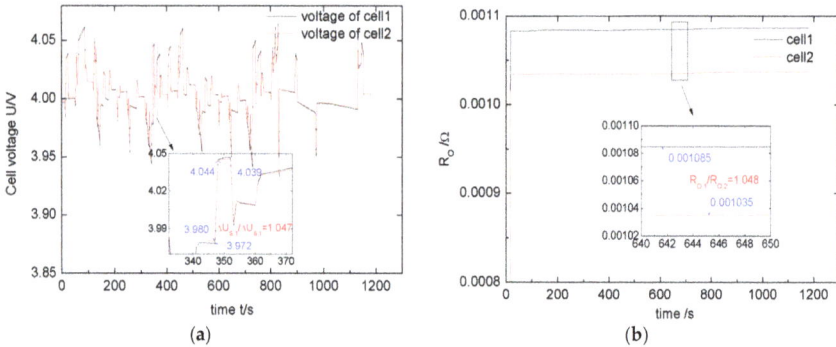

Figure 4. Voltage and ohmic resistance of the two cells: (**a**) voltage of the two cells; and (**b**) ohmic resistance of the two cells.

(3) Determination of ratio vector B

The polarization voltages of the cells can be obtained by:

$$\begin{cases} U_{TH1,k} = \exp(-\tfrac{\Delta t}{\tau_1})U_{TH1,k-1} + (1 - \exp(-\tfrac{\Delta t}{\tau_1}))R_{TH1}I_{B,k} \\ U_{TH2,k} = \exp(-\tfrac{\Delta t}{\tau_2})U_{TH2,k-1} + (1 - \exp(-\tfrac{\Delta t}{\tau_2}))R_{TH2}I_{B,k} \end{cases} \tag{16}$$

From Equation (16), we have:

$$\frac{U_{TH1,k}}{U_{TH2,k}} = \frac{\exp(-\tfrac{\Delta t}{\tau_1})U_{TH1,k-1} + (1 - \exp(-\tfrac{\Delta t}{\tau_1}))R_{TH1}I_{B,k}}{\exp(-\tfrac{\Delta t}{\tau_2})U_{TH2,k-1} + (1 - \exp(-\tfrac{\Delta t}{\tau_2}))R_{TH2}I_{B,k}} \tag{17}$$

Normally, the time constant of the $R_{TH}C_{TH}$ network is much larger than the sampling period (generally several or tens of mini seconds), thus:

$$\begin{cases} 1 - \exp(-\tfrac{\Delta t}{\tau_1}) \to 0 \\ 1 - \exp(-\tfrac{\Delta t}{\tau_2}) \to 0 \end{cases} \tag{18}$$

By substituting Equation (18) with Equation (17), we have:

$$\frac{U_{TH1,k}}{U_{TH2,k}} \approx \frac{\exp(-\frac{\Delta t}{\tau_1})U_{TH1,k-1}}{\exp(-\frac{\Delta t}{\tau_2})U_{TH2,k-1}} \tag{19}$$

Thus:

$$\frac{\exp(-\frac{\Delta t}{\tau_1})}{\exp(-\frac{\Delta t}{\tau_2})} = \frac{U_{TH1,k}/U_{TH1,k-1}}{U_{TH2,k}/U_{TH2,k-1}} \tag{20}$$

If we define:

$$\begin{cases} K_{\tau 1} = U_{TH1,k}/U_{TH1,k-1} \\ K_{\tau 2} = U_{TH2,k}/U_{TH2,k-1} \\ K_{\tau m} = U_{THm,k}/U_{THm,k-1} \end{cases} \tag{21}$$

then, from Equation (21), we can define ratio vector B as:

$$B = [\ \frac{\exp(-\Delta t/\tau_1)}{\exp(-\Delta t/\tau_m)},\quad \frac{\exp(-\Delta t/\tau_2)}{\exp(-\Delta t/\tau_m)}\ ] = [\ \frac{K_{\tau 1}}{K_{\tau m}},\quad \frac{K_{\tau 2}}{K_{\tau m}}\ ] \tag{22}$$

On the other hand, during charge and discharge, the polarization voltages can also be calculated as:

$$\begin{cases} U_{TH1} = U_{d1} - I_B R_{O1} \\ U_{TH2} = U_{d2} - I_B R_{O2} \end{cases} \tag{23}$$

in which U_d is the dynamic voltage response caused by the impedance, and can be calculated by Equation (2).

After measuring the terminal voltages of each cell and the pack, the dynamic voltage caused by the impedance can be calculated with Equation (2). Then, with Equation (23) and the identified ohmic resistances of cell 1, cell 2 and the mean cell, the polarization voltages can be obtained. Finally, the ratio vector reflecting the difference of polarization time constants, B can be calculated with Equations (21) and (22).

(4) Determination of ratio vector C

Based on Equation (17), we have:

$$\begin{cases} R_{TH1} = \frac{U_{TH1,k} - \exp(-\frac{\Delta t}{\tau_1})U_{TH1,k-1}}{(1-\exp(-\frac{\Delta t}{\tau_1}))} \\ \\ R_{TH2} = \frac{U_{TH2,k} - \exp(-\frac{\Delta t}{\tau_2})U_{TH2,k-1}}{(1-\exp(-\frac{\Delta t}{\tau_2}))} \end{cases} \tag{24}$$

If we define:

$$\begin{cases} K_{RTH1} = \frac{U_{TH1,k} - \exp(-\frac{\Delta t}{\tau_1})U_{TH1,k-1}}{(1-\exp(-\frac{\Delta t}{\tau_1}))} \\ \\ K_{RTH2} = \frac{U_{TH2,k} - \exp(-\frac{\Delta t}{\tau_2})U_{TH2,k-1}}{(1-\exp(-\frac{\Delta t}{\tau_2}))} \\ \\ K_{RTHm} = \frac{U_{THm,k} - \exp(-\frac{\Delta t}{\tau_m})U_{THm,k-1}}{(1-\exp(-\frac{\Delta t}{\tau_m}))} \end{cases} \tag{25}$$

then, the ratio vector C can be determined as:

$$C = [\ \frac{R_{TH1}}{R_{THm}},\quad \frac{R_{TH2}}{R_{THm}}\ ] = [\ \frac{K_{RTH1}}{K_{RTHm}},\quad \frac{K_{RTH2}}{K_{RTHm}}\ ] \tag{26}$$

With vector B, and the identified time constant of the mean cell τ_m, we can get τ_1 and τ_2 with Equation (22), then with Equations (25) and (26), we can finally determine vector C, as shown in Figure 5.

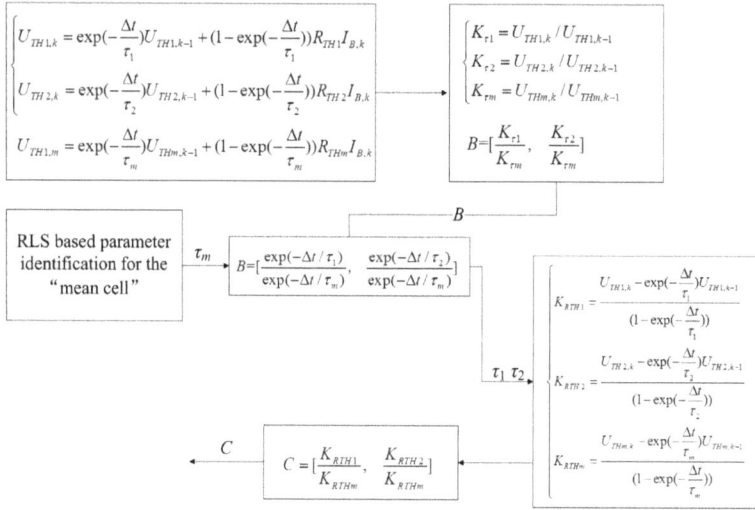

Figure 5. Calculation process of ratio vector B and ratio vector C.

(5) Online update of the ratio vectors

For a battery system composed of N series-connected battery cells, the ratio vectors reflecting the characteristics differences among the cells are:

$$
\begin{cases}
A = [\frac{\Delta U_{s1}}{\Delta U_{sM}}, \frac{\Delta U_{s2}}{\Delta U_{sM}}, \cdots, \frac{\Delta U_{sn}}{\Delta U_{sM}}] \\
B = [\frac{K_{\tau 1}}{K_{\tau M}}, \frac{K_{\tau 2}}{K_{\tau M}}, \cdots, \frac{K_{\tau n}}{K_{\tau M}}] \\
C = [\frac{K_{RTH1}}{K_{RTHM}}, \frac{K_{RTH2}}{K_{RTHM}}, \cdots, \frac{K_{RTHn}}{K_{RTHM}}]
\end{cases}
\tag{27}
$$

When using the ratio vectors to determine the parameters for each individual cell, to avoid the possible fluctuations and errors, an AF is further designed. We here take the design of the AF of vector A as an example.

According to Equation (15), at the sampling step k, the sudden voltage change vector of all the individual cells can be estimated with:

$$
\Delta U_{s,k}^{pre} = A_{k-1}\Delta U_{sM,k}
\tag{28}
$$

If the ratio vector A carries some errors, then the estimated sudden voltage changes of the individual cells should be different from the true values; then, we can adjust A with the errors as:

$$
A_k = A_{k-1} + g_A(\Delta U_{s,k} - \Delta U_{s,k}^{pre})
\tag{29}
$$

in which, g_A is an adjustment gain for vector A.

Similarly, we design the filters for vectors B and C as shown below:

$$
\begin{cases}
K_{\tau,k}^{pre} = K_{\tau M,k}B_{k-1} \\
B_k = B_{k-1} + g_B(K_{\tau,k} - K_{\tau,k}^{pre})
\end{cases}
\tag{30}
$$

$$
\begin{cases}
K_{RTH,k}^{pre} = K_{RTHM,k}C_{k-1} \\
C_k = C_{k-1} + g_C(K_{RTH,k} - K_{RTH,k}^{pre})
\end{cases}
\tag{31}
$$

in which, g_B and g_C are the adjustment gains for vectors B and C respectively.

3.2. Power Estimation Considering Cell Difference

For a battery pack consisting of N series-connected battery cells, the current limitation of each battery cell can be calculated with Equation (8) and the identified cell parameters based on the proposed identification method. The peak power estimation will be:

$$\begin{cases} P_{sysmin} = N \times \min_{i=1:N}(U_{B,i,k+m}^{chg} I_{i,min}) \\ P_{sysmax} = N \times \max_{i=1:N}(U_{B,i,k+m}^{dis} I_{i,max}) \end{cases} \tag{32}$$

In conclusion, the power estimation of the battery system considering the cell difference is illustrated in Figure 6. The main advantage of this method is that the power estimation takes the cell difference into consideration with an acceptable computation cost.

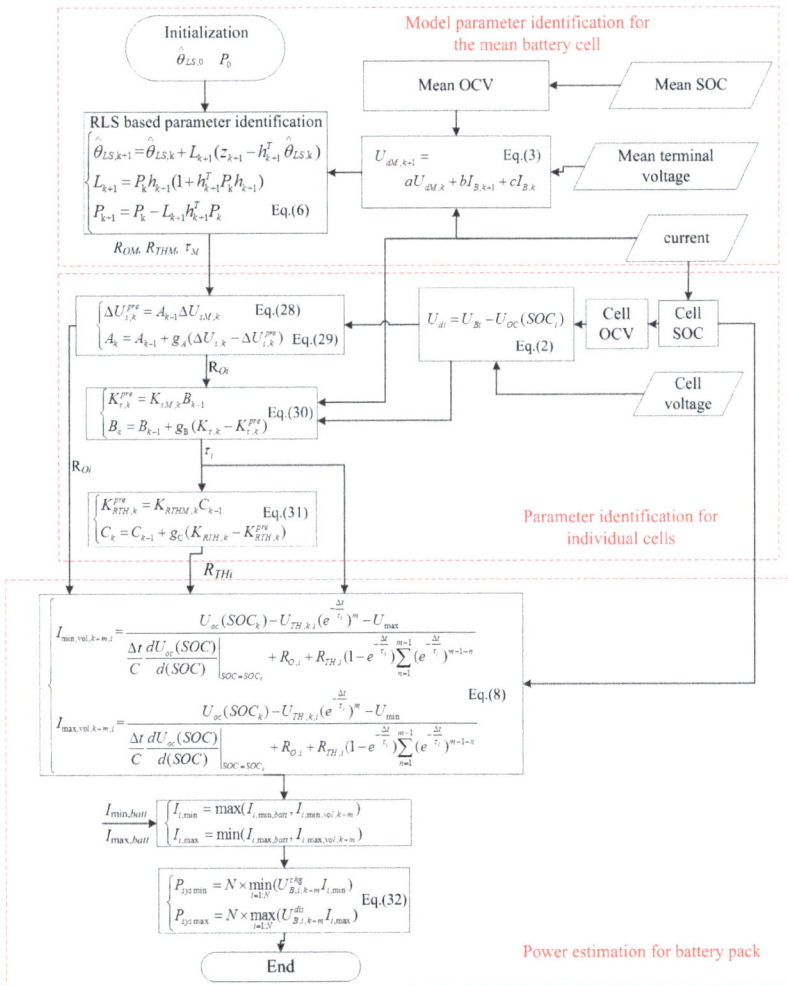

Figure 6. Power estimation of the battery pack consisting of series-connect battery cells considering cell difference.

4. Case Study

4.1. Experimental Setups

A small battery pack composed of 10 series-connected cells is constructed to validate the proposed peak power estimation method. Table 1 lists the basic information and the allowed voltage and current limitations of the battery cell suggested by the supplier at 25 °C.

Table 1. Basic information of the battery cell provided by the supplier.

No.	Parameter	Value
1	Nominal capacity (Ah)	80
2	Nominal voltage (V)	3.7
3	Discharge cut-off voltage (V)	2.8
4	Charge cut-off voltage (V)	4.2
5	Allowed maximum 30 s pulse discharge current (15%–85% SOC) (A)	480
6	Allowed maximum 30 s pulse charge current (15%–85% SOC) (A)	240

Several tests with different current profiles are designed to thoroughly investigate the performance of the method. During the tests, the battery pack is put in an environmental chamber, and the temperature is set to 25 °C. The current profiles are obtained from an EV during the J1015, New European Driving Cycle (NEDC) and Federal Test Procedure-75 (FTP-75) cycle tests, and are shown in Figure 7. The tests are then implemented by a battery tester (Arbin BTS 2000, Arbin, College Station, TX, USA) with the obtained current profiles.

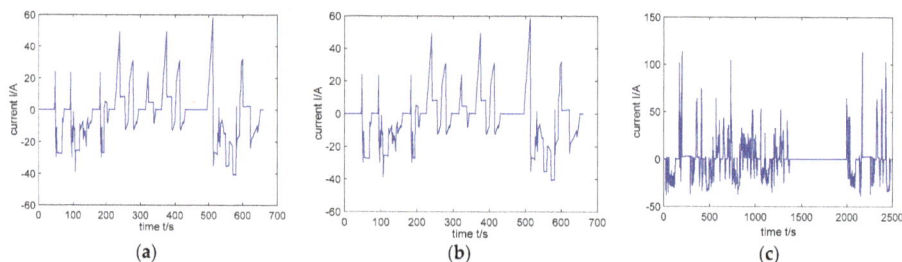

Figure 7. Current profiles of the tests: (**a**) current profile of the J1015 cycle test; (**b**) current profile of the New European Driving Cycle (NEDC) cycle test; and (**c**) current profile of the Federal Test Procedure-75 (FTP-75) cycle test.

To investigate the performance of the proposed method in different cell aging conditions and different current profiles, the battery pack is firstly tested with the NEDC current profile, followed by 50 constant full charge/discharge cycles. Then, the battery pack is tested with the J1015 current profile, followed by another 50 full charge/discharge cycles. At last, the battery pack is tested with the FTP-75 current profile. The overall test procedure is illustrated in Figure 8.

The actual battery current and voltage, and the voltages of the 10 cells during the tests are all simultaneously measured by the BMS. With all the measured currents and voltages, we obtained the parameters of each individual cell by the repetitive implementation of the RLS-based identification, and the identification results are considered as the reference values of the parameters. On the other hand, the cell parameters are also identified by the newly proposed method, and the identification results of the new method are then compared with the reference values obtained from the repetitive RLS-based identification to investigate the performance of the proposed method.

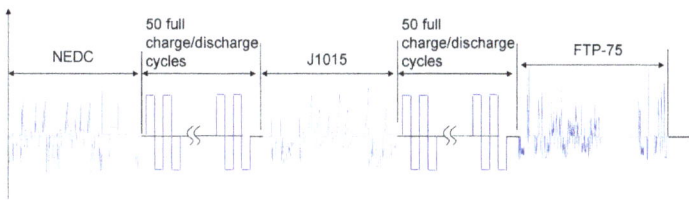

Figure 8. The overall test procedure.

Meanwhile, the power of the battery pack is estimated by the measured signals and the identified parameters. Both the peak power estimations with and without considering cell differences are obtained. The peak power estimation without considering cell differences is based on the parameters of the mean battery cell, and the process is shown in Figure 2, while the power estimation considering cell differences is realized with the method shown in Figure 6. The estimation results with and without considering cell differences are compared to investigate how cell differences affect the peak power of the battery pack. In the identification, the parameters are arbitrarily initialized with zero.

4.2. Results and Discussions

Figures 9–11 show the parameter identification results obtained by the repetitive RLS algorithm and the proposed method during the tests. In all tests, we find that the parameters obtained by the proposed method are very close to the reference values identified by the repetitive RLS algorithm, especially after the convergence of the algorithm. We also find that, because of the design of the AF of the ratio vectors, the convergence of the proposed method is a little slower than the repetitive RLS algorithm. This can be seen more clearly by a close look at the identified results during the interval from 0 s to 200 s. In this time interval, the parameters identified by the different methods show big differences. Before convergence, the parameters identified by both methods have large errors, and the errors are also different in different methods. Thus, during convergence, the parameters are not the same in different methods. However, the convergence of the repetitive RLS-based algorithm is faster than the ratio vector-based algorithm. This is because the ratio vector-based algorithm should first use the results identified by traditional RLS as the mean value of the parameters, and then determine the parameters for each individual cell with the ratio vectors. Moreover, to avoid the possible fluctuations and errors, the adaptive filters are further designed, and the adaptive filters slow down the convergence of the algorithm further. Normally, the RLS-based algorithm converges within less than 150 iteration steps, and according to the results, the presented method converges within less than 250 iteration steps.

To validate the accuracy of the parameters identified by the proposed method, we focus on the identification results at the end of the tests, which are also shown in Figures 9–11. We can see that the parameters identified by the two methods respectively are almost the same. A clear convergence of the ratio vectors can be found in the identification results. By comparing the impedance results shown in Figures 9–11, we can also conclude from the results that, during the aging process, the internal impedances of the battery cells are getting larger, regardless of the ohmic resistance or the charge transfer resistance. The larger impedance during the aging process is caused by increasing the thickness of the SEI layer, and the decreasing the conductivity of the electrolyte etc. [30,31].

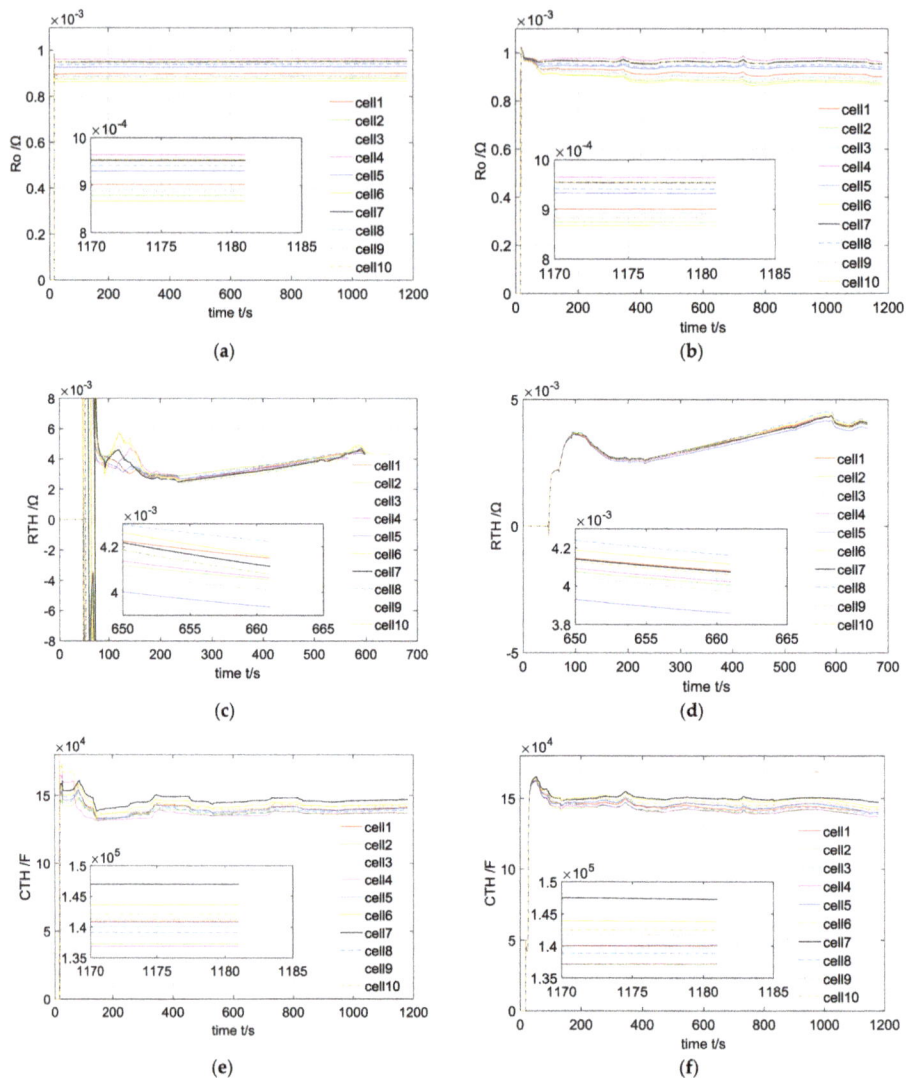

Figure 9. Identified parameters under the NEDC cycle test: (**a**) R_O identified by the repetitive recursive least squares (RLS); (**b**) R_O identified by the proposed new method; (**c**) R_{TH} identified by the repetitive RLS; (**d**) R_{TH} identified by the proposed new method; (**e**) C_{TH} identified by the repetitive RLS; and (**f**) C_{TH} identified by the proposed new method.

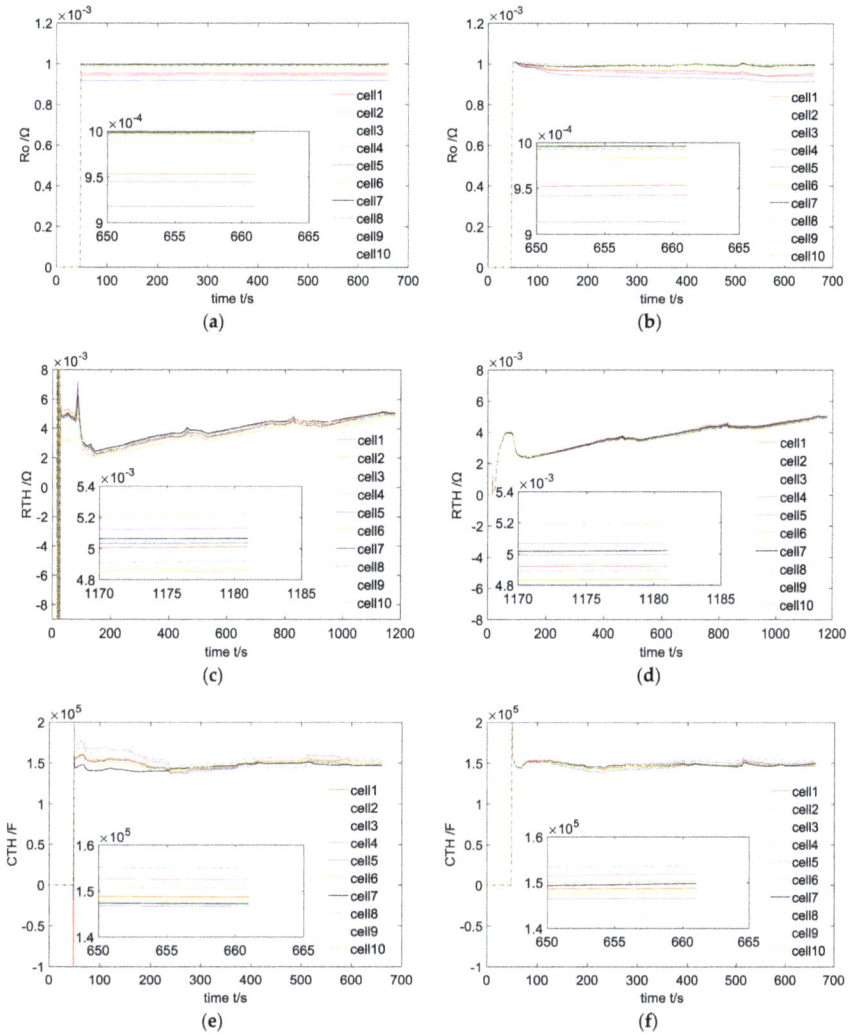

Figure 10. Identified parameters under the J1015 cycle test: (**a**) R_O identified by the repetitive RLS; (**b**) R_O identified by the proposed new method; (**c**) R_{TH} identified by the repetitive RLS; (**d**) R_{TH} identified by the proposed new method; (**e**) C_{TH} identified by the repetitive RLS; and (**f**) C_{TH} identified by the proposed new method.

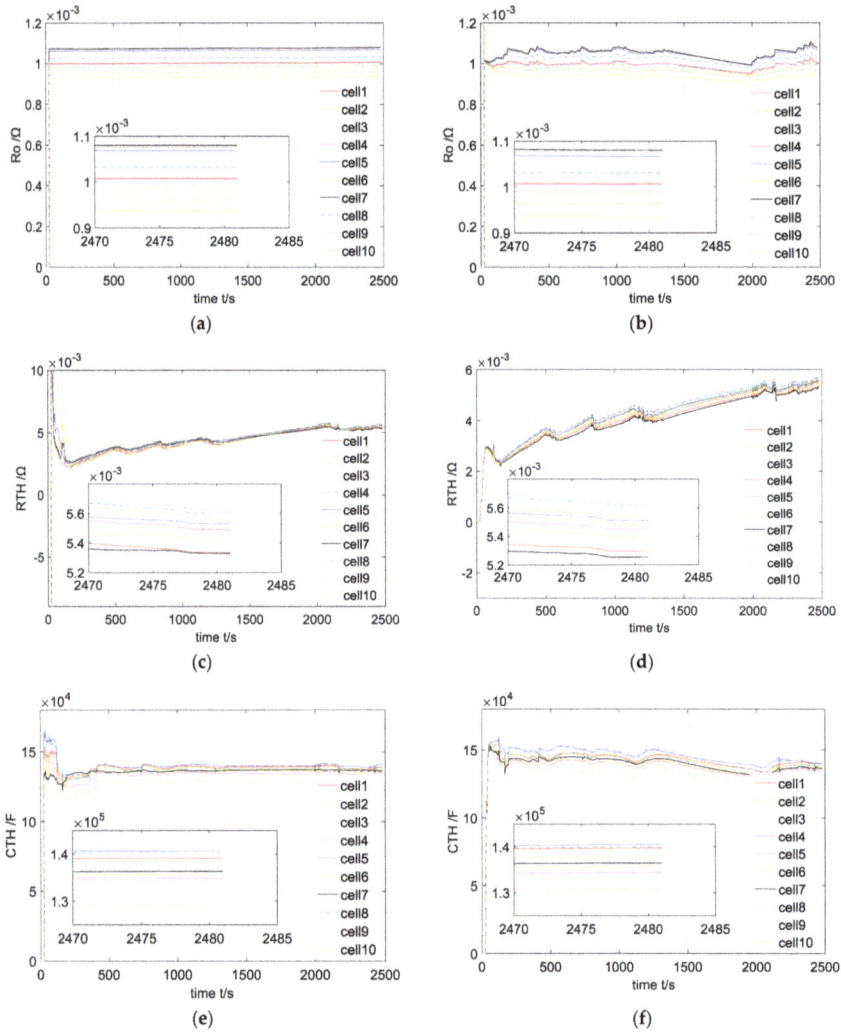

Figure 11. Identified parameters under the FTP-75 cycle test: (**a**) R_O identified by the repetitive RLS; (**b**) R_O identified by the proposed new method; (**c**) R_{TH} identified by the repetitive RLS; (**d**) R_{TH} identified by the proposed new method; (**e**) C_{TH} identified by the repetitive RLS; and (**f**) C_{TH} identified by the proposed new method.

Table 2 lists the comparison of the computation time cost by the two methods, from which we can conclude that the proposed method can achieve the same identification accuracy while evidently reducing the computation cost. Note that the time listed in Table 2 is the total computation time cost by the algorithms in different test cycles when they are executed on a PC with a 2.4 GHz CPU.

Table 2. Computation time cost of the identification algorithms.

Test Cycles	Computation time of the RLS (s)	Computation Time of the New Method (s)	Time Reduced (%)
J1015	1.46	0.67	54.1
NEDC	2.74	1.28	53.2
FTP75	5.71	2.66	53.4

Figures 12 and 13 show the peak power estimation results of the battery pack during the tests. Both the estimations with and without considering cell differences are shown in the figures. Some conclusions can be drawn from the results.

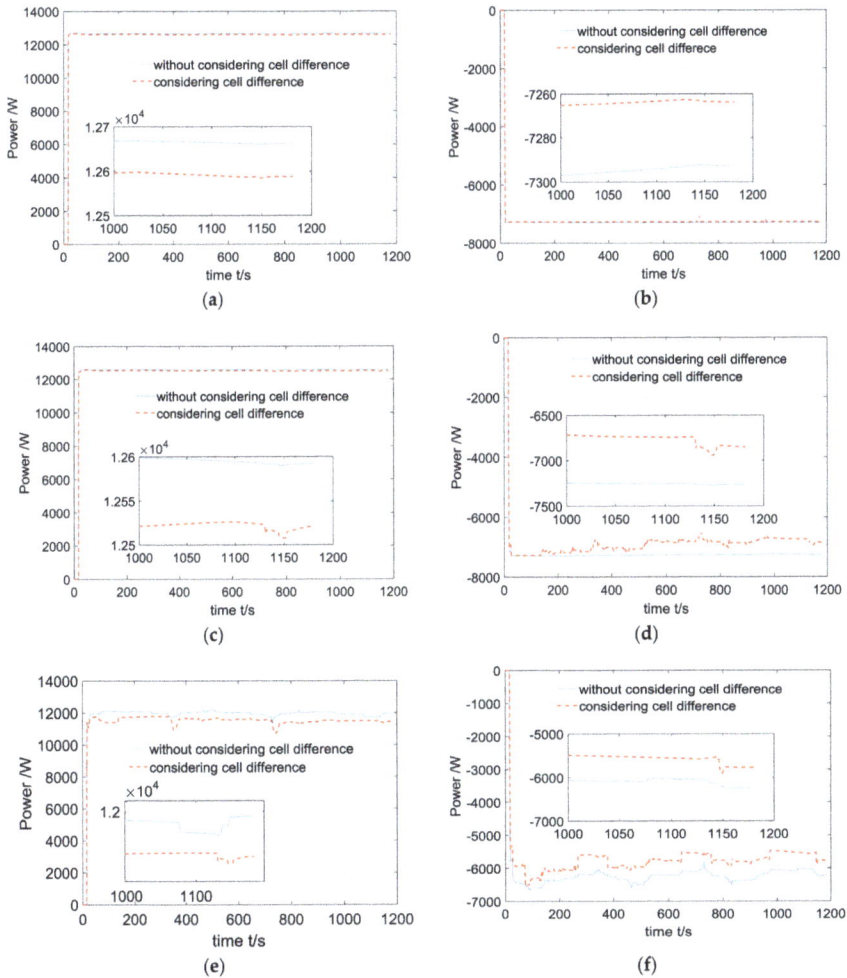

Figure 12. Power estimation results under the NEDC cycle test: (**a**) 1 s discharge power estimation result; (**b**) 1 s charge power estimation result; (**c**) 10 s discharge power estimation result; (**d**) 10 s charge power estimation result; (**e**) 30 s discharge power estimation result; and (**f**) 30 s charge power estimation result.

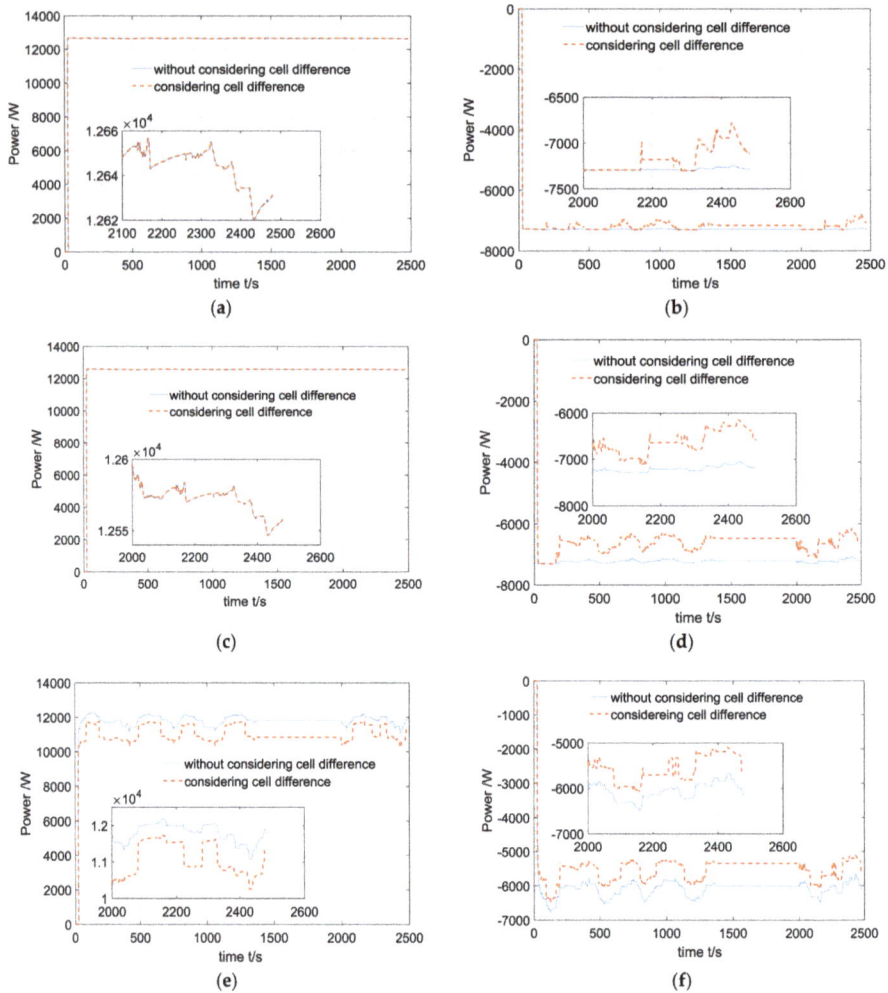

Figure 13. Power estimation results under the FTP75 cycle test: (**a**) 1 s discharge power estimation result; (**b**) 1 s charge power estimation result; (**c**) 10 s discharge power estimation result; (**d**) 10 s charge power estimation result; (**e**) 30 s discharge power estimation result; and (**f**) 30 s charge power estimation result.

(1) The power capability of the battery pack is firstly influenced by the required power duration; the longer the duration required, the smaller the power capability will be. The power capability lasting for 1 s is obviously larger than the power capabilities lasting for 10 s and 30 s. This is reasonable because during charge/discharge, the impedance caused by the internal polarization of the battery plays a more and more dominant role, which limits the power capability of the batteries. This can also be deduced from Equations (8) and (10).

(2) In the middle SOC range, the charging power capability is smaller than the discharging power capability within the same condition. From Equations (8) and (10), we can see that, if the difference between the charge and discharge impedances is neglected, the main factors affecting charging and discharging power capabilities are the cut-off voltages, U_{max} and U_{min}. Generally,

the difference between the OCV and U_{min} is much larger than the difference between the OCV and U_{max}, especially when the battery works in a middle SOC range, which makes the charging power capability smaller than the discharging power capability.

(3) During the aging process, the power capability of the battery pack is getting smaller. We can see this phenomenon by comparing the estimation results shown in Figures 12 and 13. We can clearly find that the peak power estimations during the FTP-75 test are smaller than those during the NEDC test. As introduced above, between the FTP-75 and NEDC tests, there exist 50 full charge/discharge cycles, in which the impedances of the battery cells are getting larger. The larger impedance introduces a smaller power capability.

(4) Due to cell difference, the power estimation results considering cell difference are smaller than the estimation results without considering cell difference. This result supports the statement that, because of the cell difference, the peak power of the battery pack is limited by the characteristics of the weakest cell. In the estimation without considering cell difference, an equivalent battery module that consisted of mean battery cells is used (Figure 2). The performance of the mean cell will of course be better than the worst cell in the pack.

(5) The influence of the cell difference on the pack power capability is dependent on the power duration. For the charge/discharge powers with a 1 s duration, the influence of cell difference can be neglected. This means that, for instant power capability estimations, it is not necessary to consider the influence of cell difference. However, if the estimation of continuous power capability is needed, then the consideration of cell difference is critical. We can see from the results that, in the estimations of the 30 s power capability, the results considering cell difference are much smaller than those without considering cell difference. The phenomenon can actually be deduced from Equations (8) and (10). During the process of charge/discharge, the impedance caused by the internal polarization of the battery plays a more and more dominant role, limiting the power capability of the batteries. The difference of polarization impedance among the cells then affects the power capability more and more evidently as the current excitation continues.

5. Conclusions

This paper proposed a new method to online estimate the power capability of the battery packs composed of series-connected cells. The main contribution is that the cell inconsistency is considered in the estimation by designing a novel ratio vector-based parameter identification algorithm. The main conclusions and summarizations are drawn below.

(1) A ratio vector-based parameter identification algorithm is proposed which can achieve the same identification accuracy as the repetitive RLS-based identification while evidently reducing the computation cost. This facilitates the online implementation of the algorithm.

(2) Based on the ratio vector-based parameter identification, the estimation of the power capability of the battery pack composed of series-connected cells is further developed. Validation results indicate that the proposed method is effective to estimate the power capability considering cell difference.

(3) Due to cell difference, the power estimation considering cell difference is smaller than the estimation results without considering cell difference.

(4) The influence of the cell difference on the pack power capability is dependent on the power duration. For instant power capability estimations, it is not necessary to consider the influence of cell difference. However, if the estimation of continuous power capability is needed, the consideration of cell difference is critical.

Acknowledgments: This work is financially supported by the National Natural Science Foundation of China (NSFC, Grant No. 51677136).

Author Contributions: Bo Jiang performed the experiments and wrote the paper; Haifeng Dai conceived the idea and designed the experiments; Xuezhe Wei and Letao Zhu analyzed the data; Zechang Sun contributed helpful suggestions in the design of the algorithm. All authors read and approved the manuscript.

Conflicts of Interest: The authors declare no conflict of interest.

Nomenclature

U_B	Battery terminal voltage
U_{OC}	Battery open circuit voltage
I_B	Battery working current
R_O	Ohmic resistance of the battery
U_{TH}	Voltage on network describing the charge transfer effect
R_{TH}	Resistance of charge transfer
C_{TH}	Double layer capacitor
U_d	Dynamic voltage on the impedance under current excitation
Δt	Sampling period
k	Sampling point
θ	Parameter vector of the battery model
U_{max}	Allowed maximum voltage of the battery
U_{min}	Allowed minimum voltage of the battery
$I_{max,batt}$	Allowed maximum discharge current of the battery suggested by the battery suppliers
$I_{min,batt}$	Allowed maximum charge current of the battery suggested by the battery suppliers
$I_{max,volt}$	Allowed maximum discharge current of the battery limited by voltage
$I_{min,volt}$	Allowed maximum charge current of the battery limited by voltage
R_{Om}	Ohmic resistance of the mean battery cell
R_{THm}	Charge transfer resistance of the mean battery cell
C_{THm}	Equivalent capacitor of charge transfer effect of the mean battery cell
U_O	Voltage on the ohmic resistance under current excitation
ΔU_{sm}	Sudden voltage change during current pulse of the mean battery cell
ΔU_s	Sudden voltage change during current pulse of the battery
A	Vector reflecting the difference of ohmic resistance
B	Vector reflecting the difference of time constant of charge transfer effect
C	Vector reflecting the difference of charge transfer resistance
ΔU_s^{pre}	Predicted sudden voltage change during current pulse
N	Number of the battery cells connected in series
g_A	Adjustment gain for vector A
g_B	Adjustment gain for vector B
g_C	Adjustment gain for vector C
P_{sysmin}	Maximum charge power of the battery system
P_{sysmax}	Maximum discharge power of the battery system
$K_{\tau,k}^{pre}$	Predicted ratio of the charge transfer voltages at sampling point k
K_τ	Ratio of charge transfer voltages
K_{RTH}	Ratio of the charge transfer resistance
$K_{RTH,k}^{pre}$	Predicted ratio of the charge transfer resistance at sampling point

References

1. Lu, L.; Han, X.; Li, J.Q.; Hu, J.F.; Ouyang, M. Review on the key issues for lithium-ion battery management in electric vehicles. *J. Power Source* **2013**, *226*, 272–288. [CrossRef]
2. Castaings, A.; Lhomme, W.; Trigui, R.; Bouscayrol, A. Comparison of energy management strategies of a battery/supercapacitors system for electric vehicle under real-time constraints. *Appl. Energy* **2016**, *163*, 190–200. [CrossRef]
3. *PNGV Battery Test Manual: Revision 3*; US Department of Energy: Washington, DC, USA, 2001.
4. Gregory, L.P. High-performance battery-pack power estimation using a dynamic cell model. *IEEE Trans. Veh. Technol.* **2004**, *53*, 1586–1593. [CrossRef]
5. Waag, W.; Fleischer, C.; Sauer, D.U. Adaptive on-line prediction of the available power of lithium-ion batteries. *J. Power Source* **2013**, *242*, 548–559. [CrossRef]
6. Sun, F.C.; Xiong, R.; He, H.W.; Li, W.Q.; Aussems, J.E.E. Model-based dynamic multi-parameter method for peak power estimation of lithium-ion batteries. *Appl. Energy* **2012**, *96*, 378–386. [CrossRef]

7.   Pei, L.; Zhu, C.B.; Wang, T.S.; Lu, R.G.; Chan, C.C. Online peak power prediction based on a parameter and state estimator for lithium-ion batteries in electric vehicles. *Energy* **2014**, *66*, 766–779. [CrossRef]

8.   Burgos-Mellado, C.; Orchard, M.E.; Kazerani, M.; Cardenas, R.; Saez, D. Particle-filtering-based estimation of maximum available power state in lithium-ion batteries. *Appl. Energy* **2016**, *161*, 349–363. [CrossRef]

9.   Xiong, R.; Sun, F.C.; He, H.W.; Nguyen, T.D. A data-driven adaptive state of charge and power capability a data-driven adaptive state of charge and power capability joint estimator of lithium-ion polymer battery used in electric vehicles. *Energy* **2013**, *63*, 295–308. [CrossRef]

10. Zhang, W.; Shi, W.; Ma, Z.Y. Adaptive unscented Kalman filter based state of energy and power capability estimation approach for lithium-ion battery. *J. Power Source* **2015**, *289*, 50–62. [CrossRef]

11. Malysz, P.; Ye, J.; Gu, R.; Yang, H.; Emadi, A. Battery state-of-power peak current calculation and verification using an asymmetric parameter equivalent circuit model. *IEEE Trans. Veh. Tech.* **2016**, *65*, 4512–4522. [CrossRef]

12. Wang, S.Q.; Verbrugge, M.; Wang, J.S.; Liu, P. Power prediction from a battery state estimator that incorporates diffusion resistance. *J. Power Source* **2012**, *214*, 399–406. [CrossRef]

13. Waag, W.; Fleischer, C.; Sauer, D.U. Critical review of the methods for monitoring of lithium-ion batteries in electric and hybrid vehicles. *J. Power Source* **2014**, *258*, 321–339. [CrossRef]

14. Bohlen, O.; Gerschler, J.B.; Sauer, D.U. Robust algorithms for a reliable battery diagnosis-managing batteries in hybrid electric vehicles. In Proceedings of the 22nd Electric Vehicle Symposium (EVS22), Yokohama, Japan, 23–28 October 2006.

15. Kim, D.Y.; Jung, D.Y. Method of Estimating Maximum Output of Battery for Hybrid Electric Vehicle. U.S. Patent 7,518,375 B2, 14 April 2009.

16. Duong, V.H.; Bastawrous, H.A.; Lim, K.C.; See, K.W.; Zhang, P.; Dou, S.X. Online state of charge and model parameters estimation of LiFePO$_4$ battery in electric vehicles using multiple adaptive forgetting factors recursive least-squares. *J. Power Source* **2015**, *296*, 215–224. [CrossRef]

17. Wei, Z.B.; Lim, T.M.; Skyllas-Kazacos, M.; Wai, N.; Tseng, K.J. Online state of charge and model parameter co-estimation based on a novel multi-time scale estimator for vanadium redox flow battery. *Appl. Energy* **2016**, *172*, 169–179. [CrossRef]

18. Plett, L.G. Extended Kalman filtering for battery management systems of LiPB-based HEV battery packs: Part 2. Modeling and identification. *J. Power Source* **2004**, *134*, 262–276. [CrossRef]

19. Dai, H.F.; Wei, X.Z.; Sun, Z.C. Recursive parameter identification of lithium-ion battery for EVs based on equivalent circuit model. *J. Comput. Theor. Nanosci.* **2013**, *10*, 2813–2818. [CrossRef]

20. Zou, Y.; Hu, X.S.; Ma, H.M.; Li, S.E. Combined state of charge and state of health estimation over lithium-ion battery cell cycle lifespan for electric vehicles. *J. Power Source* **2015**, *273*, 793–803. [CrossRef]

21. Nejad, S.; Gladwin, D.T.; Stone, D.A. A systematic review of lump-parameter equivalent circuit models for real-time estimation of lithium-ion battery states. *J. Power Source* **2016**, *316*, 183–196. [CrossRef]

22. Feng, T.H.; Yang, L.; Zhao, X.W.; Zhang, H.D.; Qiang, J.X. Online identification of lithium-ion battery parameters based on an improved equivalent-circuit model and its implementation on battery state-of-power prediction. *J. Power Source* **2015**, *281*, 192–203. [CrossRef]

23. Sun, F.; Xiong, R.; He, H. A systematic state-of-charge estimation framework for multi-cell battery pack in electric vehicles using bias correction technique. *Appl. Energy* **2016**, *162*, 1399–1409. [CrossRef]

24. Lim, K.C.; Bastawrous, H.A.; Duong, V.H.; See, K.W.; Zhang, P.; Dou, S.X. Fading Kalman filter-based real-time state of charge estimation in LiFePO4 battery-powered electric vehicles. *Appl. Energy* **2016**, *169*, 40–48. [CrossRef]

25. Partovibakhsh, M.; Liu, G.J. An adaptive unscented Kalman filtering approach for online estimation of model parameters and state-of-charge of lithium-ion batteries for autonomous mobile robots. *IEEE Trans. Control Syst. Technol.* **2015**, *23*, 357–363. [CrossRef]

26. Chen, C.; Xiong, R.; Shen, W. A lithium-ion battery-in-the-loop approach to test and validate multi-scale dual H infinity filters for state of charge and capacity estimation. *IEEE Trans. Power Electr.* **2017**. [CrossRef]

27. Hua, Y.; Cordoba-Arenas, A.; Warner, N.; Rizzoni, G.A. multi time-scale state-of-charge and state-of-health estimation framework using nonlinear predictive filter for lithium-ion battery pack with passive balance control. *J. Power Source* **2015**, *280*, 293–312. [CrossRef]

28. Dai, H.A.; Wei, X.Z.; Sun, Z.C.; Wang, J.Y.; Gu, W.J. Online cell SOC estimation of Li-ion battery packs using a dual time-scale Kalman filtering for EV applications. *Appl. Energy* **2012**, *95*, 227–237. [CrossRef]

Energies **2017**, *10*, 390

29. Roscher, M.A.; Bohlen, O.S.; Sauer, D.U. Reliable state estimation of multicell lithium-ion battery systems. *IEEE Trans. Energy Convers.* **2011**, *26*, 7373–7743. [CrossRef]

30. Barre, A.; Deguilhem, B.; Grolleau, S.; Gerard, M.; Suard, F.; Riu, D. A review on lithium-ion battery ageing mechanisms and estimations for automotive applications. *J. Power Source* **2013**, *241*, 680–689. [CrossRef]

31. Waag, W.; Kabitz, S.; Sauer, D.U. Experimental investigation of the lithium-ion battery impedance characteristic at various conditions and aging states and its influence on the application. *Appl. Energy* **2013**, *102*, 885–897. [CrossRef]

energies

MDPI

Article

Impact of Silicon Carbide Devices on the Dynamic Performance of Permanent Magnet Synchronous Motor Drive Systems for Electric Vehicles

Xiaofeng Ding *, Min Du, Jiawei Cheng, Feida Chen, Suping Ren and Hong Guo

School of Automation Science and Electrical Engineering, BeiHang University, Beijing 100191, China; dumin@buaa.edu.cn (M.D.); chengjiawei0218@126.com (J.C.); dege820@126.com (F.C.); rsp93_4@126.com (S.R.); guohong@buaa.edu.cn (H.G.)
* Correspondence: dingxiaofeng@buaa.edu.cn; Tel.: +86-10-8233-9498

Academic Editors: Hailong Li and Joe (Xuan) Zhou
Received: 5 January 2017; Accepted: 6 March 2017; Published: 15 March 2017

Abstract: This paper investigates the impact of silicon carbide (SiC) metal oxide semiconductor field effect transistors (MOSFETs) on the dynamic performance of permanent magnet synchronous motor (PMSM) drive systems. The characteristics of SiC MOSFETs are evaluated experimentally taking into account temperature variations. Then the switching characteristics are firstly introduced into the transfer function of a SiC-inverter fed PMSM drive system. The main contribution of this paper is the investigation of the dynamic control performance features such as the fast response, the stability and the robustness of the drive system considering the characteristics of SiC MOSFETs. All the results of the SiC-drive system are compared to the silicon-(Si) insulated gate bipolar transistors (IGBTs) drive system counterpart, and the SiC-drive system manifests a higher dynamic performance than the Si-drive system. The analytical results have been effectively validated by experiments on a test bench.

Keywords: silicon carbide (SiC) MOSFET; silicon (Si) IGBTs; permanent magnet synchronous motor (PMSM); switching characteristics; dynamic performance

---

1. Introduction

Wide band-gap power device materials, such as silicon carbide (SiC), are drawing increasing attention due to a number of superior qualities they possess, such as high switching-speed, lower specific on-resistance, and higher junction operating temperature capability [1–7]. The application of SiC in a motor drive inverter can reduce both switching and conduction losses, shorten dead time in a phase-lag, and increase switching frequency, etc. Hence, the SiC-inverter can provide higher efficiency and higher power density in comparison to its silicon (Si) inverter counterpart [8–11], which is benefit for the electric vehicles with limited capacity battery [11,12].

Aside for these inverter-level benefits, the SiC-inverter also affects features of the dynamic performance of a motor drive system, such as fast response, relative stability and robustness, etc. This has not been clearly addressed yet, therefore, this paper investigates the impact of a SiC-inverter on the performance of a motor drive system, which is unlike some previous works that solely considered SiC-inverter systems [13].

Because of its high efficiency and fast response characteristics, permanent magnet synchronous motors (PMSMs) are widely adopted in a host of high performance applications where low torque ripple, high efficiency, and remarkable dynamic response are highly demanded, such as dynamic positioning systems, machine-tool spindle electrical power steering and traction drives in electric vehicles, etc. [14]. Hence, the dynamic performance of a PMSM fed by a SiC voltage source inverter (VSI) is worthy of further investigation. Numerous research activities have been conducted to analyze different aspects of the dynamic response, such as frequency response, fast response, relative stability and robustness [15–17].

The dynamic performance analysis of a VSI-fed PMSM is based on state equations, with the *d*- and *q*-axis components of the stator currents and the rotor flux linkage as state variables [15]. A transfer function of the drive system was developed according to the state equations and control theory [18–20]. Then classical techniques such as Bode plots and Nyquist diagrams were used for analyzing the dynamic performance of the controlled system [20,21]. The traditional Bode plots were used to evaluate the frequency response in control engineering [21]. The frequency response manifests the bandwidth and the fast response capability of the system.

Not only an adequate bandwidth promises a fast response ability of a PMSM system, but also sufficient relative stability and robustness guarantee its long-term robust operation. The relative stability of a system can be evaluated by the specified gain- and phase-margins described by Nichols plots of loop transfer functions [22]. In addition, the robustness of the system can be investigated by H-infinity conditions on sensitivity functions [23,24].

The dynamic performance of a motor is mainly related with sampling delay and microprocessor calculation time, control strategies, and the characteristics of a given motor and inverter, etc. Researchers have performed a large amount of valuable work regarding the dynamic performance of the motor drive system [25–27]. In [25], the sampling and microprocessor delay were considering in a transfer function of a drive system to analyze the dynamic performance of a PMSM system. In [26], a synchronous sampling (synchronized with the pulse-width modulation (PWM) carrier) of instantaneous phase current value was adopted in a current feedback loop, which achieved a higher bandwidth than the conventional methods. In [27], the oversampling, namely sampling and controller updating done with a higher rate than the regular sampling methodology counterparts, is adopted to improve the frequency response.

Aside from the sampling delay, the computational effort on the digital signal processor (DSP) also causes a time delay [28–32]. In order to overcome the time delay, some program compensations have been done [28]. In addition to the software program improvements, faster microprocessors were adopted recently to reduce the delay time, such as field-programmable gate arrays (FPGAs), which take advantage of parallel calculations [29–32]. The use of efficient hardware and software allows the control algorithm to be run in less than 10 μs [29], which is less than the computation time of a DSP28335 (usually more than 50 μs in our previous projects).

In addition to the sampling and microprocessor analysis, more research activities are currently conducted on the applied control strategies for improving the dynamic performance of PMSM drive systems [15,28,33–37]. In [15], a new predictive direct torque control (DTC) method was developed, which is introduced to achieve the fastest dynamic response in the transient state compared to the conventional DTC method. In this proposed method only the parameters of one voltage vector need to be calculated in contrast with the previous studies where the parameters of two voltage vectors must be calculated, resulting in a lower computational burden. In [28], a proposed controller is based on a combination of deadbeat and direct predictive control techniques for a PMSM drive system. The computational delay was compensated by a modification of the control process. The proposed technique exhibited faster response as well as better robustness than the conventional proportional integral (PI) field-oriented control (FOC) technique. In [33], a per-phase control in the abc-domain was proposed, which requires only one frame transformation execution. Such a control scheme yields excellent sinusoidal current command tracking and disturbance rejection.

Although many advanced control strategies and emerging fast computation microprocessor technologies have been to improve the dynamic performance of the PMSM drive system, few investigations have been done from the viewpoint of SiC-MOSFET characteristics. Hence, the purpose of this study was to explore the impact of SiC-MOSFET switching on the dynamic performance of a PMSM drive system. The SiC-MOSFET characteristics are evaluated experimentally through a double-pulse test (DPT), taking temperature into consideration. The switching performance is applied to a physical model of voltage distortion in a SiC-inverter to quantify the distorted voltages. Then, the switching performance of a SiC-MOSFET is firstly introduced into the transfer function of the PMSM drive system. As a result, the fast response, the stability, and the robustness of the system can be conveniently investigated from a control theory point of view. Finally, the analytical results are effectively validated by experiments. All the studies of the SiC-MOSFET drive system are compared to a Si-IGBT drive system counterpart.

The remainder of this paper is organized as follows: in Section 2, the characteristics of the power devices are evaluated by DPT, and the distorted voltage of the inverter is quantitatively calculated. In Section 3, a novel PMSM transfer function is developed, taking into account the switching performance of the power device. In Section 4, the dynamic performances of the PMSM are analyzed based on transfer functions. In Section 5, the experimental setup and the validation are illustrated. Conclusions are drawn in the final section.

2. Power Devices Characteristics

In order to analyze the dynamic performance of the PMSM drive system considering the power device characteristics, in this section of the paper we investigate the switching characteristics of a SiC MOSFET (CAS300M12BM2, CREE, Durham, NC, USA) and a Si IGBT (FF400R12KE3, Infineon, Am Campeon, Neubiberg, Germany) by a double-pulse test (DPT). The DPT setup was built as shown in Figure 1, where a DC source voltage (270 V) is assigned to a leg of the inverter, and the inductors of two phases in the PMSM are adopted as the load inductors. The lower switch of phase-leg is selected as the device under test (DUT). A high precision voltage probe (P5100A, Tektronix, Beaverton, OR, USA) and a TCPA300 current probe (Tektronix, Beaverton, OR, USA) plus a TCP303 are used to measure the voltage (V_{DS}) and channel current (I_C) of the DUT, respectively.

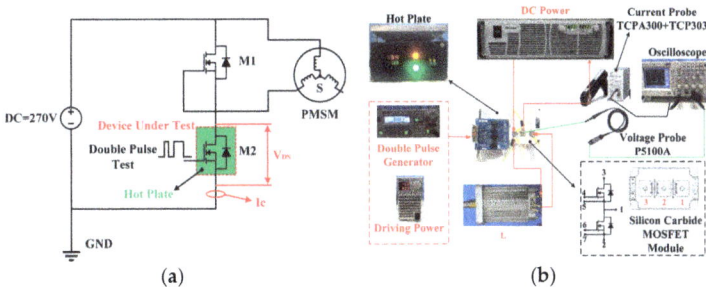

Figure 1. DPT test bench. (**a**) Simplified DPT circuit; (**b**) Actual components in the test bench.

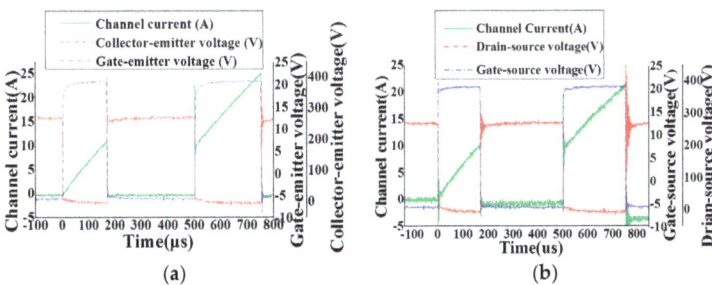

Figure 2. Total switching transitions. (**a**) Si IGBT; (**b**) SiC MOSFET.

This setup is used to characterize the on-state voltage drop, the turn-on and turn-off transient processes. A comparison of the switching waveforms for the Si IGBT and SiC MOSFET are presented in Figure 2. The enlarged figures of turn-on and turn-off transients are shown in Figure 3. The red lines show the collector-emitter voltage (V_{CE}) for IGBT or drain-source voltage (V_{DS}) for the MOSFET. The green lines represent the channel current of the power devices and the blue lines display the waveforms of the gate-drive signals.

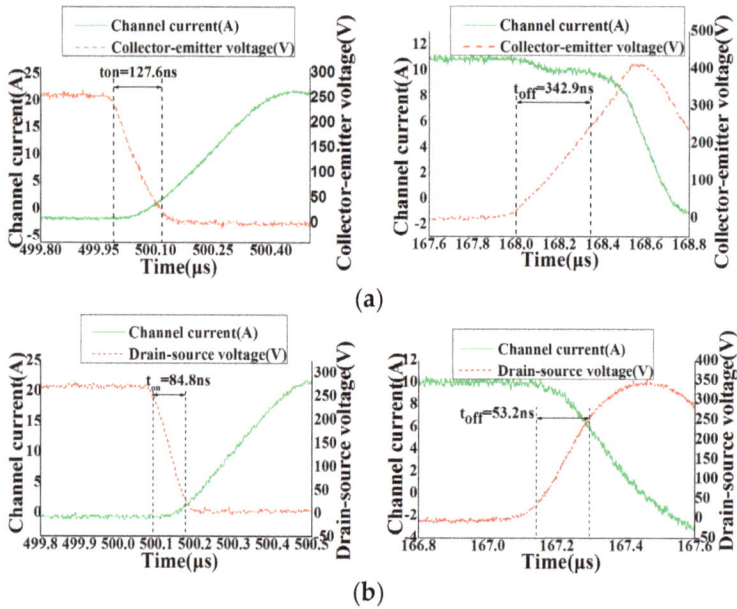

Figure 3. Turn-on and turn-off trajectories. (a) Si IGBT; (b) SiC MOSFET.

Figure 3 shows that the turn-on and turn-off times of the Si IGBT are 127.6 ns and 342.9 ns, respectively, when the ambient temperature is 25 °C and the channel current is 11A. Both the turn-on and turn-off times of the SiC MOSFET, which are 84.8 ns and 153.2 ns, correspondingly, are smaller than those of the Si counterpart.

In addition, the accumulation of system losses leads to an increase in the device junction temperature after a long operation time. The characteristics of the devices under different temperatures are systematically explored in this paper, which is helpful to analyze the properties of semiconductor devices in practical applications taking thermal effects into account. Hence, the switching times of both SiC and Si are tested at different temperatures applied by a hot plate.

Figure 4 illustrates the switching times of power devices at different temperatures. The turn-off times of Si IGBT increase as the increasing temperature. However, the switching time of SiC MOSFET remains constant under different temperatures. In addition to the switching times of the two devices, the other characteristics of the two devices are also measured as shown in Table 1. The voltage drop of SiC MOSFET is about 5% of Si IGBT. The value of SiC MOSFET is 44.2 mV and the counterpart of Si IGBT is 832.5 mV. The output capacitance of SiC MOSFET is also smaller than that of Si IGBT. These different characteristics are vital for calculating the phase voltage distortions and evaluating the dynamic performances of SiC- and Si-drive system.

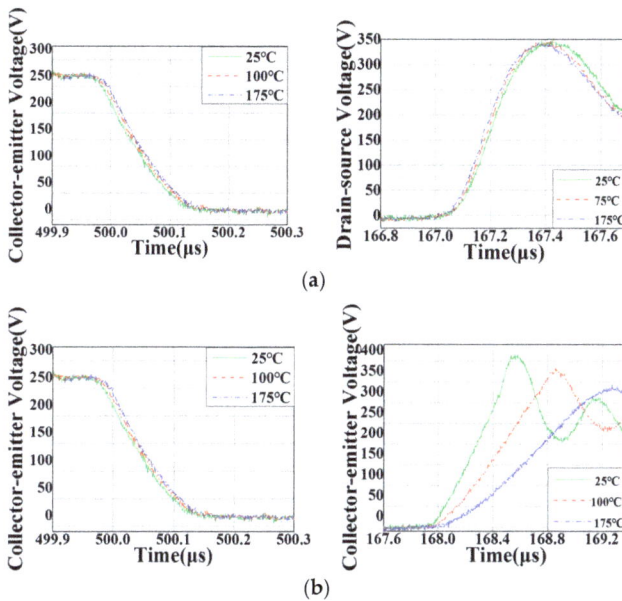

Figure 4. Switching time of power devices at different temperatures. (**a**) Si IGBT; (**b**) SiC MOSFET.

Table 1. Test parameters measured by the DPT.

Symbol	Quantity	Value	
		SiC MOSFET	Si IGBT
t_{on}	Turn on time	84.8 ns	127.6 ns
t_{off}	Turn off time	153.2 ns	342.9 ns
V_d	Voltage drop	44.2 mV	832.5 mV
C_o	Output capacitance	12.7 nF	32.7 nF
ΔV	average voltage distortion	−2.42 V	−3.28 V

In [38,39], the phase voltage distortions of the inverter are developed, taking the dead time, switching time, voltage drop and output capacitance into account. In Figure 5a, the deviation appears in the waveform of the inverter phase voltage. V_{ref} displays the ideal fundamental phase voltage of the inverter that would result if there were no distortion voltage effects. As the PMSM is an inductive load, the phase current waveform i lags behind V_{ref} by an angle θ'. Since the distortion voltage increases (decreases) the inverter phase voltage for the negative (positive) half cycle of the phase current as shown in Figure 5b, the average voltage distortion over an entire cycle could be illustrated by the square wave in Figure 5a.

The average voltage distortion is the superposition of ΔV on the ideal voltage V_{ref} shown as broken line in Figure 5a. Therefore, the fundamental phase voltage with distortion voltage V_1 is the sum of V_{ref} and ΔV_1, which is described as a heavy solid curve. When the harmonic components of the current are ignored the phase displacement between V_1 and i corresponds to the fundamental power factor angle of the load. It can be seen that the real fundamental phase voltage differs from the reference one in both the phase and magnitude. Therefore, when the fundamental phase voltage reduces and power factor angle increases, the current will increase to maintain a constant output power.

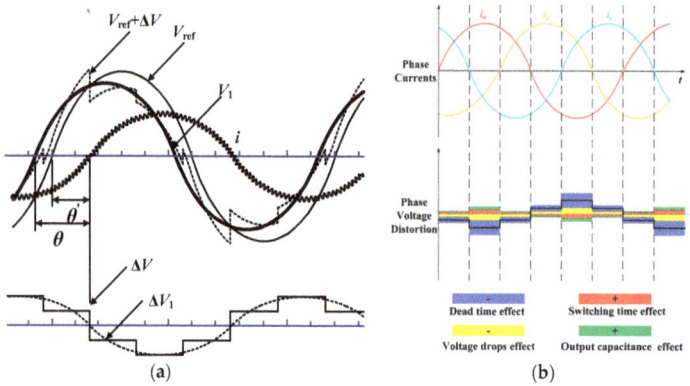

Figure 5. The voltage distortion and its impact on current. (**a**) Representation of waveforms of the voltage and current; (**b**) Average voltage distortion induced by four aspects.

According to the mathematical model of the phase voltage distortion, less switching times, smaller output capacitances and lower voltage drops of SiC MOSFET will reduce the phase voltage distortion. With a faster switching speed, the dead-time of SiC MOSFET in phase-leg configuration can be reduced, which helps to reduce the phase voltage distortion.

3. Transfer Functions of a PMSM Drive System Considering Device Characteristics

In order to conveniently investigate the impact of the power devices' characteristics on the dynamic performance of a PMSM, this paper develops a novel PMSM drive system transfer function taking into account the device characteristics. Then the impacts of SiC-MOSFETs on the fast response, relative stability and robustness of the PMSM drive system are explored based on the proposed transfer function.

3.1. Transfer Function of the Inverter

The input signal u_c of inverter originates from a microprocessor (DSP38335), which is 3.3 V space vector PWM (SVPWM) waveform. Then, the input signal is amplified to a power signal u_a via a driver broad and a power device. A phase lag between u_c and u_a is induced due to the lag of the inverter. Hence, the inverter used in motor drive system can be assumed as a black box with a gain and a phase delay. The simplified transfer function of it could be represented by a first-order lag system [40]:

$$G_1(s) = K_r/(\tau_r s + 1) \tag{1}$$

where, $K_r = \frac{2}{\pi} u_a / u_{cm}$ is the inverter gain, u_{cm} is the maximum gate drive signal voltage, u_a is the maximum phase voltage. τ_r is delay time constant, depends on the switching cycle $1/f_{sw}$, turn-on time t_{on}, turn-off time t_{off} and the dead time T_{dt}:

$$\tau_r = a_1/f_{sw} + a_2 T_{dt} + a_3 t_{on} + a_4 t_{off} \tag{2}$$

where, a_1, a_2, a_3, a_4 are the coefficients of the switching cycle, dead time, turn-on time and turn-off time, respectively.

Both the turn-on and turn-off time of SiC are smaller than those of the Si counterpart as shown in Section 2. The benefit of this is that a smaller dead time duration and a higher switching frequency can be achieved in a SiC-inverter. Hence, the delay time constant τ_{r1} of the SiC-inverter is shorter than that of a Si-inverter according to (2):

$$\tau_{r1} < \tau_{r2} \tag{3}$$

It is known that the distorted voltage of SiC-inverter is less than that of a Si-inverter counterpart as shown in Figure 5. Consequently, the amplitude of the phase voltage of SiC-inverter is higher than that of a Si-inverter when the DC sources are the same ones. Hence, it is concluded that the gain of a SiC-inverter K_{r1} is higher than that of a Si-inverter K_{r2}:

$$K_{r1} < K_{r2} \tag{4}$$

3.2. Transfer Function of the Motor and PI Regulator

The input and output are the command and output currents, respectively. The control strategy of $i_d = 0$ is adopted in this drive system. Hence, the tracking characteristic of the q-axis current i_q can reflect well the dynamic performance of the system. A simplified equation of the motor voltage is shown in the following equation:

$$u_q = Ri_q + p(L_q i_q) + \psi_f w_e \tag{5}$$

where, p is a differential operator, ψ_f is the flux linkage of the permanent magnet, w_e is the electric angular velocity of the motor. From (5), the transfer function of PMSM in q-axis frame is expressed as:

$$G_2(s) = 1/(L_q s + R_a) \tag{6}$$

where, L_q is the q-axis inductance and R_a is the stator resistance.

Besides, the equation of PI regulator in the time domain is:

$$u(t) = K_p e(t) + K_i \int e(t)dt \tag{7}$$

where, K_p and K_i are the proportional and integral coefficients, respectively and $e(t)$ is the error between the reference and feedback signals. The transfer function of the PI regulator can be obtained by a Laplace transform of (7):

$$G_3(s) = K_p(\tau_i s + 1)/(\tau_i s) \tag{8}$$

where, $\tau_i = K_p/K_i$.

In conclusion, the current control model of the PMSM drive system is shown in Figure 6. The open-loop transfer function of system can be written as:

$$G(s) = G_1(s)G_2(s)G_3(s) = \frac{K_q K_r K_p(\tau_i s + 1)}{\tau_i s(\tau_q s + 1)(\tau_r s + 1)} \tag{9}$$

where, $K_q = 1/R_a$, $\tau_q = L_q/R_a$.

The PI regulator of the q-axis current is designed by applying pole zero cancellation, namely setting $\tau_i = \tau_q$ [40]. Thus the closed-loop transfer function is obtained taking the characteristics of devices into account:

$$G_c(s) = \frac{K_q K_r K_p}{\tau_r \tau_i s^2 + \tau_i s + K_q K_r K_p} = \frac{w_n^2}{s^2 + 2\xi w_n s + w_n^2} \tag{10}$$

where, $w_n = \sqrt{K_q K_r K_p / \tau_r \tau_i}$ represents the undamped natural frequency. $\xi = \sqrt{\tau_i / 4K_q K_r K_p \tau_r}$ represents the damping ratio.

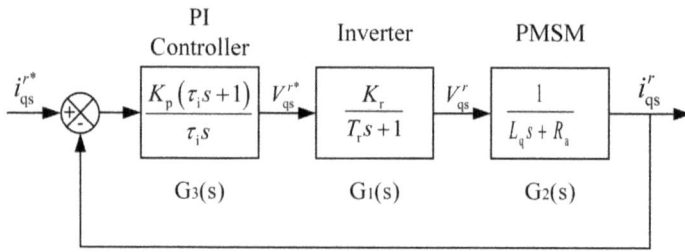

Figure 6. Schematic model of q-axis current of PMSM.

4. Dynamic Performance Analysis of the PMSM

4.1. Fast Response Performance

As a basic requirement of industrial applications, an advanced and smart inverter fed PMSM system needs fast response performance. In general, the fast response performance of the control system can be evaluated via the bandwidth and the settling time.

According to (10), the closed-loop frequency characteristic of the system can be expressed as:

$$G_c(j\omega) = \frac{i_{qs}^r(j\omega)}{i_{qs}^{r*}(j\omega)} = \frac{\omega_n^2}{(j\omega)^2 + 2\xi\omega_n(j\omega) + \omega_n^2} = M_B(\omega)e^{j\varphi_B(\omega)} \tag{11}$$

When the closed-loop amplitude-frequency characteristic $M_B(\omega)$ equals 0.707 times the amplitude at zero-frequency, the corresponding frequency is defined as cut-off frequency ω_b, i.e.:

$$1/\sqrt{\left(1 - \frac{\omega_b^2}{\omega_n^2}\right)^2 + \left(2\xi\frac{\omega_b}{\omega_n}\right)^2} = 0.707 \tag{12}$$

Hence, the relationship among the cut-off frequency ω_b, the undamped natural frequency ω_n and the damping ratio ξ can be obtained as:

$$\omega_b = \omega_n\sqrt{1 - 2\xi^2 + \sqrt{2 - 4\xi^2 + 4\xi^4}} \tag{13}$$

From the monotonicity analysis of (13), it can be concluded that ω_b decreases as the increase of ξ, and increases as ω_n increses. The following equations are obtained by substituting (3) and (4) into (10):

$$\omega_{n1} > \omega_{n2} \tag{14}$$

$$\xi_1 > \xi_2 \tag{15}$$

where, ω_{n1} and ω_{n2} represent the undamped natural frequencies of SiC- and Si-drive systems, respectively. ξ_1 and ξ_2 represent the damping ratios of SiC- and Si-drive systems, respectively.

Equations (14) and (15) are substituted into (13), resulting in:

$$\omega_{b1} > \omega_{b2} \tag{16}$$

Hence, compared with the Si-drive system, the SiC-drive system manifests a higher bandwidth leading to a fast response.

In addition, the settling time of the system t_s can be calculated as:

$$t_s \approx 4/\xi\omega_n = 4/\left(\sqrt{\tau_i/4K_qK_rK_p\tau_r} \cdot \sqrt{K_qK_rK_p/\tau_r\tau_i}\right) = 8\tau_r \tag{17}$$

From (17), the settling time t_s is directly proportional to the delay time constant τ_r of the inverter. Hence, the settling time of the SiC-drive system is shorter than that of a Si-drive system according to (3). According to the aforementioned analyzing results and the system parameters shown in Table 2, the Bode plot of the closed-loop transfer functions for the two systems was drawn, as shown in Figure 7. The bandwidth ω_b is usually defined by -3 dB amplitude response in Bode plots for a closed loop drive system. Thanks to the superior switching characteristics of SiC MOSFET, the bandwidth of the SiC-drive system is 1190 Hz, which is higher than the 1090 Hz of the Si-drive system.

Table 2. The main parameters of the PMSM drive system.

Symbol	Quantity	Value
U_{DC}	Input DC voltage	270 V
U_{cm}	Maximum gate driving voltage	20 V
L_q	q axis inductance	5.19 mH
R_a	Stator resistance	0.25 Ω
K_p	Proportional coefficient	3.8

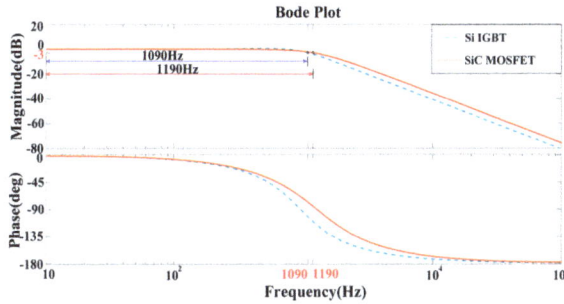

Figure 7. Bode plot of the inverter fed PMSM system with different power devices.

4.2. Relative Stability Analysis

In the process of motor control system design, stability is a necessary condition for the control system to work properly. Additionally, the control system should also have a high level of relative stability. During the operation of a PMSM system, accumulated losses will lead to an increase in temperature. Furthermore, changes in the temperature affect the stator resistance R_a and q-axis inductance L_q, which are likely to destroy the stability of the whole system.

Based on the Nyquist criterion, when the open-loop transfer function of the control system has no pole in the right part of the S-plane, and the open-loop frequency characteristic curve $G(j\omega)$ passes through the point $(-1, j0)$, the control system is at the critical stable edge. In this case, if the control system parameters drift, and it is possible to make open-loop frequency characteristic curve of the control system surround the point $(-1, j0)$, resulting in the control system instability.

In control theory, the stability margins which include the phase margin and amplitude margin are evaluation indexes of the relative stability of the system. The phase- and amplitude-margins of the system determine the stability of the dynamic performance.

From the open-loop transfer function $G(s)$ of the drive system in (9), the frequency characteristic $G(j\omega)$ can be obtained as:

$$G(j\omega) = \omega_n^2 / (j\omega(j\omega + 2\xi\omega_n)) \tag{18}$$

The amplitude-frequency characteristic $A(\omega)$ and the phase-frequency characteristic $\phi(\omega)$ are:

$$A(\omega) = \omega_n^2 / \left(\omega \sqrt{\omega^2 + (2\xi\omega_n)^2} \right) \tag{19}$$

$$\varphi(\omega) = -90° - \arctan(\omega/2\xi\omega_n) \tag{20}$$

When the open-loop amplitude-frequency characteristic of a system is equal to 1, i.e., $A(\omega) = |G(j\omega)| = 1$, the corresponding frequency is defined as the open-loop cut-off frequency ω_c. The difference between the phase angle $\varphi(\omega_c)$ and $-180°$ at the open-loop cut-off frequency represents the phase margin γ of the system, that is:

$$20\lg|G(j\omega_c)| = 0 \,(\mathrm{dB}) \tag{21}$$

$$\gamma = 180° + \varphi(\omega_c) \tag{22}$$

According to (19) and (21), the open-loop cut-off frequency can be calculated as:

$$\omega_c = \omega_n \sqrt{-2\xi^2 + \sqrt{4\xi^2 + 1}} \tag{23}$$

And the phase margin γ can be derived from (22):

$$\gamma = \arctan \frac{2\xi}{\sqrt{\sqrt{4\xi^4 + 1} - 2\xi^2}} = \arctan \frac{1}{\sqrt{-1/2 + 1/4\sqrt{4 + 1/\xi^4}}} \tag{24}$$

It is observed that the phase margin is only related to the arctangent function, the phase margin increases with an increase of damping ratio ξ. From (15), the damping ratio ξ_1 of the SiC-drive system is larger than that of the Si-drive system ξ_2, which results in a larger phase margin in SiC-drive system.

When the open-loop frequency characteristic curve of the system intersects with the negative real axis, i.e., $\varphi(\omega) = -180°$, the corresponding frequency can be obtained by (20), known as the phase cross-over frequency ω_g:

$$\omega_g = 2\xi\omega_n \cdot \tan 90° = n \cdot 2\xi\omega_n \tag{25}$$

where, n tends to infinity.

At ω_g, the reciprocal of the open-loop amplitude-frequency characteristic $A(\omega_g)$ is defined as the amplitude margin K_g of the drive system:

$$K_g = 1/A(\omega_g) = 1/|G(j\omega_g)| \tag{26}$$

According to (19) and (26), the amplitude margin of a system can be calculated as;

$$K_g = 4n\omega_n\xi^2 \sqrt{n^2 + 1} \tag{27}$$

The amplitude margin of the second-order system also tends to infinity, which makes the comparison between the two systems impossible. In order to facilitate the comparison, it is assumed that the phase cross-over frequency ω_g is the corresponding frequency of the phase-frequency characteristic $\varphi(\omega) = -179°$ replacing $\varphi(\omega) = -180°$. The amplitude margin at this scenario can be calculated as:

$$K_g \approx 4 \times 57^2 \cdot \omega_n\xi^2 \tag{28}$$

Equations (14) and (15) show that the natural frequency ω_n and the damping ratio ξ of the SiC-drive system are higher than the counterparts of the Si-drive system. Hence, it is not difficult to draw the following conclusion from (28):

$$K_{g1} > K_{g2} \tag{29}$$

where K_{g1}, K_{g2} represent the amplitude margin of SiC- and Si-drive system, respectively.

In order to figure out the phase- and amplitude-margins of the system, a Nichols plot is plotted using the transfer function of the drive system. Unlike the Bode plot, the Nichols plot presents both phase and magnitude information in one diagram.

In summary, due to the faster switching speed and lower voltage drop of the SiC MOSFET, the voltage distortion of the system is smaller. Consequently, the natural frequency and damping ratio in the second-order system transfer function of a SiC-drive system are larger than that of a Si-drive system, which yields larger phase- and amplitude-margins in the SiC-drive system.

As shown in Figure 8, the phase- and amplitude-margins of the SiC-drive system are 67.63° and 97.68 dB, which is larger than the counterparts of the Si-drive system (59.18° and 91.39 dB, respectively). Hence, the relative stability of the SiC-drive system is better.

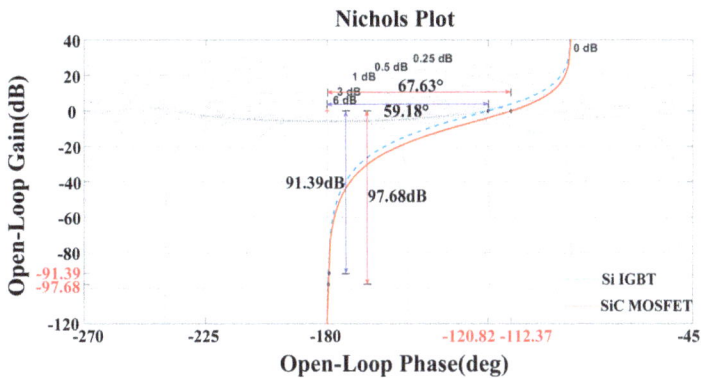

Figure 8. Nichols plot of the inverter fed PMSM system with different power devices.

4.3. Robustness Analysis

Robustness represents the sensitivity of the control system. In a case of a servo system, the robustness performance analysis is one of the fundamental issues in controller design. When the system is influenced by energy-bounded interference signals, such as system parameter variations due to varying load torque, the system robustness is a significant index to evaluate system performance.

Figure 9 shows a single-input-single-output (SISO) PMSM drive system with negative feedback-control. As the energy-bounded interference signal $u(T)$ injected into the system, the controlled plant $G_2(s)$ transforms to $\tilde{G}_2(s)$, as described by:

$$\tilde{G}_2(s) = G_2(s) + \Delta G_2(s) \tag{30}$$

where $\Delta G_2(s)$ is the uncertain part of the plant which is assumed to satisfy the following equation:

$$|\Delta G_2(j\omega)| < W_T(\omega) \tag{31}$$

where $W_T(\omega)$ is a weighting function that represents the upper bound for the plant uncertainty. If the closed-loop system remains stability under such uncertain interferences, it is assumed that the feedback-control system is robust.

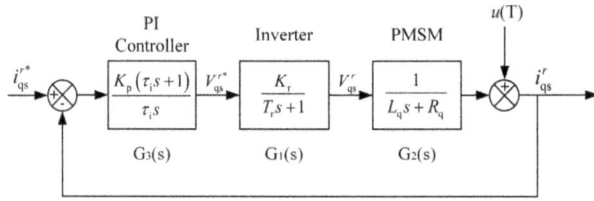

Figure 9. System block diagram with the disturbance injected.

The variation rate of the transfer function of the system could be deduced by the following procedures: the errors of the open-loop and closed-loop transfer functions caused by changes of parameters could be expressed by (32) and (33), respectively:

$$\Delta G(j\omega) = G_1(j\omega)G_3(j\omega)\Delta G_2(j\omega) \tag{32}$$

$$\Delta G_c(j\omega) = \frac{G_1(j\omega)G_3(j\omega)(G_2(j\omega)+\Delta G_2(j\omega))}{1+G_1(j\omega)G_3(j\omega)(G_2(j\omega)+\Delta G_2(j\omega))} - \frac{G_1(j\omega)G_2(j\omega)G_3(j\omega)}{1+G_1(j\omega)G_2(j\omega)G_3(j\omega)} \tag{33}$$

Since $G_1(j\omega)\cdot G_2(j\omega)\cdot G_3(j\omega) \gg G_1(j\omega)\cdot \Delta G_2(j\omega)\cdot G_3(j\omega)$, the (33) could be rewritten as:

$$\Delta G_c(j\omega) = \frac{\Delta G_2(j\omega)G_1(j\omega)G_3(j\omega)}{[1 + G_1(j\omega)G_2(j\omega)G_3(j\omega)]^2} \tag{34}$$

Combining (9), (10), (32) and (34), the variation rate of the transfer function of the system is derived, as expressed by following equation:

$$\Delta G_c(j\omega)/G_c(j\omega) = S(j\omega) \cdot \Delta G(j\omega)/G(j\omega) \tag{35}$$

$$S(s) = 1/(1 + G_1(s)G_2(s)G_3(s)) \tag{36}$$

In the robustness analysis, the sensitivity function $S(s)$ is an essential index, which reflects the ratio of the relative deviation $\Delta G(j\omega)/G(j\omega)$ of the open-loop characteristics to the gain of the closed-loop frequency characteristics $\Delta G_c(j\omega)/G_c(j\omega)$. When the value of $S(s)$ is small enough, the relative deviation of the closed-loop characteristics will be limited within the engineering allowance. Hence, the control system has a robustness capability under energy-bounded interferences.

In addition, $S(s)$ could also be defined as the ratio of the variation rate $\Delta G_c(s)/G_c(s)$ of the transfer function to the variation rate $\Delta G_2(s)/G_2(s)$ of the controlled plant $G_2(s)$, as described by:

$$\frac{\Delta G_c(s)/G_c(s)}{\Delta G_2(s)/G_2(s)} = \frac{1}{1 + G_1(s)G_2(s)G_3(s)} = S(s) \tag{37}$$

Equation (37) manifests that reducing the gain of $S(s)$ would result in the diminution of the adverse effect on control error of PMSM drive system caused by the interference $u(T)$. The sensitivity function $S(s)$ provides a general description of the impact by feedback disturbances in the control system. The $S(s)$ of the PMSM drive system in this paper could be obtained as follows by substituting (10) into (37):

$$S(s) = 1/\left(1 + K_q K_r K_p / \left(\tau_r \tau_i s^2 + \tau_i s\right)\right) \tag{38}$$

According to the switching characteristic of the SiC MOSFET and Si IGBT analyzed above, the SiC MOSFET presents higher switching speed and lower voltage drop. The phase voltage distortion of the SiC MOSFET inverter system is lower than the counterpart of the Si IGBT inverter system. Thus, the gain of SiC-drive system K_{r1} is greater than that of the Si-drive system K_{r2}. Besides, as the delay time constant of an inverter defined in (2), it could be seen that the delay time constant of SiC-inverter τ_{r1} is less than that of the Si-inverter system τ_{r2}.

In conclusion, with the greater inverter system gain K_{r1} and lower inverter delay time τ_{r1}, the sensitivity $S(s)$ of the SiC-drive system is smaller. The preferable robustness performance of the SiC-drive system is thus clarified.

5. Experimental Analysis

The impact of SiC MOSFETs on the dynamic performance of PMSM drive systems is further validated experimentally in a SiC-drive system, and the results were compared with those of the Si-drive system.

The experimental setup is illustrated in Figure 10. In order to prevent unpredictable factors from impacting the voltage and current waveforms, the same control boards based on DSP28335, the same current sensors (LEM DHAB s/14) and same PMSM with a Tamagawa resolver (TS2640N321E64) are used in the two test drive systems. An AD2S1210 resolver-digital converter with a highest accuracy of 0.24 rpm is adopted. The only difference is the power switching device: SiC MOSFET (CAS300M12BM2, Cree, 1200 V, 300 A) and Si IGBT (FF400R12KE3, Infineon, 1200 V, 400 A) are adopted.

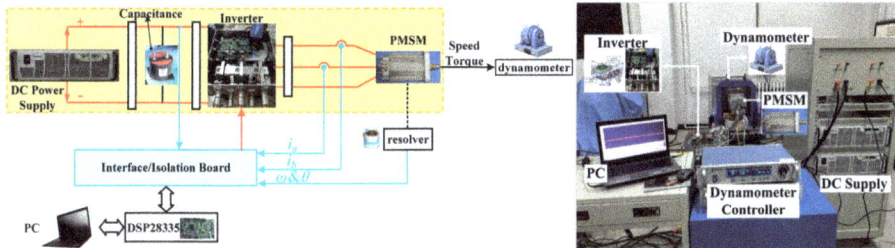

Figure 10. Experimental setup of the PMSM drive system.

5.1. Fast Response Performance Results

The same control strategy and PI parameters are set in the two systems, yielding a relative fair comparison. The speed controller is a conventional two-loop PI control. One is the inside loop, namely current loop. The corresponding K_p and K_i are 2.3 and 0.01 for the direct-axis current i_d separately. The K_p and K_i for the quadrature-axis current i_q are 2.5 and 0.005, respectively. The other one is the outside loop, namely the speed loop. The corresponding K_p and K_i are 0.042 and 0.00035, respectively. A series of bandwidths of q-axis currents are measured under different switching frequencies and dead times, and the effects of the different power devices on the bandwidth of PMSM drive system are explored. Additionally, the speed settling times of the two PMSM drive systems are investigated and compared as well.

5.1.1. The *q*-Axis Current Tracking Experiments

The references of sinusoidal current with different frequencies are assigned to the SiC- and Si-drive systems, respectively, and the loci of actual current tracking are recorded. As the frequency of the reference current increases, the phase difference between the reference and actual current increases, and the amplitude of the actual current waveform decreases. When the amplitude of actual current waveforms declines to 0.707 times the amplitude of the reference current, the corresponding frequency is defined as the bandwidth of the drive system.

Figure 11 shows the experimental results at a switching frequency of 15 kHz and a dead time of 2 μs. Figure 11a shows experimental results of the *q*-axis reference current and the actual *q*-axis current when the frequency and the amplitude of the reference current are 545 Hz and 6 A, respectively.

The actual current of the SiC-drive system presents 5.3 A amplitude and 53.8° phase lag. Meanwhile, the actual current of the Si-IGBT system has 4.2 A amplitude and 116° phase lag. The SiC-drive system manifests a faster dynamic response compared with the Si-drive system.

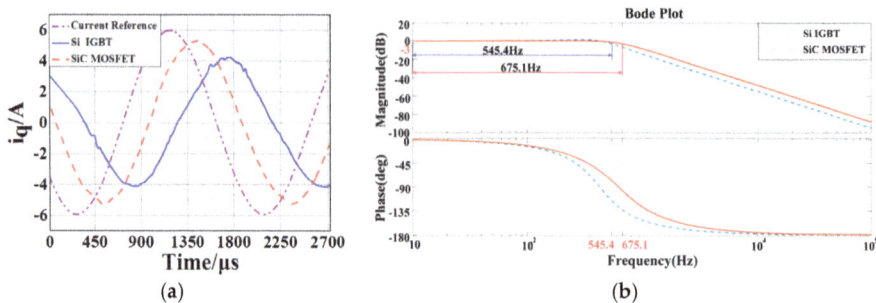

Figure 11. The *q*-axis current tracking experimental results. (**a**) Commanded and actual *q*-axis current. (**b**) Bode plot of systems at 15 kHz switching frequency and 2 μs dead time.

Besides, the loci of the current tracking of the two systems are recorded at a large scale of frequencies, then the Bode plot is obtained shown in Figure 11b. The bandwidth of the SiC-drive system is 675.1 Hz, which is higher than that of the Si-drive system (545.4 Hz). Moreover, the current bandwidths of the two systems under different switching frequencies and dead times are comprehensively compared in Table 3.

Table 3. The frequency response results of *q*-axis current with different switching frequency and dead time.

Variables		*q* Axis Current Bandwidth/Hz	
Switching Frequency/kHz	Dead Time/μs	SiC-Drive System	Si-Drive System
	2	671.5	537.4
10	3	663.4	526.3
	4	653.2	514.5
	2	675.1	545.4
15	3	668.8	531.2
	4	658.5	524.9

In accordance with the aforementioned analysis, faster switching speed, lower dead time, smaller voltage drop and output capacitance indeed result in a lower phase voltage distortion, and a higher bandwidth. Hence, the experimental results are consistent with the conclusions presented in Section 4.

5.1.2. The Step Response of the Speed Loop

The references of step speed (0 to 100 rpm) are assigned to the two PMSM drive systems under no-load conditions. The speed response loci are captured under different switching frequencies and dead times. Figure 12a,b illustrate the experimental results of the Si- and SiC-drive systems, respectively, at 15 kHz switching frequency and 2 μs dead time. The setting time of the SiC-drive system is 59.51 ms, which is less than that of the Si-drive system (68.27 ms).

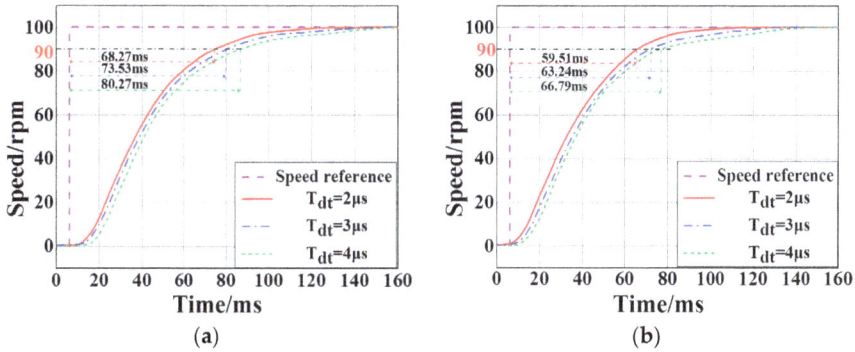

Figure 12. Step response of speed loop at 15 kHz with different dead time; (**a**) Si IGBT based system; (**b**) SiC MOSFET based system.

Furthermore, the detailed experimental step response results of speed under different switching frequencies and dead times, are summarized in Table 4. The settling time extends as the dead time increases and the switching frequency declines, which means the fast response performance of the drive system deteriorates.

Table 4. The step response results of speed loop with different switching frequency and dead time.

Variables		Settling Time t_s/ms	
Switching Frequency/kHz	Dead Time/µs	SiC-Drive System	Si-Drive System
	2	61.31	71.12
10	3	64.95	75.61
	4	68.23	82.96
	2	59.51	68.27
15	3	63.24	73.53
	4	66.79	80.27

5.2. Relative Stability Results

In this section, the relative stabilities of the two drive systems are investigated by experiments. The Nichols plot is drawn based on the transfer function of the drive system and experiment as shown in Figure 13. The phase and amplitude margins of the SiC-drive system are 62.33° and 99.82 dB, respectively, while the counterparts of the Si-drive system are 51.83° and 86.03 dB. A significant improvement in relative stability by adopting SiC MOSFETs can be observed from the results. In addition, thanks to the superior characteristics of the SiC MOSFET, the phase and amplitude margins can be further increased by decreasing the dead time and the switching period, whereas, the phase and amplitude margins of the Si-drive system are penalized by its limited switching speed and operation frequency.

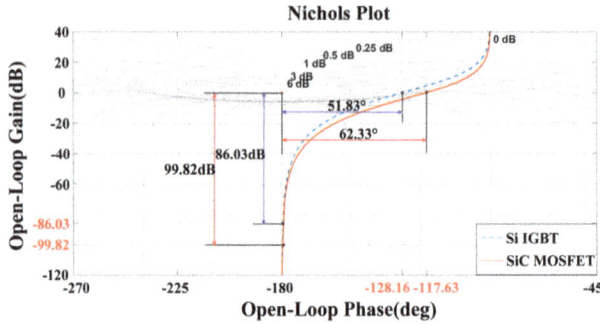

Figure 13. The Nichols plot of systems at 15 kHz switching frequency and 2 µs dead time.

5.3. Robustness Performance Results

In order to evaluate the robustness performance of the drive system, an external torque is suddenly applied to the motor while the motor is operating at a steady-state. Figure 14 shows the experimental result.

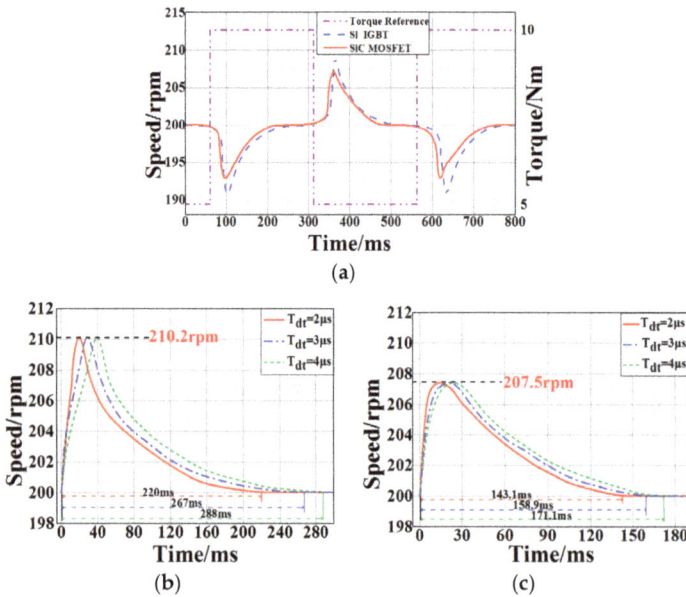

Figure 14. The real speed response curves with the external torque changes suddenly. (**a**) The fluctuations of the speeds at 15 kHz switching frequency and 2 µs dead time; (**b**) The fluctuations of the speeds of Si-drive system with different dead times; (**c**) The fluctuations of the speeds of SiC-drive system with different dead times.

The real speed response curves are measured. The fluctuations of the speeds are due to the external torque changing abruptly from 5 to 10 Nm. Both the speed deviation and regulation restoration time of the SiC-drive system are smaller than the counterparts of the Si-drive system. The different switching frequencies and dead times are adopted to comprehensively investigate the robustness performance of the two drive systems. When the switching frequency and dead time are 15 kHz and 2 µs the speed deviation of the SiC-drive system is 7.5 rpm, while the value of the Si-drive system is more than 10 rpm, as shown in Figure 14. Besides, the regulation restoration time of the SiC system is 143.1 ms, which is less than that of the Si system (220 ms). Hence, it is experimentally verified that the SiC-drive system achieves the better robustness performance to suppress the disturbance in the transient process.

It is worth mentioning that the turn-off time of the Si IGBT is increasing as temperature rises, which will increase the delay time constant τ_r and decrease the gain K_r of the Si-inverter. Hence, the fast response, relative stability and robustness capability of the Si-drive system will be worse. On the contrary, the switching time of the SiC MOSFET is almost invariable under different temperatures. As a result, the SiC-drive system can maintain superior dynamic performance for a long-time operation.

6. Conclusions

In this paper, the impact of SiC MOSFETs on the dynamic performance of PMSM drive systems has been addressed. The transfer function of SiC-drive system was first developed taking into account the switching characteristics of the SiC MOSFETs. The Bode plot, the Nichols plot and the sensitivity function were developed to display the fast response, relative stability and robustness capability of the SiC-drive system, respectively. Both the analytical and experimental results manifested that the SiC-drive system has higher dynamic performance features such as a faster response, a higher relative stability and more robustness than the Si-drive system counterpart. This study can be helpful for the applications of SiC MOSFETs in motor drive systems.

Acknowledgments: This work was supported in part by the National Natural Science Foundation of China under Project 51407004 and in part by the Aeronautical Science Foundation of China 20162851016.

Author Contributions: Xiaofeng Ding proposed the main idea for this paper and established the transfer functions of a PMSM drive system considering device characteristics with Min Du. Min Du and Feida Chen did the experiments. Min Du and Jiawei Cheng drew the figures. Hong Guo provided some useful suggestions in the construction of the paper framework. The paper was checked by Suping Ren. All authors carried out the theoretical analysis and contributed to writing the paper.

Conflicts of Interest: The authors declare no conflict of interest.

References

1. Zhang, H.; Tolbert, L.M. Efficiency impact of silicon carbide power electronics for modern wind turbine full scale frequency converter. *IEEE Trans. Ind. Electron.* **2011**, *58*, 21–28. [CrossRef]
2. Nakakohara, Y.; Otake, H.; Evans, T.M.; Yoshida, T. Three-Phase LLC Series Resonant DC/DC Converter Using SiC MOSFETs to Realize High-Voltage and High-Frequency Operation. *IEEE Trans. Ind. Electron.* **2016**, *63*, 2103–2110. [CrossRef]
3. Sung, W.; Baliga, B.J.; Huang, A.Q. Area-Efficient Bevel-Edge Termination Techniques for SiC High-Voltage Devices. *IEEE Trans. Electron Devices* **2016**, *63*, 1630–1636. [CrossRef]
4. Sarnago, H.; Lucía, Ó.; Burdío, J.M. A Comparative Evaluation of SiC Power Devices for High-Performance Domestic Induction Heating. *IEEE Trans. Ind. Electron.* **2015**, *62*, 4795–4804. [CrossRef]
5. Chen, Z.; Yao, Y.; Boroyevich, D.; Ngo, K. A 1200-V, 60-A SiC MOSFET multichip phase-leg module for high-temperature, high-frequency applications. *IEEE Trans. Power Electron.* **2014**, *29*, 2307–2320. [CrossRef]
6. Liu, S.; Gu, C.; Wei, J.; Qian, Q.; Sun, W.; Huang, A.Q. Repetitive Unclamped-Inductive-Switching-Induced Electrical Parameters Degradations and Simulation Optimizations for 4H-SiC MOSFETs. *IEEE Trans. Electron Devices* **2016**, *63*, 4331–4338. [CrossRef]
7. Sarnago, H.; Lucía, Ó.; Mediano, A.; Burdío, J.M. Design and Implementation of a High-Efficiency Multiple-Output Resonant Converter for Induction Heating Applications Featuring Wide Bandgap Devices. *IEEE Trans. Power Electron.* **2014**, *29*, 2539–2549. [CrossRef]
8. Chen, Z. *Characterization and Modeling of High-Switching-Speed Behavior of SiC Active Devices*; Virginia Polytechnic Institute and State University: Blacksburg, VA, USA, 2009.
9. Zhao, T.; Wang, J.; Huang, A.Q.; Agarwal, A. Comparisons of SiC MOSFET and Si IGBT based motor drive systems. In Proceedings of the 2007 IEEE Conference Record of the 42nd IAS Annual Meeting Industry Applications Conference, New Orleans, LA, USA, 23–27 September 2007; pp. 331–335.

10. Ding, X.; Du, M.; Zhou, T.; Guo, H.; Zhang, C. Comprehensive comparison between silicon carbide MOSFETs and silicon IGBTs based traction systems for electric vehicles. *Appl. Energy* **2016**, in press. [CrossRef]
11. Xiong, R.; Sun, F.; Chen, Z.; He, H. A data-driven multi-scale extended Kalman filtering based parameter and state estimation approach of lithium-ion polymer battery in electric vehicles. *Appl. Energy* **2014**, *113*, 463–476. [CrossRef]
12. Zhang, Y.; Xiong, R.; He, H.; Shen, W. A lithium-ion battery pack state of charge and state of energy estimation algorithms using a hardware-in-the-loop validation. *IEEE Trans. Power Electron.* **2017**, *32*, 4421–4431. [CrossRef]
13. Antonopoulos, A.; Bangtsson, H.; Alakula, M.; Manias, S. Introducing a silicon carbide inverter for hybrid electric vehicles. In Proceedings of the Power Electronics Specialists Conference, Rhodes, Greece, 15–19 June 2008; pp. 1321–1325.
14. Mattavelli, P.; Tubiana, L.; Zigliotto, M. Torque-ripple reduction in PM synchronous motor drives using repetitive current control. *IEEE Trans. Power Electron.* **2005**, *20*, 1423–1431. [CrossRef]
15. Vafaie, M.H.; Dehkordi, B.M.; Moallem, P.; Kiyoumarsi, A. Minimizing torque and flux ripples and improving dynamic response of PMSM using a voltage vector with optimal parameters. *IEEE Trans. Power Electron.* **2016**, *63*, 3876–3888. [CrossRef]
16. Holtz, J.; Quan, J.; Pontt, J.; Rodriguez, J. Design of fast and robust current regulators for high-power drives based on complex state variables. *IEEE Trans. Ind. Appl.* **2004**, *40*, 1388–1397. [CrossRef]
17. Cho, Y.; Lee, K.B.; Song, J.H.; Lee, Y.I. Torque-ripple minimization and fast dynamic scheme for torque predictive control of permanent-magnet synchronous motors. *IEEE Trans. Power Electron.* **2015**, *30*, 2182–2190. [CrossRef]
18. Guo, X.; Wen, X.; Zhao, F.; Song, X.; Zhuang, X. PI parameter design of the flux weakening control for PMSM based on small signal and transfer function. In Proceedings of the 2009 IEEE International Conference on Electrical Machines and Systems (ICEMS 2009), Tokyo, Japan, 15–18 November 2009; pp. 1–6.
19. Guo, Y.; Xi, Z.; Cheng, D. Speed regulation of permanent magnet synchronous motor via feedback dissipative Hamiltonian realisation. *IET Control Theory Appl.* **2007**, *1*, 281. [CrossRef]
20. Senroy, N.; Suryanarayanan, S.; Steurer, M.; Woodruff, S.L. Adaptive transfer function estimation of a notional high-temperature superconducting propulsion motor. *IEEE Trans. Ind. Appl.* **2009**, *45*, 651–658. [CrossRef]
21. Keel, L.H.; Bhattacharyya, S.P. A bode plot characterization of all stabilizing controllers. *IEEE Trans. Autom. Control* **2010**, *55*, 2650–2654. [CrossRef]
22. Atsumi, T.; Messner, W.C. Modified bode plots for robust performance in SISO systems with structured and unstructured uncertainties. *IEEE Trans. Control Syst. Technol.* **2012**, *20*, 356–368. [CrossRef]
23. Malwatkar, G.M.; Khandekar, A.A.; Nikam, S.D. PID controllers for higher order systems based on maximum sensitivity function. In Proceedings of the 2011 3rd International Conference on Electronics Computer Technology (ICECT), Kanyakumari, India, 8–10 April 2011; Volume 1, pp. 259–263.
24. Belove, C. The sensitivity function in variability analysis. *IEEE Trans. Reliab.* **1966**, *15*, 70–76. [CrossRef]
25. Ohmae, T.; Matsuda, T.; Kamiyama, K.; Tachikawa, M.A. Microprocessor-Controlled High-Accuracy Wide-Range Speed Regulator for Motor Drives. *IEEE Trans. Ind. Electron.* **1982**, *1*, 207–211. [CrossRef]
26. Blasko, V.; Kaura, V.; Niewiadomski, W. Sampling of discontinuous voltage and current signals in electrical drive: A system approach. *IEEE Trans. Ind. Appl.* **1998**, *34*, 1123–1130. [CrossRef]
27. Böcker, J.; Buchholz, O. Can oversampling improve the dynamics of PWM controls? In Proceedings of the 2013 IEEE International Conference on Industrial Technology (ICIT), Cape Town, South Africa, 25–28 February 2013; pp. 1818–1824.
28. Alexandrou, A.D.; Adamopoulos, N.K.; Kladas, A.G. Development of a Constant Switching Frequency Deadbeat Predictive Control Technique for Field-Oriented Synchronous Permanent-Magnet Motor Drive. *IEEE Trans. Ind. Electron.* **2016**, *63*, 5167–5175. [CrossRef]
29. Lin-Shi, X.; Morel, F.; Llor, A.M.; Allard, B. Implementation of Hybrid Control for Motor Drives. *IEEE Trans. Ind. Electron.* **2007**, *54*, 1946–1952. [CrossRef]
30. Quang, N.K.; Hieu, N.T.; Ha, Q.P. FPGA-Based Sensorless PMSM Speed Control Using Reduced-Order Extended Kalman Filters. *IEEE Trans. Ind. Electron.* **2014**, *61*, 6574–6582. [CrossRef]
31. Idkhajine, L.; Monmasson, E.; Naouar, M.W.; Prata, A. Fully Integrated FPGA-Based Controller for Synchronous Motor Drive. *IEEE Trans. Ind. Electron.* **2009**, *56*, 4006–4017. [CrossRef]

32. Horvat, R.; Jezernik, K.; Čurkovič, M. An Event-Driven Approach to the Current Control of a BLDC Motor Using FPGA. *IEEE Trans. Ind. Electron.* **2014**, *61*, 3719–3726. [CrossRef]
33. Chou, M.C.; Liaw, C.M. Dynamic Control and Diagnostic Friction Estimation for an SPMSM-Driven Satellite Reaction Wheel. *IEEE Trans. Ind. Electron.* **2011**, *58*, 4693–4707. [CrossRef]
34. Kommuri, S.K.; Defoort, M.; Karimi, H.R.; Veluvolu, K.C. A Robust Observer-Based Sensor Fault-Tolerant Control for PMSM in Electric Vehicles. *IEEE Trans. Ind. Electron.* **2016**, *63*, 7671–7681. [CrossRef]
35. El-Sousy, F.F.M. Hybrid H∞-Based Wavelet-Neural-Network Tracking Control for Permanent-Magnet Synchronous Motor Servo Drives. *IEEE Trans. Ind. Electron.* **2010**, *57*, 3157–3166. [CrossRef]
36. Calleja, C.; López-de-Heredia, A.; Gaztañaga, H.; Aldasoro, L.; Nieva, T. Validation of a Modified Direct-Self-Control Strategy for PMSM in Railway-Traction Applications. *IEEE Trans. Ind. Electron.* **2016**, *63*, 5143–5155. [CrossRef]
37. Shyu, K.-K.; Lai, C.-K.; Tsai, Y.-W.; Yang, D.-I. A newly robust controller design for the position control of permanent-magnet synchronous motor. *IEEE Trans. Ind. Electron.* **2002**, *49*, 558–565. [CrossRef]
38. Ding, X.; Chen, F.; Du, M.; Guo, H.; Ren, S. Effects of silicon carbide MOSFETs on the efficiency and power quality of a microgrid-connected inverter. *Appl. Energy* **2016**, in press.
39. Bedetti, N.; Calligaro, S.; Petrella, R. Self-Commissioning of Inverter Dead-Time Compensation by Multiple Linear Regression Based on a Physical Model. *IEEE Trans. Ind. Appl.* **2015**, *51*, 3954–3964. [CrossRef]
40. Krishnan, R. *Permanent Magnet Synchronous and Brushless DC Motor Drives*; CRC Press/Taylor & Francis: Boca Raton, FL, USA, 2010.

energies

MDPI

Article

Numerical Analysis of Shell-and-Tube Type Latent Thermal Energy Storage Performance with Different Arrangements of Circular Fins

Sebastian Kuboth *, Andreas König-Haagen and Dieter Brüggemann

Chair of Engineering Thermodynamics and Transport Processes (LTTT), Center of Energy Technology (ZET), University of Bayreuth, 95440 Bayreuth, Germany; Andreas.Koenig-Haagen@uni-bayreuth.de (A.K.-H.); brueggemann@uni-bayreuth.de (D.B.)
* Correspondence: sebastian.kuboth@uni-bayreuth.de or lttt@uni-bayreuth.de; Tel.: +49-921-55-7524

Academic Editor: Hailong Li
Received: 1 December 2016; Accepted: 20 February 2017; Published: 25 February 2017

Abstract: Latent thermal energy storage (LTS) systems are versatile due to their high-energy storage density within a small temperature range. In shell-and-tube type storage systems fins can be used in order to achieve enhanced charging and discharging power. Typically, circular fins are evenly distributed over the length of the heat exchanger pipe. However, it is yet to be proven that this allocation is the most suitable for every kind of system and application. Consequently, within this paper, a simulation model was developed in order to examine the effect of different fin distributions on the performance of shell-and-tube type latent thermal storage units at discharge. The model was set up in MATLAB Simulink R2015b (The MathWorks, Inc., Natick, MA, USA) based on the enthalpy method and validated by a reference model designed in ANSYS Fluent 15.0 (ANSYS, Inc., Canonsburg, PA, USA). The fin density of the heat exchanger pipe was increased towards the pipe outlet. This concentration of fins was implemented linearly, exponentially or suddenly with the total number of fins remaining constant during the variation of fin allocations. Results show that there is an influence of fin allocation on storage performance. However, the average storage performance at total discharge only increased by three percent with the best allocation compared to an equidistant arrangement.

Keywords: thermal energy storage; shell-and-tube; phase change material (PCM); circular fins

1. Introduction

Thermal energy storage is currently an important topic in energy science and research. Latent thermal energy storage (LTS) is of particular interest, because of its high-energy storage density. The phase change material (PCM) within the LTS stores a high amount of energy within a small temperature range by changing its phase state.

Originally used for spacecraft thermal control applications, the development and exploration of new PCM for different temperature ranges has led to a wide range of applications [1]. Latent thermal storage units can be found in heat pumps, building temperature control, off-peak electricity storage, waste heat recovery systems, the cooling of electronic devices and many other fields [1,2]. Furthermore, the combination of solar power and LTS, like solar heating systems and solar cooking, is currently being researched [1]. Malan et al. [3] examined the potential use of LTS in solar thermal power plants for performance optimisation. Due to the variety of possible applications, several reviews have been carried out inter alia by Sharma et al. [1]; Zalba et al. [4]; and Farid et al. [5].

Depending on the type of application, an adequate storage material has to be applied. Besides a suitable temperature range of the phase transition, the thermal energy capacity and the thermal conductivity are main criteria [1]. Due to the phase change enthalpy, the heat capacity of PCM is high in small temperature ranges, while the heat conductivity of PCM is generally low. Since the low heat conductivity presents a disadvantage on discharging when high heat transfer rates are required, the low thermal conductivity can limit the applications of PCM [5,6].

In order to extend the range of application and to achieve better storage performances, research activities have been carried out to examine the heat transfer process [4,7,8]; storage types and configurations like a combination of different PCMs within one storage unit [9–15]; and the integration of highly heat conductive materials into the storage material, examples of which are copper, aluminum, stainless steel or carbon fiber [16–22]. These materials can be integrated in different forms, such as fins, honeycombs, wool or brush-form. For detailed information, the reader is referred to the reviews of Fan and Khodadadi [23], as well as Jegadheeswaran and Pohekar [24].

The most common way to improve the performance of shell-and-tube LTS is to integrate longitudinal or circular fins into the storage unit [16]. Khalifa et al. [20] showed that the extracted energy per time of a LTS could be improved by 86% by adding four longitudinal fins into a shell-and-tube type LTS unit. Various articles investigate the performance enhancement by fin shape and design variation. For instance, Al-Abidi et al. [19] studied the design of shell-and-tube LTS using longitudinal fins. Furthermore, the influence of circular fin quantity and diameter have been examined by Erek et al. [25] with evenly distributed fins, which is the most common way of fin placement. Since the fin material is only relevant if highly conductive material is applied [26], fins can be the main costing of a thermal energy storage. Therefore, an increase of the number of fins is not economically viable in many cases. Furthermore, adding additional fins increases the storage weight whereas the storage capacity is generally decreased, which is a disadvantage particularly for mobile application. Therefore, the objective of this study was to investigate whether and to what extend it is possible to improve storage performance of shell-and-tube type LTS on discharge with a constant amount of fin and storage material by adjustment of fin positioning.

For this purpose, simulation models for shell-and-tube type LTS with circular fins were developed in MATLAB [27] Simulink [28]. The distribution of these fins was varied in 30 different ways and their storage power during discharging was examined. Most of the examined designs increased their fin density towards the end of the storage, whereas one geometry represented a homogeneous distribution of fins. For validation, a comparison of these models to an ANSYS Fluent [29] model was undertaken. Since conduction is the dominant heat transfer type during discharge [7,24], convectional effects in the liquid PCM were neglected for reasons of model simplification.

In the following sections, the mathematical description and numerical implementation; the validation; the boundary conditions; and the results will be demonstrated and discussed.

2. Description of the Simulation Model

2.1. Storage Setup

The setup of a shell-and-tube type LTS is shown in Figure 1. Within a container (shell), several storage elements are consistently distributed. The main component of the storage elements is a tube with heat transfer fluid (HTF) on the inside and PCM on the outside. For charging and discharging, the HTF flows through the pipes. The temperature difference between the HTF and the other parts of the storage elements causes charging or discharging of the storage system. Typically, the PCM changes its phase during these processes. On charging, the PCM melts and solidifies on discharging. Since the PCM has a comparatively low heat conductivity, fins can be included into the storage unit to improve the charge and discharge power. In the following, several different types of fin allocations and their effects on the storage performance during discharge will be examined with numerical simulations, examples of which are pictured in Figure 2. The types of analyzed fin allocations can be categorized into five groups. Each allocation consisted of a storage element with 100 fins. Mostly, the fin density (fins per axial distance) increased towards the storage unit outlet in order to adapt the fin density to the temperature difference between HTF and PCM.

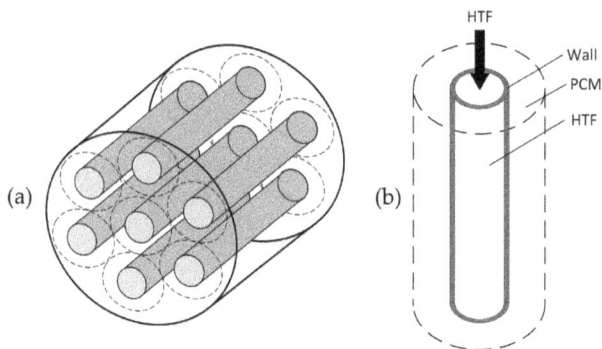

Figure 1. (**a**) shell-and-tube latent thermal energy storage (LTS) and (**b**) single storage element. HTF: heat transfer fluid; and PCM: phase change material.

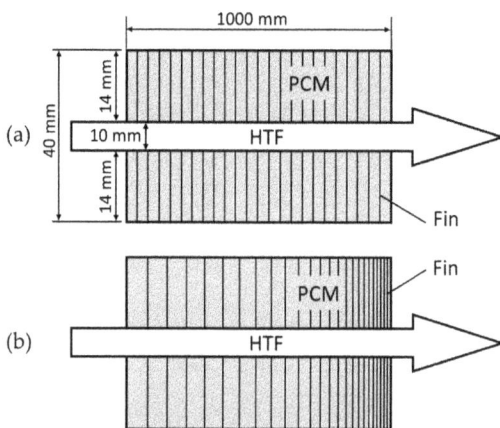

Figure 2. Single storage element with (**a**) homogeneous; and (**b**) uneven distribution of circular fins.

The examined groups of arrangements are described as follows:

- Division of the storage into three sections of the same length, with a different fin density in each section—the density ratio F_3 is calculated according to Equation (1):

$$F_3 = \frac{N_{\text{sec II}}}{N_{\text{sec I}}} = \frac{N_{\text{sec III}}}{N_{\text{sec II}}} \tag{1}$$

- Division of the storage into two sections of the same length, with a different fin density in each section—the density ratio F_2 is calculated according to Equation (2):

$$F_2 = \frac{N_{\text{sec II}}}{N_{\text{sec I}}} \tag{2}$$

- Linear increase of distances between the fins towards the storage element inlet with a minimum distance Δx_0 between two fins—the distance Δx_n between the fins n and $n+1$ is calculated in consideration of the factor for linear increasing distances F_L:

$$F_L = \frac{\Delta x_n - \Delta x_0}{n \cdot (\Delta x_1 - \Delta x_0)} \tag{3}$$

- Exponential increase of distances between the fins towards the storage element inlet—the factor for exponential increasing distances F_E is calculated according to Equation (4):

$$F_E = \frac{\Delta x_{n+1} - \Delta x_0}{\Delta x_n - \Delta x_0} \tag{4}$$

- Homogeneous arrangement

Altogether, 30 different distributions of fins were investigated with a storage containing 20 storage elements of one-meter length, an outer storage element diameter of 40 mm and an inner tube diameter of 10 mm. The pipe wall thickness and the thickness of the fins was one millimeter.

The material properties of the HTF and the PCM are listed in Table 1. The phase change of the PCM RT42 takes place within a temperature range of 4 °C. The water inlet temperature $T_{HTF,in}$ was set to 22 °C. The material properties of the pipe walls and fins corresponded to the material properties of pure copper.

Table 1. Properties of applied materials.

Material	Application	Phase Change Temperature Solid-Liquid (°C)	Latent Heat of Fusion (kJ/kg)	Liquid Heat Capacity (kJ/kg·K)	Liquid Density (kg/m³)
RT42	PCM	40–44	176	2.0	760
Water	HTF	0	334	4.18	998

2.2. Numerical Model

The model was set up in MATLAB Simulink. The simulation was based on the transport equation:

$$\frac{\partial(\rho\phi)}{\partial t} + div(\rho\phi u) = div(\Gamma \cdot grad\phi) + S_\phi. \tag{5}$$

From this equation, the energy conservation equation can be deduced:

$$\frac{\partial(\rho cT)}{\partial t} + div\left(\rho \vec{u} cT\right) = div(\lambda \cdot gradT) + S_h. \tag{6}$$

To simplify the model and to achieve a shorter computational time, the impulse and mass conservation equations were neglected and a constant HTF volume flow rate was assumed. The following adaptions have been made in relation to the PCM calculation:

- Neglect of convectional effects in the liquid PCM;
- Neglect of the temperature dependency of material properties within one phase;
- Application of an enthalpy method with apparent heat capacity;
- Integration of the phase change enthalpy according to Rösler and Brüggemann [30] by applying an apparent heat capacity.

$$c_{app} = c_{sen} + c_L \tag{7}$$

$$c_L = 4L \frac{\exp\left(-\left\{\left[\frac{4(T-T_m)}{T_{L,l}-T_{L,s}}\right]^2\right\}\right)}{(T_{L,l} - T_{L,s}) \cdot \sqrt{\pi}} \tag{8}$$

225

The governing equation for the PCM is

$$\rho c_{\text{eff}} \frac{\partial T}{\partial t} = \left(\frac{1}{r}\right) \frac{\partial}{\partial r}\left(\lambda r \frac{\partial T}{\partial r}\right) + \frac{\partial}{\partial x}\left(\lambda \frac{\partial T}{\partial r}\right). \tag{9}$$

Besides the adaptations for the PCM, the following simplifications were determined for the HTF:
- Incompressible fluid;
- One-dimensional convection (axial);
- Constant predefined velocity.

The governing equation for the HTF is:

$$\rho c \frac{\partial T}{\partial t} + u \rho \frac{\partial T}{\partial x} = \alpha \frac{A \alpha}{V}(T_F - T_{wall}) + \lambda \frac{\partial^2 T}{\partial x^2}. \tag{10}$$

In order to further reduce the computational effort, geometrical symmetry was included in the calculation. Therefore, the shell-and-tube storage system was simplified by considering only a single storage element within the simulation, as seen in Figure 1. This storage element was rotationally symmetric, thus the storage could be simulated like a two-dimensional model as shown in Figure 3.

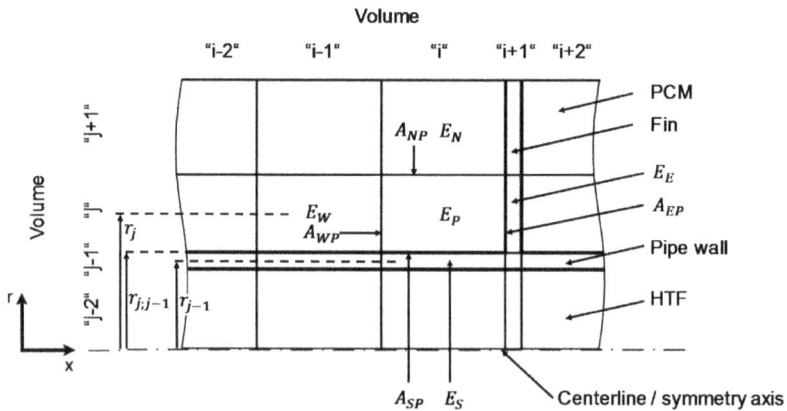

Figure 3. Section of the discretized two-dimensional model with exemplary declaration of elements, interfaces and radii.

This model was spatially discretized using the finite volume method resulting in Equation (11), while the time discretization was executed by MATLAB Simulink.

$$a_P T_P = a_E T_E^0 + a_W T_W^0 + a_N T_N^0 + a_S T_S^0 + \left(a_P^0 - a_E - a_W - a_N - a_S\right) \cdot T_P^0 \tag{11}$$

With the element declaration indices of Figure 3, the geometrical indices of the coefficients are explained in Equation (12):

$$E_P = E_j^i, E_E = E_j^{i+1}, E_W = E_j^{i-1}, E_N = E_{j+1}^i, E_S = E_{j-1}^i \tag{12}$$

The subsequent coefficients for Equation (11) are listed in the following:

$$a_E = \left(\frac{\Delta x_E}{2\lambda_E A_{EP}} + \frac{\Delta x_P}{2\lambda_P A_{EP}}\right)^{-1}, \tag{13}$$

$$a_W = \left(\frac{\Delta x_W}{2\lambda_W A_{WP}} + \frac{\Delta x_P}{2\lambda_P A_{WP}} \right)^{-1}, \tag{14}$$

$$a_N = \frac{2\pi \Delta x_P}{\frac{1}{\lambda_N} \ln\left(\frac{r_{j+1}}{r_{j+1;j}} \right) + \frac{1}{\lambda_P} \ln\left(\frac{r_{j+1;j}}{r_j} \right)}, \tag{15}$$

$$a_S = \frac{2\pi \Delta x_P}{\frac{1}{\lambda_S} \ln\left(\frac{r_{j;j-1}}{r_{j-1}} \right) + \frac{1}{\lambda_P} \ln\left(\frac{r_j}{r_{j;j-1}} \right)}, \tag{16}$$

$$a_P^0 = \rho c_{app} \frac{\Delta V_P}{\Delta t}, \tag{17}$$

$$a_P = a_P^0 \tag{18}$$

The radii, which contain a semicolon in its indices, mark the border between the associated elements, while the other radii mark the center of the elements.

In order to calculate the convection of the HTF, the coefficients of the equations calculating the tube wall and the HTF have to be adapted. The convectional heat transfer within the fluid can be considered by adjusting the coefficient a_W of Equation (11) for the HTF volumes:

$$a_W = \left(\frac{\Delta x_W}{2\lambda_W A_{WP}} + \frac{\Delta x_P}{2\lambda_P A_{WP}} \right)^{-1} + \dot{m}_F c_F, \tag{19}$$

while the convectional heat transfer between the HTF and the pipe wall was calculated by modifying the coefficient a_N of the HTF and the coefficient a_S of the pipe wall elements:

$$a_N = \left(\frac{\frac{1}{\lambda_N} \ln \frac{r_{j+1}}{r_{j+1;j}}}{2\pi \Delta x_P} + \frac{1}{\alpha_F A_{NP}} \right)^{-1}, \tag{20}$$

$$a_S = \left(\frac{1}{\alpha_F A_{SP}} + \frac{\frac{1}{\lambda_P} \ln \frac{r_j}{r_{j;j-1}}}{2\pi \Delta x_P} \right)^{-1}. \tag{21}$$

In order to account for the boundary conditions, the coefficients of the discretization equation of the boundary elements were adapted. An adiabatic boundary condition was calculated by setting the appropriate coefficient a to zero. Furthermore, losses to the surrounding were calculated by:

$$a_{E,W,N,S} = \alpha_{amb} A_{edge}, \tag{22}$$

$$T_{E,W,N,S} = T_{amb}. \tag{23}$$

Although included into the model, ambient losses were neglected within the following simulations, in order to avoid as many sources of uncertainties as possible. Similar to the ambient losses, Equation (24) was applied to include the inlet temperature:

$$T_W = T_{F,in}. \tag{24}$$

Altogether, the following starting and boundary conditions were set in relation to the coordinate system of Figure 3:

- $T_0 = 62\,^\circ\text{C}$ for every element
- $\dot{q} = 0$ for $r = 0$
- $\dot{q} = \alpha_{amb} \cdot \frac{(T_{amb} - T(r_{su}))}{N_{su}}$ for $r = r_{su}$
- $\dot{q} = 0$ for $x = 0$ and $r > r_{wall}$

- $\dot{q} = 0$ for $x = x_{su}$ and $r > r_{wall}$
- $\dot{q} = 0$ for $x = 0$ and $r_{HTF} < r \le r_{wall}$
- $\dot{q} = 0$ for $x = x_{su}$ and $r_{HTF} < r \le r_{wall}$
- $T = T_{HTF,in}$ for $x = 0$ and $r \le r_{HTF}$
- $\dot{q} = 0$ for $x = x_{su}$ and $r \le r_{HTF}$
- Simulation domain length of 1 m
- Simulation domain radius of 20 mm

The spatial discretization was defined within a s-function. To solve the system of semi-discretized ordinary differential equations, the ode113 (Adams) solver was chosen.

In each simulation, the storage was discharged completely within a physical time of 9000 s. The condition defining the state of total discharge was the remaining storage energy of 0.1% in relation to the starting conditions.

2.3. Validation

In order to validate the model, the storage unit with equidistant spacing was modeled in the commercial computational fluid dynamics (CFD) tool ANSYS Fluent [29]. Similar to the other simulation models, the storage unit was simplified by simulating only a single storage element and exploiting its geometry. Hence, the storage could be simulated using a two-dimensional model.

All material properties were set in conformity with the MATLAB Simulink model. The boundary conditions of both the ANSYS Fluent and the MATLAB Simulink model were set adiabatic to ensure heat loss independency. However, the flow speed was increased by a factor of 30 to obtain a high Reynolds number of about 2000. In this way, deviations caused by the different types of flow discretization were minimized while the flow was still laminar. The absolute output powers of both simulation environments are shown in Figure 4.

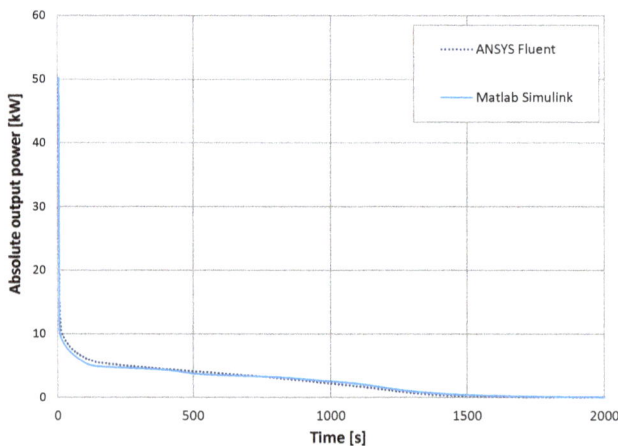

Figure 4. Comparison of equidistant allocation in ANSYS Fluent and MATLAB Simulink.

Even though the simulations do not show total congruency, the performance of both storage models are very similar and deviations are negligible. Differences between the simulations can be seen shortly after the start of the simulation. Although the Reynolds number was increased, these differences were caused by the different flow discretization types. The HTF of the Simulink model was defined to be one-dimensional, while the ANSYS Fluent model calculates two-dimensionally. Therefore, the ANSYS model had a higher flow speed in the middle of the inner pipe and a lower flow speed close to the pipe wall. Furthermore, it considers radial convection. Consequently, the output temperature of the ANSYS Fluent model drops at an earlier point in time. Since this output temperature drop involves a lower output power, the level of discharge of the ANSYS storage after 20 s is lower than that of the Simulink model. As the times of total discharge are almost the same, the output power of the ANSYS Fluent storage unit has a higher output power within a time of 50–500 s. In spite of these small deviations, the difference of the power integrals of both simulations is less than 0.5%.

Apart from the deviations caused by different types of flow discretization, the results show a very good compliance. Since the effects of the flow discretization are consistent for all MATLAB Simulink fin allocations, the results of the arrangement analysis simulation are reliable. Furthermore, simulations with increased numbers of elements in MATLAB Simulink were performed to ensure grid independency.

3. Results and Discussion

Figure 5 shows the average absolute storage power until full discharge of all fin arrangements. It can be seen that the examined allocations differ, therefore, the storage performance was affected by fin allocation. The most efficient geometry consisted of storage elements with a distribution of fins that linearly increased the fin density using a growth rate of $F_L = 10$. This geometry achieved an average output power of about 716 W until total discharge, while the least efficient arrangement with an exponential growth rate factor of $F_E = 1.035$ had an average output power of 655 W leading to a difference of 8.5%. The equidistant distribution of fins achieved an average absolute power of 695 W, which is about three percent less than the achieved maximum average power. Consequently, a homogeneous distribution of fins is not the best type of arrangement in any case considering total storage discharge. It has to be noted that three of the four kinds of uneven fin distribution pass a maximum value within the variation of growth rate. Therefore, it can be assumed that a further variation of the growth rate factors would not cause significant improvements in terms of storage performance compared to the achieved maximum values.

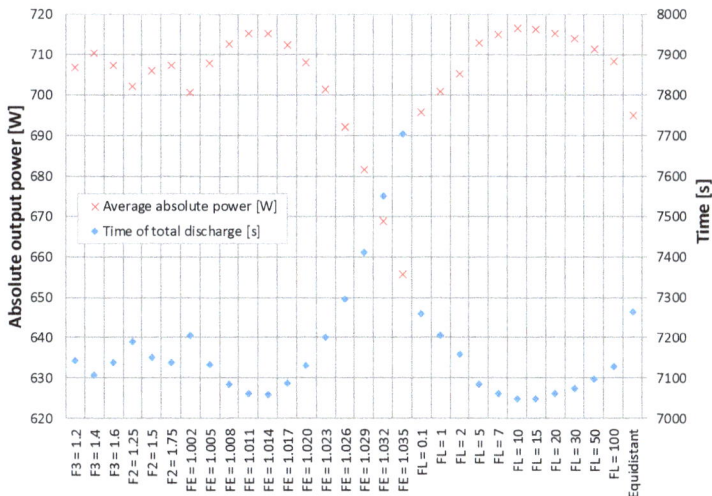

Figure 5. Absolute average power until total storage discharge of all examined arrangements of fins.

Within all different kinds of fin arrangements, a growth rate F close to the equidistant model did not lead to the shortest discharge time. The shortest times of discharge were achieved using linear growth rates of $F_L = 10$ and $F_L = 15$, followed by an exponential growth rate of $F_E = 1.014$.

The difference between these discharge times was only 12 s (0.17%). Due to this fact, both types of distribution are of interest for further research.

Since the yet discussed values only represent the final state, it was also useful to look at the discharge progress to examine intermediate results. Figure 6 depicts the progress of absolute output power of selected arrangement types. For reasons of clarity, only the most efficient storage unit in relation to the average power until total discharge and the storage unit with equidistant arrangement of fins are represented.

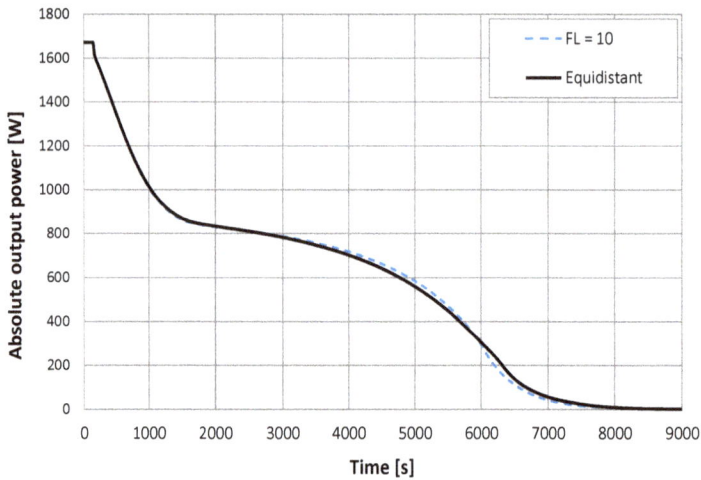

Figure 6. Storage performances of two simulated allocations of fins.

The figure shows the typical plot for LTS with regard to the performance. From 3000 s physical time until about 5900 s, the output power of the inhomogeneous distribution was higher than the arrangement with equidistant fin distribution. At 5000 s, the relative difference in output power between the allocations was about 5.2%. In addition, most of the non-equidistant arrangements provided higher average power until a simulation time of about 5900 s. At that time, the average output power of the homogeneous distribution was about 0.8% less than the maximum average power. Although the arrangement with a linear growth rate factor of $F_L = 10$ is the fastest type for total discharge, it did not provide the highest power all over the time. In the time range between 5900 s and 8000 s physical time, the storage unit using a homogeneous distribution had a higher output power.

In Figure 7, the power values for discharge times of 6300 s and 5000 s can be seen. At these times, the difference between the allocations are clearly recognizable. At 6300 s, the equidistant geometry provided about 26% more power than the allocation with the highest output power at total storage discharge. Furthermore, at 5000 s it can be seen that the higher the growth rate factor gets at that time, the more output power the storage units generate temporarily.

The differences in storage performance also caused different levels of discharge at certain times of simulation, which is also depicted in Figure 7. These differences can be explained by considering the storage temperatures. Figures 8 and 9 show the temperatures of the HTF and pipe wall elements within a storage unit with the linear growth rate of $F_L = 10$ on one hand, and the equidistant fin distribution on the other. The depicted physical times are 5000 s and 6300 s, respectively. The HTF temperature of the last element is directly proportional to the output power. It is recognizable that both fin arrangements have different material temperatures at the displayed points in time.

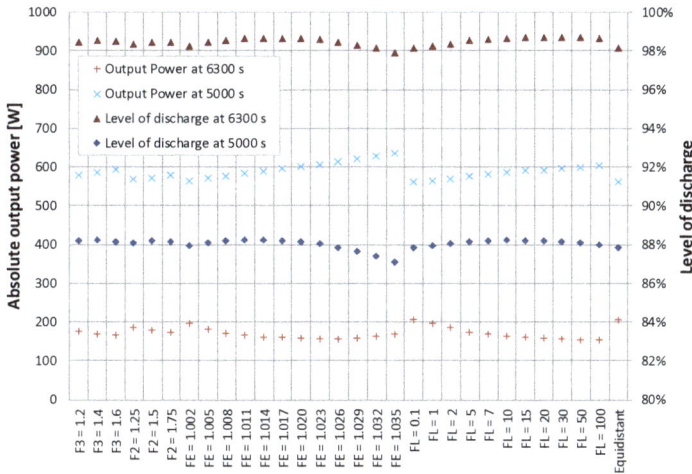

Figure 7. Comparison of absolute output power and level of discharge at 5000 s and 6300 s for all allocations.

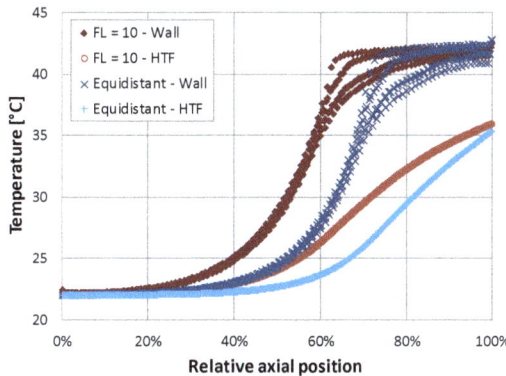

Figure 8. Heat transfer fluid (HTF) and wall element center temperatures of two distributions at 5000 s physical time.

At 5000 s of physical time, the PCM was not yet totally solidified. Due to the high heat conductivity of the fins, the pipe wall was close to the phase change temperature in a wide range at the storage outlet. In addition, high temperature deviations occurred between adjacent pipe wall elements from 60% to 100% relative axial position. Due to different material properties of the fins and the PCM, pipe wall elements adjoining fins had a higher temperature than adjacent pipe wall elements, which border the PCM elements.

Although both storage units still contained liquid PCM, the storage unit with an allocation with linear growth rate factor of $F_L = 10$ had a higher amount of PCM that was still not solidified. This led to a larger difference in temperature between the pipe wall and the HTF within a relative length of 30% to 100% when comparing the $F_L = 10$ storage unit with the one containing an equidistant arrangement of fins. Consequently, the storage with non-equidistant fin allocation had a higher output power at 5000 s of physical time.

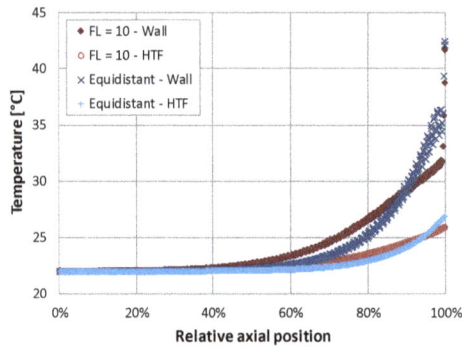

Figure 9. Heat transfer fluid (HTF) and wall element center temperatures of two distributions at 6300 s physical time.

In contrast, at 6300 s of physical time, almost all of the PCM changed its phase in the non-equidistant case, while the pipe wall of the other arrangement was still close to the phase change temperature at the outlet of the storage unit. This indicates that the $F_L = 10$ storage unit cooled more uniformly. Furthermore, the high pipe wall temperature in the end of the storage unit led to a higher HTF temperature at the outlet since the mass flow rate in the simulation was moderate. Consequently, the output power of the storage with equidistant fin allocation was higher at that time. The simulation results show that non-equidistant arrangements of fins can achieve a higher average output power than the homogeneous distribution of fins. These types of thermal energy storage units can be applied for all targeted levels of discharge. It should be noted that these results may vary using different storage types or mass flow rates, since higher heat transfer coefficients between the fluid and the tube might enhance the effect of fin arrangement modification. Further parameters, which might have an impact on the results, are the fin and the storage material, the number and width of the fins such as the outer storage element diameter. Decreasing the thermal conductivity of the storage material might increase the effect of adjusted fin distributions, whereas decreasing the latent heat of fusion might reduce the effect. Although the heat conductivity of copper is high, an even higher heat conductivity of the fins such as higher storage unit diameters, a lower number of fins such as thinner fins can also amplify the effect on storage performance. However, a modification of these parameters might also have an influence on the optimum growth rate factors. Due to the large variety of parameters, an optimization depending on the case of application is recommendable.

Since the charging process of the storage unit was not examined within this work and the influence on the system performance is small, further investigations have to be conducted with respect to technical application.

4. Conclusions

Within this study, LTS performance enhancement by varying circular fin positioning was examined. Therefore, different kinds of circular heat conduction fin distributions with a constant amount of fins and storage material were investigated numerically in order to improve the storage power at discharge. The latent thermal storage units were set up within MATLAB Simulink. The models examine shell-and- tube type LTS containing copper fins and tubes and the paraffin Rubitherm RT42 as a storage material. Different fin allocations were studied including the linearly and exponentially increased fin density towards the outlet of the storage units, the section by section change of the fin density, and the equidistant fin distribution.

In order to validate the results, appropriate thermal energy storage was set up in ANSYS Fluent. A comparison of the different simulation environments was conducted by examining the storage performance at discharge. Apart from small deviations caused by different flow discretization, the performances of both storage units show very good congruency.

The results of varying the fin arrangements within the LTS show that fin allocation affects storage performance. Non-equidistant distributions can cause a higher average output power at all levels of storage discharge than equidistant ones. At total storage discharge, the average output power of a homogeneous fin arrangement could be improved by three percent using a linear growth rate factor of 10. Exponentially increased fin densities towards the storage element outlet also showed promising results. A detailed investigation of the storage material and the HTF temperatures demonstrated that uneven distributions of fins could induce a more uniform discharge. Since the present work examined storage discharge exclusively and the achieved increase of storage performance was generally low, further work analyzing the effects of fin distribution on storage charge and the significance for applications is recommended.

Acknowledgments: The authors gratefully acknowledge financial support of the Bavarian State Ministry of Education, Science and the Arts within the framework TechnologieAllianzOberfranken (TAO). This publication was funded by the German Research Foundation (DFG) and the University of Bayreuth in the funding programme Open Access Publishing.

Author Contributions: All authors contributed to this work by collaboration. Sebastian Kuboth is the main author of this manuscript. Andreas König-Haagen had the initial idea, particularly contributed to the Simulink model and assisted in the conceptual design of the study as well as in the writing of the manuscript. The whole project was supervised by Dieter Brüggemann. All authors revised and approved the publication.

Conflicts of Interest: The authors declare no conflict of interest. The founding sponsors had no role in the design of the study; in the collection, analyses, or interpretation of data; in the writing of the manuscript, and in the decision to publish the results.

Nomenclature

Latin symbols

a	Temperature conductivity ($m^2 \cdot s^{-1}$)
A	Area (m^2)
c	Specific heat capacity ($J \cdot kg^{-1} \cdot K^{-1}$)
E	Element (-)
F	Factor for fin concentration
l	Length of the storage (m)
L	Latent heat ($J \cdot kg^{-1}$)
$\dot{m}$	Mass flow ($kg \cdot s^{-1}$)
N	Number of fins
r	Radius (m)/radial coordinate (m)
$\dot{q}$	Heat flux ($J \cdot s^{-1}$)
S	Heat source term
t	Time (s)
T	Temperature (K)
u	Flow velocity ($m \cdot s^{-1}$)
$\vec{u}$	Flow velocity vector
V	Volume (m^3)
x	Axial coordinate (m)

Greek symbols

α	Heat transfer coefficient ($W \cdot m^{-2} \cdot K^{-1}$)
Δ	Difference (-)
ρ	Density ($kg \cdot m^{-3}$)
φ	General variable
Γ	Diffusion coefficient ($m^2 \cdot s^{-1}$)
λ	Heat conductivity ($W \cdot m^{-1} \cdot K^{-1}$)

Subscripts

amb	Ambience
app	Apparent value
E	Concerning the element left of the calculation element/exponential
edge	Concerning the edge of the calculation element, that borders the ambience
EP	Concerning the element east of the calculation element (CE) and the CE
F	Heat transfer fluid
h	Concerning the specific enthalpy
HTF	Concerning the outer radius of the HTF
in	Inlet
I	Concerning the first section
II	Concerning the second section
III	Concerning the third section
j	Position indicator in radial direction
l	Liquid phase
L	Linear/latent
m	Thermodynamic mean
n	Counting variable
N	Concerning the element above the calculation element
NP	Concerning the element north of the calculation element (CE) and the CE
P	Concerning the calculation element
s	Solid phase
S	Concerning the element below the calculation element
sec	Concerning one section of the storage element
sen	Concerning the sensible heat capacity
SP	Concerning the element south of the calculation element (CE) and the CE
su	Storage unit
W	Concerning the element right of the calculation element
wall	Pipe wall
WP	Concerning the element west of the calculation element (CE) and the CE
0	Concerning minimum distance
1	First element
2	Storage element divided into two parts
3	Storage element divided into three parts
α	Convective heat transfer
φ	Concerning the general variable

Superscripts

i	Position indicator in axial direction
0	Concerning the last time step

References

1. Sharma, A.; Tyagi, V.V.; Chen, C.R.; Buddhi, D. Review on thermal energy storage with phase change materials and applications. *Renew. Sustain. Energy Rev.* **2009**, *13*, 318–345. [CrossRef]
2. Tan, F.L.; Tso, C.P. Cooling of mobile electronic devices using phase change materials. *Appl. Therm. Eng.* **2009**, *24*, 159–169. [CrossRef]
3. Malan, D.J.; Dobson, R.T.; Dinter, F. Solar thermal energy storage in power generation using phase change material with heat pipes and fins to enhance heat transfer. *Energy Procedia* **2015**, *69*, 925–936. [CrossRef]
4. Zalba, B.; Marín, J.M.; Cabeza, L.F.; Mehling, H. Review on thermal energy storage with phase change: Materials, heat transfer analysis and applications. *Appl. Ther. Eng.* **2003**, *23*, 251–283. [CrossRef]
5. Farid, M.M.; Khudhair, A.M.; Razack, S.A.K.; Al-Hallaj, S. A review on phase change energy storage: Materials and applications. *Energy Convers. Manag.* **2004**, *45*, 1597–1615. [CrossRef]
6. Sharma, S.D.; Sagara, K. Latent heat storage materials and systems: A review. *Int. J. Green Energy* **2005**, *2*, 1–56. [CrossRef]
7. Seddegh, S.; Wang, X.; Henderson, A.D. Numerical investigation of heat transfer mechanism in a vertical shell and tube latent heat energy storage system. *Appl. Ther. Eng.* **2015**, *87*, 698–706. [CrossRef]

8.   Kozak, Y.; Rozenfeld, T.; Ziskind, G. Close-contact melting in vertical annular enclosures with a non-isothermal base: Theoretical modeling and application to thermal storage. *Int. J. Heat Mass Transf.* **2014**, *72*, 114–127. [CrossRef]

9.   Adine, H.A.; el Qarnia, H. Numerical analysis of the thermal behaviour of a shell-and-tube heat storage unit using phase change materials. *Renew. Sustain. Energy Rev.* **2009**, *13*, 318–345. [CrossRef]

10. Nithyanandam, K.; Pitchumani, R. Analysis and optimization of a latent thermal energy storage system with embedded heat pipes. *Int. J. Heat Mass Transf.* **2011**, *54*, 4596–4610. [CrossRef]

11. Trp, A.; Lenic, K.; Frankovic, B. Analysis of the influence of operating conditions and geometric parameters on heat transfer in water-paraffin shell-and-tube latent thermal energy storage unit. *Appl. Therm. Eng.* **2006**, *26*, 1830–1839. [CrossRef]

12. Fang, M.; Chen, G. Effects of different multiple PCMs on the performance of a latent thermal energy storage system. *Appl. Therm. Eng.* **2007**, *27*, 994–1000. [CrossRef]

13. Wang, W.; Zhang, K.; Wang, L.; He, Y. Numerical study of the heat charging and discharging characteristics of a shell-and-tube phase change heat storage unit. *Appl. Therm. Eng.* **2013**, *58*, 542–553. [CrossRef]

14. Khalifa, A.; Tan, L.; Date, A.; Akbarzadeh, A. Performance of suspended finned heat pipes in high-temperature latent heat thermal energy storage. *Appl. Therm. Eng.* **2015**, *81*, 242–252. [CrossRef]

15. Regin, A.F.; Solanki, S.C.; Saini, J.S. Heat transfer characteristics of thermal energy storage system using PCM capsules: A review. *Renew. Sustain. Energy Rev.* **2008**, *12*, 2438–2458. [CrossRef]

16. Rozenfeld, T.; Kozak, Y.; Hayat, R.; Ziskind, G. Close-contact melting in a horizontal cylindrical enclosure with longitudinal plate fins: Demonstration, modeling and application to thermal storage. *Int. J. Heat Mass Transf.* **2015**, *86*, 465–477. [CrossRef]

17. Agyenim, F.; Eames, P.; Smyth, M. A comparison of heat transfer enhancement in a medium temperature thermal energy storage heat exchanger using fins. *Sol. Energy* **2009**, *83*, 1509–1520. [CrossRef]

18. Languri, E.M.; Aigbotsua, C.O.; Alvarado, J.L. Latent thermal energy storage system using phase change material in corrugated enclosures. *Appl. Therm. Eng.* **2013**, *50*, 1008–1014. [CrossRef]

19. Al-Abidi, A.A.; Mat, S.; Sopian, K.; Sulaiman, M.Y.; Mohammada, A.T. Internal and external fin heat transfer enhancement technique for latent heat thermal energy storage in triplex tube heat exchangers. *Appl. Therm. Eng.* **2013**, *53*, 147–156. [CrossRef]

20. Khalifa, A.; Tan, L.; Date, A.; Akbarzadeh, A. A numerical and experimental study of solidification around axially finned heat pipes for high temperature latent heat thermal energy storage units. *Appl. Therm. Eng.* **2014**, *70*, 609–619. [CrossRef]

21. Rathod, M.K.; Banerjee, J. Thermal performance enhancement of shell and tube Latent Heat Storage Unit using longitudinal fins. *Appl. Therm. Eng.* **2015**, *75*, 1084–1092. [CrossRef]

22. Fukai, J.; Kanou, M.; Kodama, Y.; Miyatake, O. Thermal conductivity enhancement of energy storage mediausing carbon fibers. *Energy Convers. Manag.* **2000**, *41*, 1543–1556. [CrossRef]

23. Fan, L.; Khodadadi, J.M. Thermal conductivity enhancement of phase change materials for thermal energy storage: A review. *Renew. Sustain. Energy Rev.* **2011**, *15*, 24–46. [CrossRef]

24. Jegadheeswaran, S.; Pohekar, S.D. Performance enhancement in latent heat thermal storage system: A review. *Renew. Sustain. Energy Rev.* **2009**, *13*, 2225–2244. [CrossRef]

25. Erek, A.; İlken, Z.; Acar, M.A. Experimental and numerical investigation of thermal energy storage with a finned tube. *Int. J. Energy Res.* **2005**, *29*, 283–301. [CrossRef]

26. Cabeza, L.F.; Mehling, H.; Hiebler, S.; Ziegler, F. Heat transfer enhancement in water when used as PCM in thermal energy storage. *Appl. Therm. Eng.* **2002**, *22*, 1141–1151. [CrossRef]

27. The MathWorks, Inc. *Release*; The MathWorks, Inc.: Natick, MA, USA, 2015.

28. The MathWorks, Inc. *Simulink Release*; The MathWorks, Inc.: Natick, MA, USA, 2015.

29. ANSYS, Inc. *Fluent Release 15.0*; ANSYS, Inc.: Canonsburg, PA, USA, 2013.

30. Rösler, F.; Brüggemann, D. Shell-and-tube type latent heat thermal energy storage: Numerical analysis and comparison with experiments. *Heat Mass Transf.* **2011**, *47*, 1027–1033. [CrossRef]

![energies logo] *energies*

MDPI

Article

A Cell-to-Cell Equalizer Based on Three-Resonant-State Switched-Capacitor Converters for Series-Connected Battery Strings

Yunlong Shang, Qi Zhang, Naxin Cui * and Chenghui Zhang *

School of Control Science and Engineering, Shandong University, Jinan 250061, China;
shangyunlong@mail.sdu.edu.cn (Y.S.); zhangqi2013@sdu.edu.cn (Q.Z.)
* Correspondence: cuinx@sdu.edu.cn (N.C.); zchui@sdu.edu.cn (C.Z.);
  Tel.: +86-531-8839-2907 (N.C.); +86-531-8839-5717 (C.Z.)

Academic Editors: Rui Xiong, Hailong Li and Joe (Xuan) Zhou
Received: 23 December 2016; Accepted: 6 February 2017; Published: 11 February 2017

Abstract: Due to the low cost, small size, and ease of control, the switched-capacitor (SC) battery equalizers are promising among active balancing methods. However, it is difficult to achieve the full cell equalization for the SC equalizers due to the inevitable voltage drops across Metal-Oxide-Semiconductor Field Effect Transistor (MOSFET) switches. Moreover, when the voltage gap among cells is larger, the balancing efficiency is lower, while the balancing speed becomes slower as the voltage gap gets smaller. In order to soften these downsides, this paper proposes a cell-to-cell battery equalization topology with zero-current switching (ZCS) and zero-voltage gap (ZVG) among cells based on three-resonant-state SC converters. Based on the conventional inductor-capacitor (LC) converter, an additional resonant path is built to release the charge of the capacitor into the inductor in each switching cycle, which lays the foundations for obtaining ZVG among cells, improves the balancing efficiency at a large voltage gap, and increases the balancing speed at a small voltage gap. A four-lithium-ion-cell prototype is applied to validate the theoretical analysis. Experiment results demonstrate that the proposed topology has good equalization performances with fast equalization, ZCS, and ZVG among cells.

Keywords: battery equalizers; battery management systems; switched-capacitor (SC) converters; zero-voltage gap (ZVG); modularization; electric vehicles (EVs)

1. Introduction

The world is being confronted with unprecedented crises, i.e., the depletion of fossil fuels and the global warming [1]. Energy conservation is becoming of paramount concern to people. In response to the crises, electric vehicles (EVs) have been implemented and are considered to be the inevitable development trend of vehicles for the future [2]. Due to high energy density, long lifetime, and environmental friendliness, lithium-based batteries have been dominating the high power battery packs of EVs [3,4]. However, the terminal voltage of a single lithium battery cell is usually low, e.g., 3.7 V for lithium-ion batteries and 3.2 V for lithium iron phosphate (LiFePO4) batteries [5,6]. In order to meet the demands of the load voltage and power, lithium batteries are usually connected in series and parallel [7]. For example, Tesla Model S uses 7616 lithium-ion 18650 cells connected in series and parallel [8]. Unfortunately, there are slight differences among cells in terms of capacity and internal resistance, which cause the cell voltage imbalance as the battery string is charged and discharged. On the one hand, this imbalance reduces the available capacity of battery packs. On the other hand, it may lead to over-charge or over-discharge for a cell in the battery pack, increasing safety risks. In fact, the most viable solution for this problem might not originate merely from the improvement in the battery chemistry. It also uses suitable power electronics topologies to prevent the cell imbalance, which is known as battery equalization.

During the last few years, many balancing topologies have been proposed, which can be classified into two categories: the passive balancing methods [7,9] and the active balancing methods [10–32]. The passive equalizers employ a resistor connected in parallel with each cell to drain excess energy from the high energy cells [7,9]. These methods have the outstanding advantages of small size, low cost, and easy implementation. However, their critical disadvantages are energy dissipation and heat management problems [7]. To overcome these drawbacks, active cell balancing topologies are proposed, which employ non-dissipative energy-shuttling elements to move energy from the strong cells to the weak ones [7], reducing energy loss. Therefore, active balancing methods have higher balancing capacity and efficiency than the passive equalization ones. They can be further divided into three groups, which are capacitor based [10–18], inductor based [19–21], and transformer based [22–32] methods. Among these active balancing topologies, switched-capacitor (SC) based solutions have the inherent advantages of smaller size, lower cost, simpler control, and higher efficiency. Ref. [10] proposes an SC equalizer for series battery packs. As shown in Figure 1a, one capacitor is employed to shift charge between the adjacent two cells. The capacitor is switched back and forth repeatedly, which diffuses the imbalanced charge until the two cell voltages match completely [10]. The main disadvantage of this structure is the high switching loss. To solve this problem, an automatic equalization circuit based on resonant SC converters is proposed in [15]. As shown in Figure 1b, an inductor L_0 is added to form a resonant inductor-capacitor (LC) converter, which operates alternatively between the charging state and discharging state with zero-current switching (ZCS) to automatically balance the cell voltages [15]. However, it is difficult to apply this topology to the systems with low voltage gap among cells. For example, the voltage difference among lithium-ion battery cells is not allowed to exceed 0.1 V [15]. This small voltage difference causes the Metal-Oxide-Semiconductor Field Effect Transistor (MOSFETs) of the equalizers to fail to conduct, which results in the inevitable residual voltage gap among cells. Moreover, the equalization current becomes smaller as the voltage gap gets smaller, resulting in a very long balancing time.

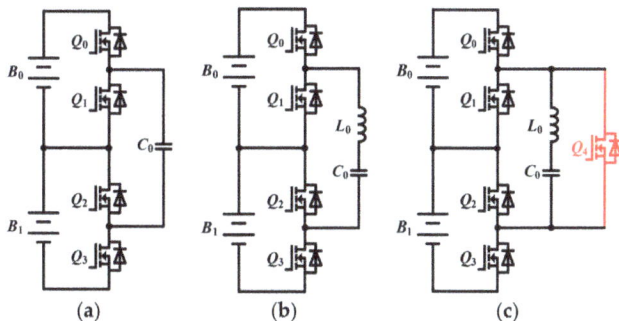

Figure 1. Battery equalizers based on switched-capacitor (SC) converters. (**a**) the classical SC equalizer [10]; (**b**) the resonant SC equalizer [15]; and (**c**) the proposed equalizer based on an inductor-capacitor-switch (LCS) converter.

In order to overcome these problems, a battery equalizer is proposed based on a resonant LC converter and boost converter that offers several major advantages, e.g., ZCS and zero-voltage gap (ZVG) among cells, etc. [16]. However, the balancing efficiency of this topology is strongly related to the voltage conversion ratio, which is expressed as $\eta_e = V_{output}/V_{in}$. The lower the conversion ratio (or the larger the voltage difference), the larger the balancing current, but the lower the balancing efficiency. This means that high efficiency cannot be achieved at a large voltage gap. Ref. [33] proposes a high-efficiency SC converter that decouples the efficiency from the voltage conversion ratio. Ref. [34] applies the switched-capacitor gyrator to photovoltaic systems, demonstrating ultimate improvement in the power harvesting capability under different insolation levels. Based on these works, the objective of this paper is to introduce an adjacent cell-to-cell battery equalization topology based on three-resonant-state LC converters, with the potential of fulfilling the expectations of high current capability, high efficiency, easy modularization, ZCS, and ZVG among cells. As shown in Figure 1c, except the classical design, an additional switch Q_4 is added to be connected in parallel with the LC tank, which is hereinafter to be referred as the inductor-capacitor-switch (LCS) converter. This structure obtains another resonant current path to release the residual energy stored in the capacitor to the inductor, which lays the foundations to achieve the bi-directional power flow and weakens the couplings of a large voltage gap with low efficiency and a small voltage gap with slow balancing speed.

2. The Proposed Equalizer

2.1. Basic Circuit Structure

As shown in Figure 2, the proposed equalizer can be easily extended to a long series battery string without limit. The architecture consists of n battery cells connected in series and $n-1$ resonant LCS tanks connected in parallel with each two adjacent battery cells, through which energy can be exchanged among all cells.

The proposed equalizer has several major advantages per the following:

(1) The proposed equalizer can achieve ZCS for all MOSFETs, and obtain ZVG among cells.
(2) Due to the other resonant current path, the balancing efficiency is improved at a large voltage gap among cells, and the balancing speed is increased at a small voltage gap.
(3) By changing the parameters of the resonant LCS converter, different balancing speeds can be achieved to meet the requirements of different energy storage devices.
(4) The concept is modular [35], and the topology can be extended to any long series-connected battery strings or individual cells without limit.

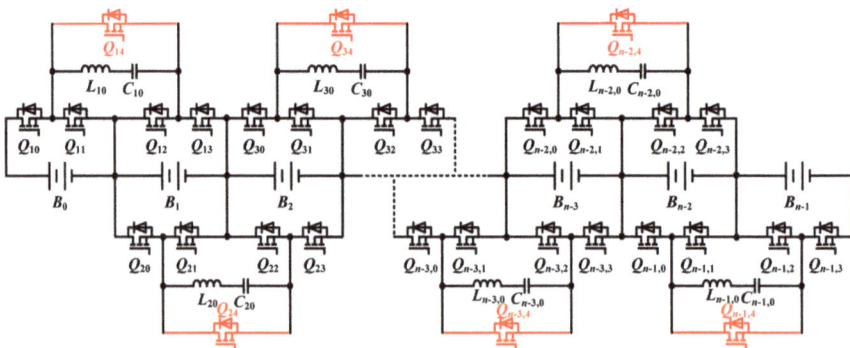

Figure 2. Schematic diagram of the proposed system for n series-connected battery cells.

2.2. Operation Principles

In order to simplify the analysis for the operation states, the following assumptions are made: the proposed equalizer is applied to two cells connected in series, i.e., B_0 and B_1, where B_0 is over-charged and B_1 is undercharged. The operation principles are shown in Figure 3. The switching sequence is set as (Q_0, Q_2), (Q_1, Q_3), and Q_4, as shown in Figure 4. Three resonant states S_1–S_3 are employed to charge, discharge, and release the LC tank, which is connected to a voltage of V_{B0}, V_{B1}, or 0 in each switching state, respectively. Figure 5 shows the theoretical waveforms of the proposed equalizer at $V_{B0} > V_{B1}$.

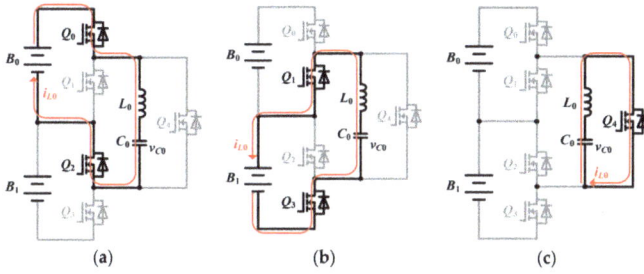

Figure 3. Operating states of the proposed equalizer at $V_{B0} > V_{B1}$. (**a**) charge state S_1; (**b**) discharge state S_2; (**c**) release state S_3.

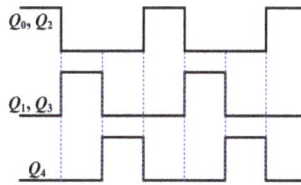

Figure 4. Switching sequences of the proposed equalizer at $V_{B0} > V_{B1}$.

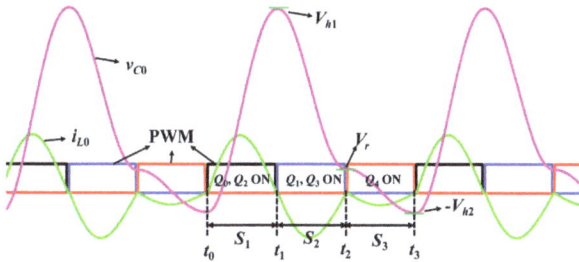

Figure 5. Theoretical waveforms of the capacitor voltage and the resonant current at $V_{B0} > V_{B1}$.

Charge State S_1 [t_0-t_1]: At t_0, switches Q_0 and Q_2 are turned ON with ZCS. The LC tank is connected with B_0 in parallel through Q_0 and Q_2, as shown in Figure 3a. B_0, L_0, and C_0 form a resonant current loop. The capacitor C_0 is charged by B_0. v_{C0} increases from $-V_{h2}$, which is a remnant of C_0 from the last period (see Figure 5). i_{L0} and v_{C0} in this state can be expressed as

$$i_{L0}(t) = \frac{V_{B0} + V_{h2}}{Z_r \cdot \sqrt{1 - \rho^2}} \cdot e^{-\rho \omega_n (t - t_0)} \cdot \sin\left[\omega_n \cdot \sqrt{1 - \rho^2} \cdot (t - t_0)\right], \tag{1}$$

$$v_{C0}(t) = -V_{h2} + (V_{B0} + V_{h2}) \cdot \left\{1 - \frac{e^{-\rho \omega_n (t - t_0)}}{\sqrt{1 - \rho^2}} \cdot \cos\left[\omega_n \cdot \sqrt{1 - \rho^2} \cdot (t - t_0)\right]\right\}, \tag{2}$$

where $Z_r = \sqrt{L_0/C_0}$, $\omega_n = 1/\sqrt{L_0 C_0}$, and $\rho = R_S/2Z_r$. R_S represents the equivalent parasitic resistance in each current path.

The charge state ends when i_{L0} crosses zero at $t = t_1$. From Equation (1), the duration of this state is determined by

$$\Delta t = t_1 - t_0 = \frac{\pi}{\omega_n \cdot \sqrt{1 - \rho^2}}. \tag{3}$$

At t_1, v_{C0} is positively charged to V_{h1}, which can be given by

$$V_{h1} = v_{C0}(t_1) = (V_{B0} + V_{h2}) \cdot \left(1 + \frac{e^{-\rho\omega_n \Delta t}}{\sqrt{1-\rho^2}}\right) - V_{h2}. \tag{4}$$

Discharge State S_2 [t_1-t_2]: At t_1, the switches Q_1 and Q_3 are turned ON with ZCS, connecting B_1 to the resonant LC tank. B_1, L_0, and C_0 form a resonant loop. B_1 is charged by C_0. i_{L0} and v_{C0} in this state are given as

$$i_{L0}(t) = -\frac{V_{h1} - V_{B1}}{Z_r \cdot \sqrt{1-\rho^2}} \cdot e^{-\rho\omega_n(t-t_1)} \cdot \sin\left[\omega_n \cdot \sqrt{1-\rho^2} \cdot (t-t_1)\right], \tag{5}$$

$$v_{C0}(t) = V_{h1} - (V_{h1} - V_{B1}) \cdot \left\{1 - \frac{e^{-\rho\omega_n(t-t_1)}}{\sqrt{1-\rho^2}} \cdot \cos\left[\omega_n \cdot \sqrt{1-\rho^2} \cdot (t-t_1)\right]\right\}. \tag{6}$$

At $t = t_2$, the discharge state ends when i_{L0} drops to zero. The voltage V_r of C_0 at $t = t_2$ is represented by

$$V_r = V_{h1} - (V_{h1} - V_{B1}) \cdot \left(1 + \frac{e^{-\rho\omega_n \Delta t}}{\sqrt{1-\rho^2}}\right). \tag{7}$$

Release State S_3 [t_2-t_3]: During this state, the resonant LC tank is short-circuited by turning on the switch Q_4 with ZCS. This releases the residual charge of the capacitor into the inductor and even charges reversely the capacitor C_0, so B_0 can charge C_0 with a large current at the beginning of S_1. This state provides the opportunity to transfer energy from a low voltage cell to a high voltage one, which lays the foundations to achieve ZVG among cells. i_{L0} and v_{C0} in this state are given by

$$i_{L0}(t) = -\frac{V_r}{Z_r \cdot \sqrt{1-\rho^2}} e^{-\rho\omega_n(t-t_2)} \cdot \sin\left[\omega_n \cdot \sqrt{1-\rho^2} \cdot (t-t_2)\right], \tag{8}$$

$$v_{C0}(t) = V_r \cdot \frac{e^{-\rho\omega_n(t-t_2)}}{\sqrt{1-\rho^2}} \cdot \cos\left[\omega_n \cdot \sqrt{1-\rho^2} \cdot (t-t_2)\right]. \tag{9}$$

The release state ends when i_{L0} crosses zero at $t = t_3$. The voltage V_{h2} of C_0 at $t = t_3$ can be expressed as

$$-V_{h2} \equiv v_{C0}(t_3) = V_r \cdot \frac{e^{-\rho\omega_n(t_3-t_2)}}{\sqrt{1-\rho^2}} \cdot \cos\left[\omega_n \cdot \sqrt{1-\rho^2} \cdot (t_3-t_2)\right] = -\lambda V_r, \tag{10}$$

where

$$\lambda = \frac{e^{-\rho\omega_n \Delta t}}{\sqrt{1-\rho^2}} = \frac{e^{-\pi\rho/\sqrt{1-\rho^2}}}{\sqrt{1-\rho^2}}. \tag{11}$$

By solving Equations (4), (7), and (10), V_{h1}, V_r, and V_{h2} can be calculated as

$$V_{h1} = \frac{V_{B0} + \lambda^2 V_{B1}}{1 - \lambda + \lambda^2}, \tag{12}$$

$$V_r = \frac{V_{B1} - \lambda V_{B1}}{1 - \lambda + \lambda^2}, \tag{13}$$

$$V_{h2} = \frac{\lambda(V_{B1} - \lambda V_{B0})}{1 - \lambda + \lambda^2}. \tag{14}$$

The operating period T is composed of three resonant states, which can be expressed as

$$T = \frac{3\pi}{\omega_n \cdot \sqrt{1-\rho^2}} = \frac{3\pi \cdot \sqrt{L_0 C_0}}{\sqrt{1-\rho^2}}. \tag{15}$$

The direction of the balancing power flowing can be changed by controlling the switching sequences. According to the above analysis, the switching sequence (Q_0, Q_2), (Q_1, Q_3), Q_4 is to deliver energy from B_0 to B_1. In the case of energy transferred from B_1 to B_0, the switching sequence is changed to (Q_1, Q_3), (Q_0, Q_2), Q_4. Figure 6 shows the three consecutive operating states of the proposed equalizer: (a) charge state; (b) discharge state; and (c) release state at $V_{B0} < V_{B1}$. Figure 7 shows the corresponding switching sequence. It can be seen that, by controlling the switching sequence, energy can be delivered between two adjacent cells arbitrarily, by which ZVG between cells can be achieved without any limit.

It is important to note that the release state can also be achieved by turning simultaneously on Q_1 and Q_2 without using Q_4, which results in a reduced MOSFET number but complex control. Figures 8 and 9 show the three consecutive operating states without using Q_4 and the corresponding switching sequences at $V_{B0} > V_{B1}$. Figures 10 and 11 show the three consecutive operating states without using Q_4 and the corresponding switching sequences at $V_{B0} < V_{B1}$. The operation principles of this system are similar to those shown in Figures 3–6 and will not be described here in detail.

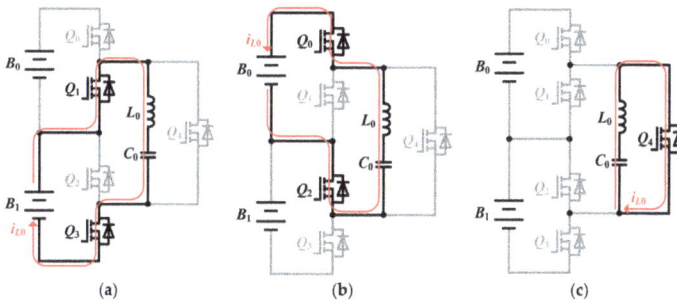

Figure 6. Operating states of the proposed equalizer at $V_{B0} < V_{B1}$. (**a**) charge state; (**b**) discharge state; (**c**) release state.

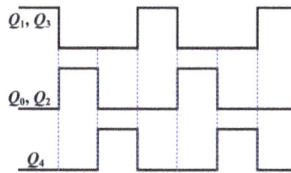

Figure 7. Switching sequences of the proposed equalizer at $V_{B0} < V_{B1}$.

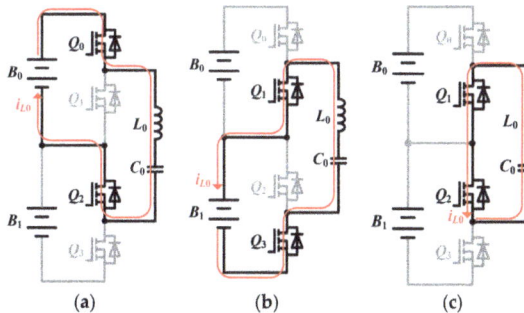

Figure 8. Operating states of the proposed equalizer without using Q_4 at $V_{B0} > V_{B1}$. (**a**) charge state; (**b**) discharge state; (**c**) release state.

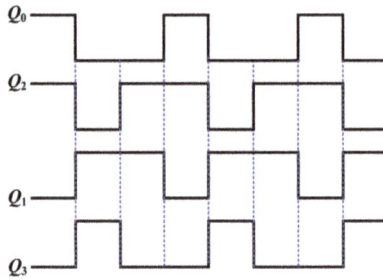

Figure 9. Switching sequences of the proposed equalizer without using Q_4 at $V_{B0} > V_{B1}$.

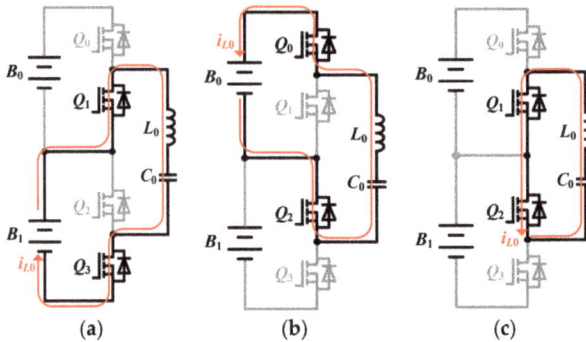

(a)   (b)   (c)

Figure 10. Operating states of the proposed equalizer without using Q_4 at $V_{B0} < V_{B1}$. (a) charge state; (b) discharge state; (c) release state.

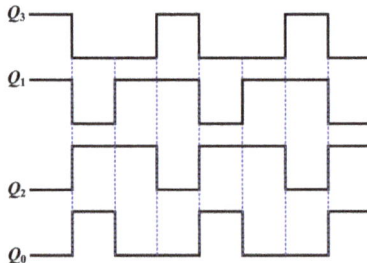

Figure 11. Switching sequences of the proposed equalizer without using Q_4 at $V_{B0} < V_{B1}$.

2.3. Equalizing Power and Efficiency

During one switching period T, the charge delivered to C_0 from B_0 is

$$\Delta Q_D = C_0 \cdot (V_{h_1} + V_{h_2}), \tag{16}$$

and the charge received by B_1 is expressed as

$$\Delta Q_R = C_0 \cdot (V_{h_1} - V_r). \tag{17}$$

Using Equations (12)–(14) and (16), the average power flowing out of B_0 is obtained as

$$P_{avg,D} = \Delta Q_D \cdot V_{B_0} = \frac{V_{B0} \cdot \sqrt{1-\rho^2}}{3\pi Z_r} \times \frac{(1+\lambda) \cdot [(1-\lambda) \cdot V_{B0} + \lambda \cdot V_{B1}]}{1 - \lambda + \lambda^2}, \tag{18}$$

and, using Equations (12)–(14) and (17), the average power flowing into B_1 is given as

$$P_{avg,R} = \Delta Q_R \cdot V_{B_1} = \frac{V_{B1} \cdot \sqrt{1-\rho^2}}{3\pi Z_r} \times \frac{(1+\lambda) \cdot [V_{B0} - (1-\lambda) \cdot V_{B1}]}{1 - \lambda + \lambda^2}. \tag{19}$$

Based on Equations (18) and (19), the equalization efficiency η_e can be calculated as

$$\eta_e = \frac{P_{avg,R}}{P_{avg,D}} = \frac{V_{B1}}{V_{B0}} \cdot \frac{V_{B0} - (1-\lambda) \cdot V_{B1}}{(1-\lambda) \cdot V_{B0} + \lambda \cdot V_{B1}} \times 100\%. \tag{20}$$

Figure 12 shows the balancing efficiency curves obtained from Equation (20) as a function of the L_0/C_0 ratio, for various R_S, under the conditions of V_{B0} = 3.3 V and V_{B1} = 3.2 V. It can be observed that the efficiency increases as the L_0/C_0 ratio increases or R_S decreases, which show how the coupling of the large voltage gap with low efficiency can be weakened by keeping R_S as low and the L_0/C_0 ratio as high as possible. However, from Equations (1) and (5), it can be concluded that the balancing current would become smaller as the L_0/C_0 ratio increases. Therefore, an appropriate L_0/C_0 ratio (e.g., L_0/C_0 = 10) should be selected in order to achieve a higher balancing efficiency and larger balancing current.

Figure 13 presents the efficiency curve as a function of power at L_0/C_0 = 10 and R_S = 0.18 Ω. The balancing efficiency rises rapidly when the power increases from 0.12 W to 0.5 W and basically stays at a high value when the power increases from 0.5 W to 0.9 W, but decreases slightly when the power increases from 0.9 W to 1.3 W. The peak efficiency of 91.5% is achieved at 0.74 W.

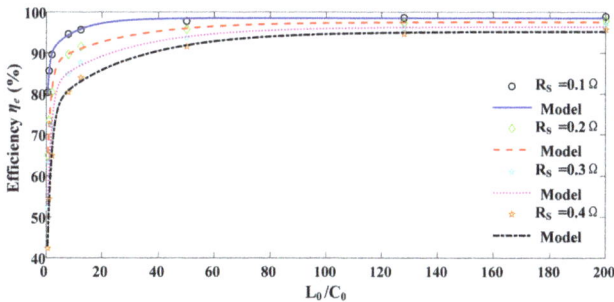

Figure 12. Theoretical efficiency η_e as a function of L_0/C_0 ratio with different R_S.

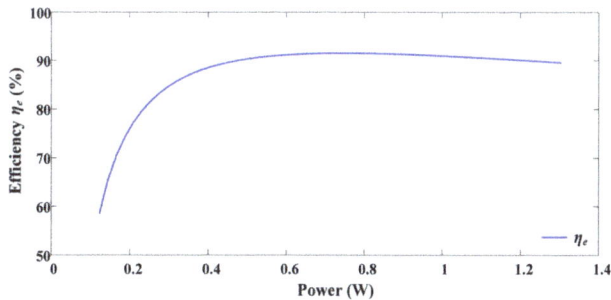

Figure 13. Theoretical efficiency η_e as a function of power at L_0/C_0 = 10 and R_S = 0.18 Ω.

3. Experimental Results

In order to verify the theoretical analysis and evaluate the equalization performance of the proposed system, a prototype for four 6200-mA·h lithium-ion cells is implemented and tested. Figure 14 shows the photographs of the experimental setup. The MOSFETs are implemented by STP220N6F7 MOSFETs with 2.4 mΩ internal resistance. The values of L_0 and C_0 are determined as 10.99 µH and 1.05 µF, respectively. The measured equivalent resistance R_S in the LC converter is about 0.18 Ω. A MicroAutoBox® II manufactured by dSPACE (Wixom, MI, USA) was used for the digital control, which can generate Pulse-Width Modulation (PWM) singles to control the MOSFETs, and receive the cell voltage information by analog-to-digital converters.

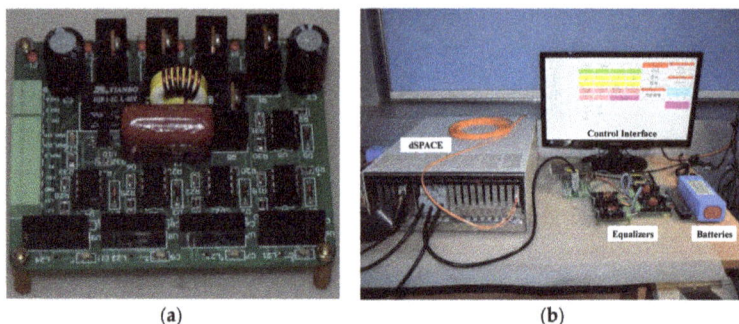

(a)   (b)

Figure 14. Photographs of the implemented engineering prototype for four lithium-ion battery cells. (a) balancing circuit; (b) experimental platform.

Figure 15 shows the experimental waveforms of resonant current i_{L0} and capacitor voltage v_{C0} with different switching sequences. It can be observed that the MOSFETs are turned ON and OFF at zero current state, thus significantly reducing the switching losses. This provides the equalizer with the potential to work at higher frequencies, leading to a small size of the proposed equalizer. From Figure 15a,b, it can be seen that controlling the switching sequence can govern the direction of the balancing power flowing. This agrees well with the theoretical waveforms.

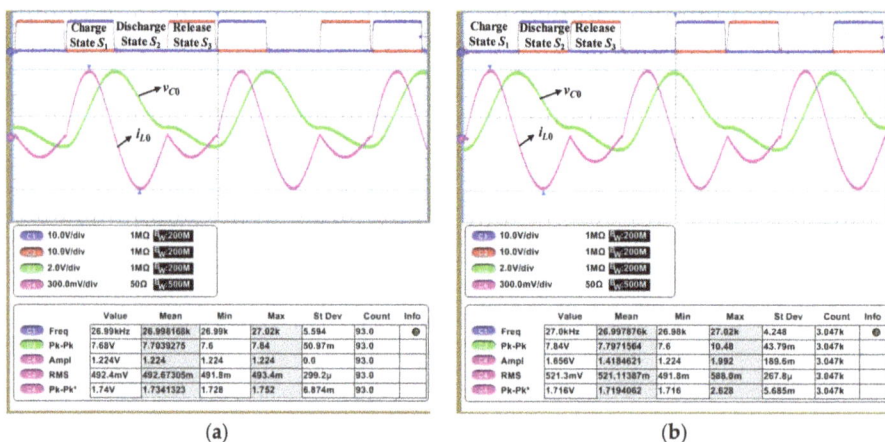

(a)   (b)

Figure 15. Experimental waveforms of the proposed equalizer with different switching sequences. (a) energy transfer from B_0 to B_1; (b) energy transfer from B_1 to B_0.

Figure 16 shows the measured efficiency η_e as a function of power at $L_0/C_0 \approx 10$. When power increases from 0.226 to 0.595 W, η_e increases from 47.7% to 89.1%. When power increases from 0.595 to 0.913 W, η_e decreases slightly from 89.1% to 81.5%. This indicates that the proposed equalizer obtains a high efficiency over a wide range of output power.

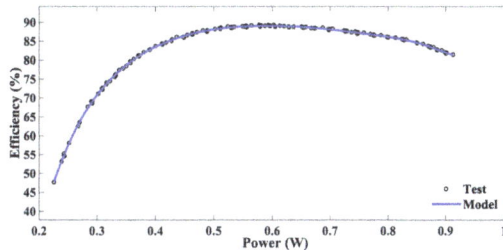

Figure 16. Measured efficiency η_e as a function of power at $L_0/C_0 \approx 10$.

Figure 17 shows the experimental results for two cells connected in series. The initial cell voltages are set as $V_{B0} = 3.240$ V and $V_{B1} = 2.574$ V, respectively. The initial maximum voltage gap is about 0.666 V. It is important to note that, in order to achieve the initial cell voltages, the battery string is not balanced until 200 s' standing. Figure 17a shows the balancing result with the classical switched capacitor. After about 8.2 h, the voltage gap between the cells is still larger than 0.109 V, which shows that the switched capacitor method cannot achieve ZVG between the two cells. Figure 17b shows the balancing result with the resonant switched capacitor. The balancing speed is increased a lot, but ZVG between cells is still not achieved after 8000 s. Figure 17c shows the balancing result with the proposed method. We observe that, after about 2056 s, the cell voltages are fully balanced to the same value of 3.171 V, showing the outstanding balancing performances (i.e., fast balancing and ZVG between cells) of the proposed scheme. Figure 17d shows the balancing result using the proposed equalizer without the release sate. It can be seen that the balancing speed becomes slow, and ZVG between cells cannot be achieved, which indicates that the release sate plays an active role in the balancing process.

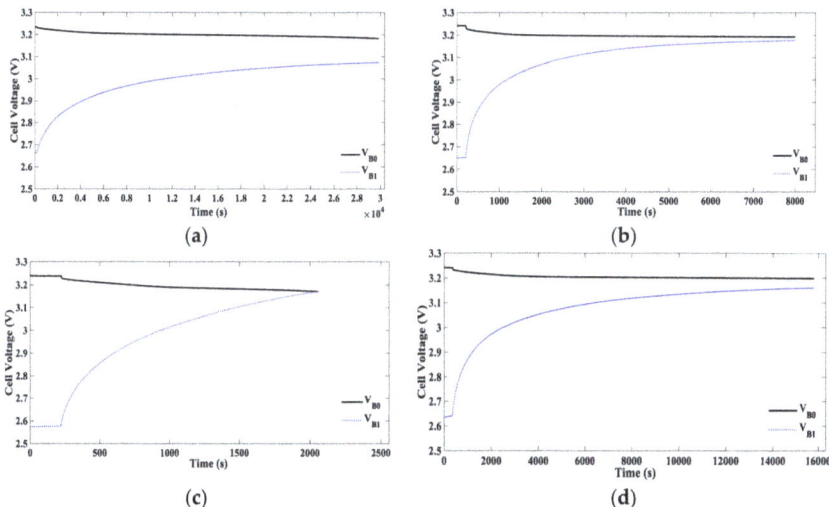

Figure 17. The voltage equalization results for two cells. (**a**) the classical SC method [10]; (**b**) the resonant SC method [15]; (**c**) the proposed method based on a LCS converter; (**d**) the proposed method based on a LCS converter without the release sate.

Figure 18 shows the experimental results for four cells connected in series. Because of the nonlinear behavior of lithium-ion batteries, it is very difficult to determine when the cell voltages are fully balanced. Thus, it is optimal to take numerous small equalization cycles to complete the energy exchange. In our method, one equalization cycle includes 10-s equalization time and 20-s standing time for the equalizer. The initial cell voltages are set as $V_{B0} = 3.216$ V, $V_{B1} = 2.783$ V, $V_{B2} = 3.233$ V, and $V_{B3} = 3.023$ V, respectively. After about 12,960 s, a balanced voltage of 3.096 V is achieved with about 178 equalization cycles.

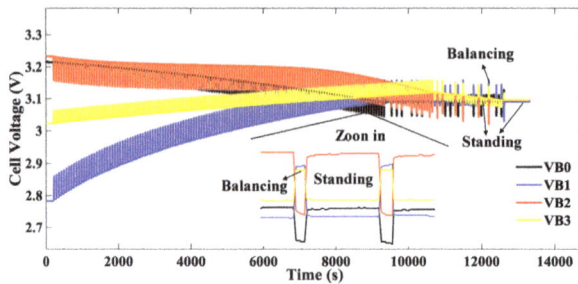

Figure 18. The voltage equalization results for four cells.

4. Comparison with Conventional Equalizers

In order to systematically evaluate the proposed scheme, Table 1 gives a comparative study with conventional battery equalizers focusing on the components, balancing speed, balancing efficiency, ZCS, ZVG among cells, and modularization. It is assumed that the battery string includes n cells connected in series, which is divided into m battery modules. Components focuses mainly on the numbers of switches (SW), resistors (R), inductors (L), capacitors (C), diodes (D), and transformers (T). The equalization speed is determined by the equalization current, the number of cells involved in balancing at the same time, and the average switching cycles to complete the charge transportation from the source cell to the target one. The balancing efficiency is evaluated according to the average energy conversion efficiency for one switching cycle and the average switching cycles to transfer energy from a cell to another one. ZCS and ZVG are evaluated according to whether the systems can achieve ZCS for all MOSFETs and obtain ZVG among cells in a battery string. Modularization is evaluated according to the implemented complexity of the equalizers when a new cell is added. These balancing performance parameters are fuzzified into three fuzzy scales, for which "H" represents the higher performance, "L" represents the lower performance, and "M" represents the medium performance, specifically, Speed (L: low, H: high), Efficiency (L: low, H: high), ZCS (L: no, H: yes), ZVG among cells (L: no, M: yes), and Modularization (L: difficult, H: easy).

All of the existing solutions provide good performance targeting. For example, the dissipative equalization method [9] has the outstanding advantages of small size, low cost, and easy implementation. However, the excess energy is consumed by the shunt resistors, resulting in a very low balancing efficiency.

SC based methods [10–14] tend to be lighter and smaller due to the absence of any magnetic components. Moreover, they have the outstanding advantages of simple control, easy modularization, and automatic equalization without cell monitoring circuits. However, the balancing efficiency is very low at a large voltage gap among cells, and the balancing speed becomes slower as the voltage gap gets smaller. In other words, these methods cannot have a high equalization efficiency and a fast balancing speed at the same time.

Table 1. Comparison of several battery equalizers.

Category	Components						Speed	Efficiency	ZCS	ZVG	Modularization
	SW	R	L	C	D	T					
Dissipative equalizer [9]	n	n	0	0	0	0	M	L	L	M	H
SC [10]	$2n$	0	0	$n-1$	0	0	L	M	L	L	H
Chain structure of SC [11]	$2(n+2m)$	0	0	$n+m$	0	0	L	M	L	L	L
ZCS SC [15]	$2n$	0	$n-1$	$n-1$	0	0	L	M	H	L	H
Single LC resonant converter [18]	$2(n+5m)$	0	m	m	0	0	M	M	H	L	M
Buckboost (multiple inductors) [19]	$2n$	0	$n-1$	0	0	0	M	M	L	M	H
Multiphase interleaved method [20]	$2(n-1)$	0	$n-1$	0	0	0	M	M	L	M	L
Optimized next-to-next balancing [21]	$4(n-1)$	0	$2(n-1)$	0	0	0	L	M	L	M	H
Flyback conversion [22]	$2(n-m)$	0	0	0	$2(n-m)$	m	M	M	L	M	L
Flyback or forward conversion [23]	$2n$	0	0	0	0	m	M	H	H	L	M
Forward conversion [24]	n	0	0	n	n	n	H	H	H	L	M
Wave-trap [28]	$2m$	0	n	n	0	0	M	M	H	M	L
Proposed equalizer with Q_4	$5(n-1)$	0	$n-1$	$n-1$	0	0	H	M	H	H	H
Proposed equalizer without using Q_4	$4(n-1)$	0	$n-1$	$n-1$	0	0	H	M	H	H	H

n is the number of cells in the battery string; m is the number of battery modules in the battery string; SC (Switched capacitor); ZCS (zero-current switching); LC (inductor capacitor).

Inductor based methods [19–21] require only inductors and MOSFETs. Therefore, the sizes of these solutions are small, and the costs are low. These approaches can also achieve automatic equalization among cells without the requirement of cell monitoring circuits. Moreover, they are easily modularized and not limited to the numbers of battery cells in a battery string. However, they work in the hard-switching mode, and the switching loss tends to be high, leading to a low balancing efficiency. Particularly, ZVG among cells cannot be achieved due to the asymmetry of inductors and the voltage drops across power electronic devices.

Transformer-based solutions [22–32] have the inherent advantages of easy isolation, high efficiency, and simple control. However, it is definitely difficult to apply a single multi-winding transformer into a long series-connected battery string because of the mismatching, bulk size, and high complexity implementation of the multi windings. Moreover, the mismatched multi windings naturally cause the imbalance voltages during the balancing. In addition, these methods need additional components for the equalization among modules, leading to bulk size and loss related to the modularization.

By using an additional switch Q_4 connected in parallel with the LC tank, the proposed solution obtains another resonant current path to release the residual energy stored in the capacitor to the inductor, which lays the foundations to achieve the bi-directional power flow and weakens the couplings of a large voltage gap with low efficiency and a small voltage gap with slow balancing speed. From Table 1, it is apparent that the size of the proposed equalizer is comparable with the existing solutions. Moreover, it has clear advantages in terms of the balancing speed, efficiency, ZCS, ZVG, and modularization, which make the proposed system be a feasible solution for EVs in the future.

5. Conclusions

In this paper, an adjacent cell-to-cell equalizer with ZCS and ZVG based on three-resonant-state SC converters is proposed. The scheme configuration, modular design, operation principles, theoretical analysis, cell-balancing performance, and comparative studies with the conventional battery equalizers are presented. The proposed scheme obtains ZCS due to the three resonant states of the LCS converter, which reduces inherently the frequency dependent switching losses, allowing efficient operation at very high switching frequencies. ZVG among cells is achieved thanks to the newly added resonant current path, which also weakens the couplings of a large voltage gap with low efficiency and a small voltage gap with slow balancing speed. A prototype with four 6200-mA·h lithium-ion cells is optimally implemented. Experiment results show that the proposed scheme exhibits good balancing performance with ZCS and ZVG, and the measured peak conversion efficiency is 89.1% at $L_0/C_0 \approx 10$.

Acknowledgments: This work was supported by the Major Scientific Instrument Development Program of the National Natural Science Foundation of China under Grant No. 61527809, the Key Project of National Natural Science Foundation of China under Grant No. 61633015, the National Natural Science Foundation of China under Grant No. 61273097, and the Major International (Regional) Joint Research Project of the National Natural Science Foundation of China (NSFC) under Grant No. 61320106011. The authors would like to thank them for their support and help. The authors would also like to thank the reviewers for their corrections and helpful suggestions.

Author Contributions: Yunlong Shang conceived this paper, designed and performed the experiments, and analyzed the data; Qi Zhang assisted the experiment and revised the paper; Naxin Cui and Chenghui Zhang revised the paper and provided some valuable suggestions.

Conflicts of Interest: The authors declare no conflict of interest.

References

1. Zhang, Y.; Xiong, R.; He, H.; Shen, W. A lithium-ion battery pack state of charge and state of energy estimation algorithms using a hardware-in-the-loop validation. *IEEE Trans. Power Electron.* **2016**. [CrossRef]
2. Sun, F.; Xiong, R.; He, H. A systematic state-of-charge estimation framework for multi-cell battery pack in electric vehicles using bias correction technique. *Appl. Energy* **2016**, *162*, 1399–1409. [CrossRef]
3. Xiong, R.; Sun, F.; Gong, X.; Gao, C. A data-driven based adaptive state of charge estimator of lithium-ion polymer battery used in electric vehicles. *Appl. Energy* **2014**, *113*, 1421–1433. [CrossRef]

4. Xia, B.; Mi, C. A fault-tolerant voltage measurement method for series connected battery packs. *J. Power Source* **2016**, *308*, 83–96. [CrossRef]

5. Lu, L.; Han, X.; Li, J.; Hua, J.; Ouyang, M. A review on the key issues for lithium-ion battery management in electric vehicles. *J. Power Source* **2013**, *226*, 272–288. [CrossRef]

6. Xiong, R.; Sun, F.; Chen, Z.; He, H. A data-driven multi-scale extended Kalman filtering based parameter and state estimation approach of lithium-ion polymer battery in electric vehicles. *Appl. Energy* **2014**, *113*, 463–476. [CrossRef]

7. Lozano, J.G.; Cadaval, E.R. Battery equalization active methods. *J. Power Source* **2014**, *246*, 934–949. [CrossRef]

8. Battery. Available online: http://batteryuniversity.com/learn/article/electric_vehicle_ev (accessed on 8 February 2017).

9. Gallardo-Lozano, J.; Romero-Cadaval, E.; Milanes-Montero, M.I.; Guerrero-Martinez, M.A. A novel active battery equalization control with on-line unhealthy cell detection and cell change decision. *J. Power Source* **2015**, *299*, 356–370. [CrossRef]

10. Pascual, C.; Krein, P.T. Switched capacitor system for automatic series battery equalization. In Proceedings of the IEEE 1997 Applied Power Electronics Conference, Atlanta, GA, USA, 23–27 February 1997; pp. 848–854.

11. Kim, M.Y.; Kim, C.H.; Kim, J.H.; Moon, G.W. A chain structure of switched capacitor for improve cell balancing speed of lithium-ion batteries. *IEEE Trans. Ind. Electron.* **2014**, *61*, 3989–3999. [CrossRef]

12. Shang, Y.L.; Xia, B.; Lu, F.; Zhang, C.H.; Cui, N.X.; Mi, C. A switched-coupling-capacitor equalizer for series-connected battery strings. *IEEE Trans. Power Electron.* **2016**. [CrossRef]

13. Ye, Y.; Cheng, K.W.E. An automatic switched-capacitor cell balancing circuit for series-connected battery strings. *Energies* **2016**, *9*, 138. [CrossRef]

14. Ye, Y.; Cheng, K.W.E. Modeling and analysis of series-parallel switched-capacitor voltage equalizer for battery/supercapacitor strings. *IEEE J. Emerg. Sel. Top. Power Electron.* **2015**, *3*, 977–983. [CrossRef]

15. Ye, Y.; Cheng, K.W.E.; Yeung, Y.P.B. Zero-current switching switched-capacitor zero-voltage-gap automatic equalization system for series battery string. *IEEE Trans. Power Electron.* **2012**, *27*, 3234–3242.

16. Shang, Y.; Zhang, C.; Cui, C.N.; Guerrero, J.M. A cell-to-cell battery equalizer with zero-current switching and zero-voltage gap based on quasi-resonant LC converter and boost converter. *IEEE Trans. Power Electron.* **2015**, *30*, 3731–3747. [CrossRef]

17. Shang, Y.; Zhang, C.; Cui, C.N.; Guerrero, J.M.; Sun, K. A crossed pack-to-cell equalizer based on quasi-resonant LC converter with adaptive fuzzy logic equalization control for series-connected lithium-ion battery strings. In Proceedings of the IEEE 2015 Applied Power Electronics Conference, Charlotte, NC, USA, 15–19 March 2015; pp. 1685–1692.

18. Lee, K.; Chung, Y.; Sung, C.-H.; Kang, B. Active cell balancing of li-ion batteries using LC series resonant circuit. *IEEE Trans. Ind. Electron.* **2015**, *62*, 5491–5501. [CrossRef]

19. Kim, M.-Y.; Kim, J.-H.; Moon, G.-W. Center-cell concentration structure of a cell-to-cell balancing circuit with a reduced number of switches. *IEEE Trans. Power Electron.* **2014**, *29*, 5285–5297. [CrossRef]

20. Mestrallet, F.; Kerachev, L.; Crebier, J.-C.; Collet, A. Multiphase interleaved converter for lithium battery active balancing. *IEEE Trans. Power Electron.* **2014**, *29*, 2874–2881. [CrossRef]

21. Phung, T.H.; Collet, A.; Crebier, J.-C. An optimized topology for next-to-next balancing of series-connected lithium-ion cells. *IEEE Trans. Power Electron.* **2014**, *29*, 4603–4613. [CrossRef]

22. Imitiaz, A.M.; Khan, F.H. "Time shared flyback converter" based regenerative cell balancing technique for series connected li-ion battery strings. *IEEE Trans. Power Electron.* **2013**, *28*, 5960–5975. [CrossRef]

23. Chen, Y.; Liu, X.; Cui, Y.; Zou, J.; Yang, S. A multi-winding transformer cell-to-cell active equalization method for lithium-ion batteries with reduced number of driving circuits. *IEEE Trans. Power Electron.* **2016**, *31*, 4916–4929. [CrossRef]

24. Li, S.; Mi, C.; Zhang, M. A high-efficiency active battery-balancing circuit using multiwinding transformer. *IEEE Trans. Ind. Appl.* **2013**, *49*, 198–207. [CrossRef]

25. Uno, M.; Kukita, A. Double-switch equalizer using parallel-or series-parallel-resonant inverter and voltage multiplier for series-connected supercapacitors. *IEEE Trans. Power Electron.* **2014**, *29*, 812–828. [CrossRef]

26. Anno, T.; Koizumi, H. Double-input bidirectional DC/DC converter using cell-voltage equalizer with flyback transformer. *IEEE Trans. Power Electron.* **2015**, *30*, 2923–2934. [CrossRef]

27. Hua, C.; Fang, Y.-H. A charge equalizer with a combination of APWM and PFM control based on a modified half-bridge converter. *IEEE Trans. Power Electron.* **2016**, *31*, 2970–2979. [CrossRef]

28. Arias, M.; Sebastián, J.; Hernando, M.; Viscarret, U.; Gil, I. Practical application of the wave-trap concept in battery-cell equalizers. *IEEE Trans. Power Electron.* **2015**, *30*, 5616–5631. [CrossRef]

29. Lim, C.-S.; Lee, K.-J.; Ku, N.-J.; Hyun, D.-S.; Kim, R.-Y. A modularized equalization method based on magnetizing energy for a series-connected Lithium-ion battery string. *IEEE Trans. Power Electron.* **2014**, *29*, 1791–1799. [CrossRef]

30. Kim, C.-H.; Kim, M.-Y.; Moon, G.-W. A modularized charge equalizer using a battery monitoring IC for series-connected Li-ion battery strings in electric vehicles. *IEEE Trans. Power Electron.* **2013**, *28*, 3779–3787. [CrossRef]

31. Park, H.-S.; Kim, C.-H.; Park, K.-B.; Moon, G.-W.; Lee, J.-H. Design of a charge equalizer based on battery modularization. *IEEE Trans. Veh. Technol.* **2009**, *58*, 3216–3223. [CrossRef]

32. Xu, A.; Xie, S.; Liu, X. Dynamic voltage equalization for series-connected ultracapacitors in EV/HEV applications. *IEEE Trans. Veh. Technol.* **2009**, *58*, 3981–3987.

33. Cervera, A.; Evzelman, M.; Mordehai Peretz, M.; Ben-Yaakov, S. A high efficiency resonant switched capacitor converter with continuous conversion ratio. *IEEE Trans. Power Electron.* **2014**, *30*, 1373–1382. [CrossRef]

34. Blumenfeld, A.; Cervera, A.; Mordechai Peretz, M. Enhanced differential power processor for PV systems: Resonant switched-capacitor gyrator converter with local MPPT. *IEEE J. Emerg. Sel. Top. Power Electron.* **2014**, *2*, 883–892. [CrossRef]

35. Dong, B.; Li, Y.; Han, Y. Parallel architecture for battery charge equalization. *IEEE Trans. Power Electron.* **2015**, *30*, 4906–4913. [CrossRef]

![energies logo] **energies**

MDPI

Article

Application of Liquid Hydrogen with SMES for Efficient Use of Renewable Energy in the Energy Internet

Xin Wang [1], Jun Yang [1,*], Lei Chen [1] and Jifeng He [2]

[1]   School of Electrical Engineering, Wuhan University, Wuhan 430072, China;
       bzzhagnxy@126.com (X.W.); stclchen1982@163.com (L.C.)
[2]   State Grid Hubei Electric Power Economic and Technology Research Institute, Wuhan 430077, China;
       18202792668@163.com
*   Correspondence: JYang@whu.edu.cn; Tel.: +86-13995638969

Academic Editor: Hailong Li
Received: 6 January 2017; Accepted: 2 February 2017; Published: 8 February 2017

Abstract: Considering that generally frequency instability problems occur due to abrupt variations in load demand growth and power variations generated by different renewable energy sources (RESs), the application of superconducting magnetic energy storage (SMES) may become crucial due to its rapid response features. In this paper, liquid hydrogen with SMES (LIQHYSMES) is proposed to play a role in the future energy internet in terms of its combination of the SMES and the liquid hydrogen storage unit, which can help to overcome the capacity limit and high investment cost disadvantages of SMES. The generalized predictive control (GPC) algorithm is presented to be appreciatively used to eliminate the frequency deviations of the isolated micro energy grid including the LIQHYSMES and RESs. A benchmark micro energy grid with distributed generators (DGs), electrical vehicle (EV) stations, smart loads and a LIQHYSMES unit is modeled in the Matlab/Simulink environment. The simulation results show that the proposed GPC strategy can reschedule the active power output of each component to maintain the stability of the grid. In addition, in order to improve the performance of the SMES, a detailed optimization design of the superconducting coil is conducted, and the optimized SMES unit can offer better technical advantages in damping the frequency fluctuations.

Keywords: superconducting magnetic energy storage (SMSE); load frequency control; generalized predictive control (GPC); energy internet

---

1. Introduction

The increasing number of renewable energy sources (RESs) and distributed generators (DGs) has become a serious challenge for the stability and reliability of the electric power system, because of the fluctuation of power supply needed to meet the demand [1,2]. With the concerns related to this and other problems, e.g., conventional energy cost, greenhouse gas emissions, security of traditional power systems [3], the concept of the energy internet is proposed [4], which is composed of numerous micro-energy grids and supports the flexible access of various RESs [5]. Therefore, energy storage technology is crucial for the energy internet to suppress power fluctuations and achieve the efficient operation of RESs by decoupling the electricity generation from demand [6,7].

Superconducting magnetic energy storage (SMES) units offer quick responses to power fluctuations and the ability to deliver large amounts of power instantaneously, while their limited storage capacity is a weak point for long term operation [8]. Liquid hydrogen (LH$_2$) storage units have the characteristics of large storage capacity [9] and economic efficiency that can make up for the disadvantages of SMES, but their response is too slow to be used as the single storage mode to support the RESs in the energy internet.

Some studies have focused on the use of LH_2 as the cooling medium for SMES [10–12] for a long period, and the concept of liquid hydrogen with SMES (LIQHYSMES) that combines the SMES with LH_2 storage units is proposed for further study [13]. The simulation and analysis of the buffering behavior of the LIQHYSMES plant model was carried out in [14] and it seems to be capable of handling even very strong variations of the imbalance between supply and demand. Also, different SMES structure designs for the 10 GJ range are compared in terms of size and ramping losses in [15], and the cost targets for different power levels and supply periods are addressed. It can be concluded from these publications that the application of LIQHYSMES are quite feasible and suitable for the energy internet.

The load frequency control (LFC) has been widely used in conventional electric power systems, and the micro-grid can maintain the stability of frequency by the optimal control, proportional integral (PI) control and other methods [16–18]. In the micro energy grid including LIQHYSMES units proposed here, the changes of the state of the system are fairly rapid so a controller with robust performance over a wide range of operating conditions is strongly needed for LFC in an isolated micro energy grid. The LFC of the micro-grid including the SMES was studied in [19], however the impacts of the parameters were not taken into account. The GPC algorithm can also be used to control isolated micro-grids with electric vehicles [20].

In this paper, a LIQHYSMES unit to be used as the energy storage system with RESs in the energy internet to solve the frequency instability problem is proposed. Based on the presentation of the LIQHYSMES characteristics, the benefits and applications to the energy internet are analyzed. Then a new coordinated LFC controller based on the GPC algorithm is proposed for the equivalent model of the micro energy grid with LIQHYSMES. Meanwhile, the optimization design of the superconducting coil parameters, including initial current, inductance and initial energy storage capacity is carried out in this paper.

The rest of this paper is organized as follows: Section 2 introduces the structure of the LIQHYSMES. In Section 3, the equivalent models of the components in the micro energy grid for LFC are constructed. Then, the coordinated LFC controller based on GPC is proposed in Section 4. In Section 5, the superconducting coil is optimized; the effectiveness and robustness of the proposed coordinated controller is demonstrated by numerical simulations on an isolated micro energy grid with LIQHYSMES in Section 6. Finally, conclusions are drawn in Section 7.

2. Liquid Hydrogen with Superconducting Magnetic Energy Storage (SMES)

The core of the energy internet in the future is the electric power system, combined with the natural gas network, transportation network and thermal network to form a comprehensive network. As shown in Figure 1, the LIQHYSMES can play an important role as the energy router in connecting, scheduling and controlling the networks concertedly in the future energy internet.

Figure 1 shows the structure of the hybrid energy storage device, which consists of three major parts: the electrochemical energy conversion (EEC), the LIQHYSMES storage unit (LSU) and the power conversion & control Unit (PCC). When a power fluctuation occurs as a result of the RES connected to the electric power system, it's prone to cause a power imbalance and affect the stability of the electric power system.

To suppress the fluctuation rapidly, the PCC will control the SMES unit to charge or discharge depending on the supply and demand imbalance of the system, in which condition the energy is stored and released by means of electric energy. As shown in Figure 2, the SMES system has a DC magnetic coil that is connected to the AC grid through a power conversion system. Meanwhile, PCC will control the LH_2 storage unit to work on the conversion of electric energy to achieve the slow suppression of the fluctuation. The surplus electric energy generated by the RESs can be converted into chemical energy by the electrolyser. Then, the gaseous hydrogen produced previously is liquefied into LH_2

for storage. On the contrary, when a power shortage occurs in the electric power system, the liquid hydrogen stored in the liquid hydrogen tank is vaporized and supplied to the gas turbine (GT), fuel cells (FC), and combined heat & power (CHP) to supply electricity and heat to the electric power system and the thermal network, respectively.

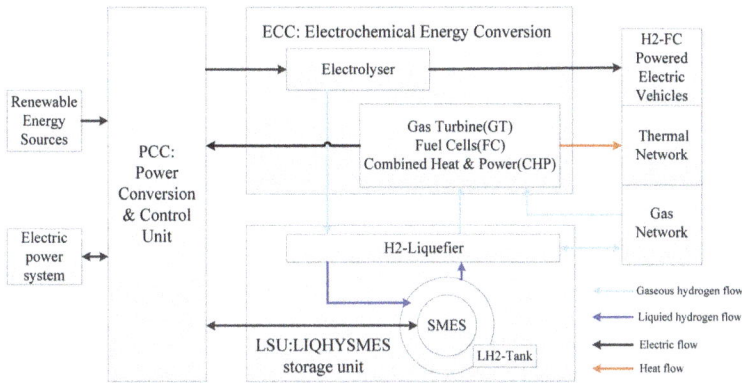

Figure 1. The liquid hydrogen with superconducting magnetic energy storage (LIQHYSMES) unit used in the energy internet.

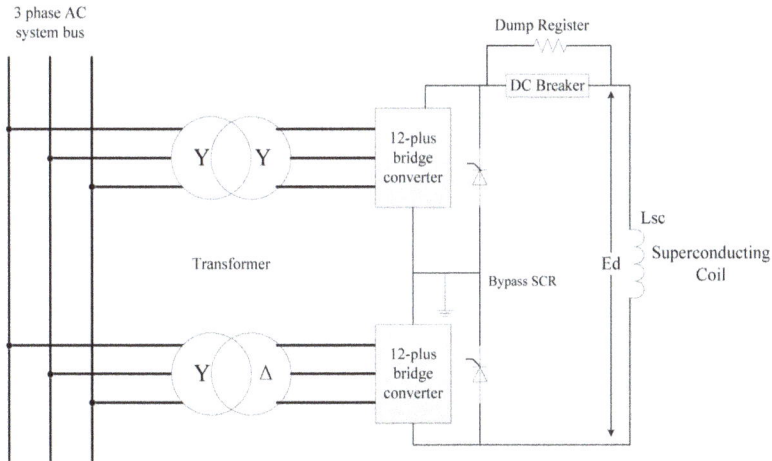

Figure 2. The schematic diagram of superconducting magnetic energy storage (SMES) connected to electric AC grid.

The schematic of a LIQHYSMES device is shown in Figure 3. It includes a multi-stage compressor, two-stage heat exchangers (HEX), liquid nitrogen or multi-component refrigerants' pre-cooling (PREC), expansion turbines and gas recycling (EXP-REC), Joule-Thomson expansion valves (JTV), a LH$_2$ storage tank and liquid nitrogen shielding. SMES based on coated conductors (based on high temperature superconductors, mostly YBaCuO and magnesium diboride (MgB$_2$) superconducting wires [21,22]) is utilized here, which could be operated in the LH$_2$ bath for sharing cooling system. In the charging process 1–4 shown in Figure 3, the gaseous hydrogen obtained after electrolysis is passed through a multistage compressor, heat exchangers and JTV1, so that most of the gaseous hydrogen is liquefied and stored in the LH$_2$ tank at 10 bar and 30 K. Besides, a small amount of unliquefied gaseous hydrogen is fed to the JTV2 as a cooling medium for the SMES coil at 1.2 bar and 20 K, and subsequently supplied to the HEX and re-compressed for another expansion cycle. From the discharge process 5 to process 8, the LH_2 stored in the LH_2 tank is sent to JTV2. Then the LH_2 is converted into gaseous hydrogen through the two HEX stages for use in GT, FC or CHP.

LIQHYSMES features the combined use of LH_2 storage and SMES to stabilize power fluctuations in the electric power system with RES. The combined use of the SMES and liquid hydrogen can help to expand storage capacity substantially. On the other hand, the liquid hydrogen is used as cooling medium for the SMES and shares the refrigeration plant with it to enhance the refrigeration efficiency and reduce the investment cost.

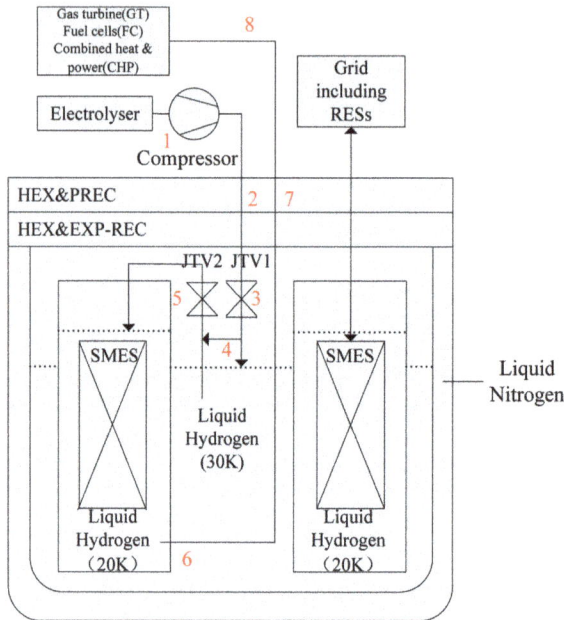

Figure 3. Schematic of the LIQHYSMES.

The use of LIQHYSMES is not subject to strict geographical restrictions, and can be applied to a variety of voltage levels in the electric power system, which are of important significance for achieving large-scale use of RES.

3. The Micro Energy Grid Including LIQHYSMES

The micro energy grid concept is consistent with the notion of the future electric power system, the characteristics of which will be profoundly different from those of the systems existing today. It represents the further development of microgrids. In a microgrid the energy is only transmitted in the form of electricity. However, in the micro energy grid with LIQHYSMES shown in Figure 1 the energy can be converted into electricity, chemical energy, thermal energy and other forms. The smart loads (SL) become controllable, and energy-storage systems, as well as vehicle-to-grid (V2G) systems, further contribute to active controllable loads. Renewable energy sources will thus be used more and more in homes, buildings, and factories [23].

The configuration architecture of the micro energy grid is presented in Figure 4. It is composed of a micro turbine (MT), DGs, an electrical vehicle station, smart loads and a LIQHYSMES unit. The micro energy grid is managed by a distribution management system (DMS). Phasor measurement units

(PMUs) are installed in this micro energy grid to measure the real-time information of the components. A large number of data from PMUs can be handled by cloud computing in the DMS [24–30]. The micro energy grid is capable to work in either the grid-connected or isolated mode, transforming from one into the other by controlling the circuit breaker 1. In the grid-connected mode, the deviation of the frequency resulting from abrupt variations in load demand growth and generated power variations from different RESs can be eliminated rapidly by the electric power system to maintain stable operation. The isolated micro energy grid on the other hand needs to control the components coordinately to maintain the stability. With the LIQHYSMES, the system inertia could be increased, thus improving the frequency stability of the isolated micro energy internet. Here the equivalent model for LFC of the isolated micro energy grid is constructed.

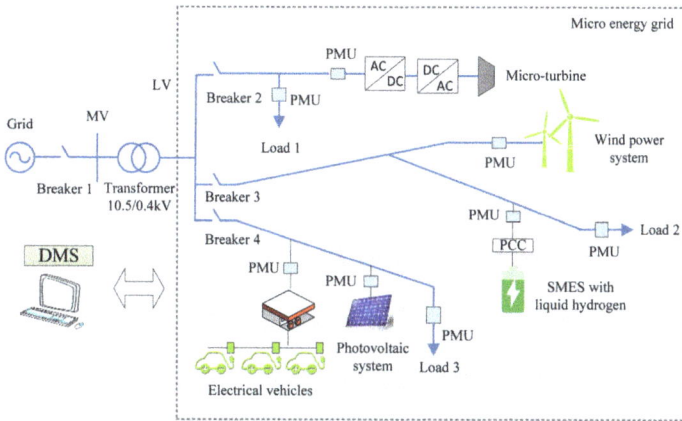

Figure 4. Schematic of a micro energy grid including the LIQHYSMES.

3.1. Model of LIQHSMES

During the LFC, the voltage and current of the superconducting coil vary with the frequency deviation to supply different amounts of power to maintain the stability of the system. When a disturbance disappears, the current of the superconducting coil should be restored to the initial value preparing for the subsequent disturbances.

The deviation of the superconducting coil (SC) voltage is given by:

$$\Delta E_d = \frac{K_{SMES}}{1 + sT_{DC}} \Delta f \tag{1}$$

The deviation of the SC current is expressed as:

$$\Delta I_d = \frac{\Delta E_d}{sL_{SC}} \tag{2}$$

The power supplying by the SMES can be obtained as follows:

$$\Delta P_{SMES} = \Delta E_d \cdot (I_{SC0} + \Delta I_d) \tag{3}$$

Therefore, the model of the SMES in LFC can be represented as in Figure 5. Herein, the GPC algorithm proposed to be used in LFC is based on the controlled auto-regressive integrated moving-average (CARIMA) model. It can identify and linearize the model of the system online, so a feedback of the deviation of the SC current is added to eliminate the error caused by linearization and achieve rapid recovery of SC current meanwhile.

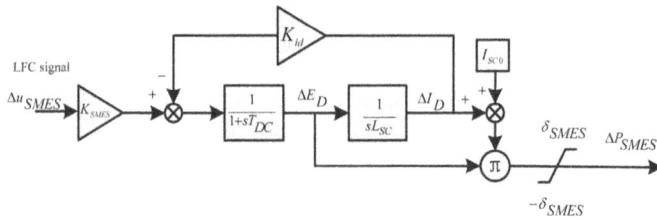

Figure 5. The transfer function model of the SMES unit for load frequency control (LFC).

The power supplied by the LH$_2$ storage unit and SMES to the micro energy grid are only controlled by PCC and are independent of each other, which means the two can be regarded as a common parallel system in the LFC. When the frequency deviation is negative, the controlled LH$_2$ storage unit provides power to compensate the power shortage by using the FC, GT or CHP. When the frequency deviation is positive, the LH$_2$ storage unit utilizes the electrolyser for consumption of excess power. The model of the LH$_2$ storage unit for LFC is shown in Figure 6. Here, the FC is used as the device to convert the LH$_2$ into electric energy. The FC constant time is set as same as the electrolyser time constant to simplify the model used in this paper.

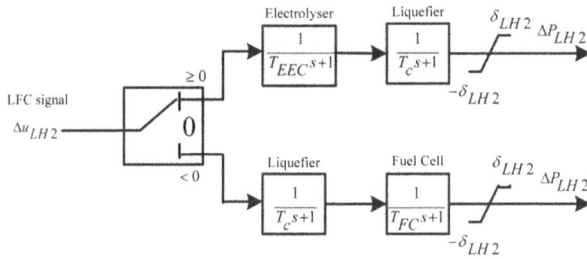

Figure 6. The transfer function model of the LH$_2$ storage unit for LFC.

3.2. Models of Other Components

Figure 7 shows the model of a micro-turbine for LFC, which simulates the dynamic process of the micro-turbine output power following the LFC signal. The model includes the governor, fuel system and gas turbine of the micro-turbine. The equivalent models of the fuel system and the turbine are represented by the first-order inertia units.

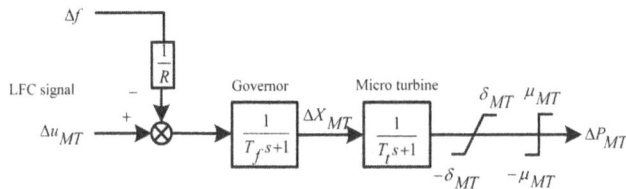

Figure 7. The transfer function model of the micro turbine for LFC.

Since there are different numbers of EVs in each EV station, the modelling of EVs could be handled by using equivalent EVs with different inverter capacities. The equivalent EV model which can be used for LFC is shown in Figure 8 [31]. The EV can be charged and discharged only within the range of $\pm\mu_e$. However, if the energy of the EV exceeds the upper limit (i.e., E_{max}), the EV can only be discharged within the range of $(0\sim\mu_e)$. Also, if the energy of the EV is smaller than the lower limit (i.e., E_{min}), the EV can only be charged within the range of $(-\mu_e\sim0)$. T_e is the time constant of EV.

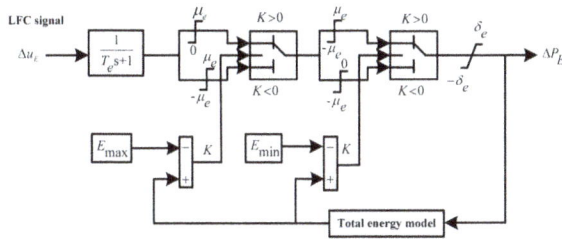

Figure 8. The transfer function model of the electric vehicle for LFC.

In the micro energy internet, the loads data computing center can calculate the total supplied power depending on the frequency deviation. Subsequently, it adjusts the amount of smart loads in need of being open or closed, although the output power of each smart load is uncontrollable. Smart loads have the advantage of rapid response for LFC and the model is shown in Figure 9.

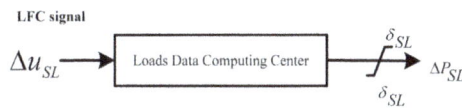

Figure 9. The transfer function model of the smart load for LFC.

Because the fluctuation of wind power and photovoltaic (PV) power output is relatively large, they can be all equalized to the disturbance sources in the LFC model [32]. The power disturbances of wind have similar responses to PV systems in the LFC model, so it is only considered here.

Based on the LFC response models of the above-mentioned components, a model of micro energy network including LIQHYSMES with load frequency controller for LFC is constructed as shown in Figure 10. ΔP_L is the load disturbance, ΔP_W is the fluctuation of the wind power generation, and H_t is the inertia constant of the micro energy internet.

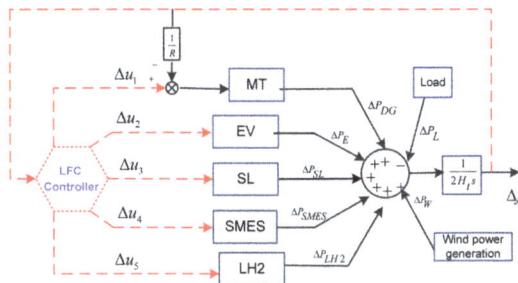

Figure 10. The control model of the micro energy grid including LIQHYSMES.

4. Generalized Predictive Control Algorithm for LFC

The principle of the GPC algorithm can be summarized as three parts: predictive model, rolling optimization and feedback compensation. In the GPC algorithm, the predictive model is described by the CARIMA model, which is suitable for unstable system and is easier to be recognized online. The LFC signals shown in Figure 10 are the multiple inputs and the frequency deviation is the single output of the predictive model. Also, the unmeasured disturbance caused by load disturbance or the fluctuation of wind power generation and measurable noise are taken into account in the model. The predictive model can be described as follows:

$$A(z^{-1})\Delta f(t) = \sum_{i=1}^{5} B_i(z^{-1})\Delta u_i + D(z^{-1})w(t) + \xi(t)/\Delta \tag{4}$$

where t is the discrete sampling time point of control, Δu_i is the LFC signal for each component in Figure 10, z^{-1} is the backward shift operator. $\Delta = 1 - z^{-1}$ is the difference operator, which represents the effect of random noise. $\xi(t)$ is a n-dimensional zero mean white noise sequence. $A(z^{-1})$, $B(z^{-1})$, $D(z^{-1})$ are the polynomial matrixes of z^{-1}:

$$\begin{cases} A(z^{-1}) = I_{n\times n} + a_1 z^{-1} + a_2 z^{-2} + \ldots + a_{n_a} z^{-n_a} \\ B_i(z^{-1}) = b_{i0} + b_{i1} z^{-1} + b_{i2} z^{-2} + \ldots + b_{in_b} z^{-n_b} \\ D(z^{-1}) = d_0 + d_1 z^{-1} + d_2 z^{-2} + \ldots + d_{n_d} z^{-n_d} \end{cases} \tag{5}$$

where $a_1, a_2 \ldots, b_{i1}, b_{i2} \ldots$ and $d_1, d_2 \ldots$ are the polynomial coefficients, n_a, n_b, n_d are the orders of the polynomial, respectively. n_a is the prediction time domain, n_b is the control time domain, where the first term can be 0, denoting the number of time delays of the response, and n_d is the interference time domain.

In order to track the set reference value $w(t + j)$ of the predicted output, we can calculate the control vectors using optimization techniques based on the following objective function:

$$J = \sum_{j=1}^{n_a} ||\hat{\Delta f}(t+j|t) - w(t+j)||_Q^2 + \sum_{i=1}^{5} (\sum_{j=1}^{n_b} ||\Delta u_i(t+j-1)||_R^2) \tag{6}$$

where Q and R are the positive definite weighting matrixes, $\hat{\Delta f}(t+j|t)$ is an optimal j-step prediction of the frequency deviation at time t. The reference value $w(t + j)$ of the j-step frequency deviation is set as constant 0. The control vectors can be obtained by many algorithms to solve this quadratic programming problem, such as sequential minimal optimization (SMO) [33–37]. Herein, the function 'quadprog' provided by Matlab is used and the first row of the vectors $\Delta u(t|t)$ is carried out as the LFC signals to eliminate the frequency deviation at the sampling time point t.

In the GPC algorithm, the recursive least squares method is used to identify the parameters of the predictive model for the LFC in the micro energy internet, which means that the polynomial matrixes $A(z^{-1})$, $B(z^{-1})$, $C(z^{-1})$ vary with the sampling time. Then the optimal control sequence can be calculated. This online identification and the control sequence correction mechanism constitute the GPC algorithm feedback correction.

5. The Optimization Design of the Superconducting Coil

The SC parameters, e.g., initial current I_{SC0}, coil inductance L_{SC} and initial stored energy E_{sc0}, are optimized to improve the control effect of the LFC model for the micro energy internet. In order to improve the stability, the following objective function can be used:

$$\text{Min } J_1 = \int_0^{t_{sim}} |\Delta f| dt \tag{7}$$

where t_{sim} is the total simulation time.

$$\Delta f(s) = (\Delta P_{MT}(s) + \Delta P_{EV}(s) + \Delta P_{SL}(s) + \Delta P_{SMES}(s) + \Delta P_{LH2}(s) - \Delta D) \cdot \frac{1}{2H_t s} \tag{8}$$

ΔD is the system uncertainty model which represents several operating conditions of unpredictable wind power and loads variation.

The response of SMES is expressed as:

$$P_{SMES}(s) = \left(\frac{R(s)}{sL_{SC}(1+sT_{DC})+1} + \frac{I_{SC0}}{s}\right)\left(\frac{sL_{SC}R(s)}{sL_{SC}(1+sT_{DC})+1}\right) \tag{9}$$

The input signal $R(s)$ is the load frequency control signal. Also the responses of other components can be expressed as the form of SMES so that the response of the frequency deviation can be obtained. Then, the numerical inversion of Laplace transform is employed for the time domain response of the frequency deviation.

Moreover, the initial stored energy is given by:

$$E_{sc0} = \frac{1}{2}L_{SC}I_{SC0}^2 \tag{10}$$

To optimize the E_{sc0}, the optimal L_{SC} and I_{SC0} is obtained by taking it into consideration. The above two parts are weighted linearly, so the optimization problem can be formulated as follows:

$$\text{Min } J = W_1 J_1 + W_2 E_{SC0} \tag{11}$$

Subject to:

$$0.001 \leq L_{SC} \leq 10H$$

$$1.5 \leq I_{SC0} \leq 4kA$$

where the weighting factors are set as $W_1 = 1$, $W_2 = 0.01$ in this paper. Then the particle swarm optimization (PSO) is applied to solve the problem. Figure 11 shows the flowchart of PSO [38].

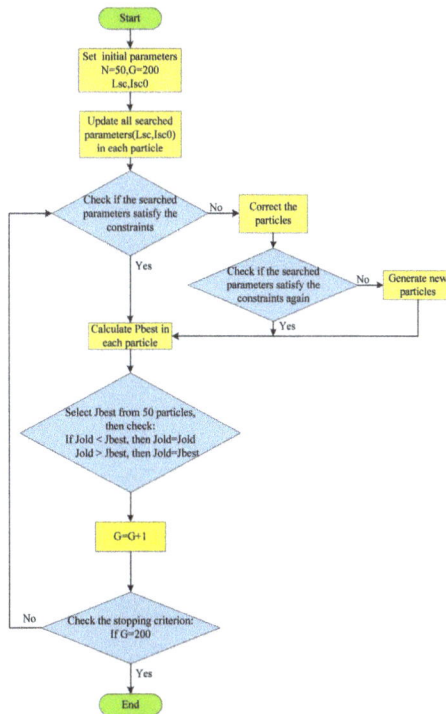

Figure 11. Flowchart of PSO.

6. Simulation Study

Simulations are carried out based on the above-mentioned model of micro energy grid, and the model parameters are shown in Table 1. Some of the values are chosen referring to [20]. Suppose that the micro energy grid is in steady isolated state at the beginning of the simulation. The wind power from an offshore wind farm in Denmark is shown in Figure 12, where $\Delta P_W = 0$ means the wind power is equal to the average power during the period.

Table 1. Parameters of the micro energy grid.

Grid Component	Parameters	Values	Unit
SMES	T_{DC}	0.03	s
	K_{SMES}	1	/
	K_{id}	1	/
	L_{SC}	5	H
	I_{SC0}	1.5	kA
	δ_{SMES}	0.15	pu·MW/s
LH$_2$ storage unit	T_{EEC}	1	s
	T_c	50	s
	T_{FC}	1	s
	δ_{LH2}	0.006	pu·MW/s
MT	T_f	0.1	s
	T_t	8	s
	R	2.5	Hz/pu·MW
	δ_{MT}	0.01	pu·MW/s
	μ_{MT}	0.04	pu·MW
EV	T_e	1	s
	δ_e	0.05	pu·MW/s
	μ_e	0.025	pu·MW
	E_{max}	0.95	pu·MWh
	E_{min}	0.80	pu·MWh
SL	δ_{SL}	0.1	pu·MW
Gird Inertia	H_t	7.11	s

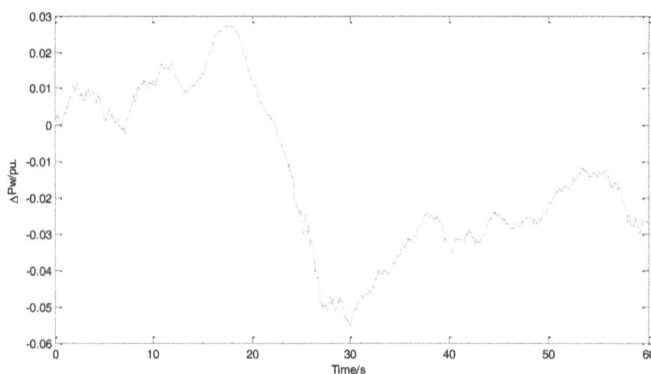

Figure 12. The power fluctuation of wind power generation.

As shown in Figure 13, it can be concluded that the frequency response of the system is better under the control of the GPC algorithm proposed in this paper than PI during the simulation period, except for the initial moment. The frequency deviation of the system is controlled generally within the range of ±0.002 Hz with LIQHYSMES, which shows that the controller based on the GPC algorithm offers better stability and robustness. The frequency oscillation is smoother and the range is smaller with the LIQHYSMES unit compared with the conditions without it. It can be seen that the LIQHYSMES unit plays a positive role in suppressing the oscillation of the load frequency caused by wind power fluctuation in an isolated micro energy grid.

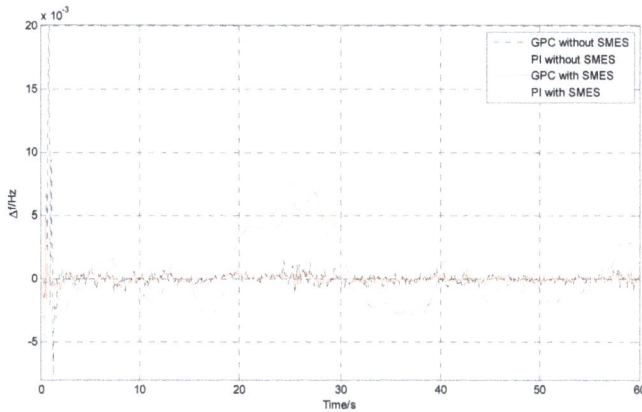

Figure 13. The frequency deviation of the micro energy grid.

At the initial moment of the simulation, the PI control is more effective than the GPC algorithm, because there is little historical data provided for the online identification of the system predictive model parameters causing the large prediction deviations and the remarkable frequency fluctuations. Actually this situation can be avoided by providing historical inputs and outputs data on the parameters of the predictive model to be pre-set before the simulation.

Figures 14 and 15 show the active output power of the components of the micro energy grid controlled by the two methods. SMES has the ability to respond quickly to the frequency oscillation, while the response speed of the LH_2 storage unit is slower due to its larger inertia. Since the output power of each smart load is not adjustable, the total active output power of the smart loads can't change smoothly. In addition, electric vehicles based on V2G technology can also be used as energy storage devices to participate in the system of load frequency control.

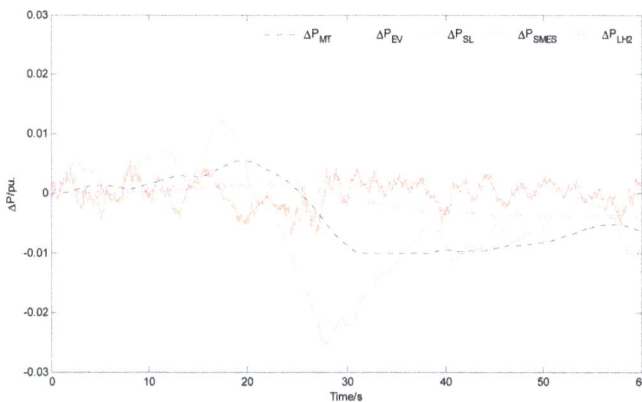

Figure 14. The output power increment of micro turbine (MT), EV, smart loads, non-optimized SMES and LH_2 storage unit controlled by PI.

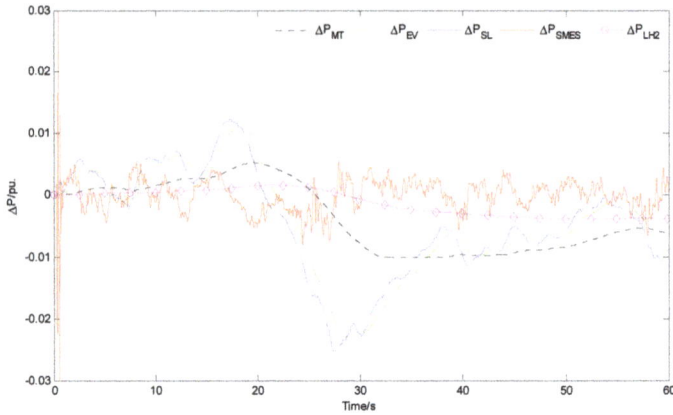

Figure 15. The output power increment of MT, EV, smart loads, non-optimized SMES and LH₂ storage unit controlled by generalized predictive control (GPC).

For an isolated micro energy grid with RESs, the abrupt change in load demand is also a challenge for the system to maintain frequency stability. Assuming that there are step disturbances in load demand ($\Delta P_L = -0.1$ pu, $\Delta P_L = 0.12$ pu, and $\Delta P_L = 0.06$ pu at $t = 5$ s, $t = 40$ s and $t = 80$ s, respectively). The fluctuation of the wind power is added to obtain the combined power disturbances shown in Figure 16.

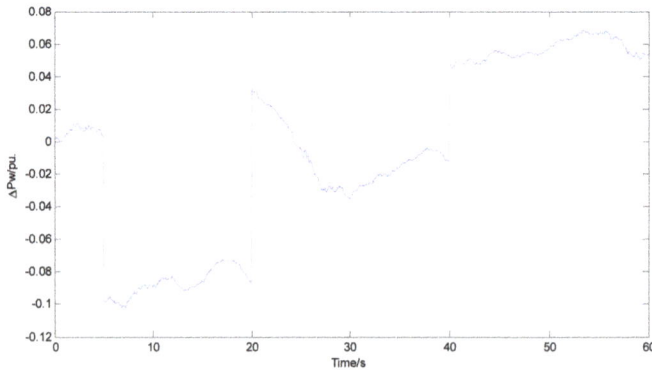

Figure 16. The power disturbances applied in the case.

Figure 17 shows the frequency deviation results when PI control and GPC algorithm are applied to the LFC. Comparing with PI control, GPC can suppress the frequency oscillation more rapidly and the peak of it is also smaller with or without the LIQHYSMES unit. In the case with LIQHYSMES unit, the advantage of the proposed GPC algorithm in suppressing the frequency oscillation is more obvious. Figures 18 and 19 show the active power contribution of the various components of the system. In the event of abrupt change in load demand, SMES can provide or consume power from the system in response to a rapid change in load frequency.

Based on the PSO algorithm, the superconducting coil parameters of LIQHYSMES are optimized, and the iterative process of optimization is shown in Figure 20. The number of particles is set as 50 and the number of the iterations is set as 200. Here, the input signal R(s) is chosen as a step signal with the amplitude of 0.1.

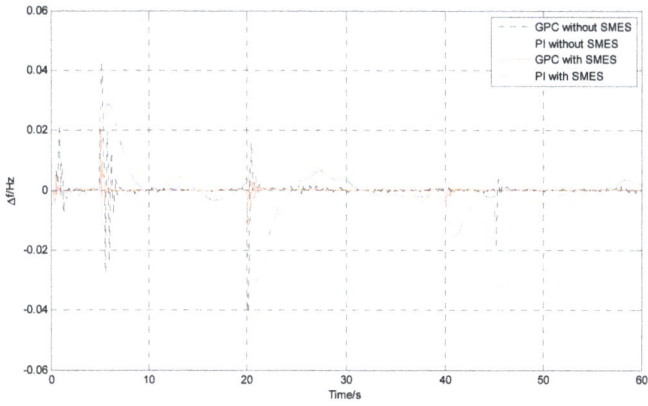

Figure 17. The frequency deviation of the micro energy grid.

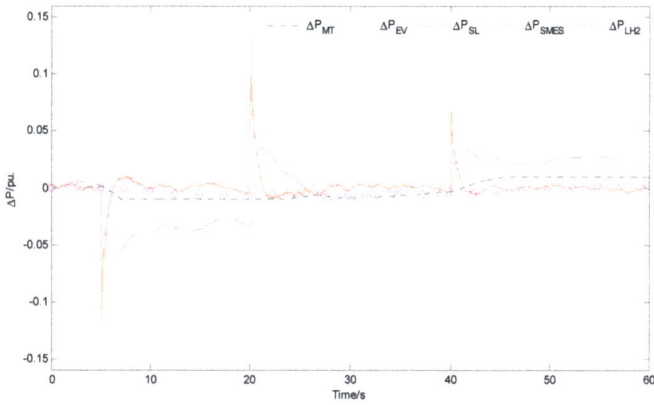

Figure 18. The output power increment of MT, EV, smart loads, non-optimized SMES and LH$_2$ storage unit controlled by PI.

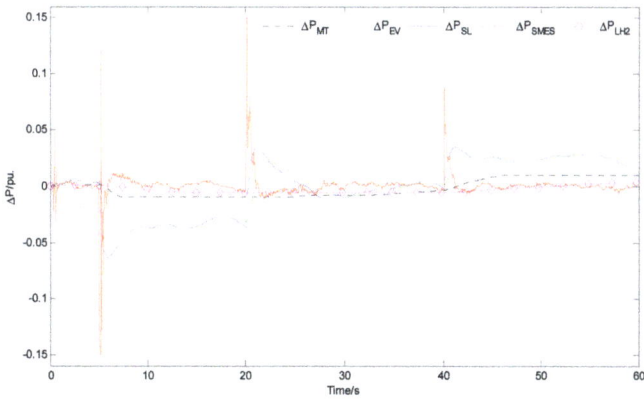

Figure 19. The output power increment of MT, EV, smart loads, non-optimized SMES and LH$_2$ storage unit controlled by GPC.

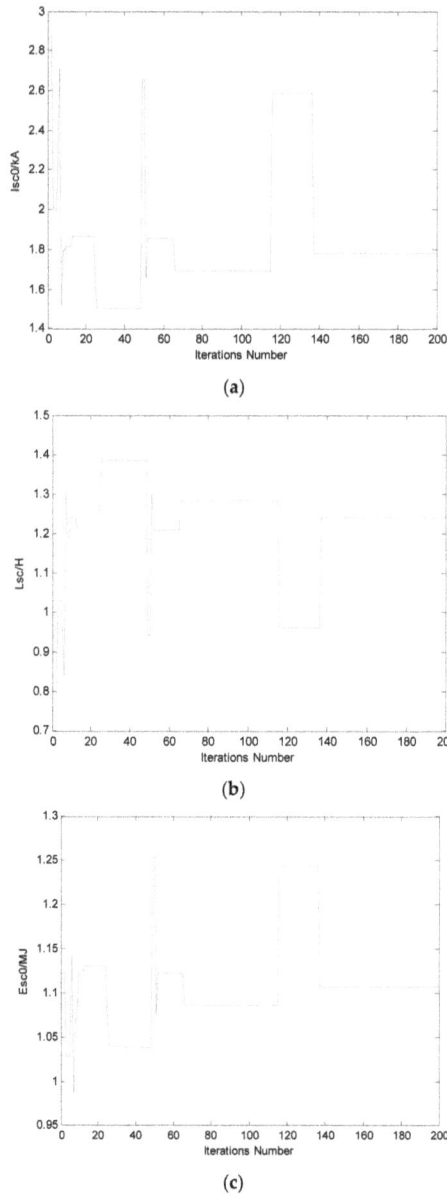

Figure 20. Iteration process, (**a**) initial current; (**b**) coil inductance; (**c**) initial stored energy.

The optimized LIQHYSMES unit is applied to the load frequency control of the micro energy grid, and the performance of the non-optimized LIQHYSMES unit is compared in Table 2.

As shown in Figures 21 and 22, the system with optimized SC is able to maintain better frequency stability when different load disturbances occur. As the wind power fluctuating, the output power of SMES unit with optimized SC increases comparing with the non-optimized SC controlled by the two methods as shown in Figures 23 and 24. The tendency is the same in the case with the combined disturbances as shown in Figures 25 and 26.

The proposed GPC algorithm utilizes CARIMA that features an easy online identification. Meanwhile, it can improve the robustness of the controller. However, as the high penetration rate of the DGs in the energy internet, CARIMA may result in prediction deviations for the LFC. Therefore, it is necessary to have stochastic studies with given confidence interval in this condition.

Table 2. The Parameters of SC.

Parameter	Non-Optimized SC	Optimized SC
I_{SC0}	1.5 kA	1.784 kA
L_{SC}	5 H	1.241 H
E_{SC0}	5.625 MJ	1.975 MJ

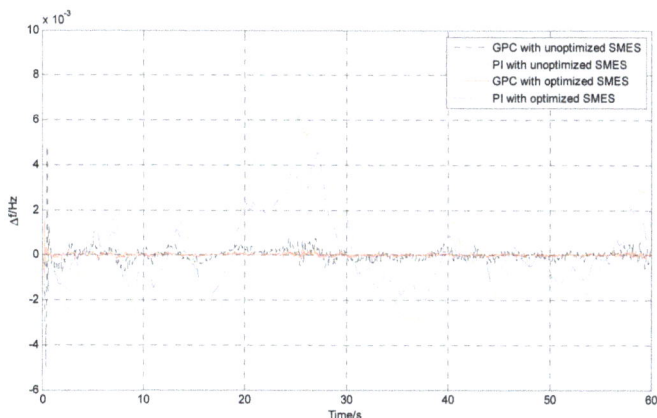

Figure 21. The frequency deviation only with wind power fluctuation.

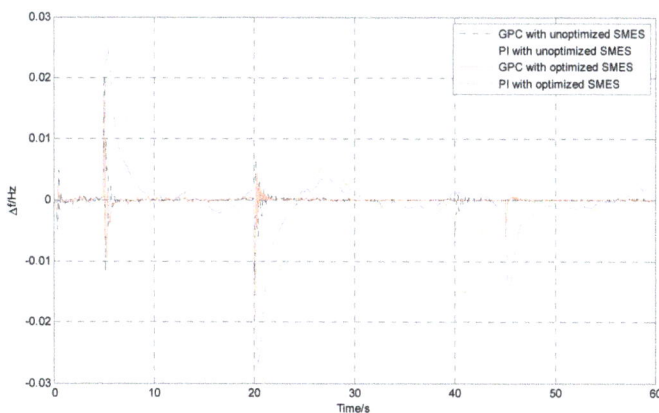

Figure 22. The frequency deviation with combined disturbances.

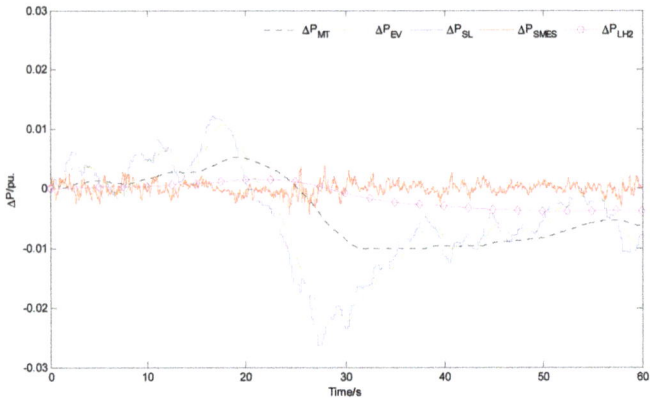

Figure 23. The output power increment of MT, EV, smart loads, optimized SMES and LH$_2$ storage unit controlled by PI only with wind power fluctuation.

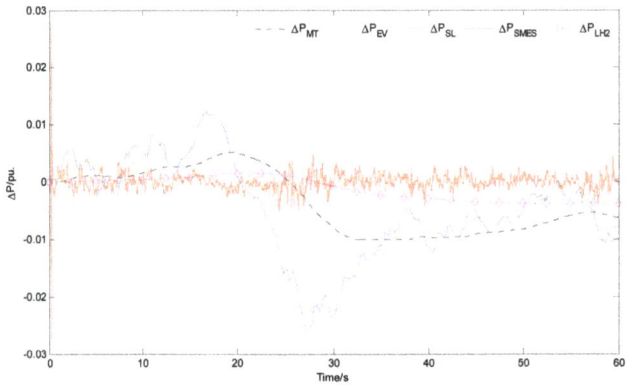

Figure 24. The output power increment of MT, EV, smart loads, optimized SMES and LH$_2$ storage unit controlled by GPC only with wind power fluctuation.

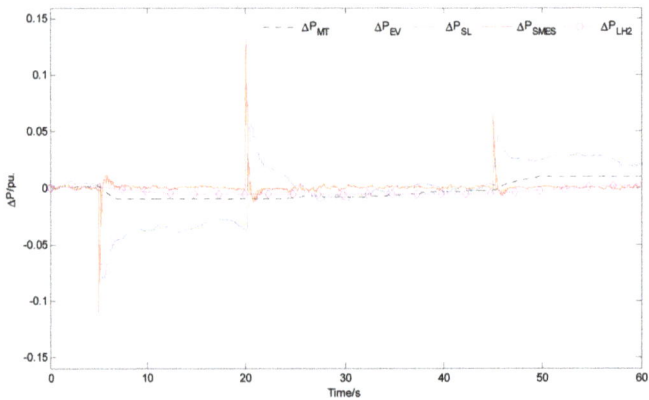

Figure 25. The output power increment of MT, EV, smart loads, optimized SMES and LH$_2$ storage unit controlled by PI with combined disturbances.

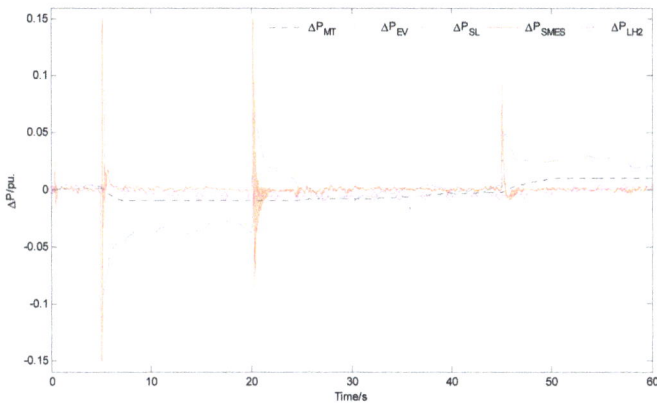

Figure 26. The output power increment of MT, EV, smart loads, optimized SMES and LH_2 storage unit controlled by GPC with combined disturbances.

7. Conclusions

Energy storage devices are necessary in the energy internet due to an increasing number of RESs are adopted in the future. The LIQHYSMES unit can obtain better economic benefits with promising applications. The LFC controller based on GPC algorithm is designed and applied to the micro grid energy including the LIQHYSMES unit. To obtain better control effect, the SC parameters are optimized as well. Simulations of the load frequency control based on the equivalent model of the micro grid energy are carried out and the results are summarized as follows.

1. The LIQHYSMES unit can be used to achieve the energy storage and transformation in the energy internet. It is also helpful for the efficient use of renewable energy through solving the frequency instability problems of the isolated micro energy grid.
2. In the isolated micro energy grid including the LIQHYSMES unit, the proposed controller based on GPC algorithm can obtain better robust performance on LFC in complex operation situations, namely, random renewable energy generations and continuous load disturbances. It plays a significant role in the load frequency control, especially in the cases where the load demand changes violently in the system with RESs.
3. The optimization SC parameters of the SMES can offer better technical advantages in alleviating the load frequency fluctuations in different cases.

For the future work, an experimental study which implements the proposed method in reality will be carried out.

Acknowledgments: The work is funded by the National Science Foundation of China (51277135, 50707021).

Author Contributions: Jun Yang and Xin Wang conceived and designed the study; Xin Wang performed the experiments; Lei Chen analyzed the experimental results; Jifeng He contributed analysis tools; Xin Wang and Jun Yang wrote the paper.

Conflicts of Interest: The authors declare no conflict of interest.

Nomenclature

ΔD	Unpredictable wind power and loads variation	T_{DC}	Converter time constant
		T_{EEC}	Electrolyser time constant
ΔE_d	Deviation of the SC voltage	T_c	Liquefier time constant
E_{SC0}	Initial stored energy of SC	T_{FC}	FC time constant
$E_{\max}$	Maximum controllable energy	T_f	Governor time constant
$E_{\min}$	Minimum controllable energy	T_t	Generator time constant
Δf	Deviation of the frequency	T_e	EV time constant
H_t	Gird inertia	Δu_{MT}	LFC signal for MT
ΔI_d	Deviation of the SC current	Δu_1	
I_{SC0}	Initial current of SC	Δu_E	LFC signal for EV
J_1, J	Objective functions	Δu_2	
K_{SMES}	Gain of the SMES	Δu_{SL}	LFC signal for SL
K_{id}	Gain of the feedback in SMES	Δu_3	
L_{SC}	Coil inductance of SC	Δu_{SMES}	LFC signal for SMES
n_a	Prediction time domain	Δu_4	
n_b	Control time domain	Δu_{LH2}	LFC signal for LH2 storage unit
n_d	Interference time domain	Δu_5	
ΔP_{SMES}	Output power increment of SMES	δ_{SMES}	Power ramp rate limit of SMES
ΔP_{LH2}	Output power increment of LH$_2$ storage unit	δ_{LH2}	Power ramp rate limit of LH$_2$ storage unit
ΔP_{MT}	Output power increment of MT	δ_{MT}	Power ramp rate limit of MT
ΔP_E	Output power increment of EV	δ_e	Power ramp rate limit of EV
ΔP_{SL}	Output power increment of SL	δ_{SL}	Power ramp rate limit of SL
ΔP_L	Load disturbance	μ_{MT}	Power increment limit of MT
		μ_e	Inverter capacity limit of EV
ΔP_W	Fluctuation of the wind power generation	W_1, W_2	Weighting factors
R	Speed regulation of MT	ΔX_{MT}	Valve position increment of the governor
$R(s)$	Input signal of LFC		
t_{sim}	Total simulation time	$\xi(t)$	White noise sequence

References

1. Carrasco, J.M.; Franquelo, L.G.; Bialasiewicz, J.T.; Galván, E.; Guisado, R.C.P.; Prats, M.Á.M.; León, J.I. Power-electronic systems for the grid integration of renewable energy sources: A survey. *IEEE Trans. Ind. Electron.* **2006**, *53*, 1002–1016. [CrossRef]
2. Georgiou, P.N.; Mavrotas, G.; Diakoulaki, D. The effect of islands' interconnection to the mainland system on the development of renewable energy sources in the Greek power sector. *Renew. Sustain. Energy Rev.* **2011**, *15*, 2607–2620. [CrossRef]
3. Sun, Q.; Zhang, Y.; He, H.; Ma, D.; Zhang, H. A novel energy function-based stability evaluation and nonlinear control approach for energy internet. *IEEE Trans. Smart Grid* **2015**, *PP*, 1–16. [CrossRef]
4. Huang, A.Q.; Crow, M.L.; Heydt, G.T.; Zheng, J.P.; Dale, S.J. The future renewable electric energy delivery and management (FREEDM) system: The energy internet. *Proc. IEEE* **2010**, *99*, 133–148. [CrossRef]
5. Zhou, K.; Yang, S.; Shao, Z. Energy internet: The business perspective. *Appl. Energy* **2016**, *178*, 212–222. [CrossRef]
6. Palizban, O.; Kauhaniemi, K. Energy storage systems in modern grids—Matrix of technologies and applications. *J. Energy Storage* **2016**, *6*, 248–259. [CrossRef]
7. Yang, J.; Zhang, L.; Wang, X.; Chen, L.; Chen, Y. The impact of SFCL and SMES integration on the distance relay. *Phys. C Supercond. Its Appl.* **2016**, *530*, 151–159. [CrossRef]
8. Hirano, N.; Watanabe, T.; Nagaya, S. Development of cooling technologies for SMES. *Cryogenics* **2016**, *80*, 210–214. [CrossRef]
9. Yang, W.J.; Aydin, O. Wind energy–hydrogen storage hybrid power generation. *Int. J. Energy Res.* **2001**, *25*, 449–463. [CrossRef]

10. Hirabayashi, H.; Makida, Y.; Nomura, S.; Shintomi, T. Liquid hydrogen cooled superconducting magnet and energy storage. *IEEE Trans. Appl. Supercond.* **2008**, *18*, 766–769. [CrossRef]

11. Nakayama, T.; Yagai, T.; Tsuda, M.; Hamajima, T. Micro power grid system with SMES and superconducting cable modules cooled by liquid hydrogen. *IEEE Trans. Appl. Supercond.* **2009**, *19*, 2062–2065. [CrossRef]

12. Hirabayashi, H.; Makida, Y.; Nomura, S.; Shintomi, T. Feasibility of hydrogen cooled superconducting magnets. *IEEE Trans. Appl. Supercond.* **2006**, *16*, 1435–1438. [CrossRef]

13. Sander, M.; Gehring, R. LIQHYSMES—A novel energy storage concept for variable renewable energy sources using hydrogen and SMES. *IEEE Trans. Appl. Supercond.* **2011**, *21*, 1362–1366. [CrossRef]

14. Sander, M.; Gehring, R.; Neumann, H. LIQHYSMES—A 48 GJ toroidal MgB2-SMES for buffering minute and second fluctuations. *IEEE Trans. Appl. Supercond.* **2013**, *23*, 5700505. [CrossRef]

15. Sander, M.; Neumann, H. LIQHYSMES—Size, loss and cost considerations for the SMES—A conceptual analysis. *Supercond. Sci. Technol.* **2011**, *24*, 105008–105013. [CrossRef]

16. Vachirasricirikul, S.; Ngamroo, I. Robust LFC in a smart grid with wind power penetration by coordinated v2g control and frequency controller. *IEEE Trans. Smart Grid* **2014**, *5*, 371–380. [CrossRef]

17. Shankar, G.; Lakshmi, S. Frequency control of hybrid renewable energy system with PSO optimized controller. In Proceedings of the 2015 International Conference on Recent Developments in Control, Automation and Power Engineering (RDCAPE), Noida, India, 12–13 March 2015; pp. 220–225.

18. Rao, C.S. Adaptive neuro fuzzy based load frequency control of multi area system under open market scenario. In Proceedings of the 2012 International Conference on Advances in Engineering, Science and Management (ICAESM), Tamil Nadu, India, 30–31 March 2012; pp. 5–10.

19. Deepak, M. Improving the dynamic performance in load frequency control of an interconnected power system with multi source power generation using superconducting magnetic energy storage (SMES). In Proceedings of the 2014 International Conference on Advances in Green Energy (ICAGE), Thiruvananthapuram, India, 17–18 December 2014; pp. 106–111.

20. Yang, J.; Zeng, Z.; Tang, Y.; Yan, J.; He, H.; Wu, Y. Load frequency control in isolated micro-grid with electrical vehicle based on multivariable generalized predictive theory. *Energies* **2015**, *8*, 2145–2164. [CrossRef]

21. Shikimachi, K.; Hirano, N.; Nagaya, S.; Kawashima, H. System coordination of 2 GJ class YBCO SMES for power system control. *IEEE Trans. Appl. Supercond.* **2012**, *19*, 2012–2018. [CrossRef]

22. Atomura, N.; Takahashi, T.; Amata, H.; Iwasaki, T.; Son, K.; Miyagi, D.; Tsuda, M.; Hamajima, T.; Shintomi, T.; Makida, Y.; et al. Conceptual design of MgB2 coil for the 100 MJ SMES of advanced superconducting power conditioning system (ASPCS). *Phys. Procedia* **2012**, *27*, 400–403. [CrossRef]

23. Wu, F.F.; Varaiya, P.P.; Hui, R.S.Y. Smart grids with intelligent periphery: An architecture for the energy internet. *Engineering* **2015**, *1*, 436–446. [CrossRef]

24. Zhi, H.X.; Xin, H.W.; Lian, G.Z.; Zhan, Q.; Xing, M.S.; Kui, R. A Privacy-preserving and Copy-deterrence Content-based Image Retrieval Scheme in Cloud Computing. *IEEE Trans. Inf. Forensics Secur.* **2016**, *11*, 2594–2608.

25. Zhang, J.F.; Xing, M.S.; Sai, J.; Guo, W.X. Towards efficient content-aware search over encrypted outsourced data in cloud. In Proceedings of the 35th Annual IEEE International Conference on Computer Communications (IEEE INFOCOM), San Francisco, CA, USA, 10–14 April 2016; pp. 1–9.

26. Qi, L.; Wei, D.C.; Jian, S.; Zhang, J.F.; Xiao, D.L.; Nigel, L. A speculative approach to spatial-temporal efficiency with multi-objective optimization in a heterogeneous cloud environment. *Secur. Commun. Netw.* **2016**, *9*, 4002–4012.

27. Zhang, J.F.; Xing, M.S.; Qi, L.; Lu, Z.; Jian, G.S. Achieving efficient cloud search services: Multi-keyword ranked search over encrypted cloud data supporting parallel computing. *IEICE Trans. Commun.* **2015**, *E98-B*, 190–200.

28. Zhang, J.F.; Xin, L.W.; Chao, W.G.; Xing, M.S.; Kui, R. Toward efficient multi-keyword fuzzy search over encrypted outsourced data with accuracy improvement. *IEEE Trans. Inf. Forensics Secur.* **2016**, *11*, 2706–2716.

29. Zhang, J.F.; Kui, R.; Jian, G.S.; Xing, M.S.; Feng, X.H. Enabling personalized search over encrypted outsourced data with efficiency improvement. *IEEE Trans. Parallel Distrib. Syst.* **2016**, *27*, 2546–2559.

30. Zhi, H.X.; Xin, H.W.; Xing, M.S.; Qian, W. A secure and dynamic multi-keyword ranked search scheme over encrypted cloud data. *IEEE Trans. Parallel Distrib. Syst.* **2015**, *27*, 340–352.

31. Singh, M.; Kumar, P.; Kar, I. Implementation of vehicle to grid infrastructure using fuzzy logic controller. *IEEE Trans. Smart Grid* **2012**, *3*, 565–577. [CrossRef]

32. Masato, T.; Yu, K.; Shinichi, I. Supplementary load frequency control with storage battery operation considering SOC under large-scale wind power penetration. In Proceedings of the IEEE Power and Energy Society General Meeting (PES), Vancouver, BC, Canada, 21–25 July 2013.

33. Bin, G.; Victor, S.S. A robust regularization path algorithm for ν-support vector classification. *IEEE Trans. Neural Netw. Learn. Syst.* **2016**, *PP*, 1–8.

34. Bin, G.; Victor, S.S.; Keng, Y.T.; Walter, R.; Shuo, L. Incremental support vector learning for ordinal regression. *Trans. Neural Netw. Learn. Syst.* **2015**, *26*, 1403–1416.

35. Bin, G.; Victor, S.S.; Zhi, J.W.; Derek, H.; Said, O.; Shuo, L. Incremental learning for ν-Support vector regression. *Neural Netw.* **2015**, *67*, 140–150.

36. Bin, G.; Victor, S.S.; Shuo, L. Bi-parameter space partition for cost-sensitive SVM. In Proceedings of the 24th International Conference on Artificial Intelligence, Buenos Aires, Argentina, 25–31 July 2015; pp. 3532–3539.

37. Bin, G.; Xing, M.S.; Victor, S.S. Structural minimax probability machine. *IEEE Trans. Neural Netw. Learn. Syst.* **2016**. [CrossRef]

38. Yan, C.X.; Tie, L.Z.; Ze, Y.D.; Chun, Y.L. The study of fuzzy proportional integral controllers based on improved particle swarm optimization for permanent magnet direct drive wind turbine converters. *Energies* **2016**, *9*, 343.

![energies logo] *energies*

MDPI

Article

Pressure Fluctuations in the S-Shaped Region of a Reversible Pump-Turbine

Zijie Wang [1], Baoshan Zhu [1,*], Xuhe Wang [1] and Daqing Qin [2]

[1] Department of Thermal Engineering, State Key Laboratory of Hydro Science and Engineering, Tsinghua University, Beijing 100084, China; 13331391026@163.com (Z.W.); wangxuhe1985@163.com (X.W.)
[2] Harbin Institute of Large Electrical Machinery, Harbin 150040, China; qindq@hec-china.com
*   Correspondence: bszhu@mail.tsinghua.edu.cn; Tel.: +86-10-6279-6797

Academic Editor: Hailong Li
Received: 11 June 2016; Accepted: 9 January 2017; Published: 13 January 2017

Abstract: Numerical simulations were performed to investigate pressure fluctuations in the S-shaped region of a pump-turbine model. Analyses focused on pressure fluctuations in the draft tube and in the gap between the guide vanes and runner. Calculations were made under six different operating conditions with a constant guide vane opening, and the best efficiency point, runaway point, and low-discharge point in the turbine brake zone were determined. The simulated results were compared with experimental measurements. In the draft tube, a twin vortex rope was observed. In the gap between the guide vanes and runner, a low frequency component was captured at both the runaway and low-discharge points in the turbine brake zone, which rotated at 65% of the runner frequency. This low frequency component was induced by the rotating stall phenomenon. At the runaway point, a single stall cell was found in the gap between the guide vanes and runner, while at the low-discharge point, four stall cells were observed.

Keywords: pump-turbine; pressure fluctuation; S-shaped region; vortex rope; rotating stall

1. Introduction

Pumped hydro energy storage (PHES) is currently the only proven large-scale (>100 MW) energy storage technology. There are great benefits in using PHES in electricity generation systems. Its flexibility can provide upregulation and downregulation. Its quick start capabilities make it suitable for black starts, and for providing spinning and standing reserves. Interest in this technology has recently been renewed because of the increasing use of renewable energy such as wind-powered electricity generation. Such technology is weather dependent, and so is highly variable over time [1,2].

A key component of PHES stations is the pump-turbine. Pump-turbines usually use a single runner to perform the functions of both pump and turbine. In order to adapt to the load changes of the power system, a pump-turbine is required to frequently switch between turbine mode and pump mode. This leads to extended operations in off-design operating conditions, such as at start-up and during load rejection processes. In the associated transient processes of pump-turbines, the S-shaped characteristics of the performance curves in turbine mode usually induce instability problems. The most common problems are difficulties in synchronizing with power grids during turbine start-ups, and unstable performance during turbine load rejections [3,4].

In recent years, both experimental and modelling studies have investigated the unstable flow characteristics of pump-turbines operating in off-design conditions [5–10]. Husmatuchi et al. [5,6] experimentally investigated pressure fluctuations in the gap between the guide vanes and runner. In their research, a low frequency component was observed under runaway conditions and turbine brake conditions. This low frequency instability was induced by one stall cell rotating inside the runner. Widmer et al. [7] performed numerical simulations of flows in a model pump-turbine. It was concluded that the unstable characteristics observed with low flow masses were induced by stationary vortex formation and rotating stall. Cavazzini et al. [8] carried out a numerical analysis of the load rejection process at a constant rate, and with a large guide vane opening of a pump-turbine in turbine mode. The flow analysis showed clearly the onset and development of unsteady phenomena progressively developing into an organized rotating stall during turbine brake operations. Both experimental and numerical studies have revealed complex flow characters in the S-shaped region of pump-turbines. Such vortexes may partially block the flow [7,9,10], or cause stall cells to rotate in the runner channels at 50%–70% of the runner rotation frequency [5,6,8].

As pump-turbines operate in or near the S-shaped region, vortex ropes are inevitably formed in the draft tubes [11]. Ruprecht et al. [12] and Kirschner et al. [13] investigated unstable characteristics in the draft tube of a pump-turbine. They reported that pressure fluctuations are mainly induced by vortex ropes in the draft tube. Usually, a single-helical vortex rope is formed in the draft tube; therefore, most research has focused on regimes with a single helical vortex rope [11–13]. However, for some hydraulic turbines, at small flow masses, the single helix is replaced by a double helix, and twin vortex ropes can be formed [11,14]. One of the first detailed reports of twin vortex ropes in draft tubes was presented by Wahl [14], who was able to measure their precession frequency.

In this paper, numerical simulations are performed to study instability near, or in, the S-shaped region of a pump-turbine model operating under off-design conditions. The model runner was designed by Wang et al. [15–17] during the development of the Liyang Pumped Storage Power Station in Jiangsu Province, China. Model tests were conducted in a stand high-head test rig at the Harbin Institute of Large Electric Machinery of China. The analyses mainly focus on the pressure fluctuations and flow characteristics in the draft tube and in the gap between the guide vanes and runner. In the draft tube of the pump-turbine model, pressure fluctuations are mainly caused by a twin vortex rope. In the gap between the guide vanes and runner, a low frequency component induced by rotating stalls is captured both at runaway and low-discharge conditions.

2. Object and Research Methods

2.1. Scaled Pump-Turbine Model

The scaled model pump-turbine runner used in this study is shown in Figure 1. The model was designed in order to conduct measurements on a standard test rig. The main parameters of the model runner are listed in Table 1. The scale of the model was one-tenth of the prototype turbines installed in aforementioned Liyang Pumped Storage Power Station. The rated specific speed of both model and prototype is $n_s = 145.80$ m · Kw. The specific parameters of the prototype pump-turbine are as follows. In turbine mode, the rated head is $H_r = 259$ m and the rated output power is $P_r = 255$ MW. The rotational speed is $n_r = 300$ rpm for both turbine and pump modes. In pump mode, the maximum and minimum heads are $H_{max} = 298$ m and $H_{min} = 239$ m, respectively.

Figure 1. Model runner.

Table 1. Parameters of pump-turbine model in turbine mode.

Parameter	Value
D_2 (runner inlet diameter in turbine mode, mm)	448.2
n (runner rotating speed, rpm)	1200
Z_b (number of runner blades)	7
Z_s (number of stay vanes)	20
Z_g (number of guide vanes)	20
H_r (rated head, m)	37.68
Q_r (rated discharge, m^3/s)	0.384
n_{11} (rated unit speed, rpm)	87.62
Q_{11} (rated unit discharge, m^3/s)	0.311
n_s (rated specific speed, m·Kw)	145.80

Model tests were conducted on a standard hydraulic machinery test rig at the Harbin Institute of Large Electric Machinery, China. Figure 2 gives a schematic diagram of the model test rig. The test rig had two test stations, A and B, of which Station A could be used to conduct the model tests for Francis turbines and reversible pump-turbines. The main specifications of the test rig and the pressure sensor (PCB112A22) used for the pressure fluctuation measurement are listed in Table 2. All measurements were conducted in accordance with International Electrotechnical Commission (IEC) Standard 60193 [18].

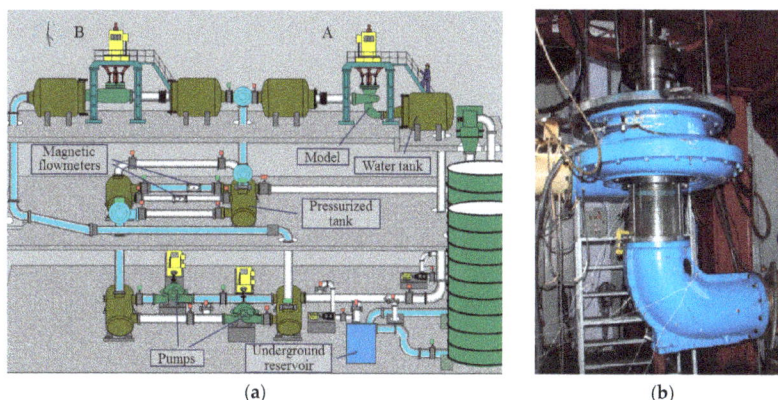

(a)   (b)

Figure 2. International Electrotechnical Commission (IEC) standard test rig. (**a**) Schematic diagram; and (**b**) test section.

Table 2. Characteristics of test rig performance and pressure sensors.

Item \ Parameter	Characteristic	Value
The test rig	Maximum head (m)	150
	Maximum discharge (m^3/s)	2.0
	Runner diameter range (mm)	300–500
	Generating power (kW)	500
	Test accuracy in efficiency (%)	±0.2
Pressure sensor	Measurement range (mPa)	0.345
	Sensitivity (mv/kPa)	14.5
	Resolution (kPa)	0.007
	Sampling frequency (kHz)	0.5–250
	Constant current excitation (mA)	2–20

In Figure 3, the measured S-shaped characteristics in some smaller guide vane openings are given. Of these, the unit speed n_{11}, unit discharge Q_{11}, and unit moment M_{11} are defined as follows:

$$n_{11} = \frac{nD_2}{\sqrt{H}}, \ Q_{11} = \frac{Q}{D_2^2\sqrt{H}}, \ M_{11} = \frac{M}{D_2^3 H} \tag{1}$$

where n stands for the runner rotating speed, D_2 for the runner inlet diameter in turbine mode, H for the head in the model tests, Q for the discharge, and M for the moment. The four-quadrant test was conducted under a constant speed of $n = 500$ rpm. In Figure 3, only the first quadrant is illustrated, in order to clearly show the S-shaped characteristics. The normal operating range of the turbine mode is indicated by the two vertical lines.

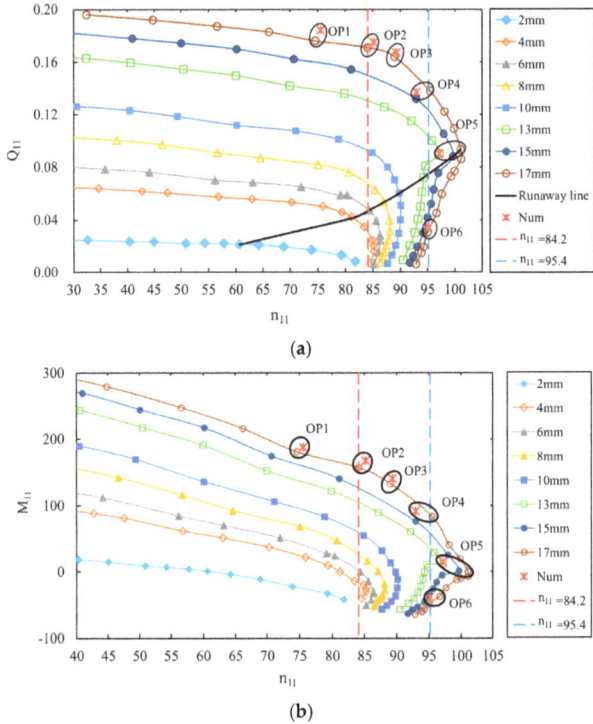

Figure 3. S-shaped curves at constant guide vane openings. (**a**) $n_{11} - Q_{11}$ characteristic curve; and (**b**) $n_{11} - M_{11}$ characteristic curve.

2.2. Numerical Solutions

Three-dimensional turbulent flow simulations were conducted for the full-passage pump-turbine. The extensively-used commercial code ANSYS CFX 15.0 (ANSYS Inc., Canonsburg, PA, USA) was used to conduct the numerical simulations.

Turbulence models are important factors in computational fluid dynamics (CFD). An advanced turbulence model, the detached eddy simulation (DES), was adopted to provide highly detailed simulations of flow patterns and unsteady phenomena. The shear stress transport $k - \omega$ model was used in the boundary layer, while the Smagorinsky-Lilly model was applied in detached regions [8].

The ANSYS ICEM (a powerful meshing software) and Turbo Grid were used for grid generation. The computational domain was divided into five parts for meshing (Figure 4), namely the spiral casing, stay vanes, guide vanes, runner and draft tube. Structured mesh was created for all these parts except for the spiral casing tongue, where unstructured mesh was employed due to the irregular structure.

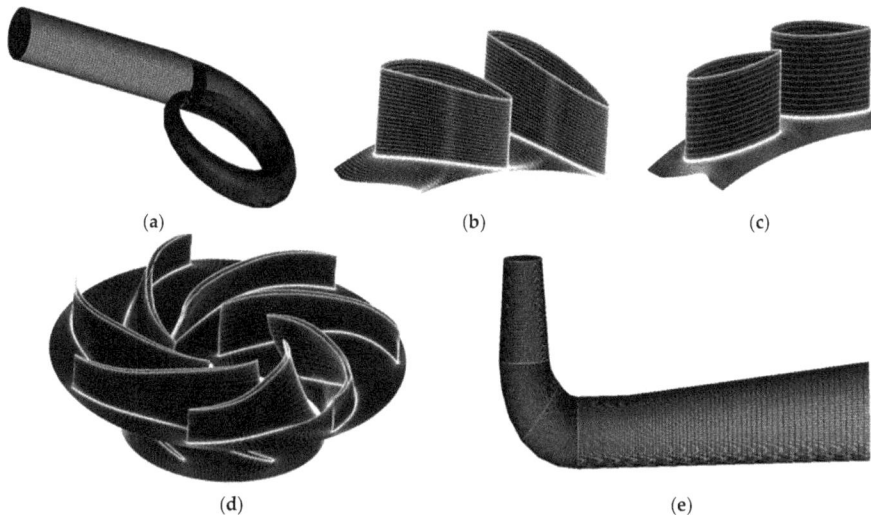

(a)　　　　　　　　　　　(b)　　　　　　　　　　　(c)

(d)　　　　　　　　　　　(e)

Figure 4. Boundary mesh for the full flow passage. (**a**) Spiral casing; (**b**) stay vanes; (**c**) guide vanes; (**d**) runner; and (**e**) draft tube.

At the inlet of the spiral casing, the mass flow rate was specified according to stochastic fluctuations of the velocities with a 5% free stream turbulent intensity. Static pressure was set at the outlet of the draft tube. The interfaces between stator-rotor blocks were set as the standard transient sliding interfaces. Walls were defined with no-slip wall boundary conditions. Runner blades, runner hub and shroud were fixed as rotating walls. The time-step size was set as 0.000139 s, such that one runner revolution was divided into 360 time steps. Second-order implicit time-stepping was adopted for time discretization. The convergence residuals for continuity and momentum equations were below 1.0×10^{-5}. All calculations were conducted in a cluster computer with eight Intel 5645 2.4 GHz processors, 96 GB RAM, and 2 TB hard drives (Dell Inc., Round Rock, TX, USA).

As shown in Figure 3, at a guide vane opening of $A = 17.0$ mm, OP1 is the best efficient point (BEP), OP2–OP4 are the operating points in the normal operating range, while OP5 is the runaway point and OP6 is the low-discharge point in the turbine brake. Usually, the guide vane opening is given as an angle; however, its width was used in this study. The opening angle was $\gamma = 18.0°$ at $A = 17.0$ mm.

Four meshes (Table 3) were used to test mesh independence. The calculated points were OP2–OP4 (Figure 4). The calculated efficiency errors for OP2–OP4 are given in Table 4. Here, the error was calculated as $Err = \left(\eta_{cal} - \eta_{exp} \right) / \eta_{num} \times 100.0\%$. In the simulations, the head cover and stay ring were not included; therefore, some leakage losses and the mechanical losses were not incorporated. The calculated efficiency η_{cal} is mainly the hydraulic efficiency, and is calculated by means of a time average after the calculated unsteady condition becomes stable. Simulated results show that mesh density has a weak influence on efficiency as the grid density is high (3.62 million). After considering the complex unsteadiness, Mesh III was chosen for the simulations. The validation of the numerical results in the next section also shows that the numerical solutions used can provide reliable results.

Table 5 gives more information on Mesh III. There were a total of 4.5 million elements in the whole domain. Figure 3 shows the boundary mesh for different components. Ten element layers with normal stretching ratios were created from the walls for improving the velocity profile from the no-slip boundaries. Near the wall, the value of no-dimensional distances was $y^+ = 50 - 180$. The mesh in the domain of the guide vanes, runner and diffuser section of the draft tube was finer than that in the other flow domain, in order to capture the complex flow characteristics in these components at higher resolution.

Table 3. Mesh densities (millions).

Component Mesh	Spiral Casing	Stay Vanes	Guide Vanes	Runner	Draft Tube	Full Domain
Mesh I	0.50	0.45	0.35	0.41	0.47	2.18
Mesh II	0.67	0.72	0.65	0.71	0.87	3.62
Mesh III	1.00	0.45	1.20	0.98	0.87	4.50
Mesh IV	1.00	1.16	1.04	1.45	1.13	5.78

Table 4. Mesh independence checks.

Operating Point Mesh	Mesh I	Mesh II	Mesh III	Mesh IV
OP2	5.51	2.67	2.43	1.92
OP3	7.88	4.54	3.25	3.07
OP4	10.63	5.49	4.37	3.86

Table 5. Statistics and quality of the mesh. BEP: best efficient point.

Component	Spiral Casing	Stay Vanes	Guide Vanes	Runner	Draft Tube
Mesh type	Hexahedral (except for small part near tongue in casing)				
Mesh density (millions)	1.0	0.45	1.2	0.98	0.87
Aspect ratio (0–100)	1–70.1	1–75.7	1–82.3	1–78.6	1–68.2
Mesh expansion factor (0–20)	0.1–15.4	0.1–18.1	0.1–12.3	0.1–10.6	0.1–11.8
Minimum orthogonality (0–90)	32.7	33.5	42.5	45.5	71.2
First layer thickness (mm)	1.6	1.6	0.2	0.2	1.6
Mesh incremental ratio	1.75	1.75	1.5	1.5	1.75
y^+ (at the calculated BEP)	165.0	172.5	80.3	54.1	140.8

2.3. Validation of the Numerical Results

Numerical results were compared with experimental measurements (Table 6). In the numerical simulations, the discharge Q, and the runner rotating speed n, were specified with the same values as those in the measurements, while the head H, and the moment M, were time-averaged values within the simulations. The Reynolds number was Re $= \omega D_2^2/\nu = 2.82 \times 10^6$, where ω is the runner rotation angular velocity, D_2 is the runner inlet diameter in turbine mode, and ν is the kinematic viscosity.

The agreement between numerical and experimental results was quite good in terms of n_{11}, Q_{11}, and M_{11}, with an error less than 5.0% at OP1 and OP2–OP4. At OP5 and OP6, there was a significant increase in error (Table 5). As regards moment M, at OP5, its values should be zero, and at OP6, its value should be a small negative value. Minor numerical overestimation in the moment will induce a significant increase in errors. Moreover, mechanical friction losses were not considered in the numerical simulations. These losses are closely related to the runner rotation rate and they remain almost unchanged with the operating condition. Therefore, the corresponding reduction of hydraulic power at these two points will induce significant increases in percentage errors, as shown in Table 6. Even at these two points, the errors in n_{11} and Q_{11} are still smaller than 5.0%. This means that the hydraulic heads are still well predicted in spite of the simplification of mechanical friction losses. The comparisons demonstrate the capability of the numerical simulations to predict the hydraulic characteristics of the pump-turbine with sufficient accuracy.

Table 6. Investigated operating points and their values of n_{11}, Q_{11}, and M_{11}.

Case	Parameter	Calculation	Experiment	Error
OP1	n_{11}	75.56	74.58	1.29
	Q_{11}	0.180	0.178	1.29
	M_{11}	188.83	185.97	1.51
OP2	n_{11}	85.27	84.24	1.21
	Q_{11}	0.172	0.170	1.21
	M_{11}	157.58	152.03	3.52
OP3	n_{11}	89.56	89.26	0.33
	Q_{11}	0.163	0.162	0.33
	M_{11}	133.63	127.89	4.29
OP4	n_{11}	93.37	95.14	1.90
	Q_{11}	0.141	0.144	1.90
	M_{11}	88.93	84.99	4.43
OP5	n_{11}	97.26	101.08	3.92
	Q_{11}	0.09	0.093	3.92
	M_{11}	16.53	0.44	97.3
OP6	n_{11}	95.56	94.98	0.60
	Q_{11}	0.032	0.031	0.60
	M_{11}	-33.65	-48.26	43.4

3. Results and Discussion

Pressure monitors in the whole water passage are presented in Figure 5. Monitor SC1 was located on the spiral casing wall. In the gap between the guide vanes and runner, 20 monitors (GV1–GV20) were arranged evenly around the circumference. On the draft tube wall there were three monitors, DT1, DT2 and DT3. Pressure fluctuations on monitors GV5, GV15, and DT1–DT3 were registered in the model test [19,20]. In a pump-turbine, pressure fluctuations in turbine mode are usually stronger than those in pump mode. Moreover, fluctuation amplitudes in the draft tube and vaneless gap between the guide vanes and runner are usually larger than those in the spiral casing and stay vane channels.

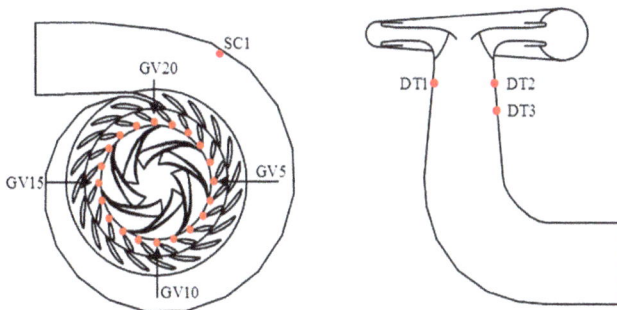

Figure 5. Pressure monitors in the whole passage.

For pressure fluctuation analyses, in frequency domains, the frequency spectrum was normalized as f/f_n, where f is the frequency component and f_n is the runner rotating frequency. In time domains, the pressure fluctuations are normalized by Equation (2).

$$\tilde{c}_p = \frac{p - \bar{p}}{\rho E} \tag{2}$$

where $\tilde{c}_p$ is the pressure fluctuation coefficient, p is the instantaneous pressure, $\bar{p}$ is the time-averaged pressure, and E is the specific energy.

3.1. Pressure Fluctuations in the Draft Tube

Three monitors, DT1–DT3, were located on the conical section in the draft tube. Table 7 shows the pressure fluctuations of monitor DT2 at six different operating conditions, along with the numerical and experimental results. The experimentally-measured pressure fluctuations are only listed in the table for OP2–OP4. This is because pressure fluctuations were not measured at operating points in the S-shaped region. Instead, pressure fluctuations were measured during the efficiency test [19,20]. Therefore, only the operating points in the normal operating ranges are given in the table.

Table 7. Amplitude and dominant frequency of pressure fluctuations at DT2.

Case	Amplitude (%)		Dominant Frequency (f_1/f_n)	
	Calculation	Experiment	Calculation	Experiment
OP1	5.57	-	0.82	-
OP2	8.12	6.95	0.84	0.82
OP3	9.13	7.94	0.86	0.89
OP4	8.88	10.09	1.18	1.16
OP5	3.39	-	1.19	-
OP6	0.91	-	0.75	-

Although the amplitudes have relative errors of a little more than 13.0% between numerical and experimental results, the first dominant frequency (f_1) is in good agreement. The single helical vortex rope in the draft tube usually rotates at 20%–40% of the runner rotating speed [11]. However, the dominant frequencies have higher values (Table 6). These higher frequencies may be induced by a twin vortex rope [14]. The pressure fluctuations at monitor DT2 at OP3 are shown in Figure 6a and the frequency spectrum is shown in Figure 6b. A total of 32 runner revolutions were calculated after the unsteady simulation stabilized (Figure 6). The first dominant frequency measured during the experiment was $0.89f_n$, compared with $0.86f_n$ in the simulation. Based on Table 7 and Figure 6, it can be seen that the pressure fluctuations in the draft tube are mainly induced by vortex ropes.

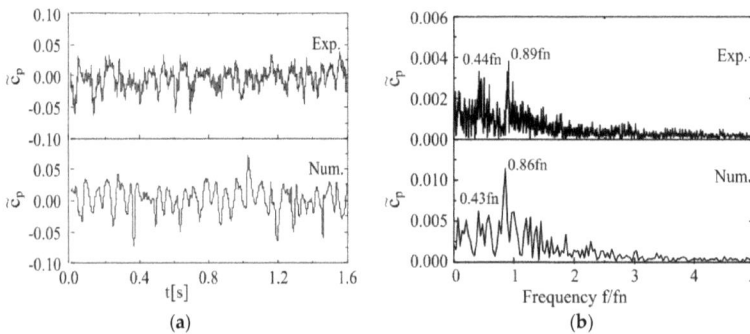

Figure 6. Measurements from monitor DT2 at OP3. (**a**) Pressure fluctuations; and (**b**) frequency spectrum.

Band-pass filtering was employed to extract the first dominant frequency component of $0.86f_n$ from the calculated pressure fluctuations of DT1, DT2 and DT3 at OP3. As shown in Figure 7, α stands for the phase difference of pressure between DT1 and DT2, and β represents the phase difference between DT2 and DT3. The α value was near zero, while β was not, which means that a twin vortex rope developed in the draft tube. If the same measurements were made at the other operating points, the twin vortex rope structure could also be detected.

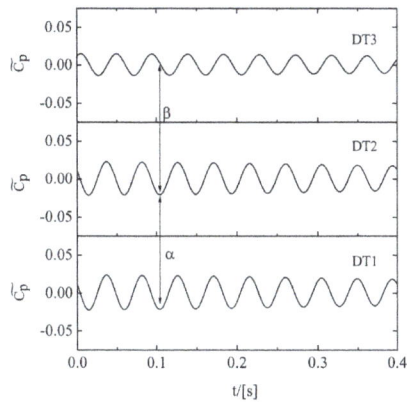

Figure 7. Band-pass-filtered pressure fluctuations at OP3.

Figure 8 shows the precession of a twin vortex rope at OP3 at a runner rotational period of T. The pressure contours at the planes where monitors DT1 and DT2 were located are also shown. The twin vortices are labelled A and B, and they rotate with the same angular velocity. This velocity corresponds to the dominant frequency of the twin vortex ropes, $0.86f_n$. Figure 9 shows a photograph of a vortex rope observed during the experiment. The twin vortex rope is generated alongside cavitation. This figure is simply used to verify the generation of twin vortex ropes because cavitation was not considered, and only a single flow field was simulated in the present study. It is well known that vortex ropes in draft tubes are usually induced by a high residual swirl in the input flow, which often forms a single helical vortex rope. Therefore, further research is required to understand the transition from a single to a double rope, and the influences of geometry on this process.

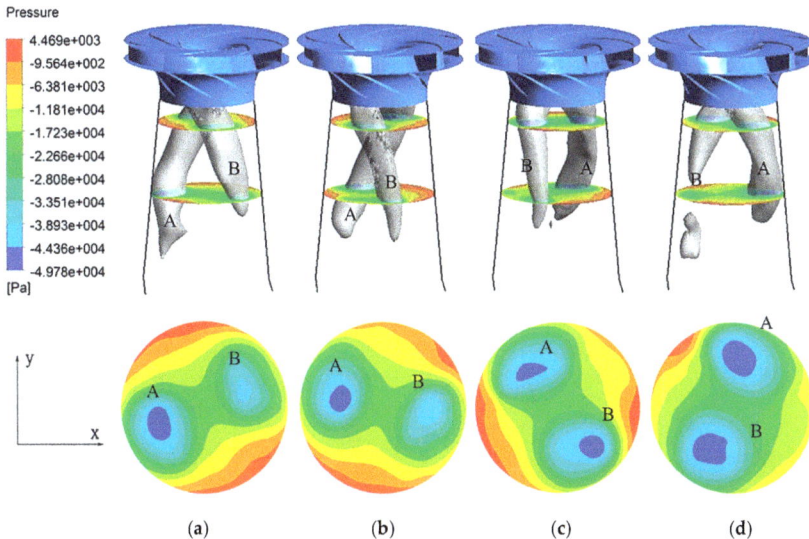

Figure 8. Twin vortex rope in the draft tube at OP3: (**a**) $t = t_0$; (**b**) $t = t_0 + 1/3T$; (**c**) $t = t_0 + 2/3T$; and (**d**) $t = t_0 + T$.

Figure 9. Photograph of a twin vortex rope taken during the experiment, with ropes 1 and 2 labelled.

3.2. Pressure Fluctuations in the Vaneless Gap

Table 8 shows the amplitude and dominant frequency of pressure fluctuations on GV5. The experimentally measured pressure fluctuations are only listed in the table for the three operation points OP2–OP4 (as per Table 7). The guide vanes and runner blades interact to generate pressure fluctuations in the gap between them. The dominant frequencies are rotor-stator interaction frequency $7f_n$ and its harmonics, for both simulated and experimental results.

Table 8. Amplitude and dominant frequency of pressure fluctuations at GV5.

Case	Amplitude (%)		Dominant Frequency (f_1/f_n)	
	Calculation	Experiment	Calculation	Experiment
OP1	5.45	-	7.00	-
OP2	5.61	6.12	7.00	7.00
OP3	8.96	10.45	7.00	7.00
OP4	35.69	30.78	7.00	7.00
OP5	35.78	-	7.00	-
OP6	28.95	-	7.00	-

The pressure amplitudes in operating conditions of low mass flow rate at OP4–OP6 were much higher than those at OP1–OP3. These increased amplitudes are believed to be induced by serious flow separation at the inlet of the runner. Figure 10 shows the distribution of streamlines at the guide vanes and runner channels under operating conditions OP1, OP5 and OP6. At OP1, the streamlines were well distributed at both guide vanes and runner channels, and there was no obvious flow separation. At OP5, flow separation occurred at the inlet of the runner, leading to backflow cells and vortices in the runner channels. However, the streamlines at the guide vanes still remained uniform. At OP6, backflow cells and vortices entirely filled the runner channels and flow separation occurred at the guide vanes.

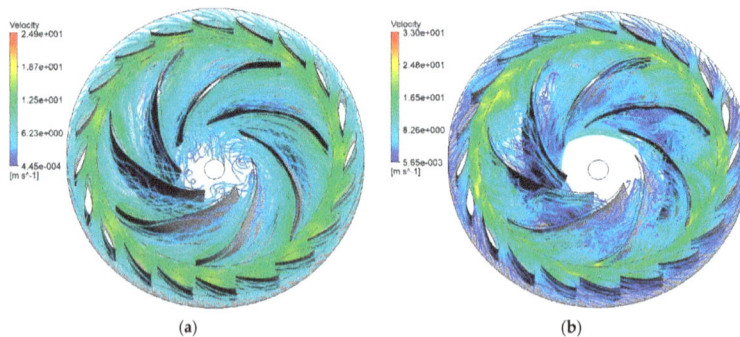

(a)   (b)

Figure 10. *Cont.*

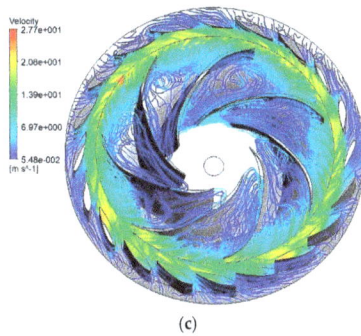

(c)

Figure 10. Streamlines in the guide vanes and runner. (**a**) Best efficiency point (OP1); (**b**) runaway point (OP5); and (**c**) low-discharge point (OP6).

Figure 11 presents the calculated pressure fluctuations and frequency spectrum at GV5 for operation conditions OP1, OP5 and OP6. For these three points, a low frequency component of $0.42f_n$ was captured at OP1, and $0.65f_n$ was well captured at OP5 as well as OP6, except for the blade passing frequency and its harmonics. The frequency component $0.42f_n$ at OP1 was about half of the vortex rope frequency, according to Table 6. Consequently, the component $0.42f_n$ was the rotation frequency of a single helical vortex rope. The frequency $0.65f_n$ was also well captured at the other 19 monitors in the gap between the runner and guide vanes at OP5 and OP6 (Figure 12). However, as shown in Table 7, the frequency of the twin vortex rope is different at OP5 and OP6. Therefore, the frequency $0.65f_n$ cannot be the rotation frequency of a single helical vortex rope at OP5 and OP6.

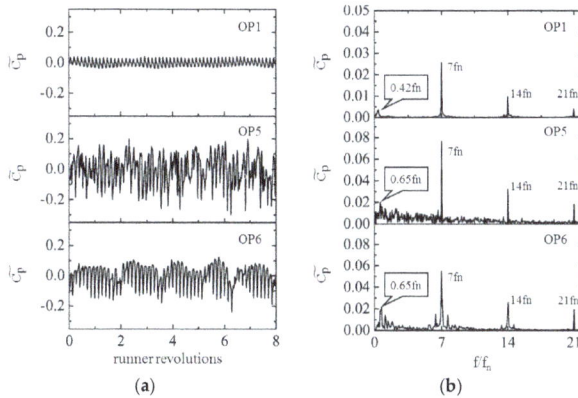

Figure 11. Measurements at monitor GV5. (**a**) Pressure fluctuations; and (**b**) frequency spectra.

Figure 12 shows the band-pass-filtered pressure signals of $0.65f_n$ for monitors GV1 to GV20 at OP5 and OP6, respectively. It is mentioned by Brennen [21] that the rotating stall phenomenon may occur in one, or in a cascade of rotor or stator blades, operating at a high incidence angle. Usually, the stall appears on a few adjacent runner channels and stall cells propagate in the circumferential direction. In Figure 12, the propagation of stall cells is represented by parallel lines, and they have the same direction as the rotating direction of the runner. At OP5, there is one single stall cell in the gap between the runner and guide vanes, while at OP6, there are four stall cells. The band-pass-filtered pressure signals for the four circumferentially-distributed monitors, GV5, GV10, GV15 and GV20, are redrawn in Figure 13 for OP5 and OP6, respectively. At OP5, the pressure signals of GV5 and GV15 had a phase difference of π, which is the same as for monitors GV10 and GV20 because of a single stall cell. At OP6, as there were four stall cells all together, and there was no definite phase difference among the four monitors.

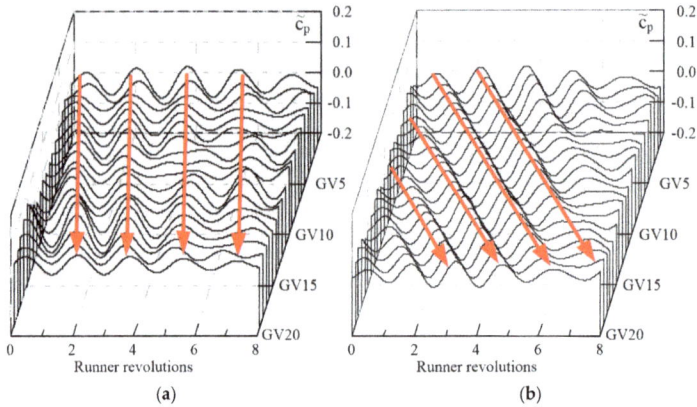

Figure 12. Propagation of rotating stall. (**a**) Runaway point OP5; and (**b**) low-discharge point OP6.

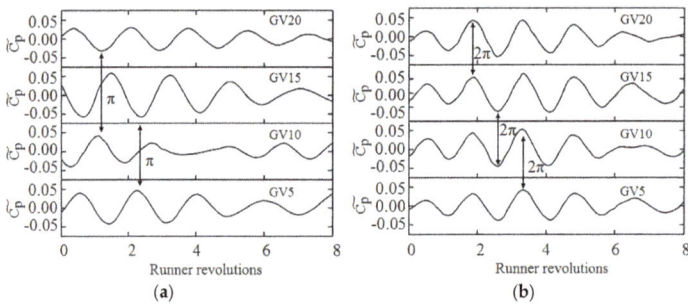

Figure 13. Band-pass-filtered pressure. (**a**) Runaway point OP5; and (**b**) low-discharge point OP6.

4. Conclusions

This study investigated flow in a reversible pump-turbine model. The unstable characteristics along a constant guide vane opening were simulated, and the numerical results were compared with experimental measurements. Analyses focused on the pressure fluctuations in the draft tube and in the gap between the runner and guide vanes. The main conclusions are summarized as:

(1) In the draft tube, a twin vortex rope formed under all the operating conditions that were investigated. The twin vortex rope rotated at 0.75–1.19 times the runner rotation speed, which was much faster than for single vortex ropes.

(2) In the gap between the runner and guide vanes, pressure fluctuations were mainly caused by rotor-stator interactions and vortices due to flow separation. A low frequency component $0.65f_n$ was well captured at both runaway and low-discharge points in the turbine brake zone, and was the effect of a rotating stall phenomenon. At the runaway point, a single stall cell was found in the gap, while at the low-discharge point, four stall cells were found.

Acknowledgments: The present work was supported by the National Natural Science Foundation of China (Grant No. 51679122). The authors would like to thank the reviewers for their constructive comments and suggestions.

Author Contributions: Zijie Wang made the computational simulations and prepared the first draft of the paper. Baoshan Zhu panned the study project, conceived, designed and performed the experiments, as well as revised the paper. Xuhe Wang contributed in the measured and calculated data analyses. Daqing Qin contributed in the experimental methodology.

Conflicts of Interest: The authors declare no conflict of interest.

References

1. Hino, T.; Lejeune, A. Pumped storage hydropower developments. *Compr. Renew. Energy* **2012**, *6*, 405–434.
2. Zhu, B.S.; Wang, X.H.; Tan, L.; Zhou, D.Y.; Zhao, Y.; Cao, S.L. Optimization design of a reversible pump-turbine runner with high efficiency and stability. *Renew. Energy* **2015**, *81*, 366–376. [CrossRef]
3. Zeng, W.; Yang, J.D.; Hu, J.H.; Yang, J.B. Guide-vane closing scheme for pump-turbine based on transient characteristics in S-shaped region. *J. Fluids Eng.* **2016**, *138*. [CrossRef]
4. Zuo, Z.G.; Fan, H.G.; Liu, S.H.; Wu, Y.L. S-shaped characteristics on the performance curves of pump-turbines in turbine mode—A review. *Renew. Sustain. Energy Rev.* **2016**, *60*, 836–851. [CrossRef]
5. Husmatuchi, V.; Farha, M. Experimental investigation of a pump-turbine at off-design operating conditions. In Proceedings of the 3rd International Meeting of the Workgroup on Cavitation and Dynamic Problems in Hydraulic Machinery and Systems, Brno, Czech Republic, 14–16 October 2009; pp. 339–347.
6. Husmatuchi, V.; Farha, M.; Roth, S.; Botero, F.; Avellan, M. Experimental evidence of rotating stall in a pump-turbine at off design conditions in generating mode. *J. Fluids Eng.* **2011**, *133*. [CrossRef]
7. Widmer, C.; Staubli, T.; Ledergerber, N. Unstable characteristics and rotating stall in turbine brake operation of pump-turbines. *J. Fluids Eng.* **2011**, *133*. [CrossRef]
8. Cavazzini, G.; Covi, A.; Pavesi, G.; Ardizzon, G. Analysis of the unstable behavior of a pump-turbine in turbine mode: Fluid-dynamical and spectral characterization of the S-shape characteristic. *J. Fluids Eng.* **2016**, *138*. [CrossRef]
9. Wang, L.Q.; Yin, J.L.; Jiao, L.; Wu, D.; Qin, D. Numerical investigation on the "S" characteristic of a reduced pump turbine model. *Sci. China Technol. Sci.* **2011**, *54*, 1259–1266. [CrossRef]
10. Sun, H.; Xiao, R.; Liu, W.; Wang, F. Analysis of the S characteristic and pressure pulsation in a pump turbine with misaligned guide vanes. *J. Fluids Eng.* **2013**, *135*, 0511011–1511016. [CrossRef] [PubMed]
11. Dörfler, P.; Sick, M.; Coutu, A. *Flow-Induced Pulsation and Vibration in Hydroelectric Machinery*; Springer: London, UK, 2014.
12. Ruprecht, A.; Helmrich, T.; Aschenbrenner, T. Simulation of vortex rope in a turbine draft tube. In Proceedings of the 21st IAHR Symposium on Hydraulic Machinery and Systems, Lausanne, Switzerland, 9–12 September 2002; pp. 259–266.
13. Kirschner, O.; Ruprecht, A.; Göde, E. Experimental investigation of pressure fluctuations caused by a vortex rope in a draft tube. *IOP Conf. Seri. Earth Environ. Sci.* **2012**, *15*. [CrossRef]
14. Wahl, T.L. Draft tube surging times two: The twin vortex phenomenon. *Hydro Rev.* **1994**, *13*, 60–69.
15. Wang, X.H.; Zhu, B.S.; Fan, H.G.; Tan, L.; Chen, Y.L.; Wang, H.M. 3D inverse design and performance investigation of a pump-turbine runner. *Trans. Chin. Soc. Agric. Mach.* **2014**, *45*, 93–98. (In Chinese)
16. Wang, X.H.; Zhu, B.S.; Cao, S.H.; Tan, L. Full 3-D viscous optimization design of a reversible pump-turbine runner. *IOP Conf. Ser. Mater. Sci. Eng.* **2013**, *52*. [CrossRef]
17. Wang, X.H.; Zhu, B.S.; Tan, L.; Zhai, J.; Cao, S.L. Development of a pump-turbine runner based on multiobjective optimization. *IOP Conf. Ser. Earth Environ. Sci.* **2014**, *22*. [CrossRef]
18. *Hydraulic Turbines, Storage Pumps and Pump-Turbines—Model Acceptance Tests*; IEC Standard 60193; International Electrotechnical Commission (IEC): Geneva, Switzerland, 1999.
19. Liu, L.; Zhu, B.; Wang, X.; Rao, C. Flow analysis of pump-turbine runner with large blade lean angle. In Proceedings of the 13th Asian International Conference on Fluid Machinery, AICFM13-127, Tokyo, Japan, 7–10 September 2015.
20. Wang, X. Multiobjective Optimization Design and Experimental Study for Reversible Pump-Turbine Runner. Ph.D. Thesis, Tsinghua University, Beijing, China, 2015. (In Chinese)
21. Brennen, C.E. *Hydrodynamics of Pumps*; Cambridge University Press: Cambridge, UK, 2011.

energies MDPI

Article

Battery Internal Temperature Estimation for LiFePO₄ Battery Based on Impedance Phase Shift under Operating Conditions

Jiangong Zhu [1,2], **Zechang Sun** [1,2], **Xuezhe Wei** [1,2,*] and **Haifeng Dai** [1,2]

[1]   Clean Energy Automotive Engineering Center, Tongji University, Shanghai 201804, China;
      zhujiangong@tongji.edu.cn (J.Z.); sunzechang@tongji.edu.cn (Z.S.); tongjidai@tongji.edu.cn (H.D.)

[2]   School of Automotive Engineering, Tongji University, Shanghai 201804, China

*   Correspondence: weixzh@tongji.edu.cn; Tel.: +86-135-0184-8129

Academic Editor: Hailong Li

Received: 17 November 2016; Accepted: 3 January 2017; Published: 6 January 2017

Abstract: An impedance-based temperature estimation method is investigated considering the electrochemical non-equilibrium with short-term relaxation time for facilitating the vehicular application. Generally, sufficient relaxation time is required for battery electrochemical equilibrium before the impedance measurement. A detailed experiment is performed to investigate the regularity of the battery impedance in short-term relaxation time after switch-off current excitation, which indicates that the impedance can be measured and also has systematical decrement with the relaxation time growth. Based on the discussion of impedance variation in electrochemical perspective, as well as the monotonic relationship between impedance phase shift and battery internal temperature in the electrochemical equilibrium state, an exponential equation that accounts for both measured phase shift and relaxation time is established to correct the measuring deviation caused by electrochemical non-equilibrium. Then, a multivariate linear equation coupled with ambient temperature is derived considering the temperature gradients between the active part and battery surface. Equations stated above are all identified with the embedded thermocouple experimentally. In conclusion, the temperature estimation method can be a valuable alternative for temperature monitoring during cell operating, and serve the functionality as an efficient implementation in battery thermal management system for electric vehicles (EVs) and hybrid electric vehicles (HEVs).

Keywords: lithium-ion battery; internal temperature estimation; impedance; phase shift; electric vehicles (EVs)

1. Introduction

Lithium-ion battery, which has been proven to be the ideal power source for electric vehicles (EVs) and hybrid electric vehicles (HEVs), strikes the best balance between power/energy density and costs for energy storage [1,2]. As safety behaviors and a longer cycle life of the battery demands a narrow temperature range, battery temperature always acts as one of the most essential operating parameters [3]. Surface mounted thermal sensors (thermistors and thermocouples) suffer from heat transfer delay due to the thermal mass of batteries. In consequence of cell thermal non-equilibrium, the internal temperature differs from the external counterpart [4]. It is complicated to directly measure the internal temperature of large format batteries in the vehicular application. An on-line detection of battery internal temperature, which is essential to facilitate operation control, can help improve the accuracy of BMS (battery management system) and the security of the power battery (battery pack). The electrochemical-thermal model [5–8] and electrical-thermal model [9–12] are widely employed to investigate the battery temperature performance during high power extraction. Inserted thermal sensors and thermal imaging are also commonly used in battery thermal research [12–14].

Nowadays, a simple technique to monitor battery internal temperature based on impedance measurement has been proposed [15–20]. Hande [15] provided a technique to estimate the cell internal temperature by measuring the pulse resistance. Srinivasan et al. [16] firstly demonstrated the intrinsic relationship between battery internal temperature and the phase shift obtained from EIS (electrochemical impedance spectroscopy). Schmidt et al. [17] introduced a sensorless temperature measurement method for a 2 Ah pouch cell via the real part of impedance spectroscopy at high frequencies with state of charge (SoC) status unknown and they also studied the influence of temperature gradient on the method by experiments. Richardson and Howey [18] proposed a one-dimensional model, which was validated utilizing internal thermocouple measurement, to estimate the temperature distribution for a 2.3 Ah LiFePO$_4$ cylindrical cell by combined the real and imaginary impedance and battery surface temperature. Further, they extended the estimation using an electrical-thermal model coupled with impedance measurement [19]. Schmidt et al. [17] and Richardson et al. [19] also showed that the estimated temperature inferred from impedance represented the equivalent uniform cell temperature. The impedance phase shift is another important parameter in the EIS test. Our previous work [20] has presented evidence for the existence of the intrinsic relationship between measured impedance phase shift and the internal cell temperature with electrochemical equilibrium, which is seldom influenced by battery degradation for an 8-Ah LiFePO$_4$ battery. Raijmakers et al. [4] also put forward an intercept frequency which was extracted from impedance spectra of a Li(NCA)O$_2$ and a LiFePO$_4$ battery and exclusively related to the internal battery temperature based on EIS. As aforementioned, the impedance-based temperature estimation methods eliminate the requirements of too many hardware temperature sensors and knowledge of the cell thermal properties [4,17–19]. Battery SoC and health often visibly change and several effective estimation strategies considering uncertain driving conditions for EVs and HEVs have been presented [21–26]. Zhang et al. [23] proposed an online battery SoC and SoE (state of energy) estimation method. This method was applied based on the hardware-in-loop setup, where the novel adaptive H infinity filter was proposed to realize the real-time estimation of battery SoC and SoE. The experiment results indicated the high estimation accuracy and strong robustness of the method to the model uncertainty and measurement noise. The impedance-based temperature estimation method reveal that there is a certain frequency which is distinctly dependent on the temperature but does not depend on SoC and battery aging state for the LiFePO$_4$ battery [4, 20], which is helpful to implement the impedance-based temperature method from lab to online application in consequence of methods capable of measuring impedance spectra using existing power electronics [27,28].

However, to satisfy the criteria of linearity and time invariance [29,30], the impedance is generally measured at an operating point with a perturbation of a small AC (alternating current) signal and long relaxation time, which hinders the practical application of the impedance-based method when the battery operates under charge and discharge conditions. The interpretation of the measured impedance under operating condition should be systematically investigated, particularly the impedance measurement under short-term relaxation time warrants further investigation for effective estimation.

In our previous study [20], the monotonic relationship between impedance phase shift and battery internal temperature, which is employed as a reference to the temperature estimator for LiFePO$_4$ battery, has been identified in the frequency range of 1–100 Hz with electrochemical equilibrium. Influence of battery SoC and aging are negligible in the selected frequency range. In this paper, we extend on our earlier work by investigating and validating the temperature estimation method with the embedded thermocouple under operating conditions. As shown in the flowchart (Figure 1), the relationship between phase shift and internal temperature is firstly reproduced with temperature homogeneities and artificial temperature gradients under the electrochemical equilibrium state. Regardless of battery internal and surface temperature gradients, the measured impedance phase shift with electrochemical equilibrium corresponds to the cell internal average temperature. Secondly, in order to promote the vehicular application, a detailed experiment is conducted to investigate the regularity of battery impedance phase shift after charge/discharge current excitations with short relaxation time. It is indicated that the impedance phase shift can also be obtained even under the current excitations. The phase shift descends as the relaxation time increases, which is considered the main contribution of this work to improve the accuracy of the estimation method. An exponential equation that accounts for both measured phase shift and relaxation time is established at 10 Hz tentatively. Furthermore, considering the effect of ambient temperature, a multivariate linear equation is derived and verified experimentally. The predicted internal temperature shows good agreement with the measured internal temperature, which guarantees a more precise assessment of the battery internal temperature.

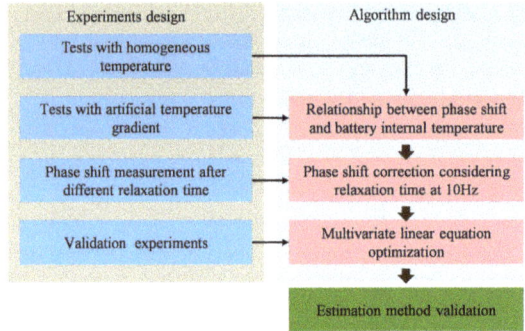

Figure 1. Implementation flowchart.

2. Experiments

The cells adopted in the experiments are commercial LiFePO$_4$ batteries with 30 Ah capacity (Shanghai Aerospace Power Technology, Shanghai, China), as depicted in Figure 2. The specifications of the lithium ion battery used are displayed in Table 1. One thermocouple is placed at the geometric center of the pouch cell in order to directly measure the battery internal temperature. Four experimental procedures are designed as shown in Figure 1, and the detailed introductions are shown in accordance with the orders of their appearance, respectively.

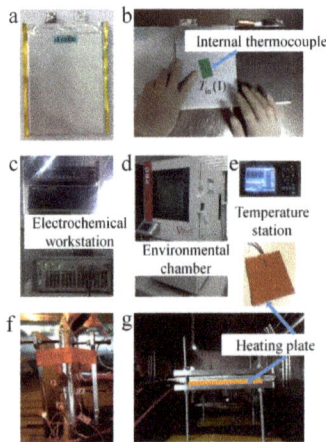

Figure 2. Battery and experimental device. (**a**) The cell sample; (**b**) the location of the internal thermalcouple; (**c**) the electrochemical workstation; (**d**) the environmental chamber; (**e**) the temperature monitor station; and (**f**,**g**) the locations of the cells during tests.

Table 1. Specifications of the lithium ion battery used.

Parameter Name	Values
Nominal voltage	3.2 V
Nominal capacity	30 Ah
Electrode chemistry	LiFePO$_4$/graphite
Internal resistance	$\leq$4 mΩ
Core size	13 mm $\times$ 132 mm $\times$ 184 mm
Storage temperature	-20–45 °C
Normal Charge voltage	3.7 V
Discharge ending voltage	2.5 V
Weight	0.675 kg
Energy density	144 Wh/kg
Manufacturer	Shanghai Aerospace Power Technology (Shanghai, China)

2.1. Impedance Phase Shift Measurement with Electrochemical Equilibrium

2.1.1. Tests with Homogeneous Temperature

The cell impedance spectra are obtained using an electrochemical workstation (Solartron SI 1287, 1255B, Solartron Mobrey, Durham, UK). The frequency range of the impedance measurement is set to span from 10 kHz to 0.1 Hz with perturbation current of 1.5 A. The ambient temperature is controlled by a Vötsch C4-180 environmental chamber (Vötsch, Germany), as displayed in Figure 2d. Measurements are made over the range -20–40 °C with an interval of 10 °C, and in the range of 0%–100% SoC. The whole test sequence is called A1 for simplification, and the detailed steps are shown in Table 2. The temperatures from all thermocouples are measured utilizing a HIOKI temperature unit (LR8510) (HIOKI, Nagano, Japan) and recorded by a HIOKI wireless logging station (LR8410-30). The tested cell is charged and discharged using an ARBIN instrument (ARBIN, College Station, TX, USA) with the test procedures listed in Table 2.

Table 2. Electrochemical impedance spectroscopy (EIS) test procedures A1 at various state of charge (SoC) and temperature. CC-CV: constant current-constant voltage.

Step No.	Type	Rate	End Condition	Set Temperature (°C)
1	Rest	0	4 h	25
2	Charge (CC-CV)	0.5 C (1 C = 30 A)	Voltage limit 3.7 V; current limit: 0.01 C	
3	Rest	0	2 h	40, 30, 20, 10, 0, -10,
4	EIS	1.5 A	10 kHz–0.1 Hz	and -20 respectively
5	Rest	0	2 h	25
6	Adjust the battery SoC, and repeat the EIS tests			
7	End			

2.1.2. Tests with Artificial Temperature Gradient

To investigate the influence of temperature gradient on EIS measurements, an artificial temperature gradient is constructed. The cell temperature gradients are controlled with the combination of the environmental chamber and a heating plate. Considering the thickness of the cell is 13 mm (Table 1), it facilitates the formation of the stabilized temperature difference, and it is beneficial to verify the relationship between the measured phase shift and the mean temperature value from the results of the embedded thermalcouple, thereby the two lateral surfaces of the cell are imposed [17]. One side of the battery cell is covered with the heat plate, and another side is exposed to the environment as depicted in Figure 3. An internal temperature gradient of the cell forms when the temperatures of the environment (T_{Am}) and heating plate (T_h) are different. To obtain a more accurate result, two thermocouples are employed to measure the actual temperatures ($T1$ and $T2$) of both side of the cell in each test, and all temperature data are logged as presented in Table 3. $T1$ is the measured surface temperature of the heating plate side, and $T2$ is the measured surface temperature without heating plate. No. 1 represents the uniform temperature of battery which is all involved in the environment chamber without heating plate. No. 2, No. 3, and No. 4 are used to describe the artificial temperature gradient sets. The influence of the internal temperature gradient on battery behavior is investigated at 50% SoC. The particular experiment procedures are illustrated in Table 4.

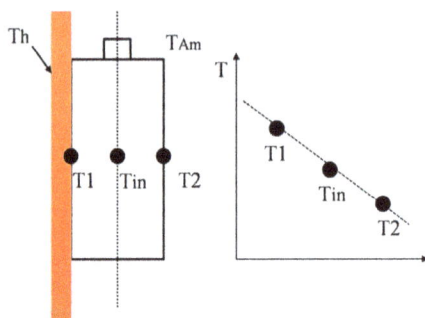

Figure 3. Artificial temperature gradient experiments diagram.

Table 3. The artificial temperature gradients.

Data Sets	T1	T2	T1	T2	T1	T2	T1	T2	T1	T2	T1	T2
No. 1	6		9		12		15		18		22	
No. 2	8	4	12	6	13	11	17	13	21	15	26	18
No. 3	11	1	15	3	16	8	20	10	25	12	29	15
No. 4	14	−3	18	−1	19	5	23	7	28	9	33	11

T1: Surface temperature with heating plate (°C); *T2*: Surface temperature without heating plate (°C).

Table 4. EIS test procedures B1 at artificial temperature gradients at 50% SoC.

Step No.	Type	Rate	End Condition	Set Temperature (°C)	
				Heating Plate	Ambient
1	Rest	0	4 h	-	
2	Charge (CC-CV)	0.5 C	Voltage limit 3.7 V; current limit: 0.01 C	-	25
3	Rest	0	2 h	-	
4	Discharge	0.5	1 h	-	
5	Rest	0	4 h	T1	T2
6	EIS	1.5 A	10 kHz–0.1 Hz		
7			End		

2.2. Impedance Phase Shift Measurement with Different Relaxation Time

2.2.1. Impedance Phase Shift Measurement after Different Relaxation Time

The experiments are designed to investigate the variation of phase shift after different relaxation time. The whole test sequence is called C1 as depicted in Table 5. The cell is first fully charged by ARBIN at room temperature. The charged cell is discharged to 50% SoC followed by a rest period of four hours at 20 °C. Then, the cell voltage and temperature are recorded with the 24 A pulse charging and discharging protocol periodically. Subsequently, relaxation time of 0 s, 10 s, 30 s, 60 s, 90 s, and 120 s is set before the EIS experiment, sequentially. After that, four hours rest is scheduled to equilibrate the cell with the chamber temperature again. The temperature control method [31], which leads to a shift of time constants and enlargements of impedance, is a powerful tool in impedance analysis. In order to distinguish each response of elemental steps in the high-medium frequency range, the expansion of impedance measurement are examined by lowering the temperature. Thus, Steps 6–12 are repeated at 10 °C and 5 °C to check the effect of relaxation time on impedance spectrum in more detail.

Table 5. The battery impedance test procedures C1 after different relaxation time.

Step No.	Type	Rate	End Condition	Set Temperature (°C)
1	Rest	0	4 h	
2	Charge (CC-CV)	0.5 C	Voltage limit 3.7 V; current limit: 0.01 C	25
3	Rest	0	2 h	
4	Discharge	0.5 C	1 h	
5	Rest	0	4 h	
6	Charge	24 A	10 s	20, 10, and 5
7	Discharge	24 A	10 s	
8	Repeat Steps 6, 7 200 times			
9	Rest	0	0 s, 10 s, 30 s, 60 s, 90 s, and 120 s respectively	20, 10, and 5
10	EIS	1.5 A	10 kHz–0.1 Hz	
11	Rest	0	4 h	
12	End			

Battery temperature first rises with the increasing pulse cycles and then gradually reaches static state because of thermal equilibrium. Following the pulse experiments described in Table 5, the maximum relaxation period of 120 s is set to ensure that thermal response of the battery to the applied 200 cycles is not obviously altered when the current excitation is switched-off. The cell internal temperature is just dropped by at most 0.3 °C during the relaxation periods monitored by the internal thermalcouple. Therefore, temperature and SoC are assumed to keep constant for all the tests to isolate the effect of relaxation time before EIS tests. The test procedures C2 in Table 6 are designed to obtain the phase shift at the homogeneous temperature with electrochemical equilibrium which refers as the static point in the following discussion. The ambient temperature values inputted in procedures C2 are calculated from battery internal and surface after the period pulse swing in procedure C1 (Table 5).

Table 6. The battery impedance test procedures C2.

Step No.	Type	Rate	End Condition	Set Temperature (°C)
1	Rest	0	4 h	
2	Charge (CC-CV)	0.5 C	Voltage limit 3.7 V; current limit: 0.01 C	25
3	Rest	0	2 h	
4	Discharge	0.5	1 h with 2.5 V	
5	Rest	0	4 h	22 (the temperature after pulse swing)
6	EIS	1.5 A	10 kHz–1 Hz	
10	Repeat Steps 5, 6 two times and Step 5 for 12.5 °C, and 8 °C respectively			
11	End			

2.2.2. Validation Experiments Design

The validation experiments are conducted using constant-current discharge and pulse swing excitation profiles. The procedures of battery charge and discharge are the main factor inducing the variation of battery internal temperature. Based on the above analysis and referring to other research methods for the battery temperature [11,14], the test programs are executed at various ambient temperatures. The discharge pulses of different current magnitudes (15 A, 20 A, and 24 A) are applied to the battery using the TOYO power booster (PBI 250-10) (TOYO Corporation, Tokyo, Japan). The impedance measurements are carried out by 10 s, 30 s and 60 s relaxations with no current load, and the surface and internal temperature are also monitored simultaneously.

3. Results and Discussion

3.1. Impedance Phase Shift with Equilibrium Temperature at 10 Hz

Many studies have elucidated that the EIS characteristics of battery are dramatically impacted by the external environment and internal conditions, especially the temperature [16–20,32]. The electrochemical reaction rate, transfer rate and diffusion rate of lithium-ion are slowed down resulting from the temperature decreasing [32], so the lower temperature enlarges the battery impedance. The EIS procedures A1 are performed and the results are illustrated in Figure 4a. As can be seen, the phase shift changes distinctly with the temperature in the whole frequency range. The relationship between impedance phase shift and battery temperature at 10 Hz is indicated in Figure 4b. A certain frequency range, which is able to exclude the influence of SoC and battery aging on the impedance phase shift, was selected in previous research [4,20]. As illustrated in Figure 4b, the phase shift does not alter with SoC at 10 Hz, which facilitates the impedance-based temperature estimator design since the SoC often visibly changes and is hard to be estimated and calculated in the vehicular application. The phenomenon mainly related to battery electrochemical reaction and diffusion process has been interpreted in our previous study from the electrochemical perspective [20]. In this study, we utilize the phase shift values at 10 Hz tentatively to track the battery internal temperature for the representative of other available frequency points, and the relationship between phase shift and temperature is employed as a reference for the estimation model in the next section.

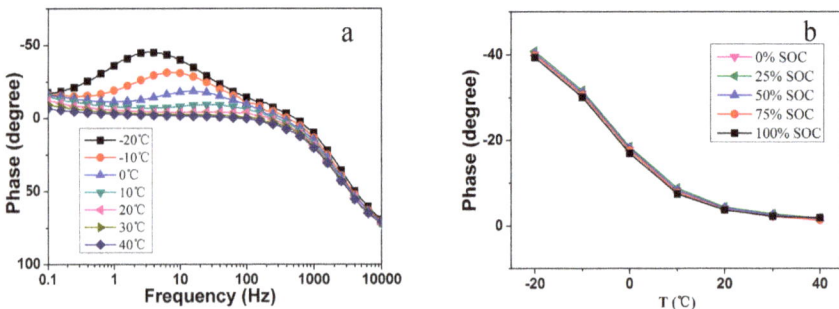

Figure 4. Relationship between impedance phase shift and temperature. (**a**) Phase shift at different frequencies and 50% SoC; and (**b**) the phase shift at various SoC and 10 Hz.

3.2. Impedance Phase Shift with Artificial Temperature Gradients

Some typical EIS measurements at 50% SoC and with different temperature gradients from procedures B1 are depicted in Figure 5. Different temperature gradients are artificially constructed by controlling the temperature of the ambient environment and the heating plate. Because of heat dissipation between cells and the environment, the battery surface temperatures are different with the setting values, so the actual measured values displayed in Table 3 are used in the paper. The internal temperatures of 6 °C, 9 °C, 12 °C, 18 °C and 22 °C were measured at the central position, and 6 °C (Figure 5a,b) and 18 °C (Figure 5c,d) are presented here for the representative. The black, red and dark cyan color lines represent the battery phase shift measured with thermal equilibrium. The gradient perpendicular to the electrodes is assumed to be linear, as shown in Figure 3. In Figure 5, the impedance spectroscopy and phase shift values are almost the same, even with different temperature gradients, and the effect of high temperature gradients on the phase shift is slightly larger in medium frequencies. It indicates that even under temperature gradients, the cell performs as under a uniform temperature with electrochemical equilibrium. The experimental results are beneficial to the impedance-based estimation method. The effect of temperature non-uniformity on the electrochemical impedance was studied by Schmidt et al. [17], who also proposed that the uniform temperature was the cell internal average temperature based on the impedance results at high frequencies. The variation of the phase shift in our paper may be related to cell impedance characteristics and the experiments schemes. Anticipated detailed interpretation for the measured results will be in next study.

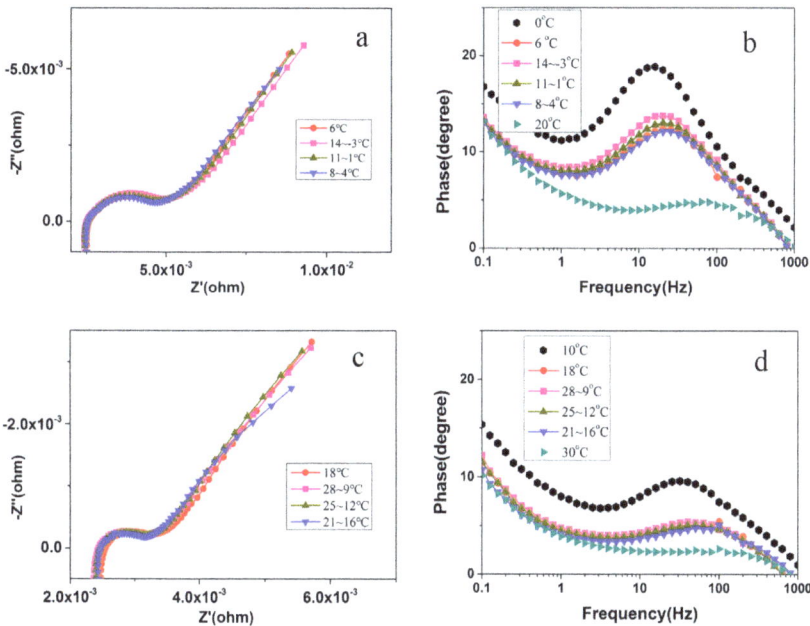

Figure 5. EIS and phase shift with artificial temperature gradients: (**a,c**) EIS results; and (**b,d**) phase shift results.

3.3. Impedance Phase Shift Correction Considering Relaxation Time

Because of the complexity of the battery charge and discharge process in practical applications, there is seldom sufficient time to satisfy the cell to reach electrochemical equilibrium. The short-term current interruption, such as waiting at red lights, may be the opportunity for AC incentives. Barai et al. [33] and Schindler et al. [34] have studied the impedance of lithium ion cell with different relaxation process between the removal of an electrical load and the impedance measurement. We find that the phase shift is also correlative to relaxation time, especially when the battery is at low temperatures. The impedance measurements are taken after current pulses; when the battery temperature reaches an approximate steady state (200 cycles), the current is switched off to allow the impedance test with different relaxation time.

The evolution of the impedance spectra and phase shift directly after switch-off the pulse current is illustrated for the tests of C1 and C2 in Figure 6. The impedance arc enlarges and the phase shift goes down with the elevation of relaxation time.

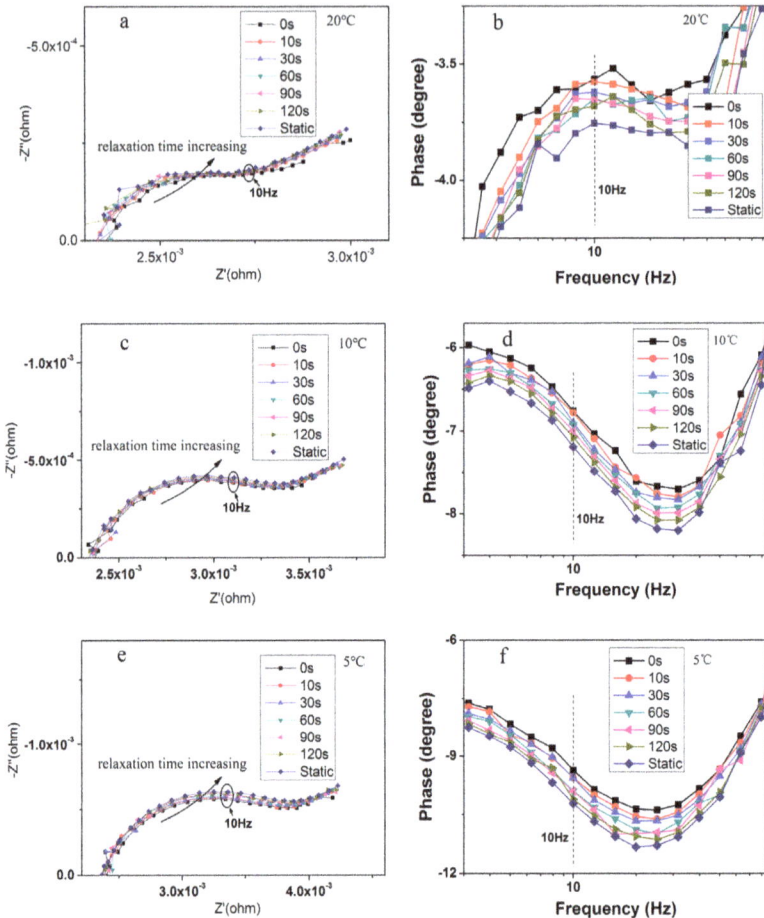

Figure 6. Impedance results at various relaxation time. (**a**) EIS results at 20 °C; (**b**) enlarged phase shift results at 20 °C; (**c**) EIS results at 10 °C; (**d**) enlarged phase shift results at 10 °C; (**e**) EIS results at 5 °C; and (**f**) enlarged phase shift results at 5 °C.

The temperature during the pulse excitation is elucidated in Figure 7. T_{in} is the battery internal temperature, and T_{surf}(1)–(5) represent the battery surface temperature monitored by the thermalcouples. The temperature first ascends with the increment of cycles and then gradually reaches static state due to the thermal equilibrium. The impedance test is performed in the relaxation break as shown the grey rectangle in Figure 7. The cell internal temperature which is monitored by the internal thermalcouple drops by at most 0.3 °C during the maximum relaxation period (120 s). The thermal response of the battery to the applied 200 cycles is not obviously altered when the current excitation is switched-off. It can be argued that the exponential decay of the phase shift is not associated with temperature decay. Therefore, temperature and SoC are kept constant for all the tests to isolate the distraction of relaxation time before cell impedance measurements.

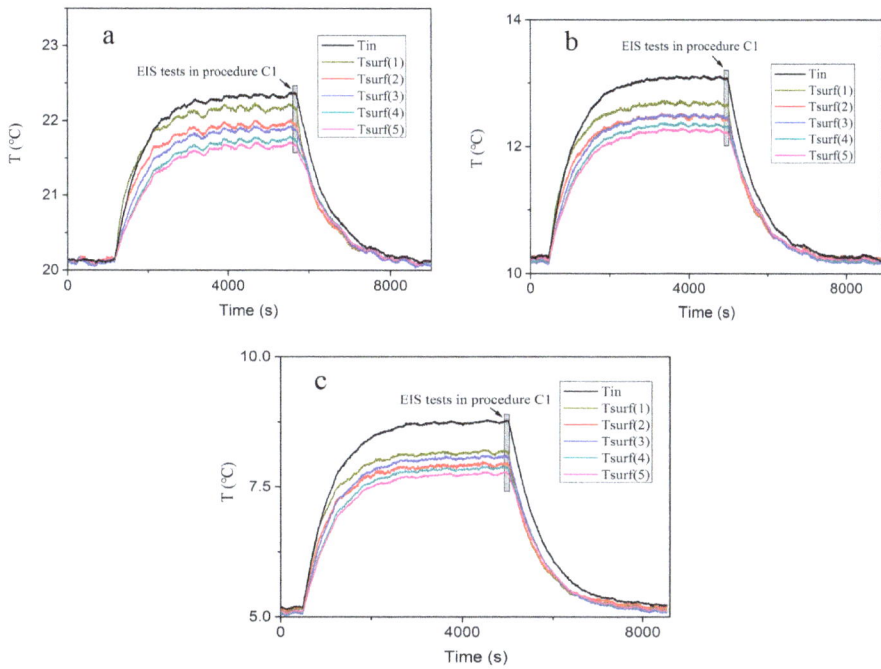

Figure 7. Battery temperatures during the test procedures C1. Battery temperatures at (**a**) 20 °C; (**b**) 10 °C; and (**c**) 5 °C.

The temperature values inputted in procedures C2 is the calculated temperature after the periodic pulse swing, and the impedance obtained in C2 is employed in Figure 6 (static scatters) for comparison. Two major features can be extracted from the results of the recorded spectra: (1) it is apparent that the impedance arc enlarges with the incremental relaxation time gradually, especially at low temperatures; and (2) the phase shift particularly slumps at a certain frequency point. In this study, the phase shift at 10 Hz for the representative of other frequency points is selected to estimate battery internal temperature tentatively, thereby the relationship between phase shift and relaxation time at 10 Hz is displayed in Figure 8. The observed variations are linked to the physical processes occurring at the cell during the relaxation period.

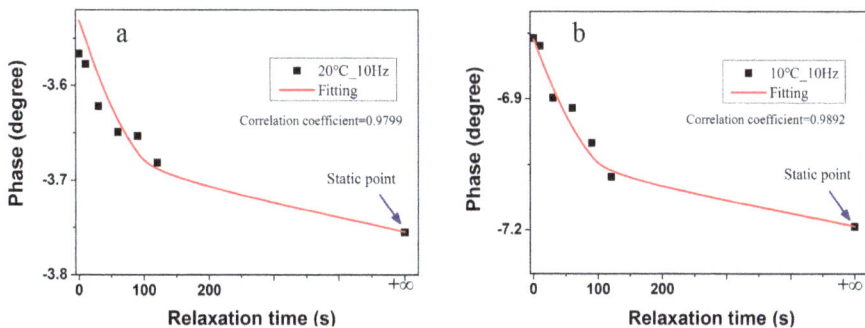

Figure 8. *Cont.*

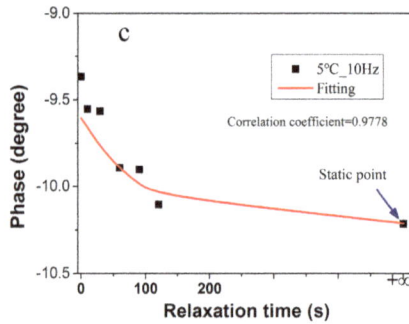

Figure 8. Experimental and fitting results of impedance phase shift at 10 Hz. Phase shift results at (a) 20 °C; (b) 10 °C; and (c) 5 °C.

The EIS is constituted with the impedance under discrete excitation frequency. The phase shift can be obtained from [20]:

$$\varphi(f_x) = \tan^{-1}\left\{\frac{-\mathrm{Im}\{Z(f_x)\}}{\mathrm{Re}\{Z(f_x)\}}\right\}, \ x \in [1, 2, \ldots, n] \tag{1}$$

where $f_x, x \in [1, 2, \ldots n]$ is with the variation of excitation frequency.

Battery relaxation is mainly dominated by diffusion processes and may take up to several hours, especially when the battery is almost empty, at low temperature, and after charging or discharging with high current rates [35]. The changes of imaginary part are mainly related to the battery capacitive component. The porosity of the electrodes confers a capacitance to the electrodes when a potential difference is applied to the cell, and the electrodes can be considered as a parallel plate capacitor as described by the Barai et al. [33]. One common observation reports that the capacitance of the cell follows an exponential decay with a subsequent continuous relaxation. The authors attribute their findings to ionic diffusion during the redistribution of ions within the electrolyte after switch-off the pulse current. This redistribution of ions declines the battery capacitance until the overall concentration of the electrolyte reaching equilibrium. As the concentration of ions at the electrolyte surface decreases, when compared to that of the cell under polarization, the total cell capacitance goes down. Therefore, one alternative explanation is that the reduction in the concentration gradients with the electrolyte induces the enlargement of the impedance arc in Figure 6. Because the variation in concentration gradient does not occur instantaneously, but rather occurs at several minutes or hours, the total cell capacitance decreases accordingly as a function of relaxation time. Similarly, the solid state diffusion will occur within the bulk of the particles during the relaxation period, which leads to a rearrangement of the lithium atoms in the electrode materials and will be also reflected in the change of observed impedance arc. S. Schindler et al. [30] also indicated the real part of medium impedance arc rises after the electrical load removed because of battery polarization by experiments.

The shrinking phenomenon of phase shift in Figure 6 should be considered in the temperature estimation method. A possible way to incorporate the information about the relaxation time is to introduce a correction phase shift factor φ', which is calculated from t_{re} and measured phase shift φ, and indicates whether the battery is completely recovered or not. An exponential function corresponding to time is numerously employed to describe the battery relaxation process, such as voltage relaxation [35], and the double layer capacitance relaxation [33]. The phase shift relaxation process is also assumed to proceed as an exponential decay with the time constant τ, and the equation can be described as:

$$\varphi' = \varphi \cdot \left(1 + a \cdot e^{\frac{-t_{re}}{\tau}}\right) \tag{2}$$

where a is the pre-exponential factor. We assume that the state of battery does not obviously alert during the frequency sweeping from 10 kHz to 10 Hz, so t_{re} represents the relaxation time, $T_{core} = f(\varphi)$ is the estimated internal temperature with measured phase shift, and $T'_{core} = f(\varphi')$ is the estimated internal temperature corresponding to correction φ'. When a equals 0.065, τ equals 85 s in the equation for the cells, the phase shift of the cell follows the exponential decay, and the experimental data (scatters) and simulation data (line) fit very well as illustrated in Figure 8. Equation (2) takes account for the effect of relaxation time on impedance phase shift, and the evaluation plays a vital role in the subsequent estimation model. In the validation experiments, a relaxation period of 60 s is adopted firstly. The cells are tested utilizing pulse swing excitation profiles above zero temperature, and constant-current discharge at subzero temperature to avoid the lithium dendrites. The temperature of the embedded thermocouple T_{in} is employed to verify the estimated results. The estimated results before and after relaxation time optimization are respectively presented in Table 7. In the results, we can find that the correction Equation (2) can improve the accuracy of the proposed model. The estimated results at a lower temperature are more accurate than that in a higher temperature, which is probably ascribed to the uncertainty of the impedance measurement as the decreasing of impedance at higher temperature. The correction considering the relaxation time, which can promote estimation accuracy, is the main contribution of this work. To further improve the accuracy of the estimator, a multivariate linear regression equation associated with environment temperature is proposed in Section 3.4.

Table 7. The verification of the temperature estimator considering the relaxation time.

T_{Am}	Current	Cycles/SoC	φ	T_{in}	T_{core}	T'_{core}	No.
20 (°C)	24 (A)	500	−2.719	26	27.94	27.42	1
		1000	−2.647	26.3	28.36	27.86	2
	20 (A)	500	−3.074	24.5	25.84	25.26	3
		1000	−3.0548	24.5	25.95	25.37	4
	15 (A)	500	−3.3313	22.5	24.32	23.69	5
		1000	−3.3303	22.7	24.33	23.69	6
10 (°C)	24 (A)	500	−6.983	12.0	12.91	12.37	7
		1000	−7.0134	11.9	12.84	12.29	8
	20 (A)	500	−7.5696	11.1	11.49	10.9	9
		1000	−7.5884	11.0	11.45	10.85	10
	15 (A)	500	−8.0317	10.3	10.37	9.891	11
		1000	−8.0522	10.3	10.30	9.797	12
0 (°C)	24 (A)	SoC I	−9.4269	7.0	8.709	8.395	13
		SoC II	−7.8341	10.2	10.85	10.24	14
	20 (A)	SoC I	−10.331	5.9	7.77	7.426	15
		SoC II	−8.7908	8.6	9.37	9.077	16
	15 (A)	SoC I	−11.294	4.8	6.77	6.393	17
		SoC I	−10.094	6.8	8.016	7.68	18
−10 (°C)	24 (A)	SoC I	−19.94	−2.5	−1.63	−2.12	19
		SoC II	−18.26	−0.5	−0.3438	−0.7924	20
	20 (A)	SoC I	−22.717	−4.7	−3.76	−4.314	21
		SoC II	−17.546	−0.7	0.28	−0.2282	22
	15 (A)	SoC I	−25.138	−6.1	−5.61	−6.227	23
		SoC II	−23.239	−4.9	−4.156	−4.727	24
−20 (°C)	24 (A)	SoC I	−33.087	−13.7	−12.4	−13.57	25
		SoC II	−31.82	−12.5	−11.3	−12.14	26

SoC I represents the 50% SoC; SoC II represents the state of battery reaching cut-off voltage.

3.4. Multivariate Linear Optimization and Validation

The environment temperature, operating time, current and other battery properties all have a great influence on battery internal temperature. Because the impedance reduces significantly at higher temperatures ($T > 10\,°C$), a multivariate linear regression equation is established to improve the battery internal temperature estimation accuracy. The relationship is expressed as:

$$T_{mult} = \beta_0 + \beta_1 \times T'_{core} + \beta_2 \times T_{Am} \tag{3}$$

To simplify the algorithm, we assume that the internal average temperature T'_{core} is related to the battery property, and the environment temperature T_{Am} is mainly pertinent to heat dissipation.

When considering the situation in which n independent multivariate observations $x_1, \ldots, x_n$ have been collected, and the number of responses measured in each observation is y, the multivariate linear regression model can be written as:

$$y = \beta_0 + \beta_1 x_1 + \cdots + \beta_n x_n \tag{4}$$

To obtain the parameter vector $(\beta_1, \ldots, \beta_n)$, N sets of observations:

$$X = \begin{bmatrix} x_{11} & x_{12} & \cdots & x_{1n} \\ x_{21} & x_{22} & \cdots & x_{1n} \\ \vdots & \vdots & \ddots & \vdots \\ x_{N1} & x_{N2} & \cdots & x_{Nn} \end{bmatrix}, \quad Y = \begin{bmatrix} y_1 \\ y_2 \\ \vdots \\ y_N \end{bmatrix}$$

n is the number of independent variables, N is the number of data sets.

Nine data sets in Table 7 (the even sequences in No. 2–18) are selected to identify the three parameter vector $(\beta_0, \beta_1, \beta_2)$. The goodness of fit ($R$ test), significance test (F test), and regression coefficient significance test (t test) are calculated. The values are all presented in Table 8. At test level $\alpha = 0.05$, all the test values (R, F, and t) prove that Equation (3) is effective and reliable to be used.

Table 8. The parameter vector values in equation.

β_0	β_1	β_2	R Test	F Test T'_{core}	T_{Am}	t Test T'_{core}	T_{Am}
1.9235	0.7408	0.1829	0.9983	206.5	14.8	10.2	2.7

3.5. Estimation Method Validation

After the obtainment of phase shift according to the correction Equation (1), the measured phase shift can be modified considering relaxation time with Equation (2). Then, the internal average temperature can be calculated from the relationship described in Figure 4. On the basis of multivariate linear equation operation, the estimated temperature T_{mult} can be observed finally.

To further verify the estimated results, a discharge profile is involved at 10 °C and 20 °C as displayed in Figure 9. The AC frequency excitations (10 Hz, dark cyan arrows) are executed after the relaxation process (10 s and 30 s). The estimated temperature results are, respectively, presented in scatters for comparison in Figure 9a,b.

In Figure 9, the estimation temperature T_{core} has larger deviation compared to T_{in}, and the estimation results T'_{core} and T_{mult} show good concordance with the measured cell internal temperature. The results with 30 s relaxation at 20 °C are more accurate than that with 10 s, as shown in Figure 9d. One interpretation is that the battery will be more stabilized and balanced with the incremental relaxation time because of the faster ions transfer and diffusion at higher temperature. Thus, the impedance can be obtained more precisely. Another interpretation could be that the model term to correct for the ambiguous error is more accurate for the 30 s relaxation, as it is already closer to the static value. When the battery operates at 10 °C, the maximum errors are 1.58 °C with Equation (2), and 0.76 °C with Equation (3). When the cells are operated at 20 °C, the errors are 2.07 °C and 0.91 °C, respectively. It indicates that the multivariate linear equation can improve the model accuracy, which mainly contains two aspects: on one hand, Equation (2) is used to modify the measuring deviation caused by electrochemical non-equilibrium. On the other hand, the ambient temperature is introduced in Equation (3) to consider battery temperature distribution due to uneven heat dissipation.

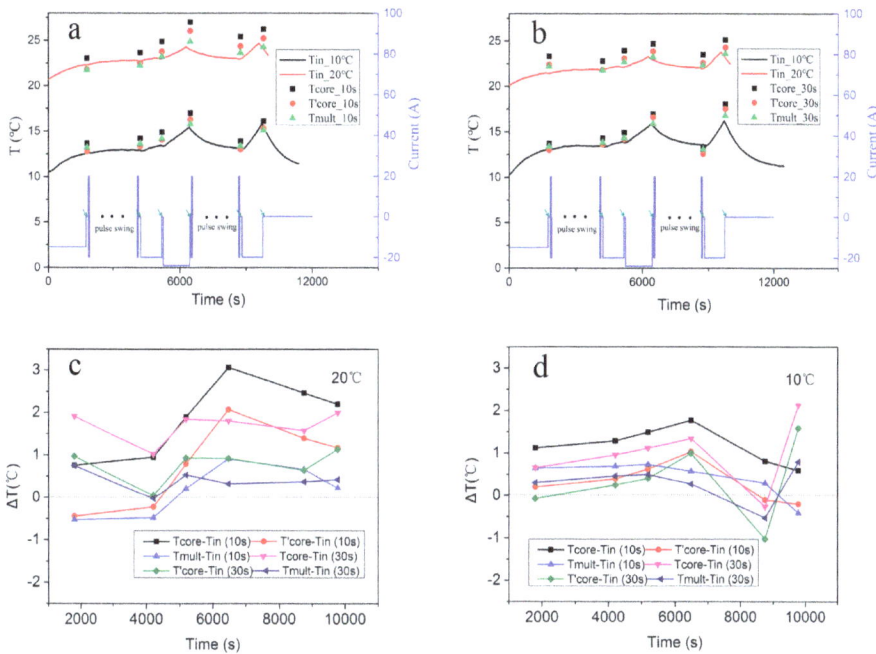

Figure 9. Estimated results with 10 s and 30 s relaxation time. Estimated results with (**a**) 10 s; (**b**) 30 s relaxation time; and temperature error at (**c**) 20 °C; (**d**) 10 °C.

The impedance changes with the degradation of the cell. Identifying ageing and degradation mechanisms in a battery is a main and most challenging goal in the implementation. L.H.J. Raijmakers et al. [4] conduct battery cyclic life tests and their temperature estimated method does not depend on the battery aging. The relationship between the phase shift and battery cyclic aging for the LiFePO$_4$ cell has been discussed in the previous study [20], which shows that the impedance magnitude varies obviously with aging, but the phase shift is not affected by the battery cyclic aging. The aforementioned research facilitates the temperature estimation method in our study. They just test the cyclic life of the cells, however, the calendar life and other complicated utilization mode, e.g., charging and discharging rates like the ones corresponding to the New European Driving Cycle or Urban Dynamometer Driving Schedule, may cause different ageing effects. Hence, validating the relationship between phase shift and other degradation mode is the next focus in our work.

4. Conclusions

Based on the monotonic relationship between impedance phase shift and battery internal temperature proposed in the previous study [20], the impedance-based temperature estimation method is further developed considering electrochemical non-equilibrium caused by current excitation. The impedance phase shift can be measured with a short-term relaxation after the current excitation switch-off. The relationship between phase shift and relaxation time at 10 Hz, which is representative of other frequency points, is investigated tentatively. The results demonstrate that the phase shift descends exponentially with the increment of the relaxation time at 10 Hz, responsible for the redistribution of ions within the electrolyte, which cause the decrease of phase shift after switch-off of the pulse current. An exponential equation is proposed to correct the measuring deviation due to electrochemical non-equilibrium. Considering the temperature inhomogeneities and uncertainty impedance measurement in higher temperature, a multivariate linear equation coupled with ambient temperature is derived. The temperature estimation method may be more accurate in low temperatures corresponding to the high resolution relationship between the temperature and the measured phase shift. The correction proposed in the study is established and verified under the excitation frequency 10 Hz. The model proposed in the paper does not rely on the battery thermal characteristics and surface temperature sensors, it can afford us much convenience in temperature monitoring during cell operating and is also functional as an efficient implementation in battery thermal management system for EVs and HEVs.

Acknowledgments: This work was supported by the National Natural Science Foundation of China (NSFC, Grant No. 51576142), Specialized Research Fund for the Doctoral Program of Higher Education (SRFDP, Grant No. 20130072110055), and International Exchange Program for Graduate Students, Tongji University.

Author Contributions: Jiangong Zhu designed the experiments, analyzed the results and finished the manuscript. Zechang Sun and Xuezhe Wei provided guidance and key suggestions. Haifeng Dai reviewed and revised the manuscript. All authors read and approved the manuscript.

Conflicts of Interest: The authors declare no conflict of interest.

Nomenclature

°C	Degree Centigrade
A	Ampere
C	Current magnitude in terms of cell capacity (1 C = 30 A)
Hz	Hertz
s	Seconds
V	Volt
h	Hour
φ	Measured phase shift
t_{re}	Relaxation time
τ	Time constant
a	Pre-exponential factor
φ'	Correction phase shift factor
t	Time (s)
T	Temperature (°C)
T_h	Temperature of heating plate
$T1$	Measured surface temperature of the heating plate side
$T2$	Measured surface temperature without heating plate
Z	Impedance
T_{Am}	Ambient temperature (°C)
T_{core}	Estimated internal temperature with measured phase shift
T'_{core}	Estimated internal temperature corresponding to correction phase shift
T_{in}	Measured internal temperature from embedded thermocouple
T_{mult}	Estimated temperature with Multivariate linear optimization
$(\beta_1, \ldots, \beta_n)$	Parameter vector
f_x	Frequency (Hz)

Acronyms

EV	Electric vehicle
HEV	Hybrid electric vehicle
BMS	Battery management system
EIS	Electrochemical impedance spectroscopy
$LiFePO_4$	Lithium iron phosphate
$Li(NCA)O_2$	Lithium cobalt aluminum nickel oxide
CC-CV	Constant charge-constant voltage
AC	Alternating current
SoC	State of charge
SoE	State of energy

Subscripts/Superscripts

re	Relaxation
h	Heat
Am	Ambient
core	Core
in	Internal
mult	Multivariate

References

1. Takahashi, M.; Tobishima, S.; Takei, K.; Sakurai, Y. Characterization of $LiFePO_4$ as the cathode material for rechargeable lithium batteries. *J. Power Sources* **2001**, *97–98*, 508–511. [CrossRef]
2. Sun, F.; Xiong, R. A novel dual-scale cell state-of-charge estimation approach for series-connected battery pack used in electric vehicles. *J. Power Sources* **2015**, *274*, 582–594. [CrossRef]
3. Yi, J.; Koo, B.; Shin, C.B. Three-dimensional modeling of the thermal behavior of a lithium-ion battery module for hybrid electric vehicle applications. *Energies* **2014**, *7*, 7586–7601. [CrossRef]
4. Raijmakers, L.H.J.; Danilov, D.L.; van Lammeren, J.P.M.; Lammers, M.J.G.; Notten, P.H.L. Sensorless battery temperature measurements based on electrochemical impedance spectroscopy. *J. Power Sources* **2013**, *247*, 539–544. [CrossRef]
5. Wu, B.; Yufit, V.; Marinescu, M.; Offer, G.J.; Martinez-Botas, R.F.; Brandon, N.P. Coupled thermal–electrochemical modelling of uneven heat generation in lithium-ion battery packs. *J. Power Sources* **2013**, *243*, 544–554.
6. Melcher, A.; Ziebert, C.; Rohde, M.; Seifert, H.J. Modeling and simulation the thermal runaway behavior of cylindrical li-ion cells—Computing of critical parameter. *Energies* **2016**, *9*, 292. [CrossRef]
7. Cai, L.; White, R.E. Mathematical modeling of a lithium ion battery with thermal effects in COMSOL Inc. Multiphysics (MP) software. *J. Power Sources* **2011**, *196*, 5985–5989. [CrossRef]
8. Song, L.; Evans, J.W. Electrochemical-thermal model of lithium polymer batteries. *J. Electrochem. Soc.* **2000**, *147*, 2086–2095. [CrossRef]
9. Capron, O.; Samba, A.; Omar, N.; van Den Bossche, P.; van Mierlo, J. Thermal behaviour investigation of a large and high power lithium iron phosphate cylindrical cell. *Energies* **2015**, *8*, 10017–10042. [CrossRef]
10. Samba, A.; Omar, N.; Gualous, H.; Firouz, Y.; Bossche, P.V.D.; Mierlo, J.V.; Boubekeur, T.I. Development of an advanced two-dimensional thermal model for large size lithium-ion pouch cells. *Electrochim. Acta* **2014**, *117*, 246–254. [CrossRef]
11. Kim, Y.; Mohan, S.; Siegel, J.B.; Stefanopoulou, A.G.; Ding, Y. The estimation of temperature distribution in cylindrical battery cells under unknown cooling conditions. *IEEE Trans. Control Syst. Technol.* **2014**, *22*, 2277–2286.
12. Sun, J.; Wei, G.; Pei, L.; Lu, R.; Song, K.; Wu, C.; Zhu, C. Online internal temperature estimation for lithium-ion batteries based on kalman filter. *Energies* **2015**, *8*, 4400–4415. [CrossRef]
13. Hong, X.; Li, N.; Feng, J.; Kong, Q.; Liu, G. Multi-electrode resistivity probe for investigation of local temperature inside metal shell battery cells via resistivity: Experiments and evaluation of electrical resistance tomography. *Energies* **2015**, *8*, 742–764. [CrossRef]

14. Forgez, C.; Do, D.V.; Friedrich, G.; Morcrette, M.; Delacourt, C. Thermal modeling of a cylindrical LiFePO$_4$/graphite lithium-ion battery. *J. Power Sources* **2010**, *195*, 2961–2968. [CrossRef]
15. Hande, A. Internal battery temperature estimation using series battery resistance measurements during cold temperatures. *J. Power Sources* **2006**, *158*, 1039–1046. [CrossRef]
16. Srinivasan, R.; Carkhuff, B.G.; Butler, M.H.; Baisden, A.C. Instantaneous measurement of the internal temperature in lithium-ion rechargeable cells. *Electrochim. Acta* **2011**, *56*, 6198–6204. [CrossRef]
17. Schmidt, J.P.; Arnold, S.; Loges, A.; Werner, D.; Wetzel, T.; Ivers-Tiffée, E. Measurement of the internal cell temperature via impedance: Evaluation and application of a new method. *J. Power Sources* **2013**, *243*, 110–117.
18. Richardson, R.R.; Howey, D.A. Sensorless battery internal temperature estimation using a kalman filter with impedance measurement. *IEEE Trans. Sustain. Energy* **2015**, *6*, 1190–1199. [CrossRef]
19. Richardson, R.R.; Ireland, P.T.; Howey, D.A. Battery internal temperature estimation by combined impedance and surface temperature measurement. *J. Power Sources* **2014**, *265*, 254–261. [CrossRef]
20. Zhu, J.G.; Sun, Z.C.; Wei, X.Z.; Dai, H.F. A new lithium-ion battery internal temperature on-line estimate method based on electrochemical impedance spectroscopy measurement. *J. Power Sources* **2014**, *274*, 990–1004.
21. Sun, F.; Xiong, R.; He, H. Estimation of state-of-charge and state-of-power capability of lithium-ion battery considering varying health conditions. *J. Power Sources* **2014**, *259*, 166–176. [CrossRef]
22. He, H.; Xiong, R.; Fan, J. Evaluation of lithium-ion battery equivalent circuit models for state of charge estimation by an experimental approach. *Energies* **2011**, *4*, 582–598. [CrossRef]
23. Zhang, Y.; Xiong, R.; He, H.; Shen, W. A lithium-ion battery pack state of charge and state of energy estimation algorithms using a hardware-in-the-loop validation. *IEEE Trans. Power Electron.* **2016**. [CrossRef]
24. Sun, F.; Xiong, R.; He, H. A systematic state-of-charge estimation framework for multi-cell battery pack in electric vehicles using bias correction technique. *Appl. Energy* **2016**, *162*, 1399–1409. [CrossRef]
25. Xiong, R.; Sun, F.; Chen, Z.; He, H. A data-driven multi-scale extended Kalman filtering based parameter and state estimation approach of lithium-ion polymer battery in electric vehicles. *Appl. Energy* **2014**, *113*, 463–476. [CrossRef]
26. Xiong, R.; Sun, F.; Gong, X.; Gao, C. A data-driven based adaptive state of charge estimator of lithium-ion polymer battery used in electric vehicles. *Appl. Energy* **2014**, *113*, 1421–1433. [CrossRef]
27. Do, D.V.; Forgez, C.; Benkara, K.E.K.; Friedrich, G. Impedance observer for a Li-ion battery using Kalman filter. *IEEE Trans. Veh. Technol.* **2009**, *58*, 3930–3937.
28. Huang, W.; Qahouq, J.A.A. An online battery impedance measurement method using dc–dc power converter control. *IEEE Trans. Ind. Electron.* **2014**, *61*, 5987–5995. [CrossRef]
29. Mcdonald, J.R. *Impedance Spectroscopy Emphasizing Solid Materials and Systems*; Wiley-Interscience: New York, NY, USA, 1987.
30. Siebert, W.M. *Circuits, Signals, and Systems*; MIT Press: Cambridge, MA, USA; McGraw-Hill Book Company: New York, NY, USA, 1986.
31. Momma, T.; Matsunaga, M.; Mukoyama, D.; Osaka, T. AC impedance analysis of lithium ion battery under temperature control. *J. Power Sources* **2012**, *216*, 304–307. [CrossRef]
32. Zhang, S.S.; Xu, K.; Jow, T.R. Electrochemical impedance study on the low temperature of Li-ion batteries. *Electrochim. Acta* **2004**, *49*, 1057–1061. [CrossRef]
33. Barai, A.; Chouchelamane, G.H.; Guo, Y.; Mcgordon, A.; Jennings, P. A study on the impact of lithium-ion cell relaxation on electrochemical impedance spectroscopy. *J. Power Sources* **2015**, *280*, 74–80. [CrossRef]
34. Schindler, S.; Bauer, M.; Petzl, M.; Danzer, M.A. Voltage relaxation and impedance spectroscopy as in-operando methods for the detection of lithium plating on graphitic anodes in commercial lithium-ion cells. *J. Power Sources* **2016**, *304*, 170–180. [CrossRef]
35. Waag, W.; Fleischer, C.; Sauer, D.U. Critical review of the methods for monitoring of lithium-ion batteries in electric and hybrid vehicles. *J. Power Sources* **2014**, *258*, 321–339. [CrossRef]

energies

MDPI

Article

A Dynamic Control Strategy for Hybrid Electric Vehicles Based on Parameter Optimization for Multiple Driving Cycles and Driving Pattern Recognition

Zhenzhen Lei [1,2], Dong Cheng [1], Yonggang Liu [1,2,*], Datong Qin [1], Yi Zhang [3] and Qingbo Xie [1]

[1] State Key Laboratory of Mechanical Transmissions & School of Automotive Engineering, Chongqing University, Chongqing 400044, China; zhenlei@umich.edu (Z.L.); chengdong_1991@163.com (D.C.); dtqin@cqu.edu.cn (D.Q.); xie_qingbo@126.com (Q.X.)
[2] Key Laboratory of Advanced Manufacture Technology for Automobile Parts, Ministry of Education, Chongqing University of Technology, Chongqing 400054, China
[3] Department of Mechanical Engineering, University of Michigan-Dearborn, Dearborn, MI 48128, USA; anding@umich.edu
* Correspondence: andylyg@umich.edu; Tel.: +86-186-9665-0900

Academic Editor: Hailong Li
Received: 5 August 2016; Accepted: 26 December 2016; Published: 5 January 2017

Abstract: The driving pattern has an important influence on the parameter optimization of the energy management strategy (EMS) for hybrid electric vehicles (HEVs). A new algorithm using simulated annealing particle swarm optimization (SA-PSO) is proposed for parameter optimization of both the power system and control strategy of HEVs based on multiple driving cycles in order to realize the minimum fuel consumption without impairing the dynamic performance. Furthermore, taking the unknown of the actual driving cycle into consideration, an optimization method of the dynamic EMS based on driving pattern recognition is proposed in this paper. The simulation verifications for the optimized EMS based on multiple driving cycles and driving pattern recognition are carried out using Matlab/Simulink platform. The results show that compared with the original EMS, the former strategy reduces the fuel consumption by 4.36% and the latter one reduces the fuel consumption by 11.68%. A road test on the prototype vehicle is conducted and the effectiveness of the proposed EMS is validated by the test data.

Keywords: hybrid electric vehicles (HEVs); energy management strategy (EMS); particle swarm optimization (PSO); multiple driving cycles; driving pattern recognition

1. Introduction

To meet user demands for vehicle power performance, the parameters of hybrid electric vehicles (HEVs) are optimized to maintain the battery state of charge (SOC) and reduce the vehicle fuel consumption. This is not only related to the design parameters of the power system, but also the control parameters of the energy management strategy (EMS). To improve HEV performance in terms of fuel economy and ensure excellent driving performance, the simultaneous optimization for the main parameters of powertrain components and control system is necessary [1]. Recently, numerous works have been proposed to find the best solution. The genetic algorithm is used for the optimization of HEV control parameters which effectively improves the fuel economy [2–5]. The energy management algorithms based on adaptive multi-operating modes proposed in [6] solve the problem that different driving cycles should be provided with different control algorithms. Besides, the matching method of the powertrain based on driving cycles is presented for fuel cell

HEVs [7]. The above optimization algorithms are used to optimize the parameters of the power system or energy control strategy of HEVs. Several algorithms have been employed to optimize the parameters of both the power system and control strategy, such as the particle swarm optimization (PSO) algorithm [8–10] and multi-objective genetic algorithm [11]. A genetic algorithm with simulated annealing is proposed in [12] to balance between economy and dynamic performance. The DIRECT algorithm global optimization method has been used for calibrating the parameters of the vehicle EMS from the perspective of fuel economy [13]. Compared with the mentioned optimization algorithms, simulated annealing particle swarm optimization (SA-PSO) has the advantages of achieving a global optimal solution [14]. It is difficult but necessary to develop a set of global optimal solutions for the simultaneous optimization of power system and control parameters.

It's well known that the effectiveness of EMS for HEVs is greatly influenced by the driving patterns. However, the optimized parameters of HEVs based on a certain driving pattern may not maintain the battery *SOC* balance in other patterns, not to mention the best fuel consumption [15]. Therefore, energy management strategies based on driving pattern recognition have recently been put forward in the literature [16,17]. To optimize the vehicle performance on a random driving pattern, a multi-mode driving control algorithm using driving pattern recognition is developed for HEVs [18,19]. An intelligent energy management for parallel HEV based on driving cycle identification is proposed using a fuzzy logic controller or fuzzy neural network [20–22]. The machine-learning methods intelligently and automatically discriminate between the driving conditions [23,24]. To solve the multi-objective optimization problem for the longevity and energy efficiency of the energy storage system, a new optimization framework for determining an instantaneously optimized power management strategy has been proposed by Zhang et al. in [25], which shows excellent real-time power optimization performance against unknown diving cycles and operating conditions.

In these studies, the parameter optimization of the energy storage system, which is also very important for the effectiveness of EMS, is not taken into consideration. As mentioned above, the advantage of SA-PSO compared with other optimization algorithms is that it can obtain global optimization results, so it is meaningful to utilize the SA-PSO to realize the parameter optimization based on multiple driving cycles. Meanwhile the EMS based on driving pattern recognition should take advantage of the optimized parameters. However, few works have comprehensively analyzed how to combine the optimized parameters with the EMS based on driving pattern recognition. Besides, the EMS based on driving pattern recognition should emphasize more the influence of the variation range of battery *SOC* while focusing on the vehicle fuel economy. In general, the simultaneous optimization for parameters of power system and control strategy on this premise of maintaining balance of the battery *SOC* is worth studying and meaningful to improve the fuel economy.

In this paper, a new methodology for parameter optimization using a SA-PSO algorithm is proposed to pursue the best fuel consumption without impairing the dynamic performance. The parameters of the power system and control strategy for HEV are both optimized based on multiple driving cycles. In addition, an algorithm of the dynamic EMS based on driving pattern recognition is proposed in this paper. Twenty-three typical driving cycles from ADVISOR (2002, National Renewable Energy Laboratory, Golden, CO, USA) have been selected and classified according to the clustering analysis method through the *Euclidean distance*. Furthermore, the *Euclid approach degree* is used to realize the driving pattern recognition. The control parameters have been optimized at each class of driving patterns based on the optimization of multiple driving cycles. The proposed energy management strategies based on parameter optimization under multiple driving cycles and driving pattern recognition are both simulated on the Matlab/Simulink (R2010a, MathWorks, Natick, MA, USA) platform under the comprehensive driving cycles. Furthermore, road tests of the prototype vehicle with the proposed control strategy are conducted. The results of both the simulation and road tests validate the effectiveness of the proposed control strategies.

2. Hybrid Electric Vehicle (HEV) Rule-Based Energy Management Control Strategy

The hybrid power system considered in this paper is a typical parallel Integrated Starter and Generator (ISG) hybrid system, as shown in Figure 1. The engine and ISG motor are connected through a master clutch, and either of them can drive the vehicle alone. The ISG motor can also be used as a generator to charge the battery.

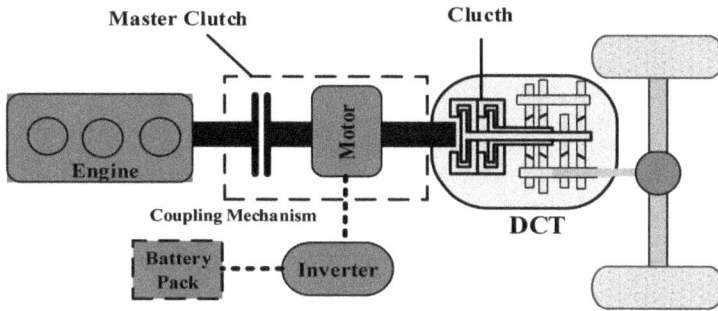

Figure 1. Configuration of the integrated starter and generator (ISG) type hybrid electric vehicle (HEV).

As shown in Figure 2, the basic control strategy in this paper is a rule-based logic threshold EMS which relies on several modes or states of operation and its decision to change modes is dependent on the power requirement of acceleration or deceleration, the *SOC* of the energy storage unit, and the vehicle speed [26,27]. In order to ensure that the engine operates more in high efficiency regions, in this paper, the coefficients of the engine torque in high efficiency regions (F_{up} and F_{low}) are designed to obtain the maximum and minimum engine torques based on the existing results presented in [28]. As shown in Figure 2a, when the battery *SOC* is higher than the low limit SOC_{low} and if the required speed is less than a certain value V_1, the vehicle will operate at pure electric mode. When the battery *SOC* is lower than SOC_{low} in Figure 2b, an additional torque T_{chg} is required from the engine to charge the battery. Therefore, the revised rule-based EMS is proposed as shown in Table 1. The parameters of the control strategy are shown in Table 2.

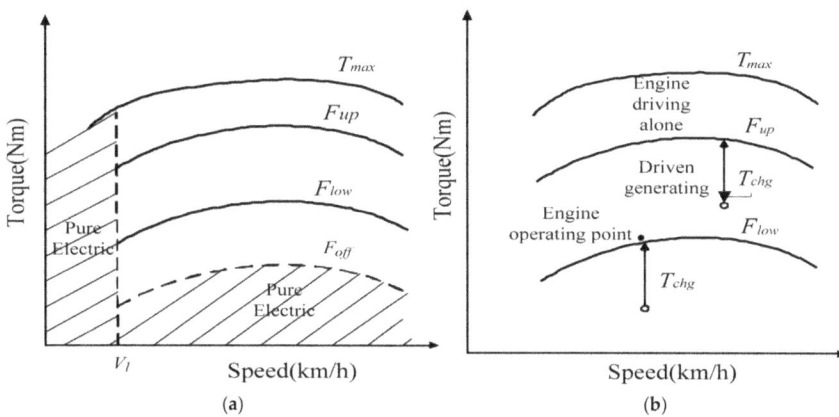

Figure 2. Logic diagram of control strategy. (**a**) $SOC > SOC_{low}$; (**b**) $SOC \leq SOC_{low}$.

Table 1. Revised rule-based energy management strategy (EMS).

Operating Mode	Constraint Condition	Torque Distribution
Electric Driving Mode	$0 < T_v \leq T_{off}$ $SOC_{up} > S\ddot{O}C > SOC_{low}; V > V_l$ $SOC_{up} > SOC > SOC_{low}; V \leq V_l$ $SOC > SOC_{up}$	$T_m = T_v; T_e = 0$
Driving & Charging Mode	$T_{off} < T_v \leq T_{low}$ $S\ddot{O}C_{low} \leq SOC \leq SOC_{up}; V > V_l$ $0 \leq T_v \leq T_{up}$	$T_e = T_{low}$ $T_m = T_v - T_{low}$ $T_e = T_{up}$ $T_m = \max(T_v - T_{up}, T_{chg\ max})$
Engine Driving Mode	$T_{low} \leq T_v \leq T_{up}$ $SOC_{up} > SOC > SOC_{low}; V > V_l$ $T_{up} \leq T_v; SOC \leq SOC_{low}$	$T_e = T_v; T_m = 0$
Motor Driving Mode	$T_{up} < T_v; SOC > SOC_{low}; V > V_l$	$T_e = T_{up}; T_m = T_v - T_{up}$
Regenerative Braking Mode	$T_v \leq 0; SOC < SOC_{up}$	$T_v = T_m + T_{mechanic}$
	$T_v \leq 0$ and $SOC > SOC_{up}$	$T_v = T_{mechanic}$

Table 2. Parameters of control strategy.

Name	Unit	Description
SOC_{up}	-	Maximum expectation of battery SOC
SOC_{low}	-	Minimum expectation of battery SOC
V	km/h	Current speed
V_l	km/h	Speed floor. When $SOC > SOC_{low}$ and $V < V_l$, pure electric mode starts
T_{max}	Nm	Maximum steady-state torque of engine
F_{off}	-	Engine off torque coefficient, $T_{off} = T_{max} \times F_{off}$
F_{low}	-	Minimum torque coefficient of engine in high efficiency regions, $T_{low} = T_{max} \times F_{low}$
F_{up}	-	Maximum torque coefficient of engine in high efficiency regions, $T_{up} = T_{max} \times F_{up}$
T_{chg}	Nm	Active charging torque of ISG motor. $T_{chg\ max}$ is the maximum charging torque of motor
T_v	Nm	Vehicle demand torque
T_m	Nm	Output torque of the ISG
T_e	Nm	Output torque of the engine
$T_{mechanic}$	Nm	Mechanic braking torque

3. Power System and Control Strategy Parameter Optimization Based on Multiple Driving Cycles

3.1. Basic Idea for Parameter Optimization

To pursue the best fuel consumption under actual driving cycle conditions, the parameter optimization of the power system and control strategy of HEV based on multiple driving cycles has been proposed. Six types of typical cycles are employed, considering the influence of urban congestion, suburban and highway conditions. The constraints of dynamic performance for the vehicle are shown in Table 3. The six types of driving cycles are shown in Table 4. The parameters of the vehicle's power system are shown in Table 5. The optimization method for the main parameters of power system and control strategy based on multiple driving cycles is generalized as follows:

(1) The assumption that the revised rule-based EMS is used for HEV.
(2) The initial parameters of power system and control strategy are selected and their values are chosen.
(3) Six types of driving cycles are selected and combined into a comprehensive driving cycle.
(4) The simultaneous optimization for the main parameters of power system and control strategy is carried out using SA-PSO algorithm with vehicle performance constraints.
(5) The optimal power system and control parameters are applied to the HEV EMS.

Table 3. Constraints of dynamic performance for the HEV.

Max. Speed		Max. Slope of Climb	Acceleration Time from 0 to 100 km/h
km/h		%	s
160 (Engine Driving Mode)	≥50 (Electric Driving Mode)	≥30 (Engine Driving Mode)	≤12 (Hybrid Driving Mode)

Table 4. Six types of typical driving cycles.

Mode	FTP	LA92	SC03	UDDS	HWFET	US06_HWY
Type	urban congestion		suburban		highway	

Table 5. Power source parameters of the HEV.

Description	Engine (P_{IC})	ISG Motor (P_{ISG})
Max Power (kW)	72	30
Max Torque (Nm)	137	115

The diagram of optimization method based on multiple driving cycles is shown in Figure 3.

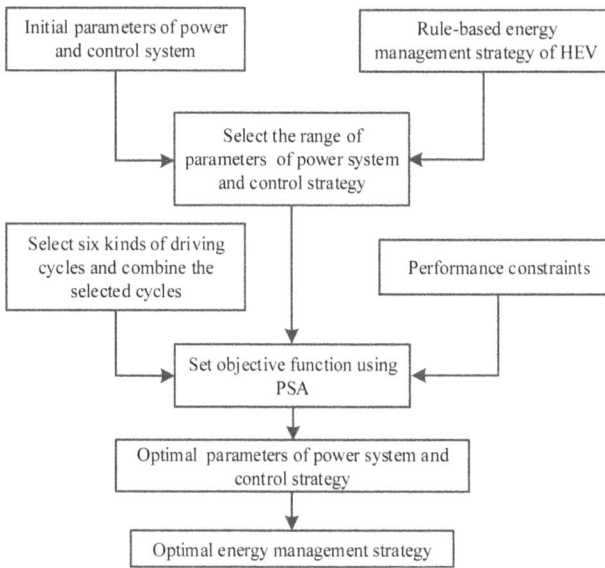

Figure 3. Diagram of optimization method under multiple driving cycles.

3.2. Parameter Definition of Power System and Control Strategy

The parameters of the power system and control strategy in terms of engine power (P_{IC}) and ISG power (P_{ISG}) are optimized in this paper to make sure that the engine and motor work in high efficiency regions on the premise of satisfying the requirements of vehicle dynamic performance. The variation of each design parameter of the power system (P_{IC} and P_{ISG}) is considered ±70% about the initial values, according to the results presented in [15]. The control parameters (F_{low}, F_{up}, F_{off}, SOC_{low}, SOC_{up} and V_l) are designed to ensure that the engine can work in high efficiency regions without interference with each other, as shown in Table 6. The initial values of selected parameters are obtained from the prototype vehicle.

Table 6. Variation of each parameter.

Optimal Variable	Initial Value	Variation Range
P_{IC} (kW)	72.0	21.6–122.4
P_{ISG} (kW)	30.0	9–51
F_{low}	0.6	0.43–0.73
F_{up}	0.9	0.75–0.93
F_{off}	0.235	0.2–0.4
SOC_{low}	0.25	0.2–0.4
SOC_{up}	0.8	0.75–0.9
V_l	32	10–50

3.3. State of Charge-Fuel Consumption Correction Method

In order to eliminate the influence of *SOC* on the vehicle fuel consumption evaluation, the battery *SOC* correction method should be used to correct fuel economy in the case initial and final battery *SOC* are not the same during a driving cycle. The *SOC*-fuel consumption correction method used in this paper is as follows:

$$\Delta fuel = \frac{\Delta SOC \cdot Q_{cap} \cdot \overline{U_{bat}} \cdot \overline{\eta_{eng\_chg}}}{1000 \cdot \rho} \tag{1}$$

where $\Delta fuel$ is the equivalent fuel consumption (L), ΔSOC is the variation of battery *SOC* between the starting and ending points, Q_{cap} is the total battery capacity (Ah), $\overline{U_{bat}}$ is the average battery bus voltage during drive cycles (V), $\overline{\eta_{eng\_chg}}$ is the average the engine power efficiency (g/kWh), and ρ is the gasoline density (g/L).

3.4. Optimization Objective Function

Taking the characteristics of different driving cycles into consideration, the target of parameter optimization of the power system and energy management control strategy is to achieve a set of optimal parameters to reduce fuel consumption as much as possible without impairing the dynamic performance. The fuel consumption is the optimization objective with the dynamic performance as the constraint. In order to prevent the excessive variation of battery *SOC* (ΔSOC), and specifically avoid exceeding the lower limit of *SOC* range, the weight coefficient of $\Delta fuel$ under different driving cycles is set to enable the motor to drive alone. The fitness function is as follows:

$$\text{Min } f(x) = \int Fuel_{use(t)} dt + \sum_{i=1}^{6} w_i \cdot |\Delta fuel_i| \tag{2}$$

$$s.t. \ u_j(x) \geq 0 \quad j = 1,2,3,...,m$$
$$x_i^l \leq x_i \leq x_i^k \quad i = 1,2,3,...,n$$

where $u_j(x)$ are the constraint conditions of vehicle dynamic performance (e.g., maximum speed and accelerating ability) as shown in Table 3, n is the number of optimization variables, which equals 8 in this study, x_i^l and x_i^k are the upper and lower bounds on the optimization variables respectively.

Considering the difference of the speed range and mileage of each driving cycle, the weight coefficients w_i of $\Delta fuel$ under driving cycles HWFET, FTP, LA92, US06_HWY, UDDS, SC03 through enumerative technique based on experience and simulation are chosen as 1.0, 1.0, 1.5, 1.3, 1.3 and 1.0, respectively.

3.5. Parameter Optimization for HEV Based on Simulated Annealing Particle Swarm Optimization Algorithm

The SA-PSO algorithm, firstly introduced by Metropolis et al. [29], is an optimization algorithm which combines the PSO with the Simulated Annealing method. This method has high efficiency in searching the global minimum value and the characteristics that it is easily realizable and has the advantages of both SA and PSO algorithms [30]. The particle swarm will gravitate towards the optimum solution after continuous iterations. All particles' positions and velocities are updated according to the following formulas:

$$v_i^{t+1} = w(t)v_i^t + c_1r_1(p_i^t - x_i^t) + c_2r_2(p_{gi}^t - x_i^t) \tag{3}$$

$$x_i^{t+1} = x_i^t + v_i^{t+1} \tag{4}$$

where p_i^t is the individual best optima for particle i after t iterations, p_{gi}^t is the group optima after t iterations, $w(t)$ is the inertia weight, c_1 and c_2 are two positive constants, $r_1 \in [0, 1]$ and $r_2 \in [0, 1]$ are two random parameters independent of each other, v_i^t is the velocity of particle i in iterative t, and x_i^t is the position of particle i in iterative t.

Based on the above analysis, the complete SA-PSO algorithm flowchart is shown in Figure 4.

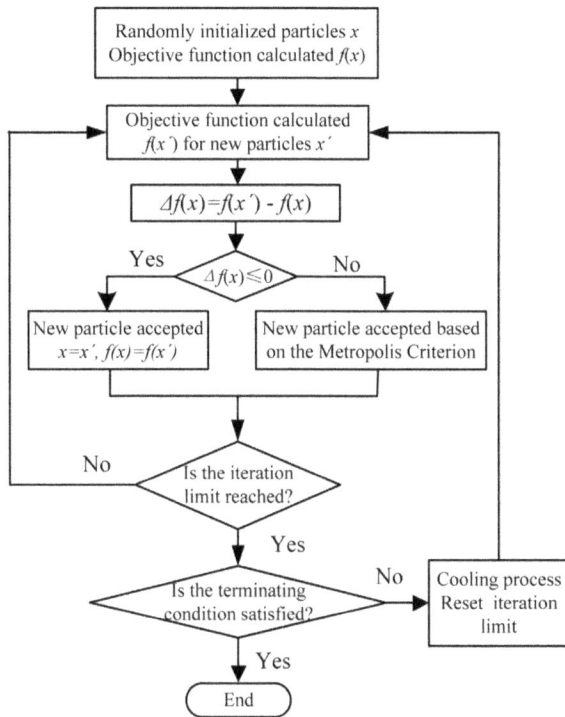

Figure 4. Optimization model based on the simulated annealing particle swarm optimization algorithm.

The detailed procedure of SA-PSO algorithm for parameter optimization is explained as follows:

Step 1: Initialize a group of random particles. The inertia should be chosen to provide a balance between the global and local exploration. The initialization consists of the following major parameters:

- Generation number: 25 Constants; c_1 and c_2: 2.05, 2.05; Initial temperature T: 9000 °C; Final temperature T_0: 0.05 °C; Anneal speed K: 0.9.

Step 2: Calculate and update the fitness function $f(x)$ of all particles. Determine p_i^t and p_{gi}^t of the current generation. Update new velocities and positions of each particle according to Equations (3) and (4).

Step 3: Calculate the difference between the optimal and non-optimal function value $\Delta f(x)$. Accept the optimal solution if $\Delta f(x)$ is greater than 0, otherwise generate a random number r within $(0, 1)$. When r is lower than $\min[1, \exp(-\Delta f(x)/t)]$, accept the optimal solution and go to Step 2, or go to the next step.

Step 4: Introduce the simulated annealing mechanism. Stop the program and output the optimal solution if the convergence criteria is satisfied, otherwise carry out the annealing process and the command "$T = 0.9 \times T$".

3.6. Simulation of Optimal Parameters Based on Multiple Driving Cycles

The simulation studies for the vehicle fuel economy are carried out using the Matlab/Simulink platform. The selected six types of driving cycles (HWFET, FTP, LA92, US06_HWY, UDDS, SC03) are successively combined into a comprehensive cycle according to driving cycles. The time-speed relationship of the comprehensive driving cycle is shown in Figure 5. The eight parameters of the power system and control strategy are optimized by the SA-PSO algorithm based on the comprehensive driving cycle, and the optimization results of the parameters are shown in Table 7.

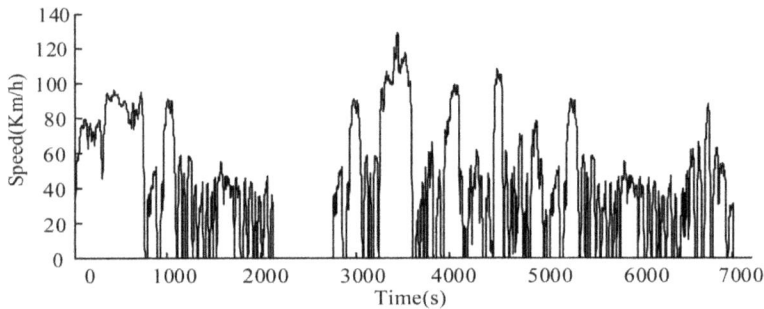

Figure 5. Time-speed relationship of the comprehensive driving cycle.

Table 7. Comparison of optimization results.

Optimal Variable	Initial Value	Optimal Value
P_{IC} (kW)	72.0	67.0
P_{ISG} (kW)	30.0	26.0
F_{low}	0.6	0.48
F_{up}	0.9	0.90
F_{off}	0.235	0.23
SOC_{low}	0.25	0.30
SOC_{up}	0.8	0.78
V_l	32	35.06

The optimized parameters satisfy the requirements of vehicle dynamic performance. The variation of the battery *SOC* and engine operation points of HEVs are simulated under the comprehensive cycle conditions, as shown Figures 6 and 7. The variation of battery *SOC* stays within 0.05 which meets the requirements for HEV in terms of the battery *SOC* consistency. Meanwhile, the battery *SOC* always fluctuates around the initial *SOC* value, which enables the battery to work in its high charging/discharging efficiency region. Furthermore, the engine can work in its high efficiency region and the vehicle can be driven in the electric driving mode with low speed and torque, which effectively improves the overall efficiency of whole system.

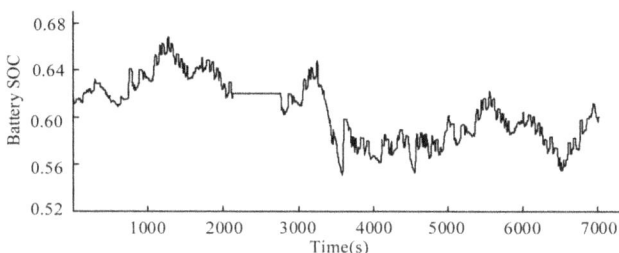

Figure 6. Variation of battery *SOC* under the comprehensive cycle conditions.

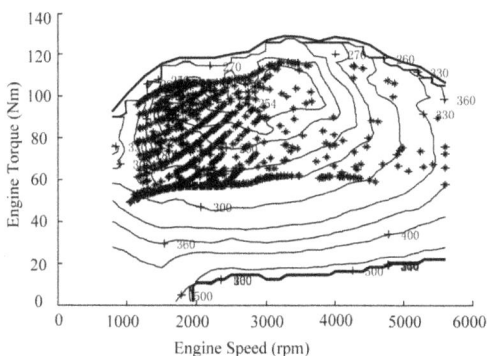

Figure 7. Engine operating points under the comprehensive cycle conditions.

4. HEV Dynamic Control Strategy Based on Driving Pattern Recognition

As mentioned above, the control parameters optimization based on multiple driving cycles is analyzed under known driving cycle conditions. However, in practice, the vehicle actual driving cycle is a random and uncertain process. In order to achieve better fuel economy, the EMS of HEVs based on driving pattern recognition is proposed after the parameter optimization under multiple driving cycles, which can optimize the control parameters in vehicle real-time control.

The diagram of EMS for HEVs based on driving pattern recognition is shown in Figure 8. Firstly, the characteristic parameters of different typical driving cycles are picked up, which are used for the clustering analysis. The control parameters of each class of the driving cycle are optimized offline based on multiple driving cycles as mentioned in Section 3. The driving pattern recognition has been realized using the *Euclid approach degree*. At last, the dynamic energy management control strategy for HEVs based on driving pattern recognition is achieved for vehicle real-time control.

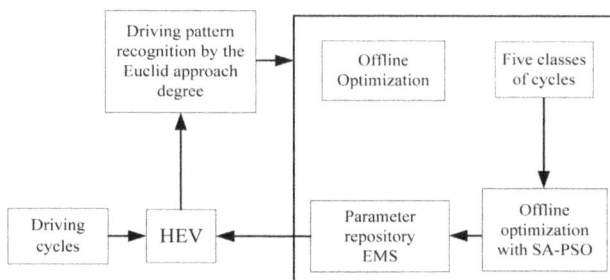

Figure 8. Diagram of EMS based on driving pattern recognition.

4.1. Selection and Classification of Characteristic Parameters for Typical Patterns

In view of the variety and complexity of vehicle driving patterns, it is significant to take all types of driving patterns into account. However, this is impractical due to the massive workload and limitation of calculation ability. Therefore, twenty-three typical driving cycles from ADVISOR are used as the research object. These driving cycles shown in Table 8 are Mode 1: JPN1015; Mode 2: ARTERIAL; Mode 3: CBD14; Mode 4: CBDTRUCK; Mode 5:COMMUTER; Mode 6: ECE_EUDC; Mode 7: HL07; Mode 8: LA92; Mode 9: MANHATTAN; Mode 10: NYCC; Mode 11: NYCCOMP; Mode 12: NYCTRUCK; Mode 13: NurembergR36; Mode 14: REP05; Mode 15: SC03; Mode 16: UDDS; Mode 17: UDDSHDV; Mode 18: US06_HWY; Mode 19: WVUCITY; Mode 20: WVUSUB; Mode 21: ARB02; Mode 22: ECE; Mode 23: IM240.

Table 8. Related characteristic parameters of twenty-three typical cycles.

Mode	v_{max}	v_{avg}	a_{max}	d_{max}	a_{avg}	d_{avg}	r_i
1	69.97	22.68	0.79	0.83	0.57	0.65	0.32
2	64.37	39.70	1.07	2.01	0.60	1.79	0.22
3	32.19	20.42	0.98	2.06	0.81	1.79	0.214
4	32.19	14.85	0.36	0.62	0.29	0.56	0.187
5	88.51	70.28	1.03	2.01	0.28	1.89	0.122
6	119.99	32.11	1.05	1.39	0.54	0.79	0.277
7	128.75	85.75	3.58	2.55	1.29	0.80	0.097
8	108.15	39.61	3.08	3.93	0.67	0.75	0.163
9	40.72	10.98	2.06	2.50	0.54	0.67	0.362
10	44.58	11.41	2.68	2.64	0.62	0.61	0.351
11	57.94	14.10	4.11	3.88	0.48	0.54	0.331
12	54.72	12.15	1.96	1.87	0.55	0.65	0.52
13	53.70	14.34	1.88	2.11	0.58	0.55	0.31
14	129.23	82.88	3.79	3.19	0.44	0.50	0.034
15	88.19	34.50	2.28	2.73	0.50	0.60	0.195
16	91.25	31.51	1.48	1.48	0.51	0.58	0.189
17	93.34	30.32	1.96	2.07	0.48	0.58	0.333
18	129.23	97.91	3.08	3.08	0.34	0.41	0.033
19	57.65	13.58	1.14	3.24	0.30	0.39	0.303
20	72.10	25.86	1.30	2.16	0.33	0.42	0.252
21	129.20	70.03	3.53	3.62	0.66	0.70	0.075
22	49.99	18.26	1.06	0.83	0.64	0.74	0.33
23	91.23	47.07	1.47	1.56	0.44	0.68	0.05

There have been some works in the literature about the selection of characteristic parameters for typical cycles [20–24]. Based on the relative importance of each parameter in driving pattern recognition, seven parameters are chosen as the characteristic parameters of driving pattern recognition in this paper. They are the maximum vehicle speed v_{max}, average vehicle speed v_{avg}, maximum acceleration a_{max}, maximum deceleration d_{max}, average acceleration a_{avg}, average deceleration d_{avg} and engine idle time ratio r_i, respectively. The *clustering analysis* is used for classification of the typical cycles. The distance between each two driving patterns of the twenty-three typical ones is calculated by the characteristic parameters with *Euclidean distance*, as expressed by Equation (5):

$$\begin{cases} \|y_i - y_j\| = \sqrt{\left(y_{i1} - y_{j1}\right)^2 + \left(y_{i2} - y_{j2}\right)^2 + \ldots + \left(y_{im} - y_{jm}\right)^2} \\ = \sqrt{\sum_{m=1}^{10} \left(y_{im} - y_{jm}\right)^2} \\ i \neq j \cap i, j \in Z^+ \cap i, j \in [1, 23] \end{cases} \quad (5)$$

Before the calculation, the feature matrix of driving cycles and under-recognition cycles should be dealt with by min–max *Normalization* due to the inconsistency between the physical dimension and quantity of feature vectors of driving cycles, as described in Equation (6):

$$y_i' = \frac{y_i - y_{min}}{y_{max} - y_{min}} \tag{6}$$

where y_i is the original variable, y_{min} is the minimum value of unscaled variable and y_{max} is the maximum value of an unscaled variable. The feature vectors are scaled to the closed interval [0, 1].

The clustering-feature tree shown in Figure 9 is obtained through Statistical Product and Service Solutions (SPSS) software (20.0, IBM SPSS, New York, NY, USA). As the clustering scale of samples decreases and the sample space is more subtly divided, the driving cycles of each category become higher. In this paper, in order to ensure the similarity of each driving cycle and the accuracy of the classes, the twenty-three types of typical driving patterns are divided into five classes when the scale of clustering distance is 0.057 (the first class includes 6, 9–13, 19; the second class includes 1, 8, 17, 20, 22; the third class includes 4, 21, 23; the forth class includes 3, 7, 14; the five class includes 5 and 18).

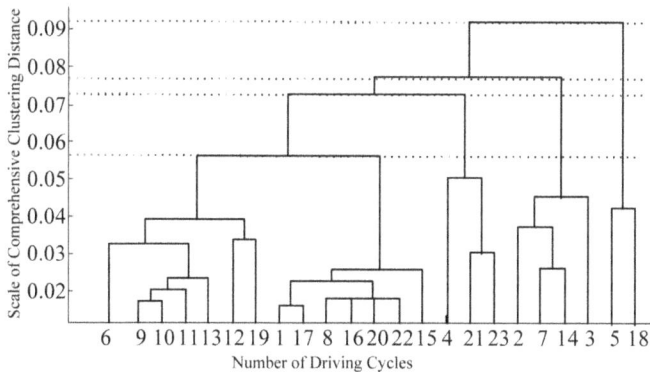

Figure 9. Clustering-feature tree of twenty-three typical patterns.

4.2. Recognition and Parameter Optimization of Driving Patterns

Although the actual vehicle driving patterns are random and uncertain, one of the twenty-three typical patterns can be selected to represent the actual driving pattern with the maximum similarity as the recognition result, and this is the basic idea of the dynamic control strategy for HEVs.

The driving pattern recognition is achieved using the *Euclid approach degree*. The representative feature vector $\mathbf{A}_n$ (n = 1, 2, ... , 23) stands for the selected twenty-three reference driving patterns, and each vector contains seven characteristic parameters of the reference driving patterns shown in Table 8. The vector $\mathbf{B}$ also contains seven characteristic parameters of the driving patterns. The distance between the feature vector of actual driving pattern and reference feature vectors is calculated by the *Euclidean distance* σ ($\mathbf{A}_n$, $\mathbf{B}$):

$$\sigma(\mathbf{A}_n, \mathbf{B}) = 1 - \frac{1}{\sqrt{m}} \left(\sum_{k=1}^{m} (\mathbf{A}_n(k) - \mathbf{B}(k))^2 \right)^{\frac{1}{2}} \tag{7}$$

where m is the number of the characteristic parameters (m = 7). In order to eliminate the deviation caused by different parameter units, parameters are standardized using the method of *Maximum magnitude* of 1.

The driving pattern showing the maximum similarity is recognized as the reference driving cycle as expressed:

$$\sigma(\mathbf{B}, \mathbf{A}_i) = \max\{\sigma(\mathbf{B}, \mathbf{A}_1), \sigma(\mathbf{B}, \mathbf{A}_2), ..., \sigma(\mathbf{B}, \mathbf{A}_n)\} \tag{8}$$

As shown in Equation (8), the result of driving pattern recognition means that the historical actual driving pattern **B** belongs to the driving pattern $\mathbf{A}_i$. In order to verify the effectiveness of driving pattern recognition, a comprehensive test driving cycle is established to represent the actual driving pattern. The comprehensive test driving cycle consists of five different types of typical cycles including NEDC, LA92, HWFET, UDDS and US06, as shown in Figure 10. An algorithm for real-time driving pattern recognition is proposed based on the assumption that the driving pattern will not change suddenly within a short period of time. This real-time driving pattern recognition algorithm can predict future driving cycles through the past sampling data analysis within a short time window. The time window for the information extraction of characteristic parameters is 120 s based on the research as presented in [31,32]. The recognition of driving patterns for each time window is realized using the *Euclid approach degree*. The result of driving pattern recognition under the comprehensive test driving cycles is shown in Figure 11.

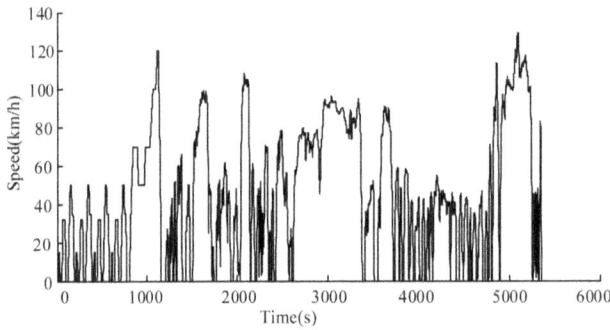

Figure 10. Time-speed relationship of the comprehensive test cycle conditions.

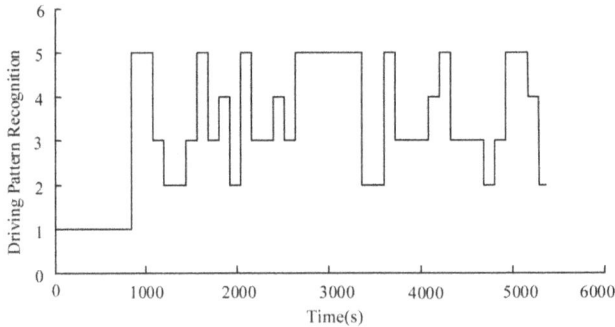

Figure 11. Result of driving pattern recognition under the comprehensive test cycle conditions.

4.3. Optimization of Control Parameters Based on Driving Pattern Recognition

In this section, the control parameters (F_{low}, F_{up}, F_{off} and V_1) of each class have been optimized based on multiple driving cycles, which has been introduced in Section 3 in detail. The optimization results of control parameters of each class are shown in Table 9.

Table 9. Optimization results of control parameters.

Classes	F_{low}	F_{high}	F_{off}	V_l
First	0.50	0.80	0.25	23.66
Second	0.48	0.90	0.23	35.06
Third	0.63	0.83	0.32	11.77
Forth	0.59	0.917	0.33	23.64
Fifth	0.68	0.80	0.40	15.98

In order to verify the effectiveness of the control parameter optimization, the driving cycles of the first class is taken as an example. The seven typical driving cycles of the first class are set as a comprehensive driving cycle (ECE_EUDC + MANHATTAN + NYCC + NYCCOMP + NYCTRUCK + NurembergR36 + WVUCITY) as shown in Figure 12. The variation of battery *SOC* (Δ*SOC*) in the first class of driving cycles is shown in Figure 13 and Table 10, where the control parameters are effective in controlling the variation of battery *SOC*.

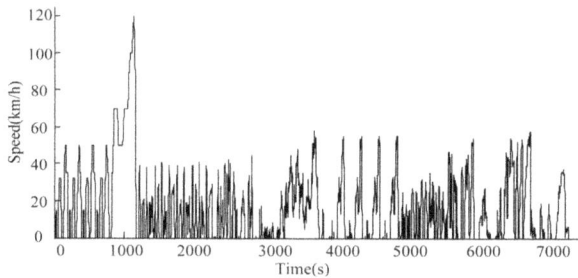

Figure 12. Time-speed relationship in the first class of driving cycles.

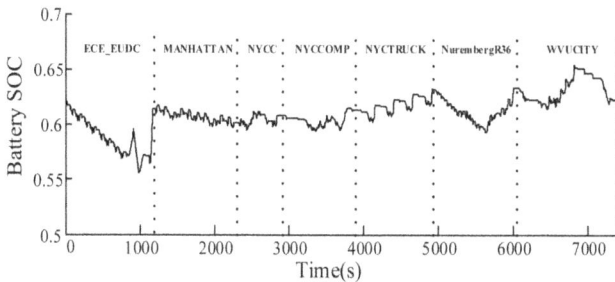

Figure 13. Variation of *SOC* in the first class of driving cycles.

Table 10. Variation of state of charge.

Mode	ECE_EUDC	MAN-HATTAN	NYCC	NYC-COMP
Δ*SOC*	−0.007	−0.017	0.01	0.009

Mode	NYC-TRUCK	NuremberR36	WVU-CITY	Comprehensive
Δ*SOC*	0.02	0.006	−0.015	0.005

5. Simulation

The proposed dynamic control strategy for HEVs based on parameter optimization at multiple driving cycles and driving pattern recognition has been simulated using the Matlab/Simulink platform under the comprehensive driving cycle (NEDC + LA92 + HWFET + UDDS + US06).

As shown in Figure 14, the variation of battery *SOC* stays within 0.01 under comprehensive driving cycle conditions with the proposed EMS based on driving pattern recognition. Meanwhile, the battery *SOC* always fluctuates around the initial *SOC* during the whole process, which enables the battery maintain to work in the high efficiency region. Compared with the EMS without driving pattern recognition, the battery *SOC* variation is more reasonable.

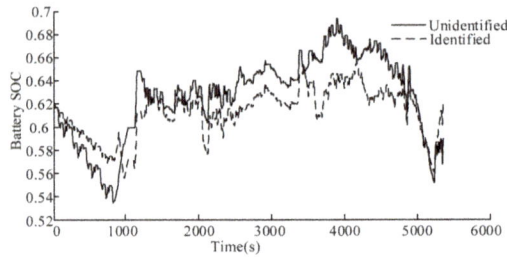

Figure 14. Variation of *SOC* under comprehensive driving cycle conditions.

Besides, the engine output power with the proposed EMS is generally larger than that with the EMS without driving pattern recognition, as shown in Figure 15. Therefore, the load of the engine is improved, which means that the engine will operate in higher efficient regions. The motor output torque at the comprehensive driving cycle is shown in Figure 16. The proposed EMS based on driving pattern recognition can adjust the control parameters to drive the vehicle in pure electric driving mode with low speed and torque, which prevents the engine from working in the low efficiency region and reduces fuel consumption. The reduction of the engine fuel consumption is shown in Figure 17.

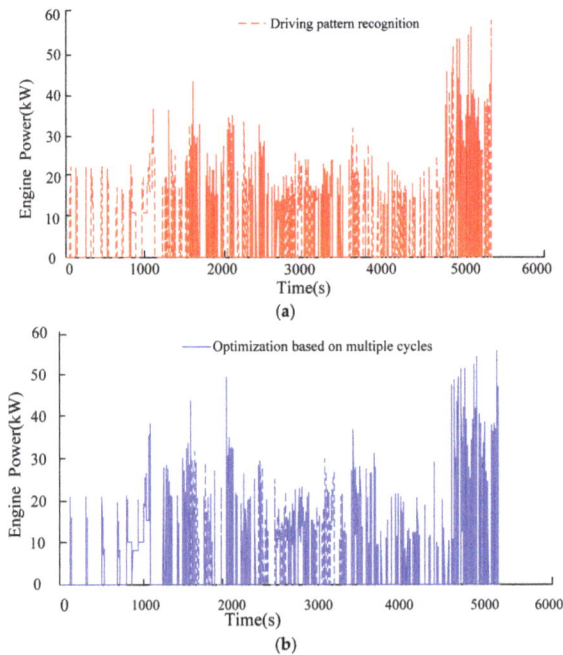

Figure 15. Engine power under the comprehensive driving cycle. (**a**) Engine power based on driving pattern recognition; (**b**) engine power based on multiple driving cycle optimization.

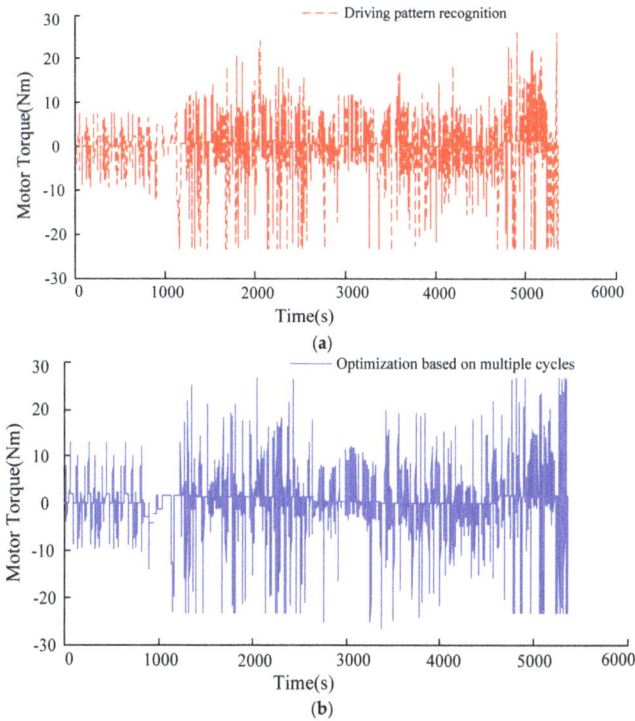

Figure 16. Motor torque under comprehensive driving cycle. (**a**) Motor power based on driving pattern recognition; (**b**) motor power based on multiple driving cycle optimization.

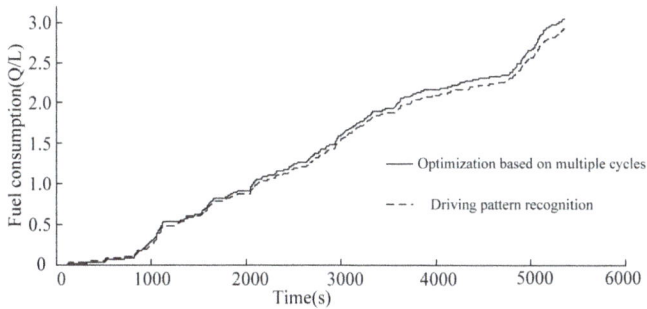

Figure 17. Fuel consumption under comprehensive driving cycle conditions.

The comparison results of fuel economy among control strategies of rule-based, multiple driving cycles optimization and driving pattern recognition are shown in Table 11 where Q_{100} is the fuel consumption of 100 km. For fuel economy comparison, *SOC* correction is very necessary. Therefore the *SOC* correction method in the SAE standards [33] is applied to compensate for the *SOC* difference. Compared with the rule-based control strategy, the fuel consumptions of energy management strategies based on multiple driving cycle optimization and driving pattern recognition are improved by 4.36% and 11.68%, respectively. Meanwhile the variation of battery *SOC* becomes smaller, which effectively improves the economic performance of the HEV vehicle.

Table 11. Fuel economy comparison results.

Factor	Rule-Based Control Strategy	Multiple Driving Cycles Optimization	Driving Pattern Recognition
Fuel Consumption (L)	3.23	3.10	2.96
Corrected Fuel Consumption (L)	3.43	3.28	3.03
Q_{100} (L/100 km)	4.75	4.55	4.36
Corrected Q_{100} (L/100 km)	5.05	4.83	4.46
Fuel Saving	-	4.24%	8.24%
Fuel Saving (*SOC* corrected)	-	4.36%	11.68%
ΔSOC	−0.28	−0.250	−0.098

6. Road Test on the Prototype Vehicle

The proposed dynamic control strategy for HEVs based on driving pattern recognition has been experimentally validated on a prototype HEV. The specifications of the prototype vehicle are shown in Table 12. The vehicle control software is developed on the Development to Production (D2P, DEV+PROD, Germany E.ON, Essen, Germany) and Matlab/Simulink platforms. The experiment is performed under the following conditions:

(1) Since the prototype HEV can only be tested on campus, for the sake of safety, the road test is only carried out at the low speed. Although the campus condition is only classified as an urban driving cycle, it is still valid to analyze the effectiveness of the proposed optimization method of HEV control strategy.
(2) The required torque during the whole test is too small compared with the maximum capacity of the HEV power system. To ensure that the vehicle operates in each mode without loss of generality, the parameters $F_{off} = 0.20$, $F_{low} = 0.44$, $F_{up} = 0.64$, $V_l = 15$ are designed as the optimal control strategy parameters according to the actual test conditions.

Table 12. Specifications of the prototype vehicle.

Main Parameter	Value
Curb weight (kg)	1350
Rated payload (kg)	1875
Effective radius (m)	0.295
Frontal area (m^2)	2.28
Maximum engine torque (Nm)	137
Nominal motor power (kW)	20
Rated voltage (V)	288

The results of the road test have been presented in Figure 18 where the vehicle speed ranges from 0 to 45 km/h. The operation modes include the electric driving mode, driving & charging mode, engine driving mode and hybrid driving mode. The prototype HEV operates at electric driving mode during the starting process, and the small required torque prevents the engine from working in the low efficiency region. The engine driving mode is mostly activated during cruising (30–35 km/h).

The effectiveness of the control strategy proposed in this paper is well verified, as seen in Figure 18. The engine can operate in the designed operating region, which effectively improves the system efficiency. Meanwhile, the battery *SOC* fluctuates smoothly and the magnitude of *SOC* variation is only 0.005, which well meets the requirements to keep the battery *SOC* as constant as possible. The engine is able to work in tandem with the motor, so as to improve the vehicle economy.

In order to show the effectiveness of the proposed algorithm better, the comparison results of road tests among different control strategies have shown in Table 13. However, during the different road rests, the vehicle can't be ensured to operate under the same working conditions among the several road tests with different control strategies. Therefore, these comparison results are roughly taken as a reference.

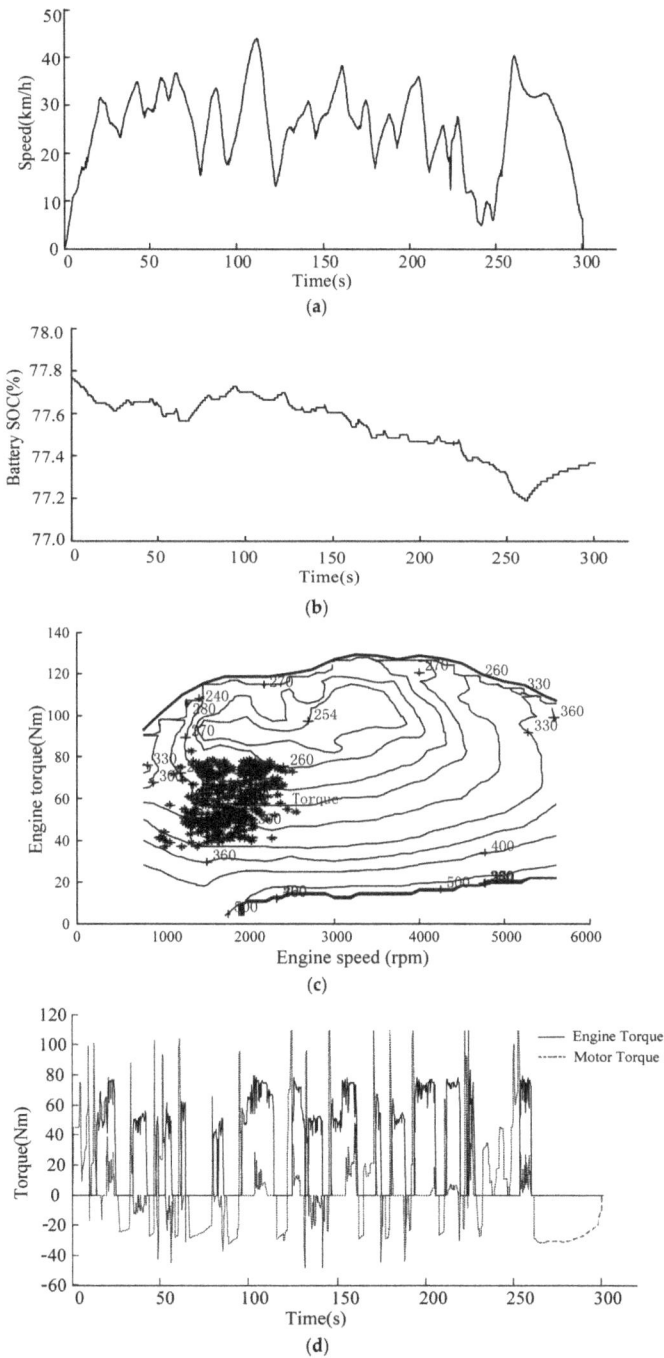

Figure 18. Road test results under the comprehensive driving cycle. (**a**) Time-speed curve of the road test; (**b**) variation of battery *SOC*; (**c**) engine operating points during testing; and (**d**) engine and motor toque distribution during testing.

Table 13. Road test comparison results.

Factor	Rule-Based Control Strategy	Multiple Driving Cycles Optimization	Driving Pattern Recognition
Total Mileage (km)	21.67	22.23	21.24
Fuel Consumption (L)	1.38	1.34	1.25
Corrected Fuel Consumption (L)	1.45	1.40	1.27
Corrected Q_{100} (L/100 km)	6.69	6.30	5.98

7. Conclusions

(1) A new methodology for parameter optimization under multiple driving cycles using SA-PSO algorithm is proposed to the simultaneous optimization for parameters of power system and control strategy. It's beneficial to achieve the best fuel consumption without impairing the dynamic performance.

(2) The EMS of HEVs based on driving pattern recognition, which optimizes the control parameters in real-time, is proposed after the parameter optimization under multiple driving cycle conditions. The proposed dynamic control strategy for HEVs based on parameter optimization under multiple driving cycles and driving pattern recognition has been simulated using Matlab/Simulink platform under the comprehensive driving cycle. Basically, the problem that the optimization based on a certain driving cycle cannot keep the battery SOC balance in other cycles has been solved in this paper.

(3) The simulation results show that compared with the original EMS, the former strategy reduces the fuel consumption by 4.36% and the latter one reduces the fuel consumption by 11.68%. The results validate the fact that the fuel consumption of EMS based on driving pattern recognition is greatly improved compared with that of the rule-based control strategy and more effective than that of multiple driving cycles. Meanwhile, the variation of battery SOC with the EMS based on driving pattern recognition is more reasonable than that of the optimization based on multiple driving cycles. It will serve as a guideline for calibrating the key parameters for road test.

(4) The proposed dynamic control strategy for HEVs based on driving pattern recognition is validated on a prototype HEV by a road test. The test results show that the EMS developed in this paper can effectively distribute the engine torque and motor torque, and significantly improve the fuel consumption of the vehicle. Furthermore, the battery SOC fluctuates smoothly and the battery SOC balance is well maintained during the test process. It will serve a reference role in dynamic control strategy for HEVs in real world.

Acknowledgments: The work presented in this paper is funded by the National Natural Science Foundation (No. 51305468), China Postdoctoral Science Foundation (No. 2016M602925XB), and the Fundamental Research Funds for the Central Universities (No. CDJZR14110005) and the Key Laboratory of Advanced Manufacture Technology for Automobile Parts, Ministry of Education (No. 2016KLMT06).

Author Contributions: Yonggang Liu provided algorithms and designed the experiments; Zhenzhen Lei wrote the paper and completed the simulation for case studies; Dong Cheng and Qingbo Xie performed the experiments and analyzed the data; Datong Qin and Yi Zhang conceived the structure and research direction of the paper.

Conflicts of Interest: The authors declare no conflict of interest.

References

1. Fang, L.; Qin, S.; Xu, G.; Li, T.; Zhu, K. Simultaneous optimization for hybrid electric vehicle parameters based on multi-objective genetic algorithms. *Energies* **2011**, *4*, 532–544. [CrossRef]
2. Montazeri-Gh, M.A. Application of genetic algorithm for optimization of control strategy in parallel hybrid electric vehicles. *J. Frankl. Inst.* **2006**, *343*, 420–435. [CrossRef]
3. Panday, A.; Bansal, H.O. Energy Management strategy for hybrid electric vehicles using genetic algorithm. *Renew. Sustain. Energy* **2016**, *8*, 741–746. [CrossRef]

4.	Wang, J.; Wang, Q.; Wang, P.; Han, B. The optimization of control parameters for hybrid electric vehicles based on genetic algorithm. In Proceedings of the SAE World Congress, Detroit, MI, USA, 8–10 April 2014.

5.	Varesi, K.; Radan, A. A Novel GA Based Technique for Optimizing Both the Design and Control Parameters in Parallel Passenger Hybrid Cars. *Int. Rev. Electr. Eng.* **2011**, *63*, 1279–1286.

6.	Wang, J.; Wang, Q.; Zeng, X.; Zhou, N.; Li, L. *The Algorithmic Research of Multi-Operating Mode Energy Management System*; SAE Technical Paper 2013-01-0988; SAE International: Warrendale, PA, USA, 2013.

7.	Gao, S.A.; Wang, X.M.; He, H.W.; Guo, H.Q.; Tang, H.L. Powertrain matching based on driving cycle for fuel cell hybrid electricvehicle. *Mech. Mater.* **2013**, *288*, 142–147. [CrossRef]

8.	Wu, J.; Zhang, C.H.; Cui, N.X. PSO Algorithn-based parameter optimization for HEV powertrain and its control strategy. *Int. J. Automot. Technol.* **2008**, *38*, 53–59. [CrossRef]

9.	Sun, F.; Xiong, R.; He, H. A systematic state-of-charge estimation framework for multi-cell battery pack in electric vehicles using bias correction technique. *Appl. Energy* **2016**, *162*, 1399–1409. [CrossRef]

10.	Chen, Z.; Xiong, R.; Wang, K.; Jiao, B. Optimal energy management strategy of a plug-in hybrid electric vehicle based on a particle swarm optimization algorithm. *Energies* **2015**, *8*, 3661–3678. [CrossRef]

11.	Fang, L.; Qin, S. Concurrent optimization for parameters of powertrain and control system of hybrid electric vehicle based on multi-objective genetic algorithms. In Proceedings of the 2006 SICE-ICASE International Joint Conference, Busan, Korea, 18–21 October 2006.

12.	Li, L.; Zhang, Y.; Yang, C.; Jiao, X.; Zhang, L.; Song, J. Hybrid genetic algorithm-based optimization of powertrain and control parameters of plug-in hybrid electric bus. *J. Frankl. Inst.* **2015**, *352*, 776–801. [CrossRef]

13.	Hao, J.; Yu, Z.; Zhao, Z.; Shen, P.; Zhan, X. optimization of key parameters of energy management strategy for hybrid electric vehicle using DIRECT algorithm. *Energies* **2016**, *9*, 997. [CrossRef]

14.	Deng, Y.W.; Chen, K.L. Simulated Annealing Particle Swarm Algorithm Based Parameters Optimization for Hybrid Electric Vehicles. *Autom. Eng.* **2012**, *34*, 580–584.

15.	Roy, H.K.; McGordon, A.; Jennings, P.A. A generalized powertrain design optimization methodology to reduce fuel economy variability in hybrid electric vehicles. *IEEE Trans. Veh. Technol.* **2014**, *63*, 1055–1070.

16.	Zhang, S.; Xiong, R. Adaptive energy management of a plug-in hybrid electric vehicle based on driving pattern recognition and dynamic programming. *Appl. Energy* **2015**, *155*, 68–78. [CrossRef]

17.	Chen, Z.; Xiong, R.; Cao, J. Particle swarm optimization-based optimal power management of plug-in hybrid electric vehicles considering uncertain driving conditions. *Energy* **2016**, *96*, 197–208. [CrossRef]

18.	Jeon, S.I.; Jo, S.; Park, Y.; Lee, J. Multi-mode driving control of a parallel hybrid electric vehicle using driving pattern recognition. *J. Dyn. Syst. Meas. Control Trans. ASME* **2002**, *124*, 141–149. [CrossRef]

19.	Lin, C.C.; Jeon, S.; Peng, H. Driving pattern recognition for control of hybrid electric trucks. *Veh. Syst. Dyn.* **2004**, *42*, 41–58. [CrossRef]

20.	Tian, Y.; Zhang, X.; Zhang, L.; Zhang, X. Intelligent energy management based on driving cycle identification using fuzzy neural network. In Proceedings of the Second International Symposium on Computational Intelligence and Design, Changsha, China, 12–14 December 2009; pp. 501–504.

21.	Dayeni, M.K.; Soleymani, M. Intelligent energy management of a fuel cell vehicle based on traffic condition recognition. *Clean Technol. Environ. Policy* **2016**, *18*, 1945–1960. [CrossRef]

22.	Wang, J.; Wang, Q.N.; Zeng, X.H.; Wang, P.Y.; Wang, J.N. Driving cycle recognition neural network algorithm based on the sliding time window for hybrid electric vehicles. *Int. J. Automot. Technol.* **2015**, *16*, 685–695. [CrossRef]

23.	Park, J.; Chen, Z.; Kiliaris, L.; Kuang, M.L.; Masrur, M.A.; Phillips, A.M.; Murphey, Y.L. Intelligent vehicle power control based on machine learning of optimal control parameters and prediction of road type and traffic congestion. *IEEE Trans. Veh. Technol.* **2009**, *58*, 4741–4756. [CrossRef]

24.	Huang, X.; Tan, Y.; He, X. An Intelligent Multifeature Statistical Approach for the Discrimination of Driving Conditions of a Hybrid Electric Vehicle. *IEEE Trans. Intell. Transp. Syst.* **2010**, *12*, 1–13. [CrossRef]

25.	Zhang, S.; Xiong, R.; Cao, J.Y. Battery durability and longevity based power management for plug-in hybrid electric vehicle with hybrid energy storage system. *Appl. Energy* **2016**, *179*, 316–328. [CrossRef]

26.	Alonso, E.; Ruiz, J.; Astruc, D. Power Management Optimization of an Experimental Fuel Cell/Battery/Supercapacitor Hybrid System. *Energies* **2015**, *8*, 6302–6327.

27.	Meintz, A.; Ferdowsi, M. Control strategy optimization for a parallel hybrid electric vehicle. In Proceedings of the 2008 IEEE Vehicle Power and Propulsion Conference (VPPC '08), Harbin, China, 3–5 September 2008.

28. Wu, L.; Wang, Y.; Yuan, X.; Chen, Z. Multiobjective optimization of HEV fuel economy and emissions using the self-adaptive differential evolution algorithm. *IEEE Trans. Veh. Technol.* **2011**, *60*, 2458–2470. [CrossRef]
29. Metropolis, N.; Rosenbluth, A.W.; Rosenbluth, M.N.; Teller, M.; Teller, E. Equation of state calculations by very fast computing machines. *J. Chem. Phys.* **1953**, *21*, 1087. [CrossRef]
30. Shieh, H.L.; Kuo, C.C.; Chiang, C.M. Modified particle swarm optimization algorithm with simulated annealing behavior and its numerical verification. *Appl. Math. Comput.* **2011**, *218*, 4365–4383. [CrossRef]
31. Johnson, V.H.; Wipke, K.B.; Rausen, D.J. HEV Control Strategy for Real Time Optimization on Fuel Economy and Emission. In Proceedings of the 2000 Future Car Congress, Arlington, VA, USA, 2–6 April 2000.
32. Pisu, P.; Rizzoni, G. A comparative study of supervisory control strategies for hybrid electric vehicles. *IEEE Trans. Control Syst. Technol.* **2007**, *15*, 506–518. [CrossRef]
33. Clark, N.; Xie, W.; Gautam, M.; Lyons, D.W.; Norton, P.; Balon, T. Hybrid Diesel-Electric Heavy Duty Bus Emissions: Benefits of Regeneration and Need for State of Charge Correction. In Proceedings of the 2000 International Fall Fuels and Lubricants Meeting and Exposition, Baltimore, MD, USA, 16–19 October 2000.

energies

MDPI

Article

Improved Battery Parameter Estimation Method Considering Operating Scenarios for HEV/EV Applications

Jufeng Yang [1,2], Bing Xia [2,3], Yunlong Shang [2,4], Wenxin Huang [1,*] and Chris Mi [2,*]

[1] Department of Electrical Engineering, Nanjing University of Aeronautics and Astronautics, Nanjing 211106, China; jufeng.yang@mail.sdsu.edu
[2] Department of Electrical and Computer Engineering, San Diego State University, San Diego, CA 92182, USA; bixia@eng.ucsd.edu (B.X.); shangyunlong@mail.sdu.edu.cn (Y.S.)
[3] Department of Electrical and Computer Engineering, University of California San Diego, San Diego, CA 92093, USA
[4] School of Control Science and Engineering, Shandong University, Jinan 250061, China
* Correspondence: huangwx@nuaa.edu.cn (W.H.); cmi@sdsu.edu (C.M.); Tel.: +86-138-5149-7182 (W.H.); +1-619-594-3741 (C.M.)

Academic Editor: Rui Xiong
Received: 3 October 2016; Accepted: 13 December 2016; Published: 22 December 2016

Abstract: This paper presents an improved battery parameter estimation method based on typical operating scenarios in hybrid electric vehicles and pure electric vehicles. Compared with the conventional estimation methods, the proposed method takes both the constant-current charging and the dynamic driving scenarios into account, and two separate sets of model parameters are estimated through different parts of the pulse-rest test. The model parameters for the constant-charging scenario are estimated from the data in the pulse-charging periods, while the model parameters for the dynamic driving scenario are estimated from the data in the rest periods, and the length of the fitted dataset is determined by the spectrum analysis of the load current. In addition, the unsaturated phenomenon caused by the long-term resistor-capacitor (RC) network is analyzed, and the initial voltage expressions of the RC networks in the fitting functions are improved to ensure a higher model fidelity. Simulation and experiment results validated the feasibility of the developed estimation method.

Keywords: lithium-ion battery; operating scenario; equivalent circuit modeling; parameter estimation

1. Introduction

Lithium-ion batteries have been widely used in the energy storage systems of hybrid electric vehicles (HEVs) and pure electric vehicles (EVs) because of their low self-discharge rate, high energy and power densities. To ensure the safe and reliable operation of lithium-ion batteries, the battery management system (BMS) is of significant importance. The main task of a BMS includes monitoring of critical states, fault diagnosis and thermal management [1–7].

1.1. Review of the Literature

The performance of a BMS is highly dependent on the accurate description of battery characteristics. Hence, a proper battery model, which can not only correctly characterize the electrochemical reaction processes, but also be easily implemented in embedded microcontrollers, is necessary for a high-performance BMS. There are two common forms of battery models available in the literature: the electrochemical model and the equivalent circuit model (ECM). The electrochemical model expresses the fundamental electrochemical reactions by complex nonlinear partial differential algebraic equations

(PDAEs) [8]. It can accurately capture the characteristics of the battery, but requires extensive computational power to obtain the solutions of the equations. Hence, such models are suitable for the battery design rather than the system level simulation. In contrast, the ECM abstracts away the detailed internal electrochemical reactions and characterizes them solely by simple electrical components; thus, it is ideal for circuit simulation software and implementation in embedded microcontrollers. The accuracy of the ECM is highly dependent on the model structure and model parameters. Theoretically, a higher order ECM can represent a wider bandwidth of the battery application and can generate more accurate voltage estimation results. However, the high order ECM can not only increase the computational burden, but also reduce the numerical stability for the further battery states' estimation [9,10]. Hence, considering a tradeoff among the model fidelity, the computational burden and the numerical stability, the second order ECM is employed in this paper [11–18]. The common structure of the second order ECM is illustrated in the top subfigure of Figure 1, where the open circuit voltage (*OCV*), which is a function of state of charge (*SoC*), stands for the open circuit voltage, R_{in} is the internal resistance, which represents the conduction and charge transfer processes [19–21], and two resistor-capacitor (RC) networks approximately describe the diffusion process. Among them, the short-term RC network models the fast dynamics diffusion process (Part A in the bottom subfigure of Figure 1), and the long-term RC network represents the slow dynamics diffusion process (Part B in the bottom subfigure of Figure 1). The above model parameters can be identified either through the time-domain or the frequency-domain parameter extraction experiments. For the time-domain parameter estimation methods, model parameters are usually identified through fitting the voltage response from the parameter extraction experiment with the exponential-based functions. The electrochemical impedance spectroscopy (EIS) test is the commonly-used frequency-domain parameter extraction experiment. Compared to the time-domain test process, one limitation of the EIS test is that the amplitude of the current excitation is so low that the battery can be considered as equalized during the whole test process, which seldom happens in HEV/EV applications. In order to overcome the above drawback, references [22–24] propose superimposing the direct current (DC) offset over the EIS signals to determine the current dependency of impedance parameters. However, since significant time is required for the EIS test, the battery *SoC* changes significantly during the test procedure if the amplitude of the superimposed current is improper. This can reduce the parameter estimation accuracy and make this method practically not applicable at moderate and high current rates [25,26]. Based on the aforementioned analysis, the second order ECM with parameters estimated by the time-domain analysis is discussed in this paper.

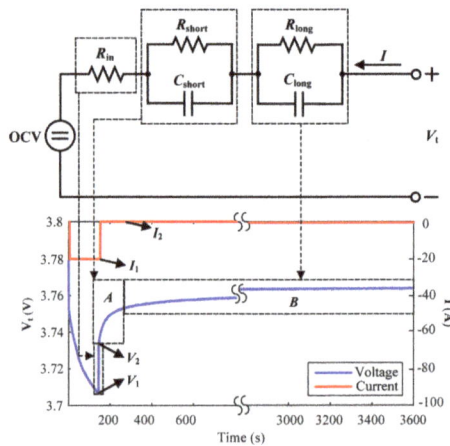

Figure 1. The second order equivalent circuit model (ECM). *OCV*, open circuit voltage.

Generally speaking, batteries usually operate in two scenarios in automotive applications: The constant-current (CC) charging scenario and the dynamic driving scenario [27]. Usually, the motions of lithium ions under the continuous external excitation (representing the CC charging scenario) and the discontinuous external excitation (representing the dynamic driving scenario) show different characteristics, and this difference is related to the diffusivity of ions. In other words, the model parameters, especially the RC network parameters, show diverse values under different operating scenarios [21,28]. Therefore, battery parameters should be identified separately according to the actual operating scenarios. Abundant research work has been conducted to seek the accurate ECM for the specific operating scenario. For the charging scenario, a universal model based on a simple mathematical equation with constant parameters is proposed [29–31]. The mathematical equations include one polynomial component and one or two exponential functions, and relevant parameters can be obtained by fitting collected charging profiles. Verification results in related literature show that the overall model output profiles match well with the experimental data, but there still exists obvious estimation errors during certain periods (at the beginning of the plateau region and the last charging region). This is mainly caused by the constant parameters during the whole charging process since the actual model parameters, such as time constants, may vary greatly at different SoC regions [32]. The works in [32–34] estimate the model parameters through the data in the rest periods of the pulse-rest test at different SoC points, and the estimated model parameters can be shown as functions of SoC. However, the charging concentration process under continuous excitation is different from the charging recovery process under the rest period [19,35]; thus, the estimated model parameters may not accurately represent the charging characteristics of the battery. For the dynamic driving scenario, many modeling approaches have been reported on the basis of the pulse discharge analysis. In [36–38], model parameters are obtained by simple algebraic operations. This is straightforward, but large estimation errors exist. A more accurate method is to fit the voltage response of the whole rest period with an exponential function [39–41]. The limitation of this method is its poor dynamic performance. In order to improve the battery model accuracy, Hu and Wang in [42] propose a two time-scale identification algorithm to separate the identifications of slow and fast battery dynamics. This method shows better frequency response matching without increasing computational complexity. Xiong in [17] uses the bias correction method to ensure the battery model prediction performance. This approach shows excellent performance and high accuracy against uncertain operating scenarios and battery packs. Instead of the conventional pulse-rest test, [43,44] propose two types of application-oriented parameter extraction tests, leading to a fast dynamics battery model with high fidelity. One major limitation of this kind of method is that the parameter extraction test corresponds to a specific operating scenario. If the actual load profiles show obviously different bandwidths under different working conditions, the parameter extraction test should be re-implemented. One solution to overcome this drawback is to conduct as many parameter extraction tests as possible to cover the typical load characteristics, but this requires an extensive amount of time and effort.

1.2. Contributions of This Paper

Based on the battery parameter estimation methods discussed above, it can be concluded that seldom does work in the previous literature discuss a battery model considering both the CC charging and dynamic driving scenarios. Hence, the focus of this paper is to propose a battery parameter estimation method, which is applicable to common operating scenarios in HEV/EV applications. The main contributions are: (1) both the constant-current charging and the dynamic driving scenarios are taken into consideration, and two separate sets of model parameters are estimated through different parts of the pulse-rest test; (2) the model parameters for the constant-current charging scenario are estimated from the data in the pulse-charging periods; (3) the model parameters for the dynamic driving scenario are estimated from the data in the rest periods, and the length of the fitted dataset is determined by the spectrum analysis of the load current; (4) the unsaturated phenomenon caused by the long-term RC network is analyzed, and the initial voltage expressions of the RC networks in the fitting functions are improved to ensure a higher model fidelity; (5) both the simulation and experiment results agree with the analysis and demonstrate the improvement of the proposed battery parameter estimation method over the existing ones.

2. Parameter Extraction Procedure

2.1. Parameter Extraction Test Design

It can be seen from Figure 1 that the second order ECM contains one *OCV-SoC* relationship and five impedance parameters (R_{in}, R_{short}, C_{short}, R_{long} and C_{long}), which need to be estimated. Theoretically, all of the impedance parameters mentioned above should be multivariable functions of *SoC*, the C-rate of the load current (C is the amplitude of the current with which the battery can be fully discharged in 1 h), temperature and cycle numbers [39,45]. These functions not only make the parameter extraction process complex and time consuming, but also increase the computational burden of the BMS. Hence, within certain error tolerance, some relationships can be simplified or ignored. Usually, aging periods are generally in the range of months to years. While for the system-level simulations of automotive applications, the time periods of interest are typically in the range of seconds to hours or days in special cases [43,45]. Hence, the long-term aging effect is usually ignored in the parameter estimation process and handled separately in most cases [39,46].

In this paper, all of the model parameters are estimated through the discharging/charging pulse-rest test at room temperature (22 °C–25 °C). A lithium-ion polymer battery with nickel-manganese-cobalt-based cathode and graphite-based anode is under test. Its specifications are given in Table 1, and the detailed experimental steps are described as follows.

Table 1. Specification of the tested battery.

Charge Capacity	40.99 Ah
Discharge capacity	40.89 Ah
Nominal voltage	3.7 V
Charge cutoff voltage	4.2 V
Discharge cutoff voltage	2.7 V

The discharging pulse-rest test starts with a fully-charged battery. In each cycle of the test, the battery is discharged at a 2% *SoC* step with C/2 constant current, then followed by a rest period. This cycle is repeated until the battery is fully discharged. Data points (including current, voltage, charging capacity and discharging capacity) are collected with the sampling frequency of 1 Hz. The relevant voltage and current profiles of the discharging pulse-rest test during the 66%–64% *SoC* interval are plotted in the bottom subfigure of Figure 1. The charging pulse-rest test is conducted similarly, that is it begins with a fully-discharged battery, then charged at a 2% *SoC* step with C/2 constant current and followed by a rest period. In order to eliminate the polarization voltage, the *OCV* values are extracted at the end of each rest period. Too short a rest time leads to a large *OCV* estimation error, whereas too long a rest time makes the whole test time consuming. It has been shown previously that for the lithium-ion polymer batteries, electrochemical reactions are negligible after a 2-h rest period [47,48]. Therefore, the rest time in this paper is predetermined as 2 h.

2.2. Parameter Estimation Algorithm

The electrical behavior of the ECM is expressed as the following state space formalism:

$$\begin{bmatrix} dV_{RC,short}/dt \\ dV_{RC,long}/dt \end{bmatrix} = \begin{bmatrix} -1/R_{short}C_{short} & 0 \\ 0 & -1/R_{long}C_{long} \end{bmatrix}\begin{bmatrix} V_{RC,short} \\ V_{RC,long} \end{bmatrix} + \begin{bmatrix} 1/C_{short} \\ 1/C_{long} \end{bmatrix}I \quad (1)$$

$$V_t = OCV(SoC) + IR_{in} + V_{RC,short} + V_{RC,long} \quad (2)$$

where Equation (1) is the state equation and Equation (2) is the output equation, $V_{RC,short}$ and $V_{RC,long}$ represent the voltages across the short-term and the long-term RC networks, respectively, *OCV(SoC)* is an eighth-order polynomial equation as a function of *SoC*, V_t is the battery terminal voltage and the positive current I represents charging. R_{in} represents the internal resistance; R_{short} and R_{long} denote the diffusion resistances; and C_{short} and C_{long} represent the diffusion capacitances. Among them, R_{in} can be directly obtained from each pulse-rest cycle through Equation (3); the corresponding four variables (V_1, V_2, I_1 and I_2) are marked in the bottom subfigure of Figure 1, and the variation of identified R_{in} with *SoC* is shown in Figure 2. *SoC* can be calculated through Equation (4), in which C_{ap} denotes the capacity of the battery in Ah.

$$R_{in} = \frac{V_2 - V_1}{I_2 - I_1} \tag{3}$$

$$SoC = SoC(0) + \frac{1}{3600 C_{ap}} \int_0^t I(\tau) d\tau \tag{4}$$

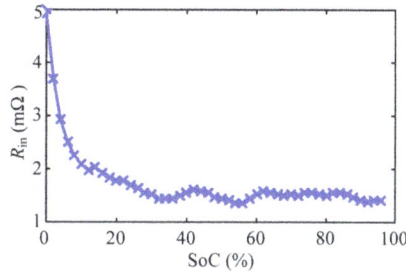

Figure 2. R_{in} variation with different state of charge (*SoC*).

For the CC operating scenario ($I \neq 0$), the analytical solutions of Equation (1) are derived as:

$$\begin{cases} V_{RC,short}(t) = V_{RC,short}(0)e^{-\frac{t}{\tau_{short}}} + IR_{short}(1 - e^{-\frac{t}{\tau_{short}}}) \\ V_{RC,long}(t) = V_{RC,long}(0)e^{-\frac{t}{\tau_{long}}} + IR_{long}(1 - e^{-\frac{t}{\tau_{long}}}) \end{cases} \tag{5}$$

where $V_{RC,short}(0)$ and $V_{RC,long}(0)$ are the initial voltages of corresponding RC networks and $\tau_{short} = R_{short}C_{short}$, $\tau_{long} = R_{long}C_{long}$, which represent the short-term and the long-term time constants, respectively.

Substituting Equation (5) into Equation (2), the output equation is rewritten as:

$$V_t(t) = OCV(SoC) + IR_{in} + V_{RC,short}(0)e^{-\frac{t}{\tau_{short}}} + V_{RC,short}(0)e^{-\frac{t}{\tau_{long}}} + IR_{short}(1 - e^{-\frac{t}{\tau_{short}}}) + IR_{long}(1 - e^{-\frac{t}{\tau_{long}}}) \tag{6}$$

During the rest period, where there is no current excitation ($I = 0$), Equation (6) can be simplified to:

$$V_t(t) = OCV(SoC) + V_{RC,short}(0)e^{-\frac{t}{\tau_{short}}} + V_{RC,long}(0)e^{-\frac{t}{\tau_{long}}} \tag{7}$$

With the knowledge of R_{in} and charging/discharging *OCV-SoC* relationships, RC network parameters (R_{short}, C_{short}, R_{long} and C_{long}) can be obtained through fitting the experimental data with relevant exponential functions, as

$$\begin{cases} y = IR_{short}(1 - e^{-\frac{t}{\tau_{short}}}) + IR_{long}(1 - e^{-\frac{t}{\tau_{long}}}) & I \neq 0 \\ y = V_{RC,short}(0)e^{-\frac{t}{\tau_{short}}} + V_{RC,long}(0)e^{-\frac{t}{\tau_{long}}} & I = 0 \end{cases} \tag{8}$$

where $y = V_t - OCV(SoC) - IR_{in}$. Since there only exists 2% *SoC* variation during each pulse-charging/discharging period, it is reasonable to make an assumption that the RC network parameters keep constant during this period. In addition, considering that the battery has converged to the steady state after a 2-h rest, $V_{RC,short}(0)$ and $V_{RC,long}(0)$ are set as zero at the beginning of the pulse-charging/discharging period.

Based on the above analysis, the RC network parameters can be estimated through fitting the experimental dataset with Equation (8). The cost function of the curve fitting method J is to minimize the sum of squared errors between the estimation results and the measured data, subjected to the following constraints:

$$\begin{cases} J = \min\limits_{r,\tau} \sum\limits_{k=1}^{n} [V_t^m(t_k) - V_t^e(r, \tau, t_k)]^2 \\ s.\,t.\ R_{short}, \tau_{short}, R_{long}, \tau_{long} > 0 \end{cases} \tag{9}$$

where t_k is the input time sequence, n is the length of the fitted experimental dataset, $r = [R_{short}, R_{long}]$, $\tau = [\tau_{short}, \tau_{long}]$, V_t^e is the model estimated voltage and V_t^m is the voltage measurements from the pulse-rest test.

3. RC Network Parameters Estimation

Based on the Introduction in Section 1, the RC network parameters show diverse values under different operating scenarios. In HEV/EV applications, batteries usually work in two typical scenarios: the CC charging scenario and the dynamic driving scenario. In the CC charging scenario, continuous external charging currents are applied to the batteries, and the transport of ions is mainly driven by the electric field. While for the dynamical driving scenario, especially for the urban driving condition, the load current has the characteristics of discontinuous amplitude values and a wide-spread frequency spectrum. In this case, besides the electric field, the gradient in concentration is also largely responsible for the transport of ions within batteries [45]. Therefore, the RC network parameters employed in different operating scenarios should be identified through different identification approaches.

3.1. RC Network Parameters for the CC Charging Scenario

The polarization voltage (V_P) is adopted to illustrate the variation of RC network parameters under the CC excitation. According to the aforementioned battery output equation, V_P can be obtained as:

$$V_P = V_{RC,short} + V_{RC,long} = V_t - OCV(SoC) - IR_{in} \tag{10}$$

The V_P-*SoC* profile during the C/2 rate CC charging process is shown in Figure 3. Since in the HEV/EV application, batteries seldom work in the extremely low or high *SoCs*, the voltage profile from 10%–90% *SoC* is covered. It can be observed from Figure 3 that the polarization voltage increases dramatically in Stage I (10%–18% *SoC*), then it declines slowly and shows a concave shape curve in Stage II, with the local minimum value at around 30% *SoC*. During Stage III (40%–70% *SoC*), the polarization voltage becomes relatively stable. After that (70%–90% *SoC*), the polarization voltage rises sharply.

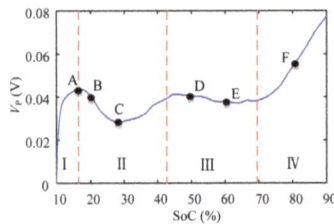

Figure 3. V_P versus *SoC* under constant-current (CC) charging.

The variation of the polarization voltage during the above *SoC* range is closely related to the internal electrochemical reaction process during charging. In the initial *SoC* region, a relatively large amount of energy is needed to form the nucleation on the surfaces of the electrodes; thus, the polarization voltage increases quickly. Once the nuclei are formed, the following lithium ions' removal process needs less energy. This explains the concave shape voltage curve occurring from 18% *SoC* to 40% *SoC*. While in the last charging stage, the lithium-ion concentration increases in the negative materials. Hence, a large amount of energy is needed to insert the lithium ions, which leads to the obvious growth of the polarization voltage in the high *SoC* region. The detailed explanation for the electrochemical reaction mechanism occurring during the CC charging process can be found in [28,32].

As mentioned in Section 2, the model parameters are estimated through fitting the measured data either from the pulse-charging period or the rest period. In order to select the proper experimental datasets that can better describe the charging characteristic of the battery, the profiles of the polarization voltage during the pulse-charging and the following rest periods, which are also calculated from Equation (10), are compared in Figure 4. Figure 4a shows the polarization voltage under the pulse-charging excitation, and Figure 4b plots the absolute values of the polarization voltage during the following rest. It can be seen from both figures that the shape of the polarization voltage curve strongly depends on the *SoC*. In Figure 4a, it is obvious that the final value of the polarization voltage obtained from 26%–28% *SoC* is the lowest, which is similar to point C in Figure 3. In addition, the final values of the voltage curves obtained from 18%–20% *SoC* and 50%–52% *SoC* are almost coincident with each other, which approximately matches the corresponding parts (point B and point D) in Figure 3. Meanwhile, the relations among the final voltage values collected from 14%–16% *SoC*, 60%–62% *SoC* and 80%–82% *SoC* are also identical to the relations among point A, point E and point F in Figure 3, respectively. Hence, it can be summarized from Figure 4a that the final values of the polarization voltage obtained from different pulse-charging periods are approximately consistent with the corresponding points in Figure 3. While in Figure 4b, the variation trend of the predicted stable voltage values differs greatly compared to the results in Figure 4a. This is because in the pulse-charging period, the ion migration is driven by external electric potential. While in the rest period, the transport of ions is mainly dominated by diffusion, owing to the concentration gradient. The detailed explanation of the electrochemical reactions occurring under different load current has been discussed in [21,45].

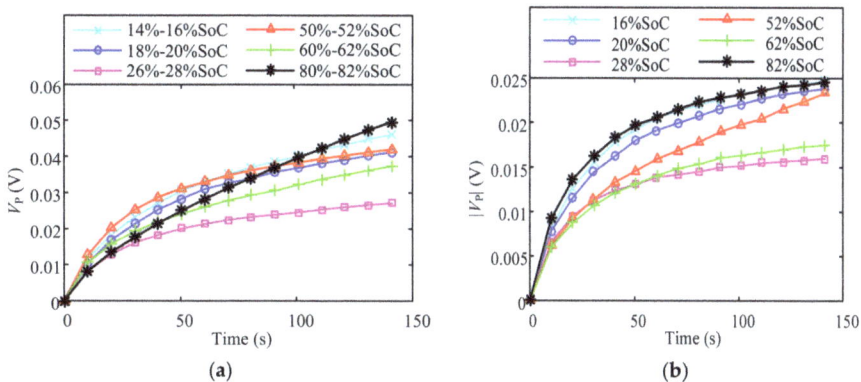

Figure 4. (a) The profiles of V_P at different *SoC* intervals during the pulse-charging period; (b) the profiles of $|V_P|$ at different *SoC* points during the rest period.

Consequently, it can be concluded that the voltage response during the pulse-charging period can better describe the characteristic of the CC charging process because of the similar current excitation.

3.2. RC Network Parameters for the Dynamic Driving Scenario

3.2.1. Typical Dynamic Driving Scenarios

For the dynamic driving scenario, especially for the urban driving scenario, vehicles accelerate and brake frequently, which cause the long lasting load current to seldom exist. There are two typical kinds of standard urban driving cycles, namely the urban dynamometer driving schedule (UDDS) and the worldwide harmonized light vehicles test procedure (WLTP), which are the American and European certification cycles, respectively. The load current profiles and the load current amplitude distributions of the two driving cycles are plotted in Figure 5. It can be observed from Figure 5a,b that both of the dynamic current profiles vary frequently over the test span. Meanwhile, from Figure 5c,d, it can be concluded that: (1) the discharging current accounts for a much larger portion, compared to the charging current during the regenerative process; (2) among the load currents, the low C-rate discharging current, particularly around zero-value amplitudes, accounts for a larger portion in both tests. Hence, the voltage response during the rest period can be employed to estimate the RC network parameters for the dynamic driving scenario.

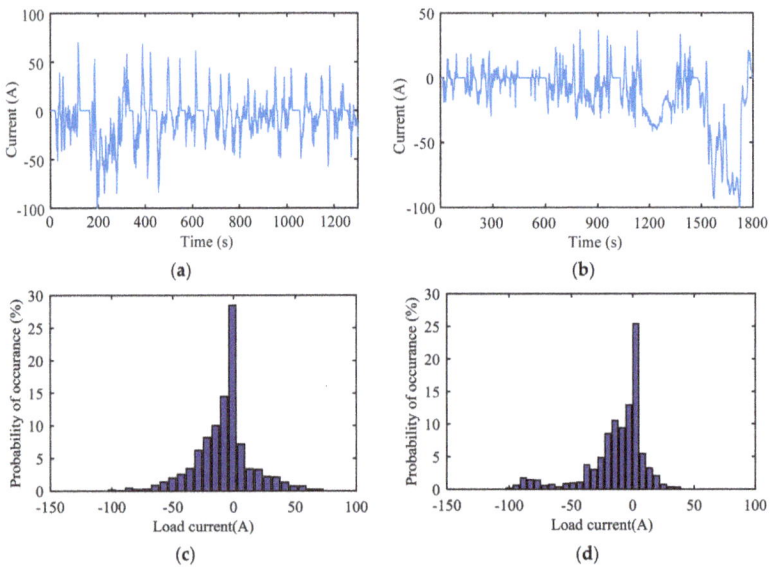

Figure 5. (a) The load current profile of the urban dynamometer driving schedule (UDDS) test; (b) the load current profile of the worldwide harmonized light vehicles test procedure (WLTP) test; (c) the load current amplitude distribution of the UDDS test; (d) the load current amplitude distribution of the WLTP test.

3.2.2. Determination of the Length of the Fitted Experimental Dataset

The diffusion process, which is caused by the gradient in concentration, plays a major role in the low C-rate load current and rest cases. Since the electrochemical reactions occurring during the diffusion process are very complex, these reactions can be accurately modeled as infinite series-connected RC networks with a wide range of time constants ($\tau_1, \tau_2, \ldots, \tau_j$). Usually, the values of time constants depend on the electrode thickness and the structure of the battery to a great extent, and typical time constants are in the range of seconds to minutes [45]. The second order RC network can only approximate the diffusion process by two parts: the fast dynamics part (the short-term RC network with τ_{short}) and the slow dynamics part (the long-term RC network with τ_{long}).

In general, the values of the two time constants are closely related to the length of the fitted experimental data Δt. When only the initial segment of the voltage response is employed in parameter estimation, such as Part A in the bottom subfigure of Figure 1, the voltages across the shorter-term RC networks have a larger degree of variability, which means that the shorter-term RC networks have a greater impact on the initial segment of the voltage response. This in turn leads to the smaller estimated time constants and subsequently ignores the slower dynamics diffusion process. On the contrary, after the initial phase of the rest period, such as Part B in the bottom subfigure of Figure 1, the voltages across the shorter-term RC networks have converged to zero; thus, the voltage variation caused by the shorter-term RC networks is negligible. Instead, the voltages across the longer term RC networks make a remarkable contribution to the total voltage response. Subsequently, it can be inferred that the measured data show a slower varying characteristic, which represent the slower dynamics diffusion process and can be modeled by the RC networks with larger time constants. Hence, if the whole voltage response of the long time rest period is adopted, data with slower varying values will account for a large portion, which will lead to the relatively larger estimated time constants. However, too large time constants will make the model output voltage severely lag behind the actual response and result in a poor dynamic performance.

In order to further illustrate the above analysis, a third order RC network circuit is simulated in MATLAB; two equivalent time constants (τ'_{short} and τ'_{long}) are estimated from the different value of Δt. In the simulation, the resistances of the three RC networks are all set as 1 mΩ, and the time constants are predetermined as $\tau_1 = 40$ s, $\tau_2 = 200$ s and $\tau_3 = 2000$ s ($\tau_3 \gg \tau_2 > \tau_1$). The applied excitation consists of a 400-s pulse-discharging current and a 2-h rest period, and the amplitude of the current is 20 A. Time constants estimated by different lengths of the voltage response are given in Table 2. It can be clearly seen from Table 2 that both τ'_{short} and τ'_{long} decrease simultaneously with the reduced value of Δt, which is consistent with the previous analysis. Hence, to obtain the appropriate values of the time constants, Δt should be predetermined properly, which is illustrated in detail as follows.

Table 2. Equivalent time constant estimation results with different values of Δt.

Δt (s)	7200	3600	1800	1400	1200	1000	900	850	800
τ'_{short} (s)	88.67	67.18	48.53	45.10	43.74	42.59	42.08	41.83	41.63
τ'_{long} (s)	971.0	484.3	284.4	256.7	245.3	235.3	230.9	228.8	226.8
k [1]	4.049×10^{-12}	4.395×10^{-5}	0.1448	0.8759	2.154	5.299	8.311	10.41	13.03

[1] k represents the degree of resistor-capacitor (RC) voltage variability; the detailed expression can referred to in Equation (13).

During Δt, the derivative of Equation (13) with respect to τ_i during the rest period is expressed as:

$$\left| \frac{dV_{RC,i}}{d\tau_i} \right| = \frac{\Delta t |V_{RC,i}(0)|}{\tau_i^2} e^{-\frac{\Delta t}{\tau_i}} \tag{11}$$

where $V_{RC,i}$ is the voltage across the i-th RC network, $i \in \{1,2,3, \dots , j\}$, $V_{RC,i}(0)$ is the corresponding initial voltage, R_i is the resistance of the i-th RC network and τ_i is the time constant of the i-th RC network, which is subject to $\tau_1 < \tau_2 < \dots < \tau_j$.

After the pulse-discharging period, $|V_{RC,i}(0)|$ can be expressed as:

$$|V_{RC,i}(0)| = |I| R_i (1 - e^{-\frac{D}{\tau_i}}) \tag{12}$$

where D denotes the length of the pulse-discharging period.

For the two well-separated time constants τ_i and τ_{i+m} ($\tau_{i+m} \geq 10\tau_i$ and $0 < m < j - i$), the voltage across the shorter term RC network $V_{RC,i}$ has a larger degree of variability when satisfying the following requirement:

$$\frac{|dV_{RC,i}/d\tau_i|}{|dV_{RC,i+m}/d\tau_{i+m}|} = k \tag{13}$$

where the constant k denotes the degree of variability, and it is subject to $k > 1$.

Substituting Equations (11) and (12) into Equation (13), the value of Δt can be derived as:

$$\Delta t = \ln\left[\frac{R_i(1 - e^{-\frac{D}{\tau_i}})\tau_{i+m}^2}{kR_{i+m}(1 - e^{-\frac{D}{\tau_{i+m}}})\tau_i^2}\right]\frac{\tau_i\tau_{i+m}}{\tau_{i+m} - \tau_i} \tag{14}$$

In Equation (14), since the values of R_i and R_{i+m} are nearly of the same order of magnitude [39,43,46], the value of R_i/R_{i+m} can be neglected when compared to the value of τ_{i+m}^2/τ_i^2; thus, Δt can be simplified as:

$$\Delta t = \ln\left[\frac{(1 - e^{-\frac{D}{\tau_i}})\tau_{i+m}^2}{k(1 - e^{-\frac{D}{\tau_{i+m}}})\tau_i^2}\right]\frac{\tau_i\tau_{i+m}}{\tau_{i+m} - \tau_i} \tag{15}$$

Equation (15) shows that k and τ_i should be determined before calculating Δt. In the aforementioned simulation, the value of k for τ_2 and τ_3 can be obtained directly from Equation (13), as shown in Table 2. This indicates that when k is larger than one, the estimated τ'_{short} and τ'_{long} are closer to τ_1 and τ_2. This is because the voltage across the RC network with τ_3 has a lower degree of variability, compared to those with τ_1 and τ_2. It can be observed from Table 2 that τ'_{short} and τ'_{long} are nearly stable when k is larger than 10. Hence, k is selected as 10 throughout the paper.

In order to set a proper τ_i in Equation (15), the discrete Fourier analysis of the load current is employed to determine the lower bandwidth limitation of the ECM. The current spectrums of UDDS and WLTP tests are shown in Figure 6. It can be observed in Figure 6a,b that there exists a large DC component (Points A and C) due to the nonzero mean value of the two current profiles. Since the characteristics of the DC component cannot be modeled by the RC circuit, they are neglected when determining the length of the fitted dataset. The major low frequency components for the two profiles are around 0.00146 Hz (point B) and 0.00138 Hz (the mean value from point D to point E), respectively. Hence, the mean value of the long-term time constant is selected as 704 s. In order to exclude the voltage variation caused by the larger time constants (larger than $10\tau_i$), the prior 1-h measured battery voltage dataset is employed to estimate the RC parameters.

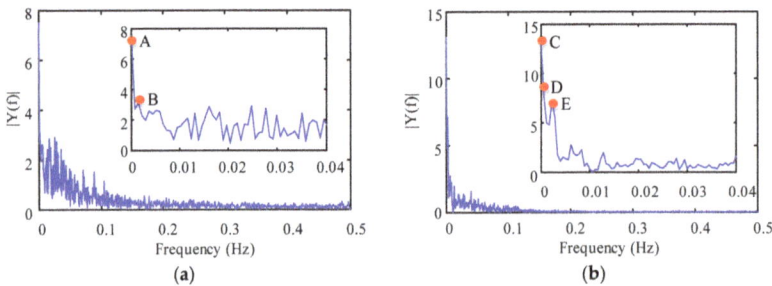

Figure 6. The spectral analysis of the load current: (**a**) the urban dynamometer driving schedule (UDDS) test; (**b**) the worldwide harmonized light vehicles test procedure (WLTP) test.

3.2.3. Improved Fitting Function

From Equations (6) and (7), it can be observed that only the initial values $V_{RC,short}(0)$, $V_{RC,long}(0)$ and time constants τ_{short}, τ_{long} can be obtained directly from the fitting results; thus, we should do the further computations to obtain the resistances and capacitances of RC networks.

In [37,39–41], two initial voltages across the RC networks are predetermined as IR_{short} and IR_{long} respectively, from which the resistances of the RC networks can be derived under the knowledge of the current value. In [49], the capacitances of the RC networks are firstly obtained from the initial voltage values. Both of the above two methods have an assumption that the capacitors of the RC networks have already converged to the steady state at the end of the pulse-discharging period.

Usually, in the parameter extraction test, in order to obtain as much data as possible at different *SoC* intervals, the length of the pulse-charging/discharging period is usually set as several minutes (resulting in 2% *SoC* variation in this paper), while the rest time is usually set as one or more hours (such as 2 h in this paper) to get an accurate *OCV* value. For the short-term RC network, the voltage can easily converge to the equilibrium state during the pulse-discharging process, which is shown in Figure 7. In other words, there is no current flowing through the capacitor branch of the short-term RC network during the last stage of the pulse-discharging period; thus, $V_{RC,short}(0)$ at the beginning of the rest period can be expressed as:

$$V_{RC,short}(0) = IR_{short} \tag{16}$$

However, for the long-term RC network, the voltage varies continuously due to a relatively large time constant, as illustrated in Figure 7. The voltage across the long-term RC network has not reached the equilibrium state at the end of the pulse-discharging period; thus, there always exists a significant proportion of the load current $I(1 - e^{-D/\tau_{long}})$ flowing through the corresponding capacitor. Consequently, $V_{RC,long}(0)$ at the beginning of the rest period should be written as:

$$V_{RC,long}(0) = IR_{long}\left(1 - e^{-\frac{D}{\tau_{long}}}\right) \tag{17}$$

where I is the value of the pulse-discharging current. Since the *SoC* variation in each test cycle is set as 2% in this paper, it can be assumed that the model parameters keep constant during the pulse-discharging period.

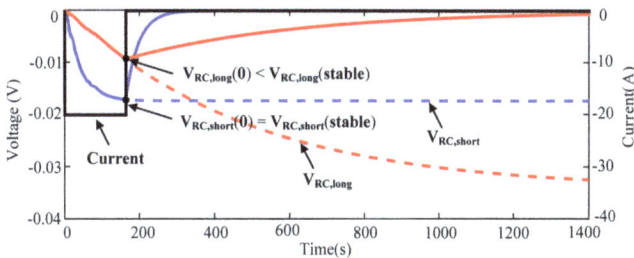

Figure 7. The voltage curve of RC networks during one cycle of the discharging pulse-rest test.

4. Experimental Results and Discussions

4.1. RC Network Parameter Estimation Results

Based on the aforementioned analysis in Section 3.1, for the case of the CC charging scenario, the charging pulse-rest test is implemented firstly. The parameters are estimated from the voltage response of the pulse-charging period, and the estimation results are shown in Figure 8. Figure 8a plots two estimated time constants; it can be seen that the general order of the magnitude of the short-term time constant is 10 s; it fluctuates greatly when the *SoC* changes, especially in the middle *SoC* region, while the order of the magnitude of the long-term time constant is 100 s; it is relatively flat during the whole *SoC* region. Figure 8b plots two estimated resistances; it can be observed that in the middle *SoC* range, the short-term resistance has a larger value, which means that the voltage across the short-term RC network accounts for more weight during this period. Hence, it can be observed from

Figures 3 and 8b that the variation tendencies of the polarization voltage and the short-term resistance are similar during the middle *SoC* range. At the end of the charging process, the short-term resistance decreases and stabilizes around a very small value, while the long-term resistance increases almost linearly after 60% *SoC*, leading to a similar variation tendency of the polarization voltage, compared to the corresponding part in Figure 3. Hence, it can be concluded that the long-term diffusion process plays a major role in this stage.

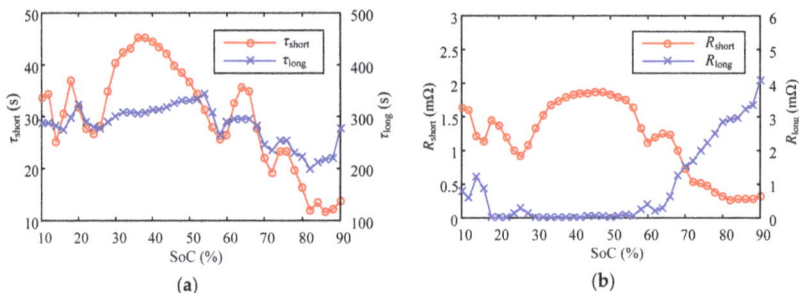

Figure 8. Parameter estimation results for the constant-current (CC) charging scenario: (a) time constant; (b) resistance.

For the case of the dynamic driving scenario, the discharging pulse-rest test is implemented, and the data from the rest periods are adopted in the parameter estimation. According to the analysis in Section 3.2.2, different time constants will be obtained from the fitted experimental datasets in different lengths. Firstly, in order to compare the best fit performances for the measured datasets in different lengths, the measured battery terminal voltage response at 60% *SoC* during a 2-h rest period is adopted, and the curve fitting results are shown in Figure 9. It can be observed from Figure 9a that the fitting result of the whole measured voltage response shows a better performance during most of the rest period, especially in the equilibrium state. Whereas for the performance of the first 200 s, the fitting result through the prior 0.5-h measured voltage response yields less errors, which is illustrated in Figure 9b. Parameter estimation results in Figure 10 show the time constants estimated from the measured voltage dataset in different lengths, ranging from 30 min–2 h with a 30-min interval. It can be observed that the time constants, both for the long term and the short term, increase simultaneously when the length of the fitted dataset increases. In addition, by comparing Figure 10 with Figure 8a, it can be concluded that the time constants applied in the CC charging scenario and the dynamic driving scenario show different variation tendencies. Hence, it is essential to adopt different sets of model parameters for different operating scenarios.

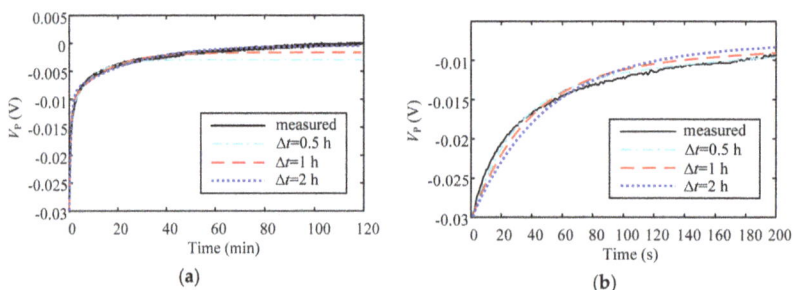

Figure 9. Curve fitting results of V_P during the rest period of the discharging pulse-rest test at 60% *SoC*: (a) the overall result; (b) a close look at the transient part at the beginning.

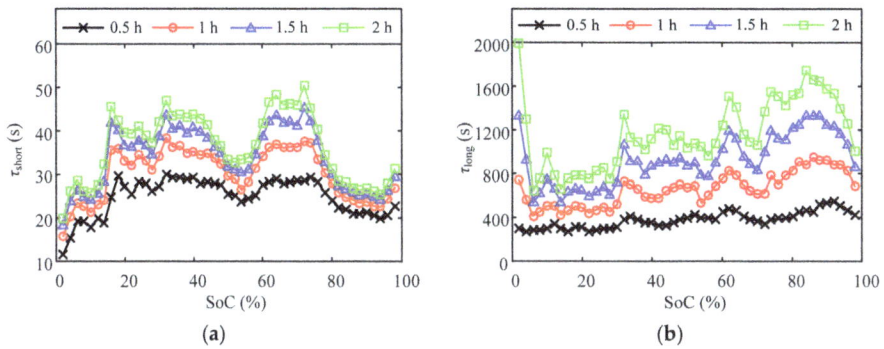

Figure 10. Time constant estimation results with different lengths of the experimental dataset: (**a**) τ_{short}; (**b**) τ_{long}.

After determining the length of the fitted experimental dataset, we can subsequently obtain the resistances. Figure 11 shows the R_{long} estimation results by the conventional fitting function and the improved fitting function. It can be concluded from Figure 11 that the R_{long} estimated by the conventional fitting function is generally less than the one estimated by the improved fitting function, because it neglects the $(1 - e^{-D/\tau_{long}})$ part. In order to demonstrate the advantage of the improved fitting function, data from the 20th cycle of the discharging pulse-rest test are adopted. In this cycle, *SoC* changes from 62% to 60% during the pulse-discharging period, then keeps the value of 60% during the following rest period. The current profile of the 20th discharging pulse-rest test is applied on the ECM MATLAB/SIMULINK model as an excitation. Figure 12a,b shows the model output voltage responses with two sets of estimated model parameters. It can be seen that the model with parameters estimated by the proposed fitting function outputs better estimation results. The lower voltage error is mainly contributed by the higher voltage drop across the long-term RC network, as plotted in Figure 12c. In addition, the root mean square errors (RMSEs) between the measured voltage and the model output voltage at different *SoCs* are given in Table 3. It can also be seen that the model parameters estimated by the proposed fitting function show a better performance for a wide range of *SoC*.

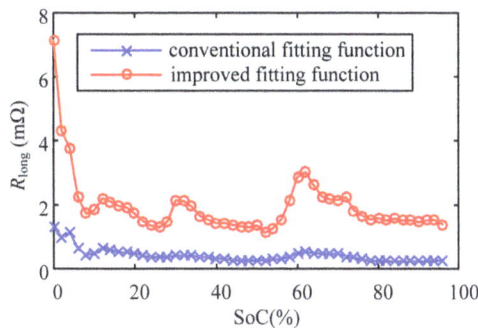

Figure 11. R_{long} estimation results.

Table 3. Comparison of RMSE at different *SoC*.

SoC (%)		10	20	30	40	50	60	70	80	90
RMSE (mV)	Conventional fitting function	1.802	1.714	2.167	1.540	1.268	2.803	2.416	1.558	1.444
	Improved fitting function	0.7658	0.7582	0.9707	0.7643	0.5000	1.202	1.242	0.7104	0.6482

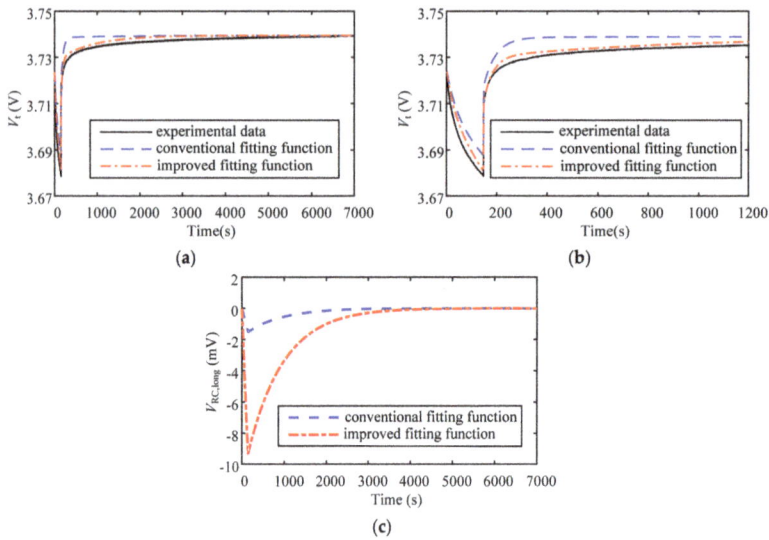

Figure 12. Voltage curves of one cycle of the discharging pulse-rest test (62%–60%): (**a**) the overview; (**b**) a close look; (**c**) the voltage across the long-term RC network.

4.2. Model Verification

In this paper, the CC charging test and the consecutive UDDS test, which respectively represent two typical operating scenarios in HEV/EV applications, are conducted separately to verify the effectiveness of the model. For the charging condition, the battery is charged from 10%–90% SoC. The typical charging current in practice varies from C/8 to 2C [50], and a C/2 rate current is employed in the charging test. The consecutive UDDS test starts from 90% SoC to 20% SoC, with a 10-min rest period in between to simulate a short parking time. In the real application, a specific set of parameters can be selected by the characteristics of the measured load current. For example, if the values of the current are approximately constant over a certain time interval, parameters estimated from the data in the pulse-charging periods are employed. On the other hand, parameters estimated from the data in the rest periods are employed when the load current shows the characteristics of high dynamics over a certain time interval.

Firstly, for the CC charging scenario, three model outputs and measured battery terminal voltage curves are plotted in Figure 13, and the corresponding RMSEs are given in Table 4. It can be observed that during the whole charging process, the model with parameters estimated from the data in pulse-charging periods outputs a voltage curve matching the measured curve better because of considering the continuous external electric driving forces. However, parameters estimated from the data in the rest periods result in relatively larger errors, especially in the high SoC region. In addition, during most part of the charging period, the model with parameters used in the dynamic driving scenarios outputs a voltage higher than the experimental voltage. Comparing the corresponding curves in Figures 8b and 11, it can be deduced that the higher estimated voltage is mainly caused by the larger value of estimated R_{long}, especially during the middle range of the SoC region.

Table 4. RMSE of model voltage estimation under the CC charging test.

Modeling Methods	Dynamic Condition	Rest-Period	Pulse-Period
RMSE (mV)	18.41	19.76	5.448

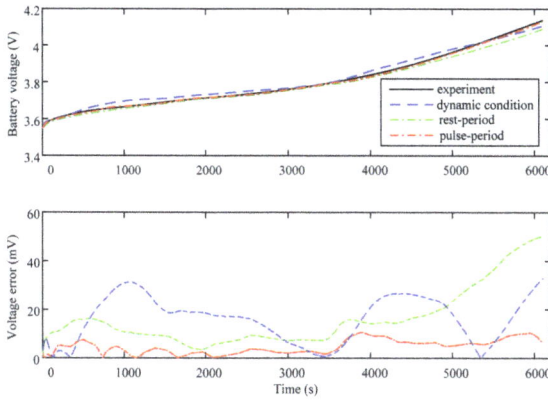

Figure 13. Verification results of different parameter estimation methods under the CC charging test.

In order to verify the robustness of the proposed parameter estimation method, the CC charging voltage profiles at different initial *SoC* are plotted in Figure 14. This shows that the estimated voltage curves match well with the measurement voltage curves, despite the different initial *SoC*.

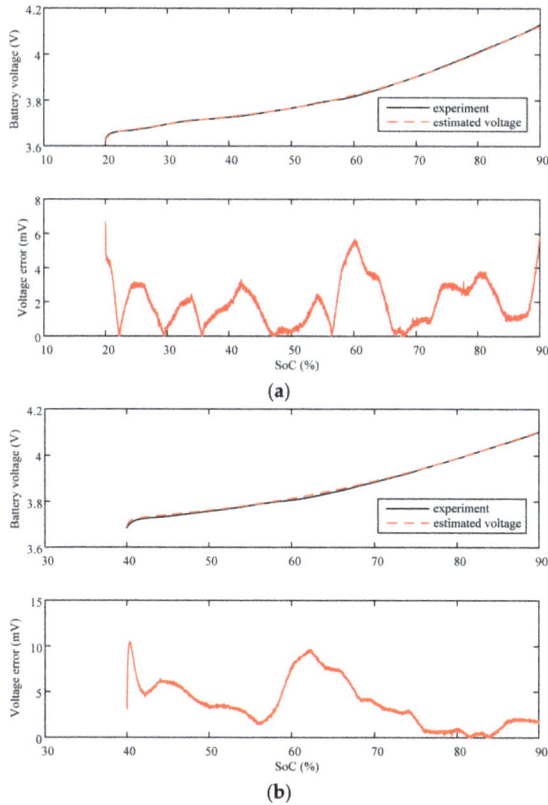

(a)

(b)

Figure 14. CC charging voltage profiles at different initial *SoC*: (a) initial *SoC* = 20%; (b) initial *SoC* = 40%.

Secondly, in order to demonstrate the improvement of the proposed battery modelling approach during the dynamic driving scenario, the model and experimental voltage outputs in the consecutive UDDS validation are plotted in Figure 15a, the corresponding calculated *SoC* profile is shown in Figure 15b, and the detailed figure from 10,000 s to 12,000 s is plotted in Figure 15c. The RMSE of the aforementioned estimation methods during the whole consecutive UDDS test are also shown in Table 5. Figure 15b shows that the consecutive UDDS test is started from 90% *SoC*, and terminated when the value of *SoC* drops below 20%. It can be observed from Figure 15c that parameters estimated by the improved fitting function generally demonstrate a better performance, especially during the dynamic period (ranging from 10,000 s to 11,400 s), because considering the unsaturated phenomenon of the long-term RC network. It can also be concluded that the model containing parameters estimated by the prior 1-h experimental data from the rest period gives voltage output with the least error, especially during the short-time rest period. In addition, it can be seen from Figure 5a that there exists a relatively long-time and high C-rate discharging current in the UDDS cycle approximately ranging from 150 s to 300 s. Since larger time constants are obtained from the data of the whole rest period, this causes the corresponding voltage output not to recover fast after a relatively long-time discharging current, which leads to an offset of voltage errors in comparison to the voltage error caused by the proposed approach.

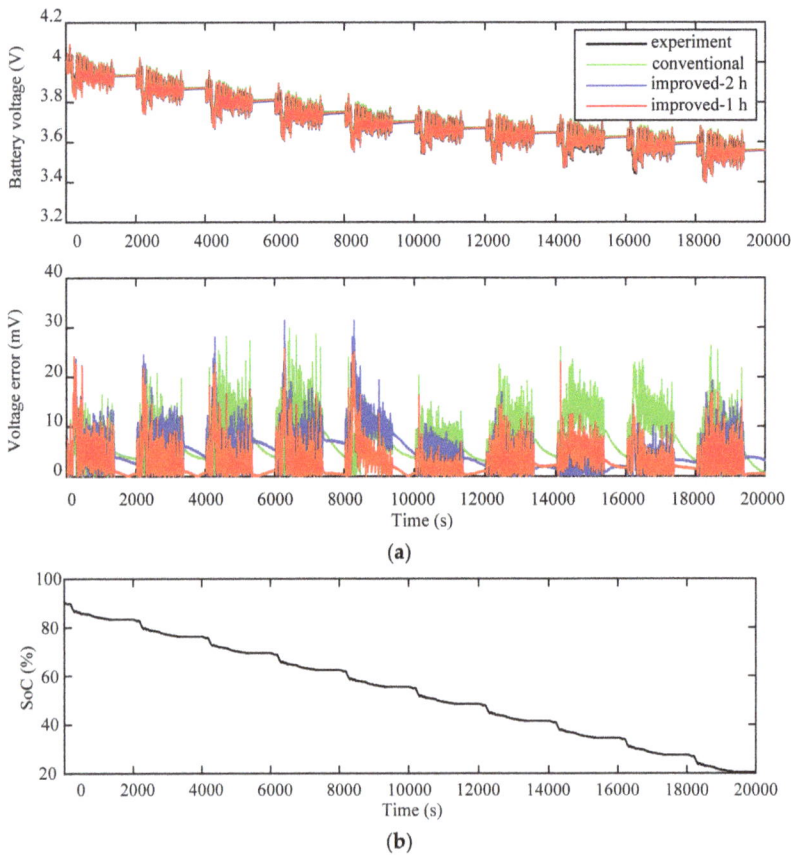

(a)

(b)

Figure 15. *Cont.*

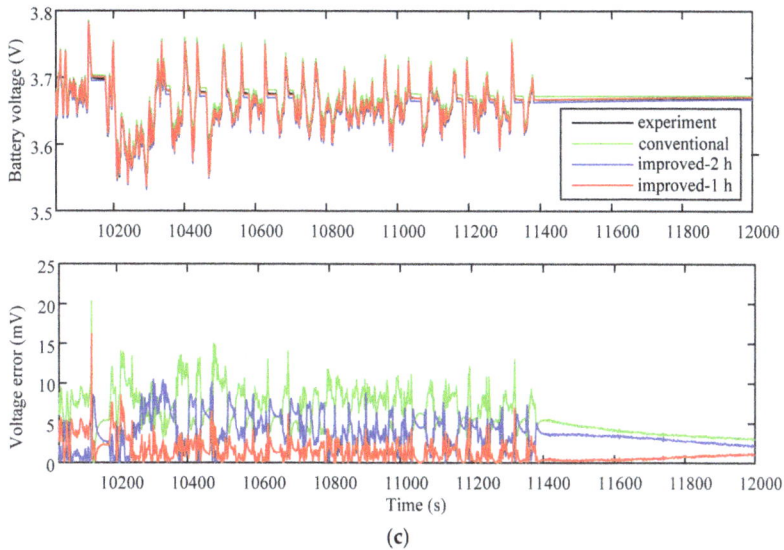

Figure 15. Verification results of different parameter estimation methods under the UDDS tests: (a) The overall look; (b) The calculated *SoC* profile (c) The close look.

Table 5. RMSE of the model voltage estimation under the urban dynamometer driving schedule (UDDS) test.

Modeling Methods	Conventional	Improved-2 h	Improved-1 h
RMSE (mV)	8.504	6.329	4.244

5. Conclusions

In this paper, an advanced battery parameter estimation method based on two general operating scenarios in HEV/EV applications is proposed. Firstly, the second order ECM is employed, and the model parameter extraction process is described in detail. Considering the typical operating scenarios in HEV/EV applications, namely the CC charging scenario and the dynamic driving scenario, two sets of model parameters are extracted from the charging/discharging pulse-rest tests. Specifically, voltage responses of the pulse-charging phases are selected to estimate model parameters applied in the CC charging scenario. For the dynamic driving scenario, the model parameters are identified through the measured data from the rest period. Instead of employing the data from the whole rest period, only the prior portion of the collected data is selected, and the length of the fitted data is determined by the frequency spectrum analysis of the load current under two typical urban driving conditions. In addition, an unsaturated phenomenon caused by the long-term RC network is analyzed in detail, and subsequently, an improved fitting equation with more accurate initial voltage expression of the RC network is adopted. Finally, verification tests simulating the CC charging scenario and the dynamic driving scenario are conducted, respectively, and comparisons between the conventional and the proposed battery parameter estimation methods are given. Experimental results show that in both cases, the voltage profiles predicted from the proposed model show a better conformity to the experimental data.

It is important to note that the proposed battery parameter estimation method for the dynamic driving scenario only considers the typical urban driving conditions at room temperature. However, the characteristics of the load current under the other special conditions (such as the highway driving condition and the extremely cold condition) will be obviously different. For the future work, the influence caused by different C-rates of the current profiles, bandwidths of the current profiles and temperature effects will be considered, and the parameter extraction test will be modified accordingly.

Acknowledgments: The authors would like to acknowledge the funding support from the China Scholarship Council (CSC); the U.S. DOE Graduate Automotive Technology Education (GATE) Center of Excellence; and Nanjing Golden Dragon Bus Co., Ltd.

Author Contributions: Jufeng Yang handled the technical modeling, drafted and revised the manuscript. Bing Xia revised the manuscript. Jufeng Yang and Bing Xia designed the experiments and analyzed the data. Yunlong Shang participated in the experiment. Wenxin Huang revised the manuscript. Chris Mi contributed the experiment platform, gave great suggestions and polished the manuscript.

Conflicts of Interest: The authors declare no conflict of interest.

References

1. Xiong, R.; Sun, F.; Chen, Z.; He, H. A data-driven multi-scale extended kalman filtering based parameter and state estimation approach of lithium-ion olymer battery in electric vehicles. *Appl. Energy* **2014**, *113*, 463–476. [CrossRef]
2. Zou, Z.; Xu, J.; Mi, C.; Cao, B.; Chen, Z. Evaluation of model based state of charge estimation methods for lithium-ion batteries. *Energies* **2014**, *7*, 5065–5082. [CrossRef]
3. Shang, Y.; Zhang, C.; Cui, N.; Guerrero, J.M. A cell-to-cell battery equalizer with zero-current switching and zero-voltage gap based on quasi-resonant lc converter and boost converter. *IEEE Trans. Power Electron.* **2015**, *30*, 3731–3747. [CrossRef]
4. Xia, B.; Mi, C. A fault-tolerant voltage measurement method for series connected battery packs. *J. Power Sources* **2016**, *308*, 83–96. [CrossRef]
5. Sun, F.; Xiong, R.; He, H. A systematic state-of-charge estimation framework for multi-cell battery pack in electric vehicles using bias correction technique. *Appl. Energy* **2016**, *162*, 1399–1409. [CrossRef]
6. Xia, B.; Shang, Y.; Nguyen, T.; Mi, C. A correlation based fault detection method for short circuits in battery packs. *J. Power Sources* **2017**, *337*, 1–10. [CrossRef]
7. Salameh, M.; Schweitzer, B.; Sveum, P.; Al-Hallaj, S.; Krishnamurthy, M. Online temperature estimation for phase change composite-18650 lithium ion cells based battery pack. In Proceedings of the 2016 IEEE Applied Power Electronics Conference and Exposition (APEC), Long Beach, CA, USA, 20–24 March 2016; pp. 3128–3133.
8. Seaman, A.; Dao, T.-S.; McPhee, J. A survey of mathematics-based equivalent-circuit and electrochemical battery models for hybrid and electric vehicle simulation. *J. Power Sources* **2014**, *256*, 410–423. [CrossRef]
9. Zou, Y.; Hu, X.; Ma, H.; Li, S.E. Combined state of charge and state of health estimation over lithium-ion battery cell cycle lifespan for electric vehicles. *J. Power Sources* **2015**, *273*, 793–803. [CrossRef]
10. Wei, Z.; Tseng, K.J.; Wai, N.; Lim, T.M.; Skyllas-Kazacos, M. Adaptive estimation of state of charge and capacity with online identified battery model for vanadium redox flow battery. *J. Power Sources* **2016**, *332*, 389–398. [CrossRef]
11. He, H.; Xiong, R.; Guo, H.; Li, S. Comparison study on the battery models used for the energy management of batteries in electric vehicles. *Energy Convers. Manag.* **2012**, *64*, 113–121. [CrossRef]
12. He, H.; Zhang, X.; Xiong, R.; Xu, Y.; Guo, H. Online model-based estimation of state-of-charge and open-circuit voltage of lithium-ion batteries in electric vehicles. *Energy* **2012**, *39*, 310–318. [CrossRef]
13. Nejad, S.; Gladwin, D.; Stone, D. A systematic review of lumped-parameter equivalent circuit models for real-time estimation of lithium-ion battery states. *J. Power Sources* **2016**, *316*, 183–196. [CrossRef]
14. Xia, B.; Zhao, X.; De Callafon, R.; Garnier, H.; Nguyen, T.; Mi, C. Accurate lithium-ion battery parameter estimation with continuous-time system identification methods. *Appl. Energy* **2016**, *179*, 426–436. [CrossRef]
15. Pérez, G.; Garmendia, M.; Reynaud, J.F.; Crego, J.; Viscarret, U. Enhanced closed loop state of charge estimator for lithium-ion batteries based on extended kalman filter. *Appl. Energy* **2015**, *155*, 834–845. [CrossRef]
16. Chen, Z.; Fu, Y.; Mi, C.C. State of charge estimation of lithium-ion batteries in electric drive vehicles using extended kalman filtering. *IEEE Trans. Veh. Technol.* **2013**, *62*, 1020–1030. [CrossRef]
17. Sun, F.; Xiong, R. A novel dual-scale cell state-of-charge estimation approach for series-connected battery pack used in electric vehicles. *J. Power Sources* **2015**, *274*, 582–594. [CrossRef]

18. Li, K.; Tseng, K.J. An equivalent circuit model for state of energy estimation of lithium-ion battery. In Proceedings of the 2016 IEEE Applied Power Electronics Conference and Exposition (APEC), Long Beach, CA, USA, 20–24 March 2016; pp. 3422–3430.

19. Fuller, T.F.; Doyle, M.; Newman, J. Relaxation phenomena in lithium-ion-insertion cells. *J. Electrochem. Soc.* **1994**, *141*, 982–990. [CrossRef]

20. Smith, K.A. Electrochemical Modeling, Estimation and Control of Lithium Ion Batteries. Ph.D. Thesis, The Pennsylvania State University, State College, PA, USA, 2006.

21. Park, M.; Zhang, X.; Chung, M.; Less, G.B.; Sastry, A.M. A review of conduction phenomena in Li-ion batteries. *J. Power Sources* **2010**, *195*, 7904–7929. [CrossRef]

22. Karden, E.; Buller, S.; De Doncker, R.W. A method for measurement and interpretation of impedance spectra for industrial batteries. *J. Power Sources* **2000**, *85*, 72–78. [CrossRef]

23. Thele, M.; Bohlen, O.; Sauer, D.U.; Karden, E. Development of a voltage-behavior model for nimh batteries using an impedance-based modeling concept. *J. Power Sources* **2008**, *175*, 635–643. [CrossRef]

24. Buller, S.; Thele, M.; De Doncker, R.; Karden, E. Impedance-based simulation models of supercapacitors and Li-ion batteries for power electronic applications. *IEEE Trans. Ind. Appl.* **2005**, *41*, 742–747. [CrossRef]

25. Waag, W.; Käbitz, S.; Sauer, D.U. Experimental investigation of the lithium-ion battery impedance characteristic at various conditions and aging states and its influence on the application. *Appl. Energy* **2013**, *102*, 885–897. [CrossRef]

26. Howey, D.A.; Mitcheson, P.D.; Yufit, V.; Offer, G.J.; Brandon, N.P. Online measurement of battery impedance using motor controller excitation. *IEEE Trans. Veh. Technol.* **2014**, *63*, 2557–2566. [CrossRef]

27. Zheng, Y.; Lu, L.; Han, X.; Li, J.; Ouyang, M. Lifepo 4 battery pack capacity estimation for electric vehicles based on charging cell voltage curve transformation. *J. Power Sources* **2013**, *226*, 33–41. [CrossRef]

28. Nakayama, M.; Iizuka, K.; Shiiba, H.; Baba, S.; Nogami, M. Asymmetry in anodic and cathodic polarization profile for LiFePO4 positive electrode in rechargeable Li ion battery. *J. Ceram. Soc. Jpn.* **2011**, *119*, 692–696. [CrossRef]

29. Musio, M.; Damiano, A. A simplified charging battery model for smart electric vehicles applications. In Proceedings of the 2014 IEEE International Energy Conference (ENERGYCON), Dubrovnik, Croatia, 13–16 May 2014; pp. 1357–1364.

30. Tsang, K.; Sun, L.; Chan, W. Identification and modelling of lithium ion battery. *Energy Convers. Manag.* **2010**, *51*, 2857–2862. [CrossRef]

31. Yao, L.W.; Aziz, J.; Kong, P.Y.; Idris, N.; Alsofyani, I. Modeling of lithium titanate battery for charger design. In Proceedings of the 2014 IEEE Australasian Universities Power Engineering Conference (AUPEC), Perth, Australia, 28 September–1 October 2014; pp. 1–5.

32. Jiang, J.; Liu, Q.; Zhang, C.; Zhang, W. Evaluation of acceptable charging current of power Li-ion batteries based on polarization characteristics. *IEEE Trans. Ind. Electron.* **2014**, *61*, 6844–6851. [CrossRef]

33. Kim, N.; Ahn, J.-H.; Kim, D.-H.; Lee, B.-K. Adaptive loss reduction charging strategy considering variation of internal impedance of lithium-ion polymer batteries in electric vehicle charging systems. In Proceedings of the 2016 IEEE Applied Power Electronics Conference and Exposition (APEC), Long Beach, CA, USA, 20–24 March 2016; pp. 1273–1279.

34. Chen, Z.; Xia, B.; Mi, C.C.; Xiong, R. Loss-minimization-based charging strategy for lithium-ion battery. *IEEE Trans. Ind. Appl.* **2015**, *51*, 4121–4129. [CrossRef]

35. Rao, R.; Vrudhula, S.; Rakhmatov, D.N. Battery modeling for energy aware system design. *Computer* **2003**, *36*, 77–87.

36. Fleischer, C.; Waag, W.; Heyn, H.-M.; Sauer, D.U. On-line adaptive battery impedance parameter and state estimation considering physical principles in reduced order equivalent circuit battery models: Part 1. Requirements, critical review of methods and modeling. *J. Power Sources* **2014**, *260*, 276–291. [CrossRef]

37. Schweighofer, B.; Raab, K.M.; Brasseur, G. Modeling of high power automotive batteries by the use of an automated test system. *IEEE Trans. Instrum. Meas.* **2003**, *52*, 1087–1091. [CrossRef]

38. Castano, S.; Gauchia, L.; Voncila, E.; Sanz, J. Dynamical modeling procedure of a Li-ion battery pack suitable for real-time applications. *Energy Convers. Manag.* **2015**, *92*, 396–405. [CrossRef]

39. Chen, M.; Rincon-Mora, G.A. Accurate electrical battery model capable of predicting runtime and iv performance. *IEEE Trans. Energy Convers.* **2006**, *21*, 504–511. [CrossRef]

40. Baronti, F.; Fantechi, G.; Leonardi, E.; Roncella, R.; Saletti, R. Enhanced model for lithium-polymer cells including temperature effects. In Proceedings of the IECON 2010—36th Annual Conference on IEEE Industrial Electronics Society, Glendale, AZ, USA, 7–10 November 2010; pp. 2329–2333.

41. Lam, L.; Bauer, P.; Kelder, E. A practical circuit-based model for li-ion battery cells in electric vehicle applications. In Proceedings of the 2011 IEEE 33rd International Telecommunications Energy Conference (INTELEC), Amsterdam, The Netherlands, 9–13 October 2011; pp. 1–9.

42. Hu, Y.; Wang, Y.-Y. Two time-scaled battery model identification with application to battery state estimation. *IEEE Trans. Control Syst. Technol.* **2015**, *23*, 1180–1188. [CrossRef]

43. Li, J.; Mazzola, M.S. Accurate battery pack modeling for automotive applications. *J. Power Sources* **2013**, *237*, 215–228. [CrossRef]

44. Widanage, W.; Barai, A.; Chouchelamane, G.; Uddin, K.; McGordon, A.; Marco, J.; Jennings, P. Design and use of multisine signals for Li-ion battery equivalent circuit modelling. Part 1: Signal design. *J. Power Sources* **2016**, *324*, 70–78. [CrossRef]

45. Jossen, A. Fundamentals of battery dynamics. *J. Power Sources* **2006**, *154*, 530–538. [CrossRef]

46. Hentunen, A.; Lehmuspelto, T.; Suomela, J. Time-domain parameter extraction method for thévenin-equivalent circuit battery models. *IEEE Trans. Energy Convers.* **2014**, *29*, 558–566. [CrossRef]

47. Petzl, M.; Danzer, M.A. Advancements in *OCV* measurement and analysis for lithium-ion batteries. *IEEE Trans. Energy Convers.* **2013**, *28*, 675–681. [CrossRef]

48. Barai, A.; Widanage, W.D.; Marco, J.; McGordon, A.; Jennings, P. A study of the open circuit voltage characterization technique and hysteresis assessment of lithium-ion cells. *J. Power Sources* **2015**, *295*, 99–107. [CrossRef]

49. Hariharan, K.S.; Kumar, V.S. A nonlinear equivalent circuit model for lithium ion cells. *J. Power Sources* **2013**, *222*, 210–217. [CrossRef]

50. Gong, X.; Xiong, R.; Mi, C.C. A data-driven bias-correction-method-based lithium-ion battery modeling approach for electric vehicle applications. *IEEE Trans. Ind. Appl.* **2016**, *52*, 1759–1765.

Article

A Generalized SOC-OCV Model for Lithium-Ion Batteries and the SOC Estimation for LNMCO Battery

Caiping Zhang [1,2], Jiuchun Jiang [1,2,*], Linjing Zhang [1,2], Sijia Liu [1,2], Leyi Wang [3] and Poh Chiang Loh [1,2]

[1] National Active Distribution Network Technology Research Center (NANTEC), Beijing Jiaotong University, Beijing 100044, China; zhangcaiping@bjtu.edu.cn (C.Z.); lj.zhang@bjtu.edu.cn (L.Z.); liusijia@bjtu.edu.cn (S.L.); bclu@bjtu.edu.cn (P.C.L.)
[2] Collaborative Innovation Center of Electric Vehicles in Beijing, Beijing Jiaotong University, Beijing 100044, China
[3] Department of Electrical and Computer Engineering, Wayne State University, Detroit, MI 48202, USA; lywang@wayne.edu
* Correspondence: jcjiang@bjtu.edu.cn; Tel.: +86-10-51683907

Academic Editors: Rui Xiong, Hailong Li and Joe (Xuan) Zhou
Received: 19 August 2016; Accepted: 26 October 2016; Published: 1 November 2016

Abstract: A state-of-charge (SOC) versus open-circuit-voltage (OCV) model developed for batteries should preferably be simple, especially for real-time SOC estimation. It should also be capable of representing different types of lithium-ion batteries (LIBs), regardless of temperature change and battery degradation. It must therefore be generic, robust and adaptive, in addition to being accurate. These challenges have now been addressed by proposing a generalized SOC-OCV model for representing a few most widely used LIBs. The model is developed from analyzing electrochemical processes of the LIBs, before arriving at the sum of a logarithmic, a linear and an exponential function with six parameters. Values for these parameters are determined by a nonlinear estimation algorithm, which progressively shows that only four parameters need to be updated in real time. The remaining two parameters can be kept constant, regardless of temperature change and aging. Fitting errors demonstrated with different types of LIBs have been found to be within 0.5%. The proposed model is thus accurate, and can be flexibly applied to different LIBs, as verified by hardware-in-the-loop simulation designed for real-time SOC estimation.

Keywords: electrochemical process analysis; SOC-OCV modeling; SOC estimation; lithium-ion batteries

1. Introduction

Lithium-ion batteries (LIBs) have been massively deployed in electric vehicles (EVs), hybrid electric vehicles (HEVs), plug-in hybrid electric vehicles (PHEVs), and stationary energy storage systems. Their attractiveness is their high voltage, high energy density, high efficiency, long cycle lifetime, and environmental benignity. Because of these advantages, their rapid growth is likely to continue with a strong likelihood of becoming the dominant storage technology. Along with this growth, accurate modeling of batteries is essential for control, optimization, energy management, diagnosis and prognosis in real time. The developed model will usually rely on the SOC-OCV relationship, which, in general, is for representing the battery electrochemical processes and thermodynamics at various SOCs. It is therefore a meaningful function needed for battery modeling, especially in the case of lumped parameter circuit models with an electromotive force (EMF) and a series of Resistance-Capacity (RC) networks [1–5]. For the RC networks, their parameters are always obtained experimentally based on data fitting, rather than specific physical principles. Similarly, the EMF is acquired by measuring the battery open-circuit terminal voltage when it reaches a steady state. It effectively reflects the concentration ratio of resultants to reactants during battery charging and discharging, and is therefore determined by the inherent electrochemical properties of the battery. This EMF is subsequently used as the approximate OCV.

The SOC-OCV function is therefore representative for a particular battery, and is generally a nonlinear monotone function between SOC and OCV for all LIBs. It is hence widely used in battery management systems (BMS) for correcting SOC calculation. Specific cases can be found in [6–9], where model based estimation of battery SOC and capacity has been developed using the SOC-OCV relationship. It has also been revealed in [10] that the accuracy of the SOC-OCV curve has great influence on the SOC value estimated. The same applies to battery capacity estimation, which has commonly been relied on for state-of-health (SOH) determination. It is consequently important to determine the SOC-OCV relationship precisely, if an accurate estimation of the battery state is necessary.

For this, some studies have proposed diversified methods for OCV modeling with each having distinctive pros and cons [11–21]. Xiong in [11] proposed a novel systematic state-of-charge estimation framework for accurately estimating SOC of the battery, where the relationship between battery SOC and OCV is highly employed. With the accurate battery model and adaptive filter based battery SOC estimator, the SOC of the battery pack can be accurately estimated. Reference [12] developed an EMF model as a function of the battery temperature, terminal voltage under open-circuit condition (not steady-state) and its slope. Its model parameters were determined from experimental data, but its accuracy gradually reduces as the battery ages. References [13,14] next use exponential and logarithmic functions for describing the relationship among OCV, EMF and time. However, like in [12], they result in battery models with high complexity, and are therefore difficult for usage in real time. Reference [15] proposed an alternative adaptive OCV estimation method based on battery diffusion principles. This method demonstrates high accuracy with its estimated SOC and capacity, but it is complex and has difficulty in online estimation because of its many coupled and non-coupled parameters. Reference [16] then employed a dynamic hysteresis model for predicting the OCV, where a hysteresis voltage has been included in the function for SOC. This model demonstrates high accuracy with $Li_4Ti_5O_{12}$ (LTO), $LiFePO_4$ (LFP), $LiMn_2O_4$ (LMO) and $LiNi_{1/3}Mn_{1/3}Co_{1/3}O_2$ (LNMCO) batteries, but its OCV hysteresis is generally not suitable for real-time model updating. Reference [17] proposed an OCV model structure in which simplified hyperbolic and exponential functions are used to represent phenomenological characteristics associated with the lithium-ion intercalation/deintercalation process. The developed SOC-OCV model applying to $LiFePO_4$ battery demonstrated higher accuracy compared to five OCV models summarized in [18]. However, its adaptability to other types of lithium-ion batteries needs to be further investigated. Reference [19] developed another type of OCV model that generates OCV vs. SOC curves based on the electrode half-cell data, which is able to be used for battery diagnostics and prognostics, and is an effective method especially for determining the degree of battery degradation in a quantitative manner. This approach requires half-cell data and thus opening the cells to reach high accuracy, which has difficulty in real-time SOC estimate applications.

This paper expands the main ideas in [10] and introduces a new model structure for the SOC-OCV relationship with some distinctive features that are especially important for model updating in real time: (1) the model uses four base functions that capture the fundamental electrochemical foundations over low, middle, and high SOC ranges; (2) it fits the experimental data for a large class of batteries of different types well, with very high accuracy; (3) it is simple and contains much fewer numbers of parameters than common existing models such as piece-wise interpolation types; (4) due to its simplicity, it becomes uniquely suitable for real-time updating on the parameter values. In other words, it is desirable for data-driven model identification, which is essential for adaptive battery management systems that can accommodate aging, environment variations, fault diagnosis, SOC estimation, and SOH monitoring.

Parameters of the generalized model must next be optimized for mapping out the SOC-OCV characteristics of different LIBs. For this, a nonlinear iterative algorithm has been developed, which is beyond the concepts presented in [10]. A real-time SOC estimation algorithm is then presented for online adaptive parameter updating needed for ensuring model accuracy throughout the battery lifetime. The updated model parameters can, in turn, be used for indicating SOH of the tested battery from the perspective of thermodynamics.

Contributions of the paper can thus be summarized as: (i) analysis of electrochemical processes during charging/discharging of commercial lithium-ion batteries, and their related function characteristics; (ii) development of a generalized SOC-OCV model that is simple, accurate and flexible for real-time battery state estimation; (iii) identification and verification of the proposed SOC-OCV model; (iv) analysis of parameter properties of the SOC-OCV model; and (v) hardware-in-the-loop (HIL) demonstration of real-time SOC estimation using the proposed SOC-OCV model.

The above contributions have been organized into five sections with Section 2 describing the SOC-OCV relations and electrochemical processes of various commercial LIBs. They include LMO, LNMCO and LFP batteries with graphite anodes, and novel batteries with LTO anodes. The generalized SOC-OCV mapping model and identification method are also introduced in this section. Section 3 then verifies the model accuracy for a variety of LIBs. The model robustness towards ambient temperature and battery aging is also analyzed in the section. Implementation issues and SOC estimation using the proposed SOC-OCV model are subsequently discussed in Section 4, where efficacy and accuracy of the estimation have been established. Finally, Section 5 summarizes the main findings of the paper, and highlights some related future issues.

2. Generalized SOC-OCV Model for Batteries

2.1. Experimental

Five types of batteries were tested to get their OCV-SOC curves. The battery types and their rated capacities are G//LMO-90 Ah, G//LNMCO-28 Ah, G//LNMCO + LMO-25 Ah, G//LFP-60 Ah, LTO//LNMCO + LMO-8.5 Ah. In the tests, the batteries were first discharged to the cutoff voltage with 1/20 C rated and rested for 2 h. Then, batteries were charged to the cutoff voltage, followed by a 2 h rest and then discharged to the cutoff voltage. Both the charge/discharge rate were 1/20 C. Arbin Instruments BT2000 test systems were used for the tests, and, during the test, batteries were put in the temperature chamber at 25 °C. $SOC = Q_{res}/Q_{max}$, where Q_{res} represents residual capacity of the battery, and Q_{max} is the maximum available capacity at the current of 1/20 C. From the test data, V-SOC curves for charge/discharge regime can be obtained, respectively, and then, by averaging the two V-SOC curves, the OCV-SOC curve of a battery is determined [22].

2.2. Electrochemical Analysis of OCV

Consider a commercial LIB with an LNMCO cathode and a graphite anode, its experimental OCV curve as a function of SOC and the corresponding dQ/dV profile is displayed in Figure 1, and its schematic presentation of the electrochemical redox reactions is illustrated in Table 1 [23]. The shown OCV behavior is caused by electrode redox reactions experienced by the cathode and anode materials. Particularly, in stage I, where the voltage gradually drops to a certain level, the main electrochemical reactions of the active materials can proceed, resulting in tardy voltage variations. This dynamic voltage decrease can appropriately be represented by a linear function that is associated with the continuous electrochemical redox reactions. In stage II, only small traces of electrochemical reactions occur due to the relatively low cell voltage. The cell voltage then suddenly drops, and can be described by a specific logarithmic function with a real (not complex) power. To better clarify this function form, the dQ/dV profile (differentiates the battery charged capacity (Q) to the terminal voltage (V)) derived from the SOC-OCV curve in Figure 1 is evaluated, beginning with the layered LNMCO cathode.

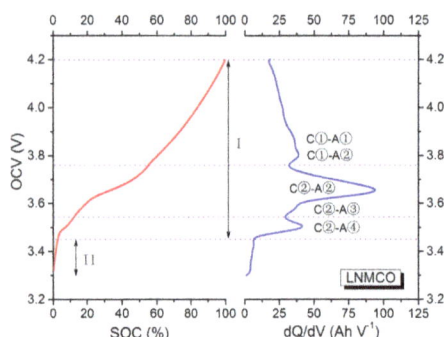

Figure 1. SOC-OCV curve (red line) and dQ/dV profile as a function of OCV (blue line) for the LNMCO/graphite lithium-ion battery.

Table 1. Schematic presentation of the electrochemical redox reactions of the LNMCO/Graphite lithium-ion battery.

Electrode	C①-A①	C①-A②	C②-A②	C②-A③	C②-A④
Cathode	C① $Ni^{4+} \leftrightarrow Ni^{3+}$			C② $Ni^{3+} \leftrightarrow Ni^{2+}$	
Anode	A① $LiC_6 \leftrightarrow LiC_{12}$	A② $LiC_{12} \leftrightarrow LiC_{18} \leftrightarrow LiC_{36}$		A③ $LiC_{36} \leftrightarrow LiC_{72}$	A④ $LiC_{72} \leftrightarrow C_6$

At the cathode, its presented voltage gradually increases/decreases throughout the whole electrochemical charging/discharging processes. When it is charged to above 4.5 V, all lithium ions deintercalated from the bulk of the material, along with the oxidation from Ni^{2+} to Ni^{4+} and from Co^{3+} to Co^{4+}. The layered LNMCO structure is thus damaged, attributed to the degraded electrochemical performance caused by the high charging voltage [24,25]. Consequently, the layered LNMCO cathode with homogenous phase reactions only works over the range of $0 \leq x \leq 2/3$ in $Li_{1-x}Co_{1/3}Ni_{1/3}Mn_{1/3}O_2$ [26]. Moreover, according to the first principle calculations reported, there are two solid state redox reactions occurring when the voltage is below 4.5 V. They are ascribed to $Li_{1/3}Co_{1/3}Ni_{1/3}Mn_{1/3}O_2$ and $Li_{2/3}Co_{1/3}Ni_{1/3}Mn_{1/3}O_2$ ranging from 4.2 V to 3.9 V (Ni^{4+}/Ni^{3+}, labeled as C①), and $Li_{2/3}Co_{1/3}Ni_{1/3}Mn_{1/3}O_2$ and $LiCo_{1/3}Ni_{1/3}Mn_{1/3}O_2$ at 3.9 V–3.7 V (Ni^{3+}/Ni^{2+}, labeled as C②), respectively.

In theory, the dQ/dV plot of the LNMCO electrode should therefore show two isolated peaks near these voltage plateaus. The two peaks are not separated, and hence appear as a broad peak in the dQ/dV plot [27]. On the other hand, for the graphite anode, there are five phase transformation stages during the charging/discharging processes over the voltage range from 0.8 to 0.01 V [28]. They include three main plateaus corresponding to the three pairs of redox reaction peaks in the dQ/dV plot of the graphite electrode. They are $LiC_6 \leftrightarrow LiC_{12}$ at 0.10/0.08 V (labeled as A①), $LiC_{12} \leftrightarrow LiC_{36}$ at 0.14/0.11 V (labeled as A②), and $LiC_{36} \leftrightarrow LiC_{72}$ at 0.22/0.20 V (labeled as A③), respectively. In addition, an unobvious plateau (or peak), $LiC_{72} \leftrightarrow C_6$ above 0.3 V (labeled as A④), usually appears in the SOC-OCV curve and dQ/dV plot.

The overall SOC-OCV curve of the cell in Figure 1 can be obtained using the approach reported in [29]. In other words, when the cathode is on a steady electrochemical reaction plateau, an additional peak will emerge at each distinct anode phase transformation plateau. Therefore, each peak in the dQ/dV plot of the cell can be distinctively identified. For example, C①-A② represents the state in which the cathode is on its first plateau and the anode is on its second plateau. The peak voltage in the dQ/dV plot is the difference in plateau voltage between the cathode and the anode [23]. It is also informative to point out that the number, location and shape of the peaks in the dQ/dV plot usually vary with operating conditions and degradation of the LIBs. It is therefore possible to identify performance decline origins of many cells based only on their dQ/dV variations.

Similarly, in the case of the spinel LMO cathode shown in Figure 2 and Table 2, it exhibits two electrochemical plateaus at 4.1 V (labeled as C′①) and 3.95 V (labeled as C′②). They correspond to the two phase transformations notated as $Mn_2O_4/Li_{0.5}Mn_2O_4$, and $Li_{0.5}Mn_2O_4/LiMn_2O_4$ [30,31]. Likewise, for the olive LFP cathode shown in Figure 3 and Table 3 [29], it manifests a unique voltage plateau at 3.45 V (labeled as C″①), which has been verified as the phase transformation between $FePO_4$ and $LiFePO_4$ [32]. In addition, unlike the LNMCO/graphite battery, the SOC-OCV curves of the LMO/graphite and LFP/graphite systems can be divided into three stages. In stage I, redox reactions basically completely with only slight traces of them continuing. The main process in this stage is also charge accumulation, whose effect is a rapid decrease of OCV value that can nicely be described by an exponential function. The latter two stages (II and III) are identical to the two stages (I and II) of LNMCO/graphite batteries, and are for representing the main electrochemical reaction and charge accumulation stages.

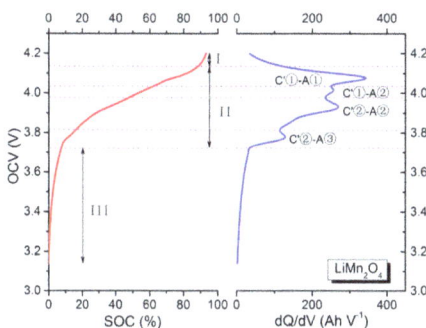

Figure 2. SOC-OCV curve and dQ/dV profile of the LMO/graphite lithium-ion battery.

Table 2. Schematic presentation of the electrochemical redox reactions of the LMO/Graphite lithium battery.

Electrode	C′①-A①	C′①-A②	C′②-A②	C′②-A③
Cathode	C′①		C′②	
	$Mn_2O_4 \leftrightarrow Li_{0.5}Mn_2O_4$		$Li_{0.5}Mn_2O_4 \leftrightarrow LiMn_2O_4$	
Anode	A①	A②		A③
	$LiC_6 \leftrightarrow LiC_{12}$	$LiC_{12} \leftrightarrow LiC_{18} \leftrightarrow LiC_{36}$		$LiC_{36} \leftrightarrow LiC_{72}$

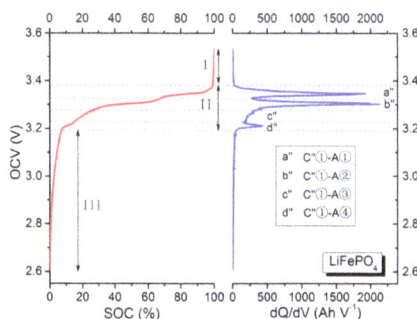

Figure 3. SOC-OCV curve and dQ/dV profile of the LFP/graphite lithium-ion battery.

Table 3. Schematic presentation of the electrochemical redox reactions of the LFP/Graphite lithium-ion battery.

Electrode	a″ C″①-A①	b″ C″①-A②	c″ C″①-A③	d″ C″①-A④
Cathode	C″① FePO$_4$↔LiFePO$_4$			
Anode	A① LiC$_6$↔LiC$_{12}$	A② LiC$_{12}$↔LiC$_{18}$↔LiC$_{36}$	A③ LiC$_{36}$↔LiC$_{72}$	A④ LiC$_{72}$↔C$_6$

For example, the voltage of the LFP batteries rapidly decreased in stage I because the main component is basically the FePO$_4$ without any more phase transformation. Upon reaching stage II, the voltage of the LFP battery remains relatively steady. The low voltage is attributed to the extraction/insertion reactions of the LFP cathode, assigned to the first-order phase transition. A flat voltage plateau is thus produced with two phase regions corresponding to FePO$_4$ and LiFePO$_4$. Taking the stage I into consideration, it may therefore introduce some differences when determining parameters for the SOC-OCV model, as compared to the earlier LNMCO battery. Figure 4 shows the SOC-OCV curve and corresponding dQ/dV profile of the novel lithium-ion batteries with NMC+LiCoO$_2$ (LCO) cathode and LTO anode, and the phase transformation is relatively uncomplicated. Table 4 reports its schematic presentation of the electrochemical redox reactions [33]. For the LTO anode [34], the phase transformation between Li$_7$Ti$_5$O$_{12}$ and Li$_4$Ti$_5$O$_{12}$ occurs near 1.55 V, showing a long plateau. For the LiCoO$_2$ cathode, it exhibits two very weak peaks at 4.19/4.13 V and 4.06/4.03 V (labeled as C‴①and C‴②) and a pair of strong redox peak at 3.97/3.85 V (labeled as C‴③), corresponding to the reduction/oxidation reactions of Co^{4+}/Co^{3+} [35], respectively. In view of the flat plateau of the LTO anode, the dQ/dV curve primarily reflects the characteristics of the NMC+LCO cathode and the LTO anode. However, due to the approximate phase transformation voltage of LCO and NMC, overlaps of peaks will appear in the dQ/dV curve of the LTO lithium-ion battery. The three stages of the SOC-OCV curve of the NMC+LCO/LTO batteries are similar to those of LMO/graphite and LFP/graphite batteries.

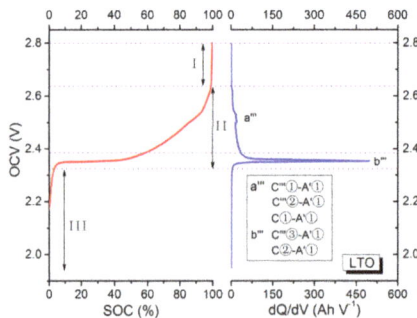

Figure 4. SOC-OCV curve and dQ/dV profile of the NMC+LCO/LTO lithium-ion battery.

Table 4. Schematic presentation of the electrochemical redox reactions of the NMC+LCO/Graphite lithium-ion battery.

Electrode	a‴ C‴①-A′①/C‴②-A′①/C①-A′①			b‴ C‴③-A′①/C②-A′①	
LCO + NMC cathode	C‴① Co^{3+}↔Co^{4+}	C‴② Co^{3+}↔Co^{4+}	C① Ni^{2+}↔Ni^{3+}	C‴③ Co^{3+}↔Co^{4+}	C② Ni^{3+}↔Ni^{4+}
LTO anode	A′① Li$_7$Ti$_5$O$_{12}$↔Li$_4$Ti$_5$O$_{12}$				

It is confirmed from electrochemical analysis of different types of LIBs that a three-segment SOC-OCV model is more potentially suitable for capturing all characteristic features of the described electrochemical processes. This model structure and its suitability for real-time implementation are described next.

2.3. Generalized SOC-OCV Model

The proposed generalized SOC-OCV model is shown in Equation (1), where a logarithmic function with real (not complex) power, a linear function, and an exponential function with a shifted exponent can clearly be seen:

$$V_{OCV} = a + b \cdot (-\ln s)^m + c \cdot s + d \cdot e^{n(s-1)} \tag{1}$$

where V_{OCV} and s represent the OCV and SOC of the battery, respectively, and $0 \le s \le 1$, $m > 0$ and $n > 0$.

To match the processes described in Section 2.2, the logarithmic function must be tuned to play a predominant role at low SOC, where charge accumulations on the surfaces of the active materials happen within the LIB. The linear function, in turn, dominates the middle SOC range, where primary phase transformation of the active materials occurs. The last exponential function then contributes to the high SOC behavior, where both partial redox reaction and charge accumulation occur. The three functions in Equation (1) are therefore essential, and will interact with each other to form the generalized OCV model over the whole SOC range.

Compared with [10], coefficients m and n included in Equation (1) are for adapting the model to match with different types of LIBs, since they depend on active materials of electrodes used in the LIBs. Their specific values and properties will be discussed in Section 3, where the hypothesis of m and n being invariant for a specific type of LIBs will also be proved. In other words, it will be proved that both m and n will not change with temperature and aging. Consequently, the SOC-OCV model is reduced to a sum of only four proportionally scaled terms, whose coefficients are a, b, c and d in Equation (1). Only these four coefficients require tuning in real-time to arrive at the desired SOC-OCV mapping. The complexity of realizing Equation (1) has therefore been considerably reduced, which is certainly encouraged for real-time state estimation.

2.4. Recursive Parameter Identification

For a given class of LIBs, its six model parameters in Equation (1) must collectively be determined from offline experimental data. This determination is nonlinear because of the logarithmic and exponential functions included in Equation (1). Some amount of complexity may therefore be involved, but will subsequently be proven to be otherwise since m and n are fixed for a given class of LIBs, and hence do not need real-time updating. They must, however, be determined for once at the beginning of real-time execution. Typical non-linear algorithm for optimization is thus needed, and is usually run iteratively. One possibility is the gradient based iterative search method, which mathematically, relies on the following equation, expressed in terms of the unknown parameter vector $\theta = (a, b, c, d, m, n)$:

$$V_{OCV} = a + b \cdot (-\ln s)^m + c \cdot s + d \cdot e^{n(s-1)} \ (0 \le s \le 1, m > 0, n > 0), \tag{2}$$

$$y = F(s; \theta) = a + b \cdot (-\ln s)^m + c \cdot s + d \cdot e^{n(s-1)}. \tag{3}$$

The obtained N sets of experimental data $(s(k), y(k))$, $k = 1, \ldots, N$, when substituted in Equation (3), further lead to the following equations:

$$\begin{aligned}
&y(k) = F(s(k); \theta) + e(k), \ k = 1, \ldots, N \\
&Y = F(S; \theta) + E, \\
&\text{where } Y = [y(1), \ldots, y(N)]^T, S = [s(1), \ldots, s(N)]^T, E = [e(1), \ldots, e(N)]^T
\end{aligned} \tag{4}$$

A nonlinear least square estimation problem has thus been formed. The purpose is to find the parameter vector that can minimize the following expression:

$$\min_{\theta} \quad \varepsilon(\theta) = \tfrac{1}{2}(Y - F(S;\theta))^T(Y - F(S;\theta))$$

$$\text{Let } G(\theta) = \left(\frac{\partial F(S;\theta)}{\partial \theta}\right)^T =$$

$$\begin{bmatrix} 1 & (-\ln s_1)^m & s_1 & e^{n(s_1-1)} & b \cdot [\ln(-\ln s_1)] \cdot (-\ln s_1)^m & d \cdot (s_1 - 1) \cdot e^{n(s_1-1)} \\ \vdots & \vdots & \vdots & \vdots & \vdots & \vdots \\ 1 & (-\ln s_N)^m & s_N & e^{n(s_N-1)} & b \cdot [\ln(-\ln s_N)] \cdot (-\ln s_N)^m & d \cdot (s_N - 1)e^{n(s_N-1)} \end{bmatrix}^T . \quad (5)$$

The process must usually be executed iteratively, using the gradient based algorithm provided as follows:

$$\theta_{j+1} = \theta_j + \mu_j G\left(\theta_j\right)\left(Y - F\left(S;\theta_j\right)\right), \quad (6)$$

where the step size μ_j must be selected to ensure algorithm convergence.

The algorithm is stopped only when $\|\theta_{j+1} - \theta_j\|$ is smaller than a pre-defined small threshold.

3. Verification of OCV Model

3.1. Estimation Accuracy Analysis with Different LIBs

To validate the proposed OCV model, a series of experiments were performed on different classes of LIBs to obtain SOC-OCV mapping data. Each experiment was performed with the battery charged from its fully discharged state at a current of 0.05 C. The charging continued until the terminal voltage of the battery reached the charging voltage limit. The battery was subsequently open-circuited for 2 h, after being discharged at 0.05 C until the battery terminal voltage reached the discharge voltage limit. Taking the average potential between the charge and discharge branch at C/20 and the normalized C/20 capacity, the voltage and its corresponding SOC can be regarded as OCV versus SOC curve [22].

The estimated and experimental SOC-OCV mapping for LMO, LNMCO, LNMCO and LMO, and LTO and LFP LIBs can eventually be illustrated in Figure 5a. Figure 5b shows the relative estimated and experimental error of each SOC-OCV mapping, which clearly indicates their close fitting except at both ends of the curve where there are a few points with larger errors. Other than those, the estimation errors have been kept within 0.5% for the LIBs except for the LFP battery when their SOCs are kept between 15% and 95%. The estimation errors are kept within 0.5% for the LFP battery when its SOC is kept between 15% and 90%. This is, in fact, the most widely used SOC region found in EVs. The larger errors at both ends of each SOC curve are thus not critical, but can still be explained from two aspects. The first is related to the model accuracy at both ends, where the more sensitive logarithmic and exponential functions are used. The second is related to polarization and Ohmic resistance, which are remarkably enlarged around 0% SOC and 100% SOC. The outcome is an enlarged random fluctuation of measured voltage, which can, no doubt, result in OCV measurement inaccuracy.

Different from other types of LIBs, the LFP LIB is attributed to the first-order phase transition mentioned in Section 2. Its freedom degree related to its terminal voltage is thus tiny, leading to very flat voltage plateaus in the middle SOC range. However, the change rate increases significantly at both ends of the SOC curve, which will then bring larger estimation errors to the LFP LIB than other types of LIBs when represented by the same proposed model. Despite that, the model errors are still well kept within 1% throughout the entire SOC range. The proposed generalized OCV model is thus accurate for representing different types of LIBs because of the presence of coefficients m and n. It is also more accurate than the model presented in [10] for the LNMCO battery. This can clearly be read from Figure 6, where the maximum estimation error for the LNMCO battery is noted to be below 0.5% when the proposed OCV model is used. The same battery will have an estimation error of 1% when the model in [10] is used. Precision enhancement is thus doubled with the proposed model, in addition to its flexible adaption to other types of LIBs.

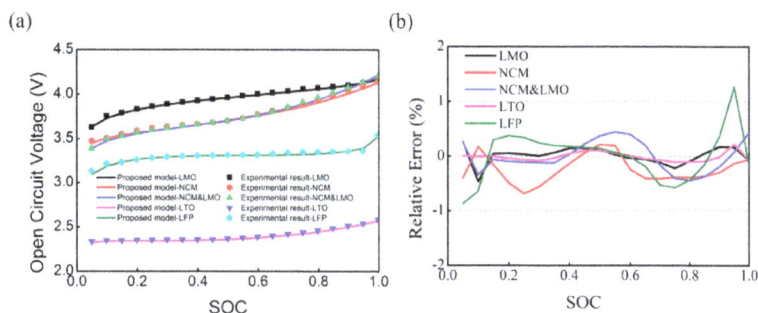

Figure 5. (**a**) the estimated and experimental SOC-OCV mapping for various LIBs; and (**b**) relative error of the OCV estimation.

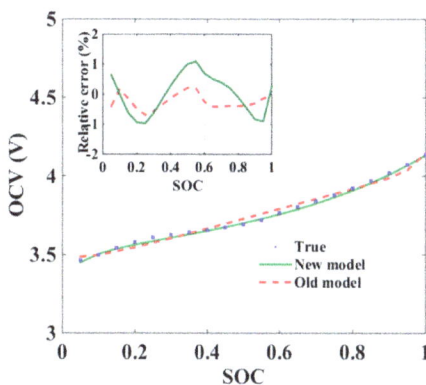

Figure 6. Comparison results of the old model and improved model for the LNMCO battery.

Moreover, it should be highlighted that parameters used with the SOC-OCV model are closely correlated to the intrinsic characteristic features of the LIBs. For example, parameter a is noted to relate to the voltage plateau with its value becoming bigger as the plateau rises higher. This can noticeably be seen in Table 5 and Figure 5. Similarly, parameter c is noted to relate to the rate of increase of voltage in the linear function, included for demonstrating the phase transformation processes. Its value will rise as the voltage increases faster, which certainly matches well with analytical results presented in this subsection.

Table 5. Model parameters for various LIBs.

Battery Type *	Parameters					
	a	b	c	d	m	n
LMO	3.875	−0.335	−0.5332	0.8315	0.653	0.6
LNMCO	3.5	−0.0334	−0.106	0.7399	1.403	2
LNMCO&LMO	3.6	−0.111	−0.5	1.113	1.093	1.9
LFP	3.135	−0.685	−1.342	1.734	0.478	0.4
LTO	2.235	−0.00132	−0.3503	0.6851	2.964	1.6

* LMO: G//LMO battery; LNMCO: G//LNMCO battery; LNMCO&LMO: G//LNMCO+LMO battery; LFP: G//LFP battery; LTO: LTO// LNMCO+LMO battery.

Two OCV models were selected in the literature, taking NCM and LFP battery, for example, which is compared to the proposed OCV model. All of the parameters in the two OCV models are refitted for the data illustrated in Figure 5 using the Matlab curve fitting toolbox (version, Manufacturer, City, US State abbrev. if applicable, Country), the results with their root mean square (RMS) and maximum errors are shown in Table 6. It can be seen that Model #1 has better accuracy for the LFP battery, while it is poor for the LNMCO battery. The fitting results of Model #2 are in contrast to Model #1. It should be noted that neither of them has good accuracy for both types of batteries. The proposed model has acceptable precisions for both NCM and LFP batteries, manifesting better adaptability compared to the other two models. It therefore can be regarded as a generalized model of commercial lithium-ion batteries.

Table 6. Compared fitting results of OCV models.

#	OCV Model	Reference	RMS Error for LNMCO Battery (mV)	Max Error for LNMCO Battery (mV)	RMS Error for LFP Battery (mV)	Max Error LFP Battery (mV)
1	$V_{OC}(s) = K_0 - K_1/s - K_2 s + K_3 \ln(s) + K_4 \ln(1-s)$	[20]	16.6	36.5	6.2	14.8
2	$V_{OC}(s) = K_0 + K_1 e^{-a(1-s)} - K_2/s$	[21]	9.7	21.8	34.9	141
3	$V_{OC} = a + b \cdot (-\ln s)^m + c \cdot s + d \cdot e^{n(s-1)}$	Proposed	13.0	20.6	15.3	27.3

3.2. Sensitivity Analysis of the Proposed Model

Case I Sensitivity towards temperature variation

The estimated and experimental results for the LNMCO battery at different temperatures are shown in Figure 7a–c. Their relative errors are shown in Figure 7d. Parameters of the model in Equation (1) used are shown in Table 7, where it has been noted that coefficients m and n have been kept constant regardless of temperature. Despite this, it can be seen from Figure 7 that the proposed OCV model fits the measured values well at 10 °C, 25 °C, and 45 °C, respectively. Their relative errors over a wide span of the SOC range are, in fact, always smaller than 0.5%. The proposed model is thus robust since its accuracy is not degraded by battery temperature variation. Moreover, coefficients m and n for the specific battery type have been fixed without affecting the model accuracy. Model updating has therefore been done by adjusting parameters a, b, c and d in Equation (1) only, which is dramatically simpler since they are simply proportional gains.

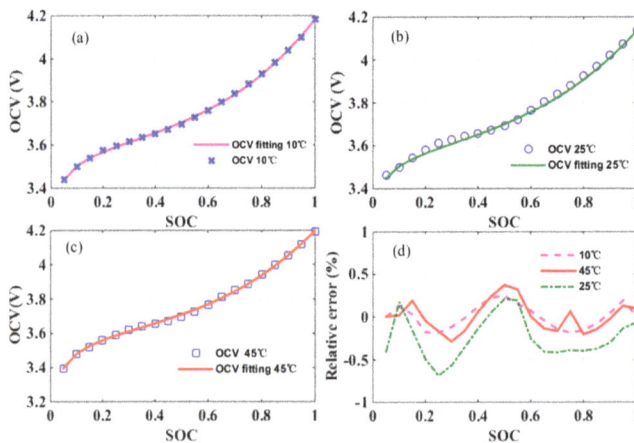

Figure 7. The estimated and experimental results for the LNMCO battery at different temperatures: (a) 10 °C; (b) 25 °C; (c) 45 °C; and (d) relative errors.

Table 7. Model parameters of the LNMCO Battery at different temperatures.

Temperature (°C)	Parameters					
	a	b	c	d	m	n
10	3.517	−0.0439	−0.2493	0.9134	1.4	2
25	3.5	−0.0334	−0.106	0.7399	1.4	2
45	3.535	−0.0571	−0.2847	0.9475	1.4	2

Case II Sensitivity to battery aging

Conventionally, the battery OCV will change as it ages even at the same SOC. This is caused by variation of battery thermodynamics, which was affected in experiments by cycling the charging and discharging processes for evaluating the proposed OCV model. The estimated and experimental results obtained for the LMO battery at different degradation states are shown from Figure 8a–d, while the model parameters used at different battery aging stages are listed in Table 8. Obviously, coefficients *m* and *n* have been kept unchanged after being first determined. Despite this, Figure 8 shows the estimated and experimental results matching well at various battery aging stages. Their relative error is, in fact, within 0.2%, except at a few individual SOC points where the relative error has been bigger at 0.5%. The proposed OCV model is thus robust against battery aging.

From Table 8, it can also be seen that the model parameters change monotonically as the battery degrades. Taking the linear function of Equation (1), for example, its parameters *a* and *c* (considering the negative sign of *c*) increase as the battery ages. These parameters can thus be employed for SOH estimation, but can be rather complex since the SOC-OCV model in Equation (1) manifests the comprehensive effects of three functions (logarithmic, linear and exponential). Estimation of battery degradation mechanism is hence rather complex with the origin of model parameter variations caused by electrochemical dynamics needing to be investigated first. This is, however, beyond the scope of the paper.

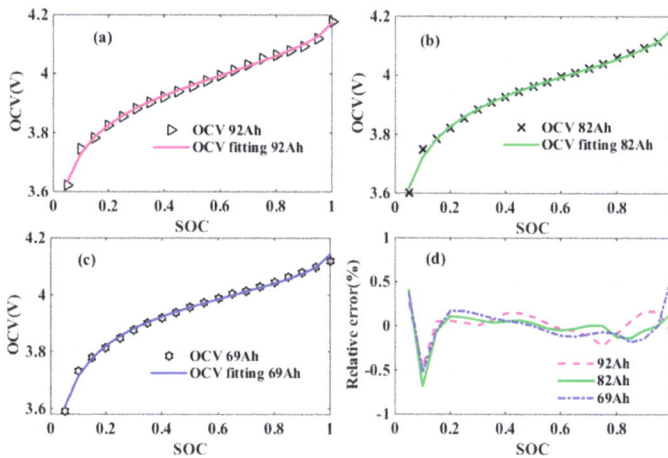

Figure 8. Estimated and experimental results for the LMO battery at different aging states: (**a**) 92 Ah; (**b**) 82 Ah; (**c**) 69 Ah; and (**d**) relative errors.

Table 8. Model parameters of the LMO battery at different aging states.

Capacity (Ah)	Parameters					
	a	*b*	*c*	*d*	*m*	*n*
92	3.875	−0.3351	−0.5332	0.8315	0.6537	0.6
82	4.061	−0.3683	−0.4946	0.5933	0.6537	0.6
69	4.132	−0.3838	−0.4912	0.5016	0.6537	0.6

4. Application of SOC Estimation

An HIL simulation platform has been set up for validating the proposed generalized OCV model and its feasibility for SOC estimation. In relation to SOC estimation, a first-order equivalent circuit and the proposed OCV model have been used for simulating the battery dynamics. An estimation algorithm based on a proportional-integral (PI) observer has also been employed for determining the estimated SOC for comparison with the actual SOC measured from the tested battery group. Principle of the PI observer has been explained in [36], and will hence not be discussed with the HIL platform described below. It is noticeable that the parameter uncertainties of the equivalent circuit model have an impact on SOC estimation. The detailed theoretical and quantitative analysis can be found in another paper [37].

4.1. HIL Simulation Platform

The HIL simulation platform is shown in Figure 9, where the central processing system is a computer for controlling the hardware experiment, obtaining voltage/current data, and executing the real-time simulated model. Specific hardware used includes the current/voltage acquisition board and the CAN bus to TCP (Transmission Control Protocol) conversion card. Software includes the Matlab/Simulink model, and driver for the CAN-TCP data acquisition card. The CAN-TCP conversion module is for receiving the CAN bus data frame. The real-time data is used by the simulated model in the central processing system for estimating the battery SOC. The estimated SOC will then be compared with the true SOC measured by the Arbin Instruments BT2000. This instrument is inter-faced to the thermal chamber, where the tested battery is placed for verification purposes.

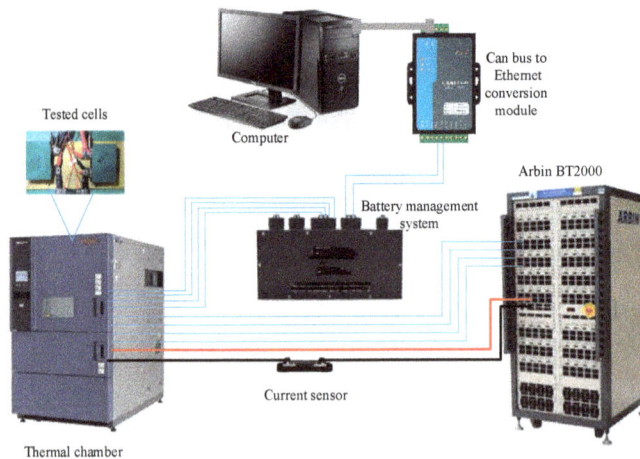

Figure 9. Hardware-in-loop simulation platform.

4.2. Accuracy Verification

Two LNMCO cells in series have been used for forming the battery group with a total nominal capacity of 28 Ah. SOCs of each cell in the group have also been estimated by the PI observer and the first-order circuit model, in which the proposed SOC-OCV model has been employed. From the cell SOCs, SOCs of the battery group can then be calculated by using the following formula [38]:

$$
\begin{aligned}
SOC^B &= \frac{Q^B_{rem}}{Q^B_{max}} \times 100\% \\
&= \frac{\min(Q_{max}[1] \times SOC[1],...Q_{max}[n] \times SOC[n])}{\min(Q_{max}[1] \times SOC[1],...Q_{max}[n] \times SOC[n]) + \min(Q_{max}[1] \times (1 - SOC[1]),...Q_{max}[n] \times (1 - SOC[n]))} \\
&= \frac{Q_{max}[i] \times SOC[i]}{Q_{max}[i] \times SOC[i] + Q_{max}[j] \times (1 - SOC[j])} \times 100\%,
\end{aligned}
\tag{7}
$$

where Q^B_{max} means the maximum available capacity of the pack, and Q^B_{rem} represents the residual capacity or maximum discharge capacity of the group. It should be noticeable that temperature will affect the SOC estimation, especially at low temperatures. The model parameters will vary with the battery temperature. Therefore, two typical temperatures at 5 °C and 25 °C have been tested with DST (dynamic stress test) profiles included for verifying SOC estimations. The model parameters have also been updated online to ensure estimation accuracy as temperature changes.

The estimated and true SOCs obtained from the HIL simulation platform at DST profiles are successively shown in Figure 10a–d, where the columbic counting results from the Arbin test instrument was regarded as the true SOC. Figure 10a–b show the estimated results at 5 °C, while Figure 10c,d show the estimated results at 25 °C. The estimated error is obviously large at the beginning because of overshooting of the PI observer, but will eventually converge to a very small value as time progresses. The estimated error is also noted to be large at partial SOC, where one possible reason is related to the assumption of resistance and capacitance of the first-order RC model being constant. Such an assumption has, no doubt, simplified calculation, but does not reflect the actual scenario, where resistance and capacitance of the RC model at partial SOC are considerably different from those at low and high SOCs. The other possible reason is related to inherent errors that may occur within the OCV model. These errors will, no doubt, affect estimation accuracy, but will usually not be as significant. Estimation error after the initial overshoot period of the battery group can thus be controlled within 3% in Figure 10 for both 5 °C and 25 °C. In other words, the proposed OCV model is suitable for SOC estimation with high precision and good adaptability demonstrated.

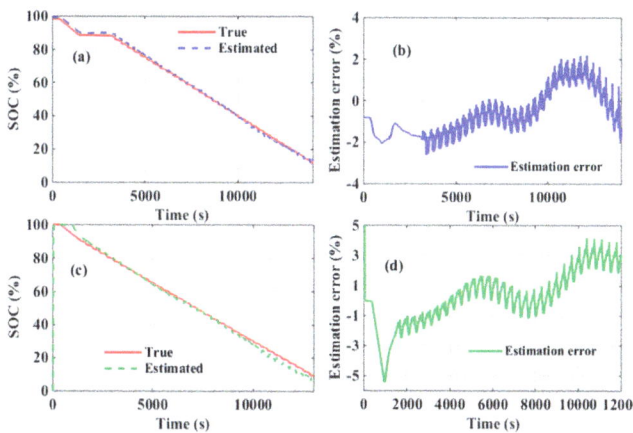

Figure 10. Estimation and true values of SOC at DST profilesat different temperatures, (**a**) SOC estimation at 5 °C; (**b**) estimation error at 5 °C; (**c**) SOC estimation at 25 °C; (**d**) estimation error at 25 °C.

5. Conclusions

From analyzing electrochemical processes of common LIBs, the charge accumulation, primary phase transformation, and partial redox reaction and charge accumulation processes have been identified. To account for these three processes, a generalized OCV model with a logarithmic, a linear and an exponential function has been developed. The model includes six parameters, but with only four proportional parameters requiring real-time updating. The other two parameters, which contribute to its nonlinear characteristics, can be kept constant after first being determined for a particular type of LIB. Complexity is thus reduced, but without affecting precision since the estimation error can still be controlled within 0.5% for all common types of LIBs. Efficacy of the model for SOC estimation has also been proven through HIL simulation and experiments, with errors of smaller than 3% observed at different temperatures and stages of degradation. The proposed model is thus robust and accurate, while retaining simplicity for real-time implementation. Further extension of the model for SOH estimation is also possible but will only be more thoroughly investigated in the future.

Acknowledgments: The authors are grateful for financial support via the National Natural Science Foundation of China (Grant No. 51477009) and the Fundamental Research Funds for the Central Universities (Grant Number E16JB00060).

Author Contributions: In this paper, author contributions are as follows: Caiping Zhang, Jiuchun Jiang and Le Yi Wang conceived the main points of this paper. Linjing Zhang analyzed the electrochemical properties of OCV, Caiping Zhang and Sijia Liu designed and performed the experiments; Poh Chiang Loh contributed materials/analysis tools.

Conflicts of Interest: The authors declare no conflict of interest.

Abbreviations

The following abbreviations are used in this manuscript:

OCV	Open Circuit Voltage
SOC	State of Charge
LIB	Lithium-Ion Battery
BMS	Battery Management System
SOH	State of Health
HIL	Hardware in Loop
LFP	$LiFePO_4$
LMO	$LiMn_2O_4$
LNMCO	$LiNi_{1/3}Mn_{1/3}Co_{1/3}O_2$
LTO	$Li_4Ti_5O_{12}$

References

1. Hu, X.; Li, S.; Peng, H. A comparative study of equivalent circuit models for Li-ion batteries. *J. Power Sources* **2012**, *198*, 359–367. [CrossRef]
2. Xiong, R.; He, H.; Sun, F.; Liu, X.; Liu, Z. Model-based State of Charge and peak power capability joint estimation of Lithium-Ion battery in plug-in hybrid electric vehicles. *J. Power Sources* **2012**, *229*, 159–169. [CrossRef]
3. Castano, S.; Gauchia, L.; Voncila, E.; Sanz, J. Dynamical modeling procedure of a Li-ion battery pack suitable for real-time applications. *Energy Convers. Manag.* **2015**, *92*, 396–405. [CrossRef]
4. Samadania, E.; Farhadb, S.; Scottc, W.; Mastali, M.; Gimenez, L.E.; Fowler, M.; Fraser, R.A. Empirical modeling of lithium-ion batteries based on electrochemical impedance spectroscopy tests. *J. Electrochim. Acta* **2015**, *160*, 169–177. [CrossRef]
5. Gagneur, L.; Driemeyer-Franco, A.L.; Forgez, C.; Friedrich, G. Modeling of the diffusion phenomenon in a lithium-ion cell using frequency or time domain identification. *Microelectron. Reliab.* **2013**, *53*, 784–796. [CrossRef]
6. Xiong, R.; Sun, F.C.; Gong, X.Z.; He, H.W. Adaptive state of charge estimator for lithium-ion cells series battery pack in electric vehicles. *J. Power Sources* **2013**, *242*, 699–713. [CrossRef]

7.   Liu, L.Z.; Wang, L.Y.; Chen, Z.Q.; Wang, C.; Lin, F.; Wang, H. Integrated system identification and state-of-charge estimation of battery systems. *IEEE Trans. Energy Convers.* **2013**, *28*, 12–23. [CrossRef]

8.   Rahimi-Eichi, H.; Baronti, F.; Chow, M.Y. Online adaptive parameter identification and state-of-charge co-estimation for Lithium-Polymer battery cells. *IEEE Trans. Ind. Electron.* **2014**, *61*, 2053–2061. [CrossRef]

9.   Wang, W.G.; Chung, H.S.H.; Zhang, J. Near-Real-Time Parameter Estimation of an Electrical Battery Model with Multiple Time Constants and SOC-Dependent Capacitance. *IEEE Trans. Power Electron.* **2014**, *29*, 5905–5920. [CrossRef]

10. Zhang, C.; Wang, L.Y.; Li, X.; Chen, W.; Yin, G.; Jiang, J. Robust and Adaptive Estimation of State of Charge for Lithium-Ion Batteries. *IEEE Trans. Ind. Electron.* **2015**, *62*, 4948–4957. [CrossRef]

11. Sun, F.; Xiong, R.; He, H. A systematic state-of-charge estimation framework for multi-cell battery pack in electric vehicles using bias correction technique. *Appl. Energy* **2016**, *162*, 1399–1409. [CrossRef]

12. Hoenig, S.; Singh, H.; Palanisamy, T.G. Method for Determining State of Charge of a Battery by Measuring Its Open Circuit Voltage. U.S. Patent 6,366,054 B1, 2 April 2002.

13. Yang, Y.P.; Liu, J.J.; Tsai, C.H. Improved estimation of residual capacity of batteries for electric vehicles. *J. Chin. Inst. Eng.* **2008**, *31*, 313–322. [CrossRef]

14. Pop, V.; Bergveld, H.J.; Danilov, D.; Regtien, P.P.L.; Notten, P.H.L. *Battery Management Systems: Accurate State-of-Charge Indication for Battery-Powered Applications*; Springer: London, UK, 2008.

15. Waag, W.; Sauer, D.U. Adaptive estimation of the electromotive force of the lithium-ion battery after current interruption for an accurate state-of-charge and capacity determination. *Appl. Energy* **2013**, *111*, 416–427. [CrossRef]

16. Barai, A.; Widanage, W.D.; Marco, J.; McGordon, A.; Jennings, P. A study of the open circuit voltage characterization technique and hysteresis assessment of lithium-ion cells. *J. Power Sources* **2015**, *295*, 99–107. [CrossRef]

17. Weng, C.; Sun, J.; Peng, H. A unified open-circuit-voltage model of lithium-ion batteries for state-of-charge estimation and state-of-health monitoring. *J. Power Sources* **2014**, *258*, 228–237. [CrossRef]

18. Hu, X.; Li, S.; Peng, H.; Sun, F. Robustness analysis of State-of-Charge estimation methods for two types of Li-ion batteries. *J. Power Sources* **2012**, *217*, 209–219. [CrossRef]

19. Dubarry, M.; Truchot, C.; Liaw, B.Y. Synthesize battery degradation modes via a diagnostic and prognostic model. *J. Power Sources* **2012**, *219*, 204–216. [CrossRef]

20. Plett, G.L. Extended Kalman filtering for battery management systems of LiPB-based HEV battery packs: Part 2. Modeling and identification. *J. Power Sources* **2004**, *134*, 262–276. [CrossRef]

21. Neumann, D.E.; Lichte, S. A multi-dimensional battery discharge model with thermal feedback applied to a lithium-ion battery pack. In Proceedings of the NDIA Ground Vehicle Systems Engineering and Technology Symposium-Modelling & Simulation, Testing and Validation (MSTV) Mini-Symposium, Dearborn, MI, USA, 9–11 August 2011.

22. Dubarry, M.; Svoboda, V.; Hwu, R.; Liaw, B.Y. Capacity loss in rechargeable lithium cells during cycle life testing: The importance of determination state-of-charge. *J. Power Sources* **2007**, *174*, 1121–1125. [CrossRef]

23. Dubarry, M.; Truchot, C.; Cugnet, M.; Liaw, B.Y.; Gering, K.; Sazhin, S.; Jamison, D.; Michelbacher, C. Evaluation of commercial lithium-ion cells based on composite positive electrode for plug-in hybrid electrical vehicle applications. Part I: Initial characterizations. *J. Power Sources* **2011**, *196*, 10328–10335. [CrossRef]

24. Shin, Y.J.; Choi, W.J.; Hong, Y.S.; Yoon, S.; Ryu, K.S.; Chang, S.H. Investigation on the microscopic features of layered oxide Li[Ni$_{1/3}$Co$_{1/3}$Mn$_{1/3}$]O$_2$ and their influences on the cathode properties. *Solid State Ion.* **2006**, *177*, 515–521. [CrossRef]

25. Myung, S.T.; Ogata, A.; Lee, K.S.; Komaba, S.; Sun, Y.K.; Yashiro, H. Structural, electrochemical, and thermal aspects of Li[(Ni$_{0.5}$Mn$_{0.5}$)$_{1-x}$Co$_x$]O$_2$ ($0 \leq x \leq 0.2$) for high-voltage application of lithium-ion secondary batteries. *J. Electrochem. Soc.* **2008**, *155*, A374–A383. [CrossRef]

26. Yabuuchi, N.; Makimura, Y.; Ohzuku, T. Solid-state chemistry and electrochemistry of LiCo$_{1/3}$Ni$_{1/3}$Mn$_{1/3}$O$_2$ for advanced lithium-ion batteries. *J. Electrochem. Soc.* **2007**, *154*, A314–A321. [CrossRef]

27. Kim, S.K.; Jeong, W.T.; Lee, H.K.; Shim, J. Characteristics of LiNi$_{1/3}$Co$_{1/3}$Mn$_{1/3}$O$_2$ Cathode Powder. *Int. J. Electrocehm. Sci.* **2008**, *3*, 1504–1511.

28. Ohzuku, T.; Lwakoshi, Y.; Sawai, K. Formation of Lithium-Graphite Intercalation Compounds in Nonaqueous Electrolytes and Their Application as a Negative Electrode for a Lithium Ion (Shuttlecock) Cell. *J. Electrochem. Soc.* **1993**, *140*, 2490–2498. [CrossRef]

29. Dubarry, M.; Liaw, B.Y. Identify capacity fading mechanism in a commercial LiFePO$_4$ cell. *J. Power Sources* **2009**, *194*, 541–549. [CrossRef]

30. Moorhead-Rosenberg, Z.; Allcorn, E.; Manthiram, A. In Situ Mitigation of First-Cycle Anode Irreversibility in a New Spinel/FeSb Lithium-Ion Cell Enabled via a Microwave-Assisted Chemical Lithiation Process. *Chem. Mater.* **2014**, *26*, 5905–5913. [CrossRef]

31. Goodenough, J.B.; Kim, Y. Challenges for Rechargeable Li Batteries. *Chem. Mater.* **2010**, *22*, 587–603. [CrossRef]

32. Padhi, A.K.; Nanjundaswamy, K.S.; Goodenough, J.B. Phospho-olivines as Positive-Electrode Materials for Rechargeable Lithium Batteries. *J. Electrochem. Soc.* **1997**, *144*, 1188–1194. [CrossRef]

33. Devie, A.; Dubarry, M.; Liaw, B.Y. Overcharge study in Li4Ti5O12 based lithium-ion pouch cell, I. Quantitative diagnosis of degradation modes. *J. Electrochem. Soc.* **2015**, *162*, A1033–A1040. [CrossRef]

34. Wang, Y.; Liu, H.; Wang, K.; Eiji, H.; Wang, Y.; Zhou, H. Synthesis and electrochemical performance of nano-sized Li$_4$Ti$_5$O$_{12}$ with double surface modification of Ti(III) and carbon. *J. Mater. Chem.* **2009**, *19*, 6789–6795. [CrossRef]

35. Pentyala, N.; Guduru, R.K.; Mohanty, P.S. Binder free porous ultrafine/nano structured LiCoO$_2$ cathode from plasma deposited cobalt. *Electrochim. Acta* **2011**, *56*, 9851–9859. [CrossRef]

36. Xu, J.; Mi, C.C.; Cao, B.G.; Deng, J.; Chen, Z.; Li, S. The state of charge estimation of lithium-ion batteries based on a proportional-integral observer. *IEEE Trans. Veh. Technol.* **2014**, *63*, 1614–1621.

37. Zhao, T.; Jiang, J.C.; Zhang, C.P.; Bai, K.; Li, N. Robustness online state of charge estimation of lithium-ion battery pack based on error sensitivity analysis. *Math. Probl. Eng.* **2015**, *2015*, 573184.

38. Wen, F. Study on Basic Issues of the Li-Ion Battery Pack Management Technology for Pure Electric Vehicles. Ph.D. Thesis, Beijing Jiaotong University, Beijing, China, 2009.

energies

MDPI

Article

Modeling of a Pouch Lithium Ion Battery Using a Distributed Parameter Equivalent Circuit for Internal Non-Uniformity Analysis

Dafen Chen [1,2], Jiuchun Jiang [1,2,*], Xue Li [1,2], Zhanguo Wang [1,2] and Weige Zhang [1,2]

[1]   National Active Distribution Network Technology Research Center, Beijing Jiaotong University, No. 3 Shang
     Yuan Cun, Haidian District, Beijing 100044, China; 11117360@bjtu.edu.cn (D.C.); 13117372@bjtu.edu.cn
     (X.L.); zhgwang@bjtu.edu.cn (Z.W.); wgzhang@bjtu.edu.cn (W.Z.)
[2]   Collaborative Innovation Center of Electric Vehicles in Beijing, Beijing Jiaotong University, No. 3 Shang Yuan
     Cun, Haidian District, Beijing 100044, China
*   Correspondence: jcjiang@bjtu.edu.cn; Tel.: +86-10-5168-4056; Fax: +86-10-5168-3907

Academic Editor: Rui Xiong
Received: 18 August 2016; Accepted: 10 October 2016; Published: 25 October 2016

Abstract: A battery model that has the capability of analyzing the internal non-uniformity of local state variables, including the state of charge (SOC), temperature and current density, is proposed in this paper. The model is built using a set of distributed parameter equivalent circuits. In order to validate the accuracy of the model, a customized battery with embedded T-type thermocouple sensors inside the battery is tested. The simulated temperature conforms well with the measured temperature at each test point, and the maximum difference is less than 1 °C. Then, the model is applied to analyze the evolution processes of local state variables' distribution inside the battery during the discharge process. The simulation results demonstrate drastic distribution changes of the local state variables inside the battery during the discharge process. The internal non-uniformity is originally caused by the resistance of positive and negative foils, while also influenced by the change rate of open circuit voltage and the total resistance of the battery. Hence, the factors that affect the distribution of the local state variables are addressed.

Keywords: lithium ion battery; distributed parameter equivalent circuit model; internal non-uniformity

1. Introduction

The lithium ion battery is one of the most promising candidates for the energy storage system (ESS) in electrical vehicles. To reduce the cost, prolong the life and ensure the safety of the ESS, the optimal design of the battery cell and pack with robust battery management system (BMS) is essential. Two kinds of battery models are proposed to fulfill the optimal design and to be implanted in the BMS, respectively. The first kind of model is physics based. A model based on the theory of intercalation electrodes and concentrated solutions was proposed by Doyle et al. [1]. Lou et.al. [2] built an extended single particle model for higher rates simulation. Kim et al. [3,4] developed a multi-scale multi-domain model framework, which resolved electrochemical, thermal and electrical coupled physics at varied length scales in the lithium ion battery. Three-dimensional electrochemical-thermal models were proposed and widely used by researchers [5–8] to analyze heat generation and temperature distribution in the lithium ion battery. A two-dimensional model was reported by Shin's group [9–11] for the thermal behavior and scale-up design of the lithium-polymer battery. These models are usually computationally complex and time consuming. In addition, a large number of parameters is needed for model inputs, and some of the parameters are extremely hard to obtain. Thus, such models are only suitable for cell designing and analyzing. For example, a three-dimensional electrochemical model was proposed by Kohneh et al. [12] and used to examine the effect of parameters, such as current collector thickness and tab location, on reducing non-uniform voltage and current distribution in the cell. Liu et al. [13] developed an integrated computational method that considered the mechanical, electrochemical and thermal behaviors of the lithium ion battery to study the nail penetration problem. Samba et al. [14] used the electrical-thermal model to design the battery thermal management system. Though simplified approaches, such as the simplified electrochemical multi-particle model and homogenous pseudo two-dimensional model, were developed by Mastali et al. [15] to decrease the computational time, the speed and simplicity of three-dimensional electrochemical-thermal models are still of concern. The second kind of models is the equivalent circuit model (ECM). Here [16–20], the battery is usually regarded as a mass point. Only a few parameters are in the model and can be derived from the external characteristics of the cell [21,22]. Therefore, they are suitable to be implanted in the battery management system (BMS) for the state of charge or the state of health estimation [23–26]. Lumped-parameter thermal models were added to ECM to predict the thermal characteristics of the cell by Lin et al. [27] and Forgez et al. [28], which made the model more comprehensive.

Meanwhile, in order to reach higher power and energy density, the size of the battery cell is growing. This leads to the imbalance of the potential, current density, temperature and state of charge (SOC) becoming significant, so the battery can no longer be treated as a mass point. Large SOC differences up to 5.3% were reported by Fleckenstein et al. [29] in a LiFePO$_4$ cell. Taking the computational cost into consideration, a new model to predict the internal imbalance of the battery is desired at this stage. Therefore, the distributed parameter equivalent circuit model (DPECM) is proposed in our former works [30]. However, the former proposed model could only solve the electric related variables, and the model is only validated by the terminal voltage. As the temperature of lithium ion batteries is vital for their life span, safety and performance, thermal models need to be established and implemented into the former proposed model, to analyze the distribution and variation trend of internal current, voltage and temperature.

On the other hand, the validation of those models that could obtain the internal variables is somehow technically difficulty, because the distribution of the internal potential, current density or SOC cannot be measured directly. Wang's group [31,32] proposed an in situ measurement method for current distribution in a lithium ion battery. The structure of the cell was changed because the positive electrode was segmented along the length of the electrode sheet; while temperature is much easier to measure compared to other variables inside a battery. Forgez et al. [28] and Li et al. [33] successfully embedded thermocouples into cylindrical and pouch cells, respectively. They proved that the cell characteristics affected by the embedded sensors were negligible. To validate the proposed model in our research, a customized cell with nine T-type thermocouples embedded inside the battery was provided by the battery manufacture. The customized battery was tested for the parameter identification and validation of the temperature distribution inside the battery.

The remaining parts of this paper are arranged as follows. Section 2 describes the experimental setup, cell dimensions and temperature sensor locations. Section 3 introduces the modeling method and validation. The internal non-uniformity is discussed in Section 4. The conclusions are summarized in Section 5.

2. Experiments

The cell studied in this paper is a 35 Ah LiNi$_{1/3}$Co$_{1/3}$Mn$_{1/3}$O$_2$ (NCM) pouch cell, which has a total of 36 laminated layers. Each laminated layer is composed of positive foil, positive material, separator, negative material, and negative foil, and the characteristics of each component are presented in Table 1. The top view from cell thick direction and the cell dimensions are shown in Figure 1. The length and width of the main body are 195 mm and 165 mm, respectively. Positive and negative tabs locate at two opposite sides of the cell, whose length and width are 40 mm and 80 mm, respectively. The total thickness of the cell is only 14 mm, which is much smaller than the length and width of battery, so the differences of local SOC, current density, and potential along thick direction are ignored in this paper. In order to acquire internal temperature reliably, nine T-type thermocouples are embedded inside the cell during the manufacturing process of the cell using the same method proposed by Li et al. [33]. The thermocouples were sandwiched by two pieces of separators and then placed in the middle of the thickness before vacuuming and sealing of the cell. Nine thermocouples are uniformly distributed on the middle plane.

Table 1. Properties of each battery component.

Component	Thickness (μm)	Heat Conductivity Coefficient ($w \cdot m^{-1} \cdot k^{-1}$)	Specific Heat Capacity ($J \cdot kg^{-1} \cdot k^{-1}$)	Electrical Resistivity ($\Omega \cdot m$)
Positive foil	20	238	903	2.83×10^{-8}
Positive material	160	3.9	839	—
Separator	40	0.33	1978	—
Negative material	110	3.3	1064	—
Negative foil	10	398	385	1.75×10^{-8}

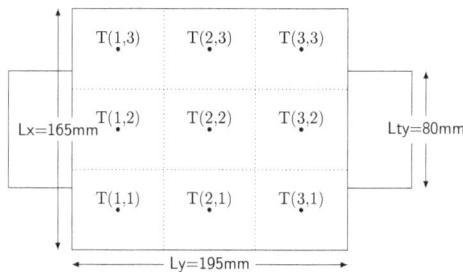

Figure 1. Cell dimensions and thermocouple locations.

In order to obtain the model input parameters and verify the accuracy of the model, two sets of experiments are executed. One is for parameter estimation, and the other is for validation.

2.1. Parameters Identification Experiments

2.1.1. Parameters Identification of the Lumped First-Order Resistor-Capacitor (RC) Model

A typical lumped first-order resistor-capacitor (RC) model is shown in Figure 2. It consists of a voltage source, a resistor and a parallel branch of the resistor and capacitor. The voltage source represents the open circuit voltage (OCV) of the cell. Resistor R_Ω represents the ohmic resistance in the cell. The parallel branch is utilized to model the chemical diffusion in the cell. In order to obtain the parameters of the first-order RC lumped model, pulse discharge is a widely-used method for lithium ion batteries. In this paper, the cell was put into an environment chamber during the test to keep the ambient temperature constant, and then, discharge pulses were applied to the cell by battery test equipment: Arbin BT-2000. Each discharge pulse depleted 5% of the total capacity and was followed by 2 hours of rest. To make the parameters suitable for different current rates, the discharge pulse at each SOC contained multiple current rates, including the current fragments of 0.5C, 1C and 2C, as shown in Figure 3a. The corresponding voltage response of the cell is shown in Figure 3b.

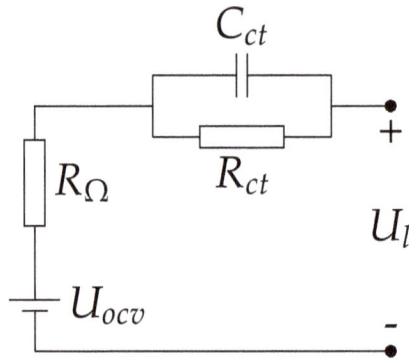

Figure 2. Lumped first-order RCmodel.

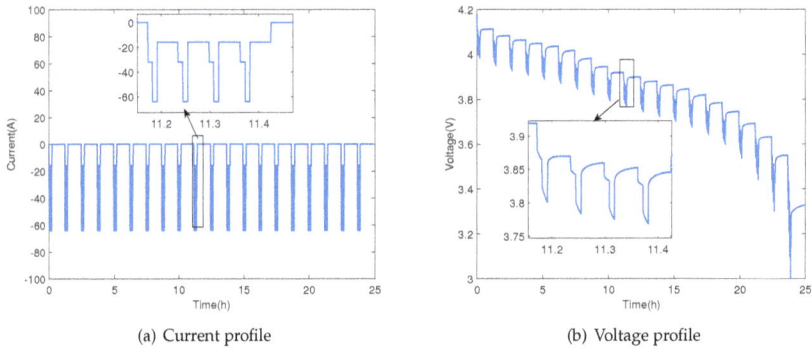

(a) Current profile

(b) Voltage profile

Figure 3. Voltage and current profiles for parameter estimation.

The lumped first-order RC model parameters can be estimated according to the current and voltage curve using the method proposed by Huria et al. [34,35]. The parameters were determined using the parameter estimation tool in Simulink Design Optimization. The tool iteratively simulated the discharge voltage profile while comparing the simulation results with experimental data. The nonlinear least squares algorithm was used to minimize the sum of square error. The fitting result of the discharge pulse at SOC = 0.55 is presented in Figure 4. The fitted results agree well with the experimental data. The estimated parameters of U_{ocv}, R_{Ω}, R_{ct} and C_{ct} are shown in Figure 5.

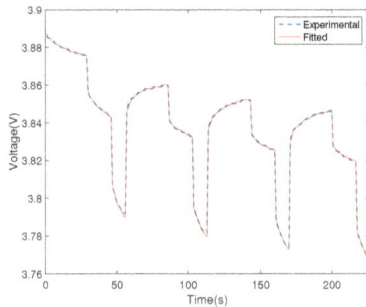

Figure 4. Fitting results of the pulse discharge at SOC = 0.55.

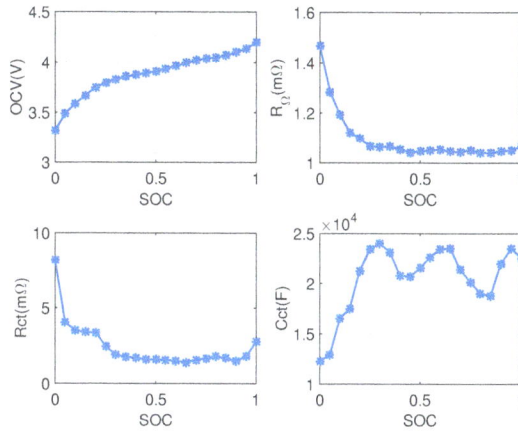

Figure 5. Estimation results of the first-order RC model parameters at 25 °C.

2.1.2. Entropy Change Measurement

The entropy change of the lithium ion battery can describe the generated reversible heat. It can be calculated using Equation (1):

$$\Delta S = F \frac{dU}{dT} \tag{1}$$

where ΔS is the entropy change of the battery; F is the Faraday constant; U is the equilibrium potential of the battery; T is the temperature.

The entropy change could be calculated through the OCV variation with temperature. The method proposed by Forgez et al. [28] was used in this paper to obtain the coefficient of OCV variation with temperature. The cell was discharged to an objective SOC and rested to the steady state. Then, the temperature of the environment chamber was changed to different values, as shown in Figure 6a. In order to extract the temperature coefficient from these data, the voltage was fitted by the function $V(t, T) = A + BT + Ct$ with A, B and C as constants and B corresponding to the temperature coefficient $\partial U / \partial T$. The fitting results for $\partial U / \partial T$ estimation are shown in Figure 6b. The $\partial U / \partial T$ results are shown in Figure 7.

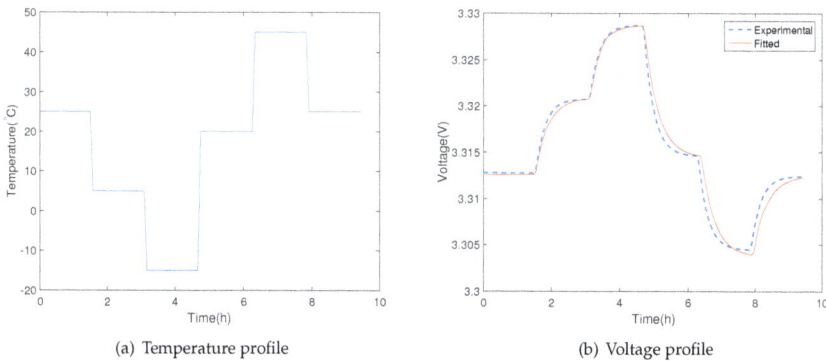

(a) Temperature profile

(b) Voltage profile

Figure 6. Temperature and voltage profiles for $\partial U / \partial T$ estimation at SOC = 0.

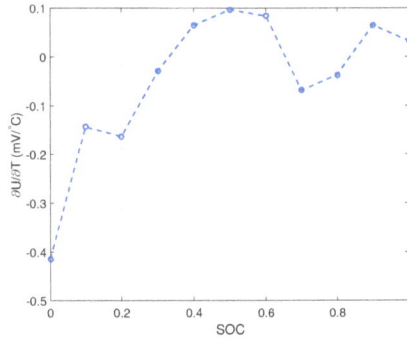

Figure 7. $\partial U/\partial T$ estimation results.

2.2. Validation Experiments

In order to analyze the voltage and temperature characteristics, constant current discharge at 1C, 2C and 3C rates was performed using Arbin BT-2000. The temperature was measured by an Agilent 34901A. The voltage and temperature during the discharge process with different rates are shown in Figure 8a. Figure 8b shows the temperature measured by the internal thermocouples. The thermocouples at $T(1, 1)$ were broken, so the data are not shown here. At the end of discharge, temperature at the center of the cell ($T(2, 2)$) is the highest. The center temperature close to the positive tab ($T(1, 2)$) is higher than that close to the negative tab ($T(3, 2)$), which might be caused by the different properties of positive and negative tabs. Although the geometries of positive and negative tabs are the same, the positive aluminum tab has a smaller heat conductivity coefficient and larger electrical resistivity compared to the negative Copper tab. Therefore, more heat is accumulated around the positive tab, leading to a higher temperature around positive tab. The temperature values of $T(2, 1)$ and $T(2, 3)$ are almost the same, owning to the symmetric cell structure.

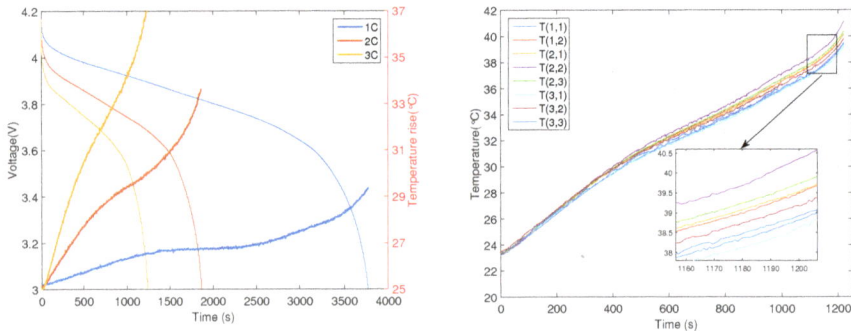

(a) Terminal voltage and temperature of $T(2,2)$ at different discharge rates

(b) Temperature results at 3C discharge

Figure 8. Voltage and temperature experimental results.

3. Modeling and Validation

The model is comprised of two parts. One is the electrical part, which is used to solve the electrical-related variables, such as the potential, current density and SOC. The other is the thermal part to solve the thermal-related variables, including heat generation, conduction and dissipation.

3.1. Electrical Model

The schematic of the electrical part in DPECM is shown in Figure 9. According to the layered structure and electrochemical principles of pouch batteries [36–38], the proposed model consists of five parts, including the positive and negative tabs, the positive and negative foils and the main material body. The current flow of positive and negative tabs only exists in the horizontal plate. The positive and negative foils not only have current flowing in the horizontal plate, but also have current pouring into or pumping out from the sub-models. The main material body, including the positive and negative materials, electrolyte and separator, etc., is divided into a group of sub-models. Each sub-model is represented by a first-order RC equivalent circuit. The first-order RC model is adopted in this paper, because Hu et al. [39] indicated that the first-order RC equivalent circuit model was the most suitable candidate for an NCM cell after comparing twelve battery models. One branch of the sub-models was magnified in the right side of Figure 9 to show the details of first-order RC model. It consists of a voltage source, a resistor and a parallel branch of the resistor and capacitor.

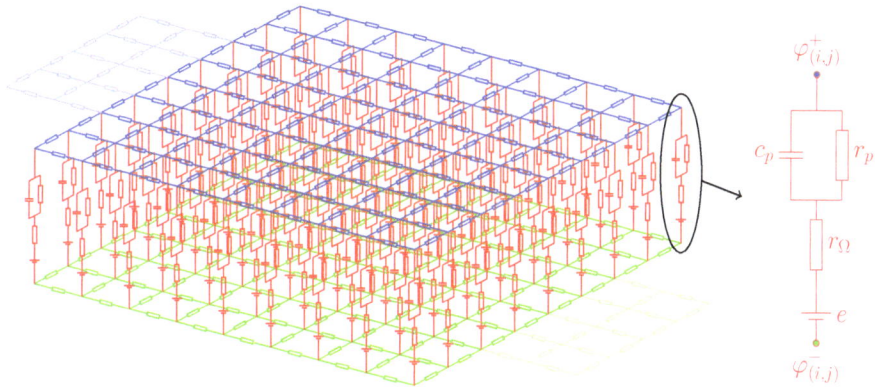

Figure 9. Distributed parameter equivalent circuit model (DPECM) structure.

The current of the sub-model located at the *i*-th row and *j*-th column in Figure 9 could be calculated as Equation (2):

$$i_{(i,j)} = \frac{e_{(i,j)} - (\varphi^+_{(i,j)} - \varphi^+_{(i,j)}) - u_{p(i,j)}}{r_{\Omega(i,j)}} \tag{2}$$

where the subscript (i,j) represents the node location; i is the current of the sub-model; e is the equilibrium potential of the sub-model; φ^- is the node potential on the negative foil connected with the sub-model; φ^+ is the node potential on the positive foil connected with the sub-model; r_Ω is the ohmic resistance of the sub-model; u_p is the polarization voltage of the sub-model, which could be calculated as:

$$c_p \frac{du_p}{dt} = i - \frac{u_p}{r_p} \tag{3}$$

where u_p is the initial polarization voltage; r_p is the polarization resistance of the sub-model; c_p is the polarization capacitor of the sub-model. The resistance value r_p in the sub-model is calculated using the equivalent resistance of the lumped first-order RC model multiplied by sub-model numbers. The capacitance value c_p in the sub-model is calculated using the equivalent capacitance of the lumped first-order RC model divided by sub-model numbers.

In this paper, u_p is iteratively updated. Therefore, Equation (3) needs to be discretized. The discretized result of Equation (3) using the Euler method is:

$$u_{p,t+\Delta t} = u_{p,t} + \frac{i_t r_{p,t} - u_{p,t}}{r_{p,t} c_{p,t}} \Delta t \tag{4}$$

where Δt is the step time; $u_{p,t}$ is the polarization voltage at instant t; $u_{p,t+\Delta t}$ is the polarization voltage at instant $t + \Delta t$; $r_{p,t}$, $c_{p,t}$ and i_t are the polarization resistance, polarization capacitance and current of the sub-model at instant t, respectively.

Assume that there are N nodes in total on the positive foil and tab, as shown in Figure 9. According to Kirchhoff's current law, the sum of currents flowing through each node is zero. Therefore, a typical node equation is:

$$i_{(i,j)} = (\varphi_{(i-1,j)}^+ - \varphi_{(i,j)}^+)/r_{+(i-1,j)(i,j)} + (\varphi_{(i+1,j)}^+ - \varphi_{(i,j)}^+)/r_{+(i+1,j)(i,j)} \\ + (\varphi_{(i,j-1)}^+ - \varphi_{(i,j)}^+)/r_{+(i,j-1)(i,j)} + (\varphi_{(i,j+1)}^+ - \varphi_{(i,j)}^+)/r_{+(i,j)(i+1,j)} \tag{5}$$

where the subscript (i,j) represents the node location; $\varphi_{(i,j)}^+$ is the potential of the node located at (i,j); $r_{+(i-1,j)(i,j)}$, $r_{+(i+1,j)(i,j)}$, $r_{+(i,j-1)(i,j)}$ and $r_{+(i,j+1)(i,j)}$ are resistances between two adjacent nodes. As the locations of all nodes are even in this paper, all of the resistances between two adjacent nodes are equal on the positive foil. Therefore, Equation (5) could be simplified as:

$$-4\varphi_{(i,j)}^+ + \varphi_{(i-1,j)}^+ + \varphi_{(i-1,j)}^+ + \varphi_{(i,j-1)}^+ + \varphi_{(i,j+1)}^+ = r_+ i_{(i,j)} \tag{6}$$

where r_+ represents the resistance between two adjacent nodes.

Define a connectivity function as:

$$f_c(a,b) = \begin{cases} 1 & \text{(when node } a \text{ and node } b \text{ are connected)} \\ 0 & \text{(when node } a \text{ and node } b \text{ are disconnected)} \end{cases} \quad a,b \in 1,2,...,N \text{ and } a \neq b \tag{7}$$

The relationship between the current of all sub-models and the potential of all nodes located on the positive foil and tab is:

$$\begin{bmatrix} -\sum_{i=1}^{N,i\neq 1} f_c^+(1,i) & f_c^+(1,2) & \cdots & f_c^+(1,N) \\ f_c^+(2,1) & -\sum_{i=1}^{N,i\neq 2} f_c^+(1,i) & \cdots & f_c^+(2,N) \\ \vdots & \vdots & \ddots & \vdots \\ f_c^+(N,1) & f_c^+(N,2) & \cdots & -\sum_{i=1}^{N,i\neq N} f_c^+(N,i) \end{bmatrix} \begin{bmatrix} \varphi_1^+ \\ \varphi_2^+ \\ \vdots \\ \varphi_N^+ \end{bmatrix} = r_+ \begin{bmatrix} i_1 \\ i_2 \\ \vdots \\ i_N \end{bmatrix} \tag{8}$$

where $f_c^+(a,b)(a,b \in 1,2,...,N$ and $a \neq b)$ is the connectivity function of the positive foil and tab; $\varphi_n^+ (n \in 1,2,\cdots,N)$ is the potential of the n-th node on the positive foil and tab; $i_n (n \in 1,2,\cdots,N)$ is the current of the n-th sub-model.

Define:

$$\mathbf{A}_+ = \begin{bmatrix} -\sum_{i=1}^{N,i\neq 1} f_c^+(1,i) & f_c^+(1,2) & \cdots & f_c^+(1,N) \\ f_c^+(2,1) & -\sum_{i=1}^{N,i\neq 2} f_c^+(1,i) & \cdots & f_c^+(2,N) \\ \vdots & \vdots & \ddots & \vdots \\ f_c^+(N,1) & f_c^+(N,2) & \cdots & -\sum_{i=1}^{N,i\neq N} f_c^+(N,i) \end{bmatrix}$$

364

$$\boldsymbol{\Phi}_+ = \begin{bmatrix} \varphi_1^+ & \varphi_2^+ & \cdots & \varphi_N^+ \end{bmatrix}' \text{ and } \mathbf{I} = \begin{bmatrix} i_1 & i_2 & \cdots & i_N \end{bmatrix}'$$

Therefore, the node potential equations of the positive foil and tab could be written as:

$$\mathbf{A}_+ \boldsymbol{\Phi}_+ = r_+ \mathbf{I} \tag{9}$$

where $\mathbf{A}_+$ is an $N \times N$ matrix representing the node connectivities of the positive foil and tab; $\boldsymbol{\Phi}_+$ is an N-dimensional node potential vector of the positive foil and tab; $\mathbf{I}$ is an N-dimensional sub-model current vector. Those nodes on the positive tab do not have sub-models, so the currents of the sub-models in the corresponding places of $\mathbf{I}$ are set to zero.

Similarly, the node potential equations of the negative foil and tab could be deduced as:

$$\mathbf{A}_- \boldsymbol{\Phi}_- = -r_- \mathbf{I} \tag{10}$$

where $\mathbf{A}_-$ is an $N \times N$ matrix representing node connectivities of the negative foil and tab; $\boldsymbol{\Phi}_-$ is an N-dimensional node potential vector of the negative foil and tab; r_- is the resistance between two adjacent nodes on the negative foil. Those nodes on the negative tab do not have sub-models, so the currents of the sub-models in the corresponding places of $\mathbf{I}$ are set to zero.

The current of all sub-models could be organized as:

$$\mathbf{R}_\Omega \mathbf{I} = \mathbf{E} - (\boldsymbol{\Phi}_+ - \boldsymbol{\Phi}_-) - \mathbf{V}_P \tag{11}$$

where $\mathbf{E}$ is the N-dimensional equilibrium potential vector of sub-models; $\mathbf{V}_P$ is the polarization voltage vector of sub-models; $\mathbf{R}_\Omega$ is an $N \times N$-dimensional diagonal matrix of sub-model ohmic resistances that:

$$\mathbf{R}_\Omega = diag(r_{\Omega 1}, r_{\Omega 2}, \cdots, r_{\Omega N}) \tag{12}$$

The resistance is set to be infinity for those nodes that do not have sub-models. Define:

$$\mathbf{G}_\Omega = \mathbf{R}_\Omega^{-1} \tag{13}$$

In order to solve the coupled sub-model currents and node potentials, combine Equations (9)–(11), and the following equation is obtained:

$$\begin{cases} \mathbf{A}_+ \boldsymbol{\Phi}_+ / r_+ &= \mathbf{G}_\Omega (\mathbf{E} - \boldsymbol{\Phi}_+ + \boldsymbol{\Phi}_- - \mathbf{V}_P) \\ \mathbf{A}_- \boldsymbol{\Phi}_- / r_- &= -\mathbf{G}_\Omega (\mathbf{E} - \boldsymbol{\Phi}_+ + \boldsymbol{\Phi}_- - \mathbf{V}_P) \end{cases} \tag{14}$$

It could be simplified as:

$$\begin{bmatrix} \mathbf{A}_+ / r_+ - \mathbf{G}_\Omega & \mathbf{G}_\Omega \\ \mathbf{G}_\Omega & \mathbf{A}_- / r_- - \mathbf{G}_\Omega \end{bmatrix} \begin{bmatrix} \boldsymbol{\Phi}_+ \\ \boldsymbol{\Phi}_- \end{bmatrix} = \begin{bmatrix} \mathbf{G}_\Omega (\mathbf{E} - \mathbf{V}_p) \\ -\mathbf{G}_\Omega (\mathbf{E} - \mathbf{V}_p) \end{bmatrix} \tag{15}$$

In order to solve Equation (15) to obtain $\boldsymbol{\Phi}_+$ and $\boldsymbol{\Phi}_-$, boundary conditions and initial values are needed. In this paper, the boundary conditions are: the node potentials at the edge of the negative tab are set to zero; the nodes at the edge of the positive tab share the external current evenly. The initial value of $\mathbf{V}_p$ is set to $\mathbf{0}$. The initial value of each element in $\mathbf{E}$ is equal to the equilibrium potential when SOC = 1. The initial values of r_Ω, r_p and c_p are also set to be the corresponding values when SOC = 1. The currents of sub-models $\mathbf{I}$ could be obtained by substituting the $\boldsymbol{\Phi}_+$ and $\boldsymbol{\Phi}_-$ into Equation (11).

The current flow in the positive foil and tab could be calculated as:

$$\vec{\mathbf{J}}_+ = -\gamma_+ \nabla \boldsymbol{\Phi}_+ \tag{16}$$

where $\vec{\mathbf{J}}_+$ is the current flow in the positive foil and tab; γ_+ is the electrical conductivity of the positive foil and tab.

Similarly, the current flow in the negative foil and tab could be calculated as:

$$\vec{J}_- = -\gamma_- \nabla \Phi_- \tag{17}$$

where $\vec{J}_-$ is the current flow in the negative foil and tab; γ_- is the electrical conductivity of the negative foil and tab.

3.2. Thermal Model

The schematic of the thermal part in DPECM is shown in Figure 10. Each sub-model consists of a parallel branch of the current source and capacitor, a resistor and a voltage reference [28]. The current source represents the heat generation source. The capacitor represents the heat capacitance of the sub-model. The voltage reference provides the ambient temperature of the sub-model. The resistor represents the resistance of heat conduction and convention from the sub-model to the ambient environment. The heat conduction in the cell is modeled by the resistors between two adjacent nodes. The values of the resistors are calculated through the thermal properties of each layer using the method proposed by Chen et al. [40].

The heat generation of each sub-model is composed of irreversible and reversible heat as Equation (18):

$$\dot{q}_{g,i} = i(e - (\varphi^+ - \varphi^-)) + iT\frac{\partial e}{\partial T} \tag{18}$$

where $\dot{q}_{g,i}$ is the heat generation rate of the i-th sub-model; T is the temperature of the sub-model.

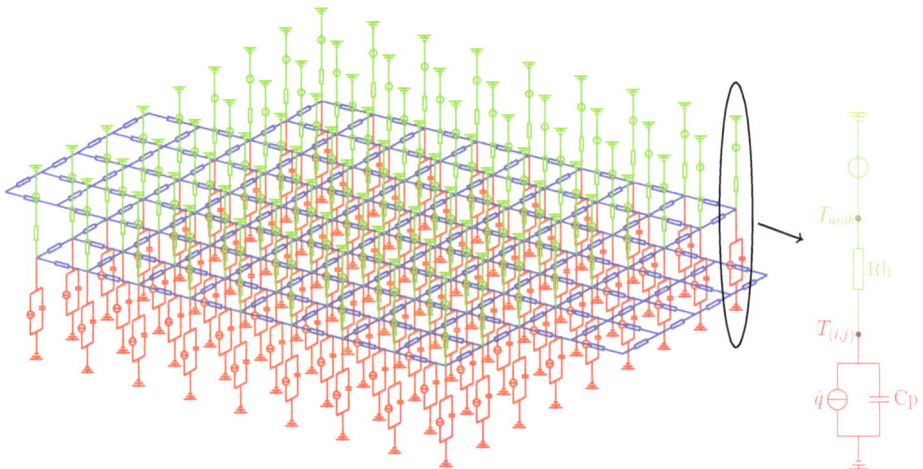

Figure 10. Thermal model.

Assume that the number of thermal nodes in Figure 10 is M. As the positive and negative tabs do not have connected sub-models, where the generated heat of those nodes are set to zero, so the heat generation rate of all of the sub-models could be described as:

$$\dot{Q}_g = \left[\underbrace{0 \quad 0 \quad \cdots \quad 0}_{M-N} \quad \underbrace{\dot{q}_{g,1} \quad \dot{q}_{g,2} \quad \cdots \quad \dot{q}_{g,2N-M}}_{2N-M} \quad \underbrace{0 \quad 0 \quad \cdots \quad 0}_{M-N} \right]' \tag{19}$$

The heat conducted from other nodes could be calculated as:

$$\dot{Q}_c = A_t T \tag{20}$$

where $\dot{Q}_c$ is an M-dimensional vector representing the conductive heat; **T** is an M-dimensional vector representing temperatures of all nodes; A_t is an $M \times M$ matrix representing the thermal connectivities of nodes in the thermal model, which is defined as:

$$
A_T = \begin{bmatrix}
-\sum\limits_{i=1}^{M,i\neq 1} f_c^T(1,i) & f_c^T(1,2) & \cdots & f_c^T(1,M) \\
f_c^T(2,1) & -\sum\limits_{i=1}^{M,i\neq 2} f_c^T(1,i) & \cdots & f_c^T(2,M) \\
\vdots & \vdots & \ddots & \vdots \\
f_c^T(M,1) & f_c^T(M,2) & \cdots & -\sum\limits_{i=1}^{M,i\neq M} f_c^T(M,i)
\end{bmatrix}
$$

where $f_c^T(a,b)(a,b \in 1,2,...,M$ and $a \neq b)$ is the connectivity function of the nodes in the thermal model.

The heat generation rate of the i-th node in the positive foil and tab is:

$$
\dot{q}_i^+ = \sum_{j=1}^{N} f_c^+(i,j)(\varphi_i - \varphi_j)^2/(2r_+) \tag{21}
$$

Therefore, the heat generation rate of all of the nodes on the positive foil and tab is:

$$
\dot{Q}_+ = \begin{bmatrix} \underbrace{\dot{q}_1^+ \quad \dot{q}_2^+ \quad \cdots \quad \dot{q}_M^+}_{N} & \underbrace{0 \quad 0 \quad \cdots \quad 0}_{M-N} \end{bmatrix}' \tag{22}
$$

Similarly, the heat generation rate of all of the nodes on the negative foil and tab is:

$$
\dot{Q}_- = \begin{bmatrix} \underbrace{0 \quad 0 \quad \cdots \quad 0}_{M-N} & \underbrace{\dot{q}_1^- \quad \dot{q}_2^- \quad \cdots \quad \dot{q}_M^-}_{N} \end{bmatrix}' \tag{23}
$$

The temperatures of all of the nodes can be calculated as Equation (24):

$$
T = T_0 + \Delta t(\dot{Q}_+ + \dot{Q}_- + \dot{Q}_c + \dot{Q}_g)/(mC) \tag{24}
$$

where T_0 is the initial temperature vector; Δt is the step time; m is the equivalent mass of the battery block corresponding to the sub-model; C is the equivalent specific heat capacity of the battery.

3.3. Solution

The above proposed model is built and solved using MATLAB language (M-code). The solution of the potential and current density of each sub-model can be obtained by Equations (11) and (15). The advantage of this method is that the variables in the electrical part are coupled directly. Then, the V_p of the next step is predicted. At the same time, the outputs of the electrical model are used to calculate the heat generation rate, so the thermal part of the model can be solved. The flow chart of the solving method is shown in Figure 11.

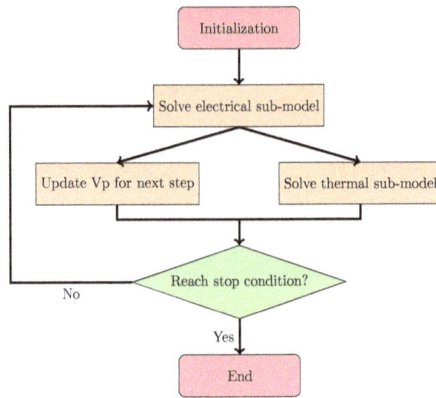

Figure 11. The flow chart of the solving method.

4. Results and Discussion

The simulated and experimental temperatures at the cell center ($T(2,2)$) and positive side center ($T(1,2)$) during 1C, 2C and 3C discharge are shown in Figure 12. The simulated temperature conforms with the experimental data very well, and the simulated temperature error is less than 1 °C. The accurate temperature results indicate that the proposed model has the capability of predicting internal situations in the cell.

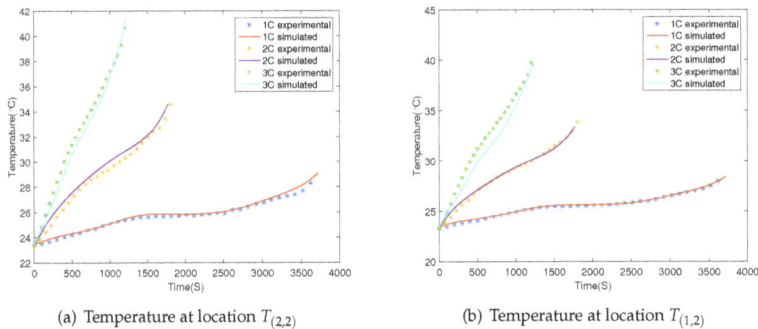

(a) Temperature at location $T_{(2,2)}$

(b) Temperature at location $T_{(1,2)}$

Figure 12. Temperature validation at different discharge rates.

4.1. Non-Uniformity Distribution at the Initial State

During discharge, according to the structure shown in Figure 9, the current from sub-models flows through the positive foil and then is collected at the positive tab end. After the current passes the external circuit and comes back to the negative tab end, it is allocated to sub-models through the negative foil. At the beginning of battery charge/discharge, SOC and temperature are uniform in all sub-models since the battery is rested to steady state, but the current distribution becomes non-uniform when the current flows to relatively narrow tabs. Figure 13a,b shows the current density in the positive and negative current collectors, respectively. The direction of arrows represents the current direction, and the length of arrows represents the magnitude of the current flow. Therefore, the current close to tabs is larger than that in the opposite side. Figure 13c,d shows the potential of the positive and negative collectors, respectively. Since the thickness of each foil is uniform, large potential drops are produced close to tabs because of larger current flow in those areas. Figure 13e shows the current density in sub-models. The potential differences between two end nodes ($\varphi^+ - \varphi^-$) of sub-models close to tabs are smaller than that of the central part. At the same time, the polarization voltages of all of the sub-models are zero, and the equivalent potentials of all of the sub-models are equal at the beginning of the discharge. Therefore, the current flows through the sub-models, decided by Equation (2), close to tabs are higher at the beginning of the discharge process. The temperature close to tabs, shown in Figure 13f, is higher due to higher current density in foils and sub-models.

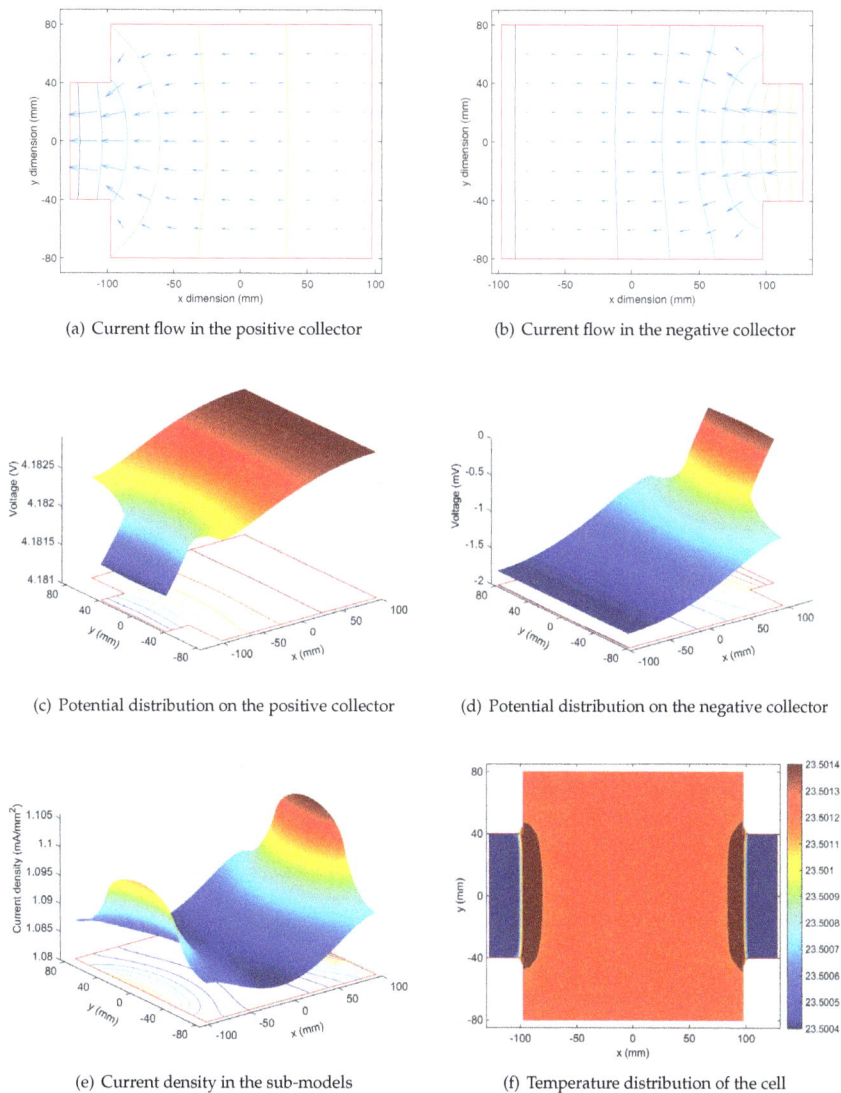

(a) Current flow in the positive collector

(b) Current flow in the negative collector

(c) Potential distribution on the positive collector

(d) Potential distribution on the negative collector

(e) Current density in the sub-models

(f) Temperature distribution of the cell

Figure 13. Non-uniformity at the beginning of the 1C discharge.

4.2. Non-Uniformity Evolution During Discharge

The non-uniformity becomes more complex after the initial state. It changes with the current profile and ambient temperature. The distribution of the current density and temperature in the sub-models could change dramatically, which sometimes even turns into an inverse distribution compared with the initial state.

The distributions of current flowing through sub-models and temperature at the end of the 1C discharge are displayed in Figure 14. Figure 14a indicates that the current density close to tabs becomes smaller than those at the cell center, which is inverse compared to the situation at the initial state, as shown in Figure 13e. This is associated with the non-uniform current accumulation and potential change in the sub-models. The current of the sub-model is calculated by $(e - (\varphi^+ - \varphi^-) - u_p)/r_\Omega$, so the current is decided by both the potential (e) and the voltage difference between positive and negative foils ($\varphi^+ - \varphi^-$). The voltage difference between positive and negative foils close to tabs is always smaller than that at the cell center during discharge because of the existence of foil resistance and the current flow direction. While the accumulation of larger current in the sub-model causes lower local SOC during the discharge process, the local SOC of the sub-models close to tabs becomes lower than that at the cell center. Therefore, the potential difference of sub-models close to tabs and the cell center increases gradually. The influence of the potential difference becomes stronger along with the accumulation of non-uniform current distribution and finally dominates the current of the sub-models, which inverses the current density distribution. Figure 14b indicates that the temperature at the cell center becomes the highest, which is mainly due to the poor heat dissipation at the center.

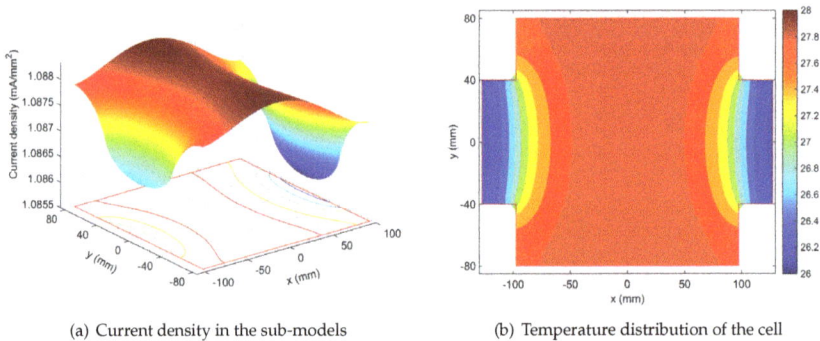

(a) Current density in the sub-models   (b) Temperature distribution of the cell

Figure 14. Non-uniformity at the end of the 1C discharge.

To further illustrate the evolution of internal non-uniformity during discharge, three points (P(1,2), P(2,2) and P(3,2) in Figure 1) are selected to be analysis points. The current density and local SOC at those three points are calculated and presented in Figure 15. Figure 15a indicates that the difference of the local current density declines at the beginning of the discharge. The current at P(3,2) is the largest, and the current at P(2,2) is the smallest. The current becomes uniform at around SOC = 0.38. After the current crosses the point at SOC = 0.38, the current difference rises again, leading to the current at P(2,2) being the largest. The current difference decreases again after SOC around 0.2.

The current non-uniformity will cause a local SOC difference of each sub-model as shown in Figure 15b. The local SOC difference increases before the current cross point and decreases later, which can be explained by its definition. The SOC, defined as:

$$SOC = SOC_0 + \frac{\int idt}{3600C} \tag{25}$$

is decided by the initial state and current flow through the sub-model. The local SOC difference reaches the maximum value at the current cross point and then is eliminated by the inversed current density distribution. The appearance of the current cross point is caused by the relatively flat characteristics of $de/dSOC$ and $dr/dSOC$ between $0.8 > SOC > 0.2$, as shown in Figure 16. The effects of resistance in foils are counteracted by the parameter difference caused by the local SOC difference at the current cross point. After the current cross point in Figure 15a, the increase of the $de/dSOC$ difference becomes larger and dominates the current distribution, which enlarges the difference of the current density. It is worth noting that the direction of currents in sub-models remains the same, while the shape of the current density distribution changes from the bowl-shaped distribution to the peak-shaped distribution. The rapid increase in cell impedance after $SOC < 0.2$ makes the difference in current density decrease again.

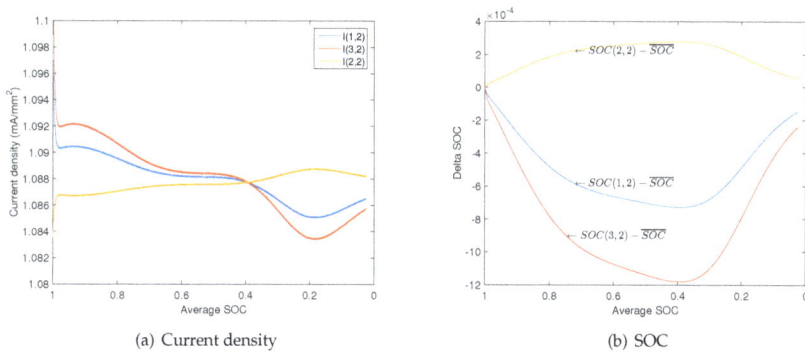

(a) Current density

(b) SOC

Figure 15. Non-uniformity evolution during discharge.

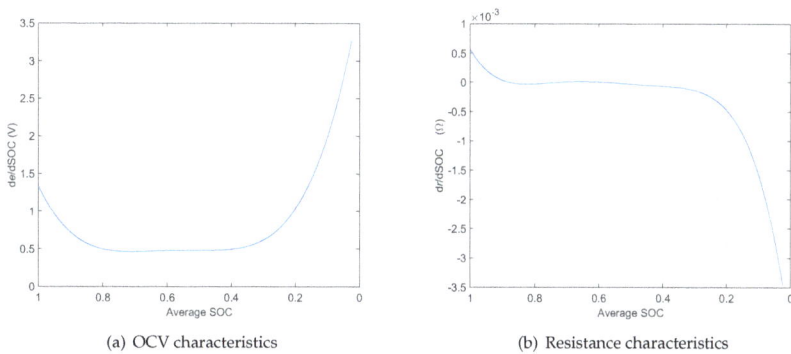

(a) OCV characteristics

(b) Resistance characteristics

Figure 16. Parameter characteristics. OCV, open circuit voltage.

4.3. Limitation of Technique

As mentioned in Section 1, the distributions of internal potential, current density or SOC can hardly be measured directly without changing the geometry and characteristics of the commercial lithium ion battery. Though the proposed model has the capability of predicting internal temperature, current density and potential, only the distributions of internal temperatures can be measured even using a customized battery. Therefore, only the internal temperatures predicted by the proposed model are validated at the present stage.

5. Conclusions

In this paper, a DPECM model consisting of a directly-coupled electrical model and a thermal model for a lithium ion battery is proposed. The model is simpler than the 3D electrical-chemical thermal model, and the parameters are easier to obtain, which is similar to the lumped ECM model. However, the proposed model has the capability of simulating the internal distribution of the current density, potential and temperature inside a cell. The model is validated by comparing the simulated temperature with measured results at test points in a customized cell that is embedded with nine T-type thermocouples. The simulated temperature error is less than 1 °C, indicating that the model can be applied to predict the internal characteristics of the battery. Summarizing the simulation results, the following conclusions can be obtained:

1. The initial non-uniformity of current density inside a 35-Ah NCM pouch cell is caused by the electrode foil resistance and relatively narrow tab.
2. The current flow through the sub-models close to tabs is higher than that at the center at the beginning of discharge, because the potential drops in positive and negative foils.
3. A current cross point exists during the constant discharge process. The local SOC difference increases with the reduction of the average SOC until the average SOC arrives at that point and then decreases until the average SOC arrives at zero.
4. The current flow through the sub-models close to tabs becomes lower than that at the center at the end of discharge due to the accumulation of the local SOC difference, rapid OCV drop and rapid resistance increase.

Acknowledgments: This work was supported by the National Natural Science Foundation of China (Grant No. 51477009).

Author Contributions: Xue Li supported the experiments. Jiuchun Jiang, Weige Zhang and Zhanguo Wang gave advice for the modeling.

Conflicts of Interest: The authors declare no conflict of interest.

References

1. Doyle, M.; Fuller, T.F.; Newman, J. Modeling of galvanostatic charge and discharge of the lithium/polymer/insertion cell. *J. Electrochem. Soc.* **1993**, *140*, 1526–1533.
2. Luo, W.; Lyu, C.; Wang, L.; Zhang, L. A new extension of physics-based single particle model for higher charge–discharge rates. *J. Power Sources* **2013**, *241*, 295–310.
3. Kim, G.H.; Smith, K.; Lee, K.J.; Santhanagopalan, S.; Pesaran, A. Multi-domain modeling of lithium-ion batteries encompassing multi-physics in varied length scales. *J. Electrochem. Soc.* **2011**, *158*, A955–A969.
4. Lee, K.J.; Smith, K.; Pesaran, A.; Kim, G.H. Three dimensional thermal-, electrical-, and electrochemical-coupled model for cylindrical wound large format lithium-ion batteries. *J. Power Sources* **2013**, *241*, 20–32.
5. Huo, W.; He, H.; Sun, F. Electrochemical–thermal modeling for a ternary lithium ion battery during discharging and driving cycle testing. *RSC Adv.* **2015**, *5*, 57599–57607.
6. Mazumder, S.; Lu, J. Faster-Than-Real-Time Simulation of Lithium Ion Batteries with Full Spatial and Temporal Resolution. *Int. J. Electrochem.* **2013**, *2013*, doi:10.1155/2013/268747.
7. Pals, C.R.; Newman, J. Thermal modeling of the lithium/polymer battery I. Discharge behavior of a single cell. *J. Electrochem. Soc.* **1995**, *142*, 3274–3281.
8. Anwar, S.; Zou, C.; Manzie, C. Distributed thermal-electrochemical modeling of a lithium-ion battery to study the effect of high charging rates. *IFAC Proc. Vol.* **2014**, *47*, 6258–6263.
9. Kwon, K.H.; Shin, C.B.; Kang, T.H.; Kim, C.S. A two-dimensional modeling of a lithium-polymer battery. *J. Power Sources* **2006**, *163*, 151–157.
10. Kim, U.S.; Shin, C.B.; Kim, C.S. Effect of electrode configuration on the thermal behavior of a lithium-polymer battery. *J. Power Sources* **2008**, *180*, 909–916.
11. Kim, U.S.; Shin, C.B.; Kim, C.S. Modeling for the scale-up of a lithium-ion polymer battery. *J. Power Sources* **2009**, *189*, 841–846.

12. Kohneh, M.M.M.; Samadani, E.; Farhad, S.; Fraser, R.; Fowler, M. *Three-Dimensional Electrochemical Analysis of a Graphite/LiFePO₄ Li-Ion Cell to Improve Its Durability*; SAE International: Warrendale, PA, USA, April 2015.

13. Liu, B.; Yin, S.; Xu, J. Integrated computation model of lithium-ion battery subject to nail penetration. *Appl. Energy* **2016**, *183*, 278–289.

14. Samba, A. Battery Electrical Vehicles-Analysis of Thermal Modelling and Thermal Management. Ph.D. Thesis, LUSAC (Laboratoire Universitaire des Sciences Appliquées de Cherbourg), Université de caen Basse Normandie; MOBI (the Mobility, Logistics and Automotive Technology Research Centre), Vrije Universiteit Brussel, Brussels, Belgium, 2015.

15. Mastali, M.; Samadani, E.; Farhad, S.; Fraser, R.; Fowler, M. Three-dimensional Multi-Particle Electrochemical Model of LiFePO₄ Cells based on a Resistor Network Methodology. *Electrochim. Acta* **2016**, *190*, 574–587.

16. Tong, S.; Klein, M.P.; Park, J.W. Comprehensive Battery Equivalent Circuit Based Model for Battery Management Application. In Proceedings of the ASME 2013 Dynamic Systems and Control Conference, American Society of Mechanical Engineers, Stanford, CA, USA, 21–23 October 2013; V001T05A005.

17. Sun, F.; Xiong, R.; He, H. A systematic state-of-charge estimation framework for multi-cell battery pack in electric vehicles using bias correction technique. *Appl. Energy* **2016**, *162*, 1399–1409.

18. Yazdanpour, M.; Taheri, P.; Mansouri, A.; Schweitzer, B. A circuit-based approach for electro-thermal modeling of lithium-ion batteries. In Proceedings of the 2016 32nd Thermal Measurement, Modeling & Management Symposium (SEMI-THERM), San Jose, CA, USA, 14–17 March 2016.

19. Knap, V.; Stroe, D.I.; Teodorescu, R.; Swierczynski, M.; Stanciu, T. Electrical circuit models for performance modeling of Lithium-Sulfur batteries. In Proceedings of the 2015 IEEE Energy Conversion Congress and Exposition (ECCE), Montreal, QC, Canada, 20–24 September 2015.

20. Blanco, C.; Sanchez, L.; Gonzalez, M.; Anton, J.C.; Garcia, V.; Viera, J.C. An Equivalent Circuit Model with Variable Effective Capacity for Batteries. *IEEE Trans. Veh. Technol.* **2014**, *63*, 3592–3599.

21. Fok, C.W.E. Simulation of Lithium-Ion Batteries Based on Pulsed Current Characterization. Ph.D. Thesis, University of British Columbia, Vancouver, BC, Canada, 2016.

22. Chen, X.K.; Sun, D. Modeling and state of charge estimation of lithium-ion battery. *Adv. Manuf.* **2015**, *3*, 202–211.

23. Xiong, R.; Sun, F.; Gong, X.; Gao, C. A data-driven based adaptive state of charge estimator of lithium-ion polymer battery used in electric vehicles. *Appl. Energy* **2014**, *113*, 1421–1433.

24. Zhang, S.; Xiong, R.; Cao, J. Battery durability and longevity based power management for plug-in hybrid electric vehicle with hybrid energy storage system. *Appl. Energy* **2016**, *179*, 316–328.

25. Xiong, R.; Sun, F.; Chen, Z.; He, H. A data-driven multi-scale extended Kalman filtering based parameter and state estimation approach of lithium-ion polymer battery in electric vehicles. *Appl. Energy* **2014**, *113*, 463–476.

26. Sun, F.; Xiong, R. A novel dual-scale cell state-of-charge estimation approach for series-connected battery pack used in electric vehicles. *J. Power Sources* **2015**, *274*, 582–594.

27. Lin, X.; Perez, H.E.; Siegel, J.B.; Stefanopoulou, A.G.; Li, Y.; Anderson, R.D.; Ding, Y.; Castanier, M.P. Online parameterization of lumped thermal dynamics in cylindrical lithium ion batteries for core temperature estimation and health monitoring. *IEEE Trans. Control Syst. Technol.* **2013**, *21*, 1745–1755.

28. Forgez, C.; Do, D.V.; Friedrich, G.; Morcrette, M.; Delacourt, C. Thermal modeling of a cylindrical LiFePO₄/graphite lithium-ion battery. *J. Power Sources* **2010**, *195*, 2961–2968.

29. Fleckenstein, M.; Bohlen, O.; Roscher, M.A.; Bäker, B. Current density and state of charge inhomogeneities in Li-ion battery cells with LiFePO₄ as cathode material due to temperature gradients. *J. Power Sources* **2011**, *196*, 4769–4778.

30. Chen, D.; Jiang, J.; Wang, Z.; Duan, Y.; Zhang, Y.; Shi, W. Research on Distribution Parameters Equivalent-Circuit Model of Power Lithium-Ion Batteries. *Trans. China Electrotechn. Soc.* **2013**, *7*, 025.

31. Zhang, G.; Shaffer, C.E.; Wang, C.Y.; Rahn, C.D. In-situ measurement of current distribution in a Li-Ion cell. *J. Electrochem. Soc.* **2013**, *160*, A610–A615.

32. Zhang, G.; Shaffer, C.E.; Wang, C.Y.; Rahn, C.D. Effects of non-uniform current distribution on energy density of Li-ion cells. *J. Electrochem. Soc.* **2013**, *160*, A2299–A2305.

33. Li, Z.; Zhang, J.; Wu, B.; Huang, J.; Nie, Z.; Sun, Y.; An, F.; Wu, N. Examining temporal and spatial variations of internal temperature in large-format laminated battery with embedded thermocouples. *J. Power Sources* **2013**, *241*, 536–553.

34. Huria, T.; Ceraolo, M.; Gazzarri, J.; Jackey, R. High fidelity electrical model with thermal dependence for characterization and simulation of high power lithium battery cells. In Proceedings of the 2012 IEEE International Electric Vehicle Conference (IEVC), Greenville, SC, USA, 4–8 March 2012.

35. Ceraolo, M.; Lutzemberger, G.; Huria, T. *Experimentally-determined Models for High-power Lithium Batteries*; SAE International: Warrendale, PA, USA, April 2011.

36. Xiao, M.; Choe, S.Y. Dynamic modeling and analysis of a pouch type $LiMn_2O_4$/Carbon high power Li-polymer battery based on electrochemical-thermal principles. *J. Power Sources* **2012**, *218*, 357–367.

37. Samba, A.; Omar, N.; Gualous, H.; Capron, O.; Van den Bossche, P.; Van Mierlo, J. Impact of Tab Location on Large Format Lithium-Ion Pouch Cell Based on Fully Coupled Tree-Dimensional Electrochemical-Thermal Modeling. *Electrochim. Acta* **2014**, *147*, 319–329.

38. Fu, R.; Xiao, M.; Choe, S.Y. Modeling, validation and analysis of mechanical stress generation and dimension changes of a pouch type high power Li-ion battery. *J. Power Sources* **2013**, *224*, 211–224.

39. Hu, X.; Li, S.; Peng, H. A comparative study of equivalent circuit models for Li-ion batteries. *J. Power Sources* **2012**, *198*, 359–367.

40. Chen, S.; Wan, C.; Wang, Y. Thermal analysis of lithium-ion batteries. *J. Power Sources* **2005**, *140*, 111–124.

![energies logo] *energies*

MDPI

Review

Seasonal Thermal-Energy Storage: A Critical Review on BTES Systems, Modeling, and System Design for Higher System Efficiency

Michael Lanahan and Paulo Cesar Tabares-Velasco *

Department of Mechanical Engineering, Colorado School of Mines, Golden, CO 80401, USA;
mlanahan@mines.edu
* Correspondence: tabares@mines.edu; Tel.: +1-303-273-3980

Academic Editor: Rui Xiong
Received: 24 February 2017; Accepted: 14 May 2017; Published: 25 May 2017

Abstract: Buildings consume approximately $\frac{3}{4}$ of the total electricity generated in the United States, contributing significantly to fossil fuel emissions. Sustainable and renewable energy production can reduce fossil fuel use, but necessitates storage for energy reliability in order to compensate for the intermittency of renewable energy generation. Energy storage is critical for success in developing a sustainable energy grid because it facilitates higher renewable energy penetration by mitigating the gap between energy generation and demand. This review analyzes recent case studies—numerical and field experiments—seen by borehole thermal energy storage (BTES) in space heating and domestic hot water capacities, coupled with solar thermal energy. System design, model development, and working principle(s) are the primary focus of this analysis. A synopsis of the current efforts to effectively model BTES is presented as well. The literature review reveals that: (1) energy storage is most effective when diurnal and seasonal storage are used in conjunction; (2) no established link exists between BTES computational fluid dynamics (CFD) models integrated with whole building energy analysis tools, rather than parameter-fit component models; (3) BTES has less geographical limitations than Aquifer Thermal Energy Storage (ATES) and lower installation cost scale than hot water tanks and (4) BTES is more often used for heating than for cooling applications.

Keywords: borehole thermal energy storage; seasonal thermal energy storage; BTES; ground source heat pump (GSHP) transient system simulation tool (TRNSYS); EnergyPlus; diurnal storage; solar thermal; solar-coupled GSHP; system modeling; component modeling

---

1. Introduction

Optimizing the performance of a sustainable and renewable grid is becoming an increasingly important topic. Societal dependence upon energy has increased significantly in the last several decades; from 10 billion MWh in 1950 to 28.5 billion MWh in 2013, totaling a 280% increase in total energy consumption in the United States [1]. Population has grown from 150 million to 316 million during the same period, indicating an energy use per capita increase of 33%. Fossil fuels generate 72% of the electricity produced in the United States, negatively impacting air quality and contributing to global warming [2–5].

Buildings consume approximately $\frac{3}{4}$ of total electricity generated in the United States and represent about 40% of the primary energy use. Building heating, ventilation and, air-conditioning (HVAC) systems are also major energy users and drivers of electric peak demand [3]. Electric utilities meet peak demand with fossil fuel energy sources because of convenient storage and quickly accessible energy [6–8]. Peak demand from buildings therefore drives fossil fuel-based pollution. To minimize pollution and building energy use, investigation of non-fossil fuel energy sources in both grid independent and grid-connected capacities is vital [9–13]. However, renewable energy sources are highly variable because of their dependence upon weather [11,12]. Energy storage is one solution for increasing grid flexibility and facilitating greater penetration of renewables. Sovacool et al. stated that the United States' grid cannot accommodate wind and solar penetrations higher than 35% without failure if solely dependent on renewable technologies with no method for storage due to their intermittent nature [11]. Others agree with this study, finding penetrations of 30–33% plausible with no energy storage [12,13]. Thermal energy storage at the building level can relieve electric peak demand and fossil fuel emissions.

A majority of renewable grid solutions consist of distributed generation (DG) with energy storage and smart-grid control [4,5,9,14–18]. A number of studies indicate that a diverse portfolio of different energy management techniques, including energy storage, are necessary for sustainable and reliable energy use [11,12,19,20]. Jacobson et al. provide a thorough economic feasibility analysis implementing wind, water, and solar (WWS) renewable energy generation with the grid, primary reliability stemming from energy distribution algorithms coupled with various methods of energy storage [12,19]. Mason et al. demonstrate a similar renewable energy analysis in New Zealand, concluding that generation mixing, combined with both hydro and virtual energy storage, as well as load shifting allow for a 100% renewable energy grid [20]. Becker et al. illustrate that expensive energy storage can be minimized by selecting the right combination of energy generation depending upon the transmission grid [21]. Others conclude that grid flexibility and energy storage are required to achieve higher renewable energy penetrations with larger grid sizes [4,22,23]. Nordell concludes that varying solar intensity, a primary energy source results in the need for seasonal storage in conjunction with short term storage [24]. According to Hyman, installation of thermal storage results in optimal outcomes with time variant loads, time dependent energy costs, and previously required equipment or system upgrades [25]. Marnay et al. claim that decoupling thermal energy and electrical energy requirements is potentially cost effective because it allows for the charging and discharge of energy storages during cost effective periods for otherwise unrelated loads [26].

Among different storage technologies, thermal energy storage nears 100% round trip efficiency, compared to the 80% efficiency batteries possess [27–29]. Using thermal storage for viable solar energy utilization through solar thermal panels to meet building heat loads becomes an important discussion [13,28,30,31]. Excess thermal energy generated throughout the day can be stored for either short or seasonal periods [32,33]. Since seasonal storage might have slow charging or discharging rates, coupling seasonal storage with diurnal storage might bridge this gap. Diurnal thermal energy storage takes the forms of chilled water and ice storage for cooling, and hot water tank storage for heating with greater energy transfer rates [30,32,34–37]. Seasonal thermal storage stores thermal energy when solar radiation or other energy sources are abundant or inexpensive to avoid energy shortages during periods of limited sun exposure or high energy cost [30,31,34,36,38–41]. The practices of using water tanks as a diurnal buffer in conjunction with solar collectors, and ice storage with conventional chillers are well documented [13,25,30,32,33,35]. Seasonal storage for both heating and cooling applications remains an emerging technology [30,31,34,39,41–46]. Therefore, coupling solar energy with sensible storage for diurnal and seasonal periods is a logical next step for DG and higher renewable energy penetrations, especially with thermal energy end use [9,14,35,41,46–49].

Thermal energy generation is readily implemented with DG mini-grids because thermal energy supports higher roundtrip efficiencies [8,15]. However, solar heating systems present the paradox of being available during the day when the sun is visible and remaining offset from peak demand periods [30,43]. This mismatch between utility energy demand and renewable energy supply is dubbed the "duck curve" [6,7]. Storage thus becomes a necessary consideration when implementing solar energy in the smart grid discussion [11,12]. Thermal storage can manifest in many different forms, which will be discussed throughout the paper [25,30,34,38,45,50–53]. Because building HVAC systems provide a major draw on the electrical grid, addressing HVAC loads with thermal energy is a practical grid decentralization solution, with solar thermal panels readily implementable at the demand side [8,27,38,46,50]. This study reviews seasonal subsurface thermal energy storage systems that accommodate entire load or partial (peak) load demands. Concentrated solar power plants are not included in the review, as the focus of this review is the system demand side [28]. A brief discussion of other seasonal energy storage techniques is shown in Section 2. Modeling techniques and tools, with advantages and shortcomings are considered in Section 3. An overview of diurnal thermal energy storage is provided.

2. Seasonal Thermal Energy Storage for Meeting Demand Side Space Heating, Cooling and Domestic Hot Water (DHW) Loads

Sensible thermal storage collects energy by increasing (or reducing) the temperature of a medium with finite heat capacitance (typically water) [30,54,55]. Seasonally, it is stored in a variety of mediums for use during periods of higher demand and/or limited energy availability [30,39,44,55–60]. The most prominent modes of storage found by this literature review are: (1) Hot Water Energy Storage (HWTES); (2) Gravel-Water Thermal Energy Storage (GWTES); (3) Aquifer Thermal Energy Storage (ATES); and (4) Borehole Thermal Energy Storage (BTES) [34,44]. Each storage method presents various advantages such as cost, location, capacity, and energy discharge capability. Among these four, BTES is the most flexible energy storage technique [61] and therefore is the primary focus of this analysis because of universal demand-side energy storage and resulting peak-load grid draw mitigation [33,39,41,57,61]. Other thermal storage methods are briefly described below.

2.1. Hot Water Thermal Energy Storage (HWTES)

Hot Water Thermal Energy Storage functions similarly to a hot water boiler: it uses heated water contained in tanks, well insulated to reduce heat losses and extend the effective storage period of the tank [30,55,57]. Hot water tanks are not commonly integrated with the surrounding geometry [38,46]. However, Dincer and Rosen present a buried concrete tank case study, despite significantly higher installation costs [32]. Similar thermal properties of the tank cement and surrounding soil provide additional heat capacity and a greater quantity of working fluid [32].

2.2. Gravel Water Thermal Energy Storage (GWTES)

Gravel water thermal energy storage units are comprised of a water gravel mixture insulated on the top and sides in a tank [62,63]. The specific heat of this mixture is lower than pure water [30]. As a result, the container must then be larger than a water-only storage tank to store comparable amounts of thermal energy [44,64]. Figure 1 below is a schematic of a gravel-water tank.

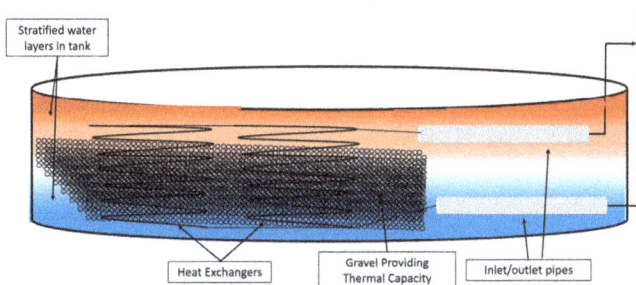

Figure 1. Gravel water thermal energy storage concept [57].

2.3. Waste Snow Pits and Ice-Pond Seasonal Thermal Storage

The primary methods of storing cooling capacity energy for seasonal periods of time are: (1) waste snow pits/warehouses and (2) ice-pond seasonal cooling storage [65]. Historically, snow and ice have been stored by Scandinavian cultures, insulated in shelters termed *Fabrikaglace* [66]. While snow utilizes latent heat storage, large volumes of storage are still required for adequate cooling capacity [67]. Taylor presented the first ice pond for air conditioning at Princeton university in 1979, producing annual savings of 31,000 $/year [68]. Yan et al. optimize a seasonal cold energy storage to supplement capacity to existing chillers with a payback time of 6 years [69]. Skosberg and Nordell describe a snow storage system in Sundsvall, Sweden for a hospital hosting a well-insulated pit equal to 2000 MWh of cooling capacity [45]. Efforts to develop seasonal cooling storage methods are most notably made in Japan and Norway [45,65,69–71].

2.4. Aquifer Thermal Energy Storage (ATES)

Similar to GWTES, Aquifer Thermal Energy Storage collects energy in a mixture of water and earth, but utilizes natural formations [30,58]. Aquifer energy storage provides an alternative to the previously mentioned storage systems due to ideal combinations of the high specific heat provided by water as well as lower cost attained from the absence of a tank [59,60]. However, ATES requires specific considerations to ensure proper performance. For example, ATES must use benign working fluids to minimize the risk of aquifer contamination with hazardous chemicals [72]. Thus, water is usually the working fluid due to mild environmental impact in comparison to other high specific heat fluids such as glycol mixtures and hydrocarbon oils [37,44]. Aquifers work with a heat source, charged by heated fluid from solar collectors, and a heat sink linked through a heat exchanger to heat the fluid required for DHW or space conditioning end use. ATES can be used for heating or cooling purposes: during summer, cool water temperatures are used for cooling, and during the winter, warm water (solar heated or not) is used for heating purposes [32,61,73]. ATES is characterized by a defining layer of non-porous rock between two volumes of water at different temperatures [30,36,57,60]. Water thermal pollution can have a negative impact on the environment, harmful to many species, and must be mitigated [74]. To reduce the impact of heated groundwater, water used in ATES is isolated, with surrounding rock possessing little porosity in order to prevent heat contamination [34]. It is important to note that the plausibility of this technology is strictly limited to preexisting aquifer formation. Figure 2 details a simplified ATES schematic.

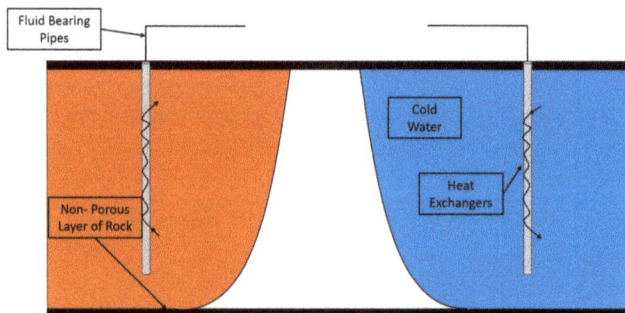

Figure 2. Aquifer thermal energy storage concept sketch [30].

The lack of insulation in this system is an important design consideration. To avoid excessive heat losses, the maximum volume to surface area ratio should be achieved through optimal borehole depth for the fluid bearing pipes [30,32,58,60]. In ATES storage, the thermal front is important for determining storage efficiency [60]. A thermal front characterizes the temperature profile between injected water into ATES, for storage, which if allowed to reach the production well will result in greater heat loss [60]. Rock-cavern thermal energy storage, or CTES, is an energy storage method similar in concept to ATES [31,39,61]. CTES functions by using heat exchangers to exchange heat with a water storage medium, contained by an artificial underground spaces (the distinction from natural aquifers), which can be very expensive to construct [31,39,61]. Because of this expensive construction cost, existing spaces are often utilized such as abandoned mines or areas previously used for underground oil storage [31,39]. Drilling cost for ATES systems range broadly from 200 $/ft to 970 $/ft [75,76]. Due to the nature of ATES open-loop configuration, typically only two boreholes are need in comparison to many for a comparable energy storage system of BTES variety and may cost significantly less. These systems offer high extraction and injection rate, and are used in some cases as both diurnal and seasonal storage, simplifying overall design [31,77].

2.5. Borehole Thermal Energy Storage

The primary seasonal thermal energy storage for heating presented in this review is BTES [43,78]. The underlying principle of the technology is consistent with the previous methods, BTES stores thermal energy utilizing soil and rock as a thermal medium [30,34,43,64,78]. BTES is a prevalent choice of seasonal storage because of its universal applicability, not limited to specific formations as with ATES and GWTES [30,32,33,36,46,48]. However, variations in climate can impact the performance of BTES systems [79]. Limitations of BTES include the comparatively large amount of heat loss compared to insulated water tank or gravel tank systems [30,56]. ATES and CTES systems also see an added advantage of combined short and seasonal time scale storage by combining large storage space and water as the storing medium [24]. A final major concern for BTES installation is the drilling cost associated with the borehole field, considerably more than in ATES configurations. A typical borehole design can be seen in Figure 3 below.

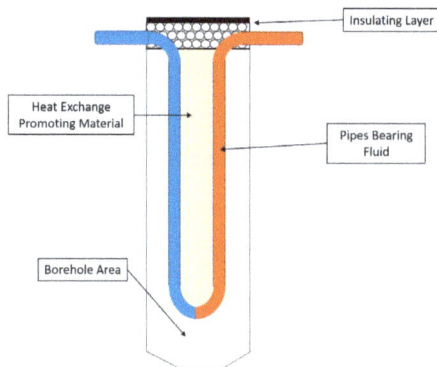

Figure 3. Single borehole design concept [61].

Despite high drilling cost thermal energy storage using boreholes is still a cost effective option. In comparison to thermal energy storage, batteries, a competing mode of energy storage, offer an attractive energy storage solution because of reduced unit storage size. Despite this advantage, BTES storage possesses a number of promising assets. BTES systems offer increasing energy return throughout their lifespan, while battery longevity is limited by the chemical reactions utilized [41,80]. The cost of batteries ranges from $300/kWh, to $400/kWh for medium and large size storage applications such as the Tesla Powerwall [80–83]. BTES energy storage at Drake Landing has a capital cost of $2.6/kWh (thermal) [43]. BTES stores thermal energy and not electrical energy which represent significantly different capital costs. A qualitative table is supplied below in Table 1. The intent of this table is to impart a comparative sense of the key advantages and disadvantages of various energy storage methods. Capacity values for snow waste pit are around half of liquid water due to the significantly lower density of snow compared to liquid water or ice. In contrast, ice ponds offer higher storage capacity than water due to latent heat of fusion and a density similar to liquid water.

Table 1. Comparsion of seasonal thermal energy storage.

Seasonal Storage	Cost (Quantitative)	Geographical Limitations	Energy Recovery	Heating/Cooling Application	Storage Capacity Compared to 1 m³ Water
HWTES	High due to manufacturing costs	None	High (~90–98%) [41,79]	Heating	1 m³
GWTES	High due to manufacturing costs	None	Lower than HWTES because of greater thermal conductivity [62]	Heating	1.5 m³ [55,57]
Snow Waste Pit/ Ice Pond	Low installation cost (~2–5$/kWh) and operating costs	Regions with high annual snowfall or ponds	Free cooling	Cooling	0.67–2 m³ [45,51,69]
ATES	Low initial drillling and equpment, with high maitenence costs [46,56,62]	Limited to aquifers	Medium (65–95%) [73,76]	Heating and Cooling	1.5–2.5 m³ [62]
BTES	High drilling cost, low maitenence and manufacturing costs, modular construction [41,48,57]	Harder rock may increase drilling costs	Low (~70–90% efficieny) [62]	Heating and Cooling	2–4 m³ [62]

BTES is the focus of this review and the principles of construction, component modeling, system working principles, and integration into systems will be discussed in the following sections.

3. BTES Principles

BTES effectively provides a large amount of heat storage despite reduced specific heat of the storage medium because of an easily increased storage volume [30,34,64]. Ground source heat pumps (GSHPs) can be coupled with BTES technology in two distinct manners. A passive GSHP system extracts energy from the ground when heating is needed, using the higher ground temperatures during the cold season [84,85]. These systems can utilize the ground as a heat sink during the summer season, combining both heating and cooling; the cooling heat rejection in this way can act as a charging source [46,62,84,85]. He and Lam demonstrate heating and cooling using a single system with energypiles in place of ground loop heat exchangers, simulated in TRNSYS [86]. The second type of system is the variety implemented at Drake Landing Solar Community (DLSC), featuring seasonal "charging" of the borehole with excess solar energy input to the ground [43,48,87–89]. However, DLSC is unique in the fact that heating is provided by water to air heat exchange fan coils located in each home [43]. The higher temperature of the systems ensure the longevity and efficiency of the BTES and GSHP system in colder climates [47,89]. Nam et al. find that GSHP systems coupled with solar thermal energy can maintain better soil temperature balances to perform at higher COPs over the lifetime of the system [79]. Sliwa and Rosen summarize a number of natural and artificial heat regeneration options for BTES alternative to solar thermal coupling [90]. Higher temperatures of the borehole after solar charging result in higher GSHP COP's and thus less electrical energy use overall. The second system is of interest in achieving higher renewable energy penetration.

3.1. BTES Construction

BTES works by entrenching a series of vertically orientated pipes with a u-tube structure in the soil, passing a working fluid through a heat exchanger, and transferring heat between the working fluid and the surrounding soil. Supplementary heat storage is easily implemented by drilling additional holes for heat exchangers [30,44,48,88]. Certain systems exist with buried horizontal piping, where lower burial depths produce lower cost and more flexible installation options [61,84]. Seibertz et al. ascertained that the lifetime and efficiency of shallow, geothermal systems is lengthened by allowing for regeneration and efficiency rather than simply heating and cooling [59]. However, a large volume of thermal storage material to the total surface area ratio is critical in order to minimize heat loss; thus horizontal orientation can often be detrimental to the system's ability to retain heat [84]. Likewise, Lee finds vertical piping advantageous due to higher temperatures at lower depths in the winter and lower temperature in the summer [61]. The comparatively large ground area required for horizontal trenching is also not inherent to a building site, and seasonal ground temperatures can fluctuate significantly at relatively low burial depths [84]. As a consequence of the presented disadvantages of horizontally entrenched pipes, vertical pipes are the more prevalent selection [64].

Stored energy in BTES is extracted when needed by the pipe-soil interface acting as a heat exchanger [43,64]. The design of BTES can vary in size, ranging from two pipes (one home) [91] up to more than 500 pipes for large scale community systems [43,78]. Large communities such as Drake Landing or Neckarsulm require larger BTES volumes for greater total storage in contrast to single building applications [47,50,64,92]. This is because heating and cooling loads impact the sizing of a borehole field [30,43,64,78]. Başer et al. conclude that undersized BTES volume will result in greater heat losses and inefficiencies in their study; this is due to greater temperature gradients per unit energy of storage, resulting in greater heat transfer rate [93]. A characteristic system design for a community scale borehole is shown below in Figure 4. This type of system is representative of a solar-coupled BTES system with GSHP, and is based on the design provided by Nussbicker et al. in their study in Crailshem, Germany [48]. Solar collectors collect energy when solar radiation is present, and depending upon the system demand either circulate water to meet heating demand or transport the heated fluid to the short term storage tank [30,38,39,78,94,95]. Thermal energy stored in the water tank is dispensed during the evening to meet peak demand or sent to the BTES if unneeded. BTES functions in either charging or discharging mode, pumping water from hot tanks to the centre of the borehole field to inject energy or pumping cool water from the outside of the borehole field inward to extract the energy [30,43,78,96].

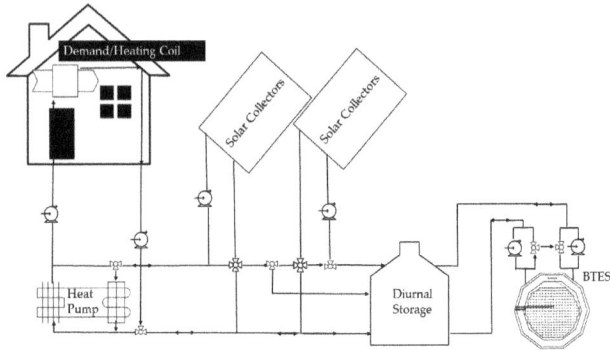

Figure 4. Concept diagram of solar-coupled GSHP with integrated BTES and diurnal storage [48].

A borehole field would ideally be insulated at the boundaries to minimize heat transfer or mass transport in undesirable directions [30,43,47,78,91]. However, insulating the sides is usually infeasible because of exponentially increasing cost. Excavation costs in borehole construction are already normally between 24 and 40% of installation totals [30,33,34,36,97].

3.2. BTES Performance Metrics

Measuring the performance of BTES systems can be done in many ways. For example, the COP of the GSHP used in the system (if a heat pump is the primary driver for the system) is a useful metric if the desired goal of the installation is to improve heat pump performance by raising evaporator temperatures [47,79,84,91]. A more common measure of system efficiency remains the BTES efficiency, which is a measure of the total heat extracted ($Q_{extracted}$) divided by the total heat injected into the storage ($Q_{injected}$), as shown in below equation [43,48,91,98]:

$$\eta_{BTES} = \frac{Q_{extracted}}{Q_{injected}} \tag{1}$$

This metric is directly impacted by the properties of the soil porosity, conductivity, water table presence, and groundwater flow [43,78,98,99] and is reported in [43,64,98]. High temperature BTES storage with direct heat exchange coupled to air units rather than heat pump assisted systems exist in installations such as DLSC. It is concluded that efficiency performance metric will likely be lower due to higher ΔT and subsequently greater heat losses. Also, due to a "warmup" period for borehole temperatures to reach target operating temperatures, less heat may be extracted during this warmup period than normal operation leading to a misleadingly low efficiency measure [43,96]. This definition of borehole efficiency is used in [43,47,48,50,64,91,98,100].

Solar fraction refers to the amount of heating demand met with solar energy [43,101]. However, other studies have used fraction of collected solar energy from the total available radiative energy available [102]. The former definition is much more useful in energy supply as it explicitly states the amount of heating energy that is provided from solar energy. Sweet and McLeskey define internal system efficiency as the heat provided to the home divided by the total solar energy collected, thus incorporating all system losses into their metric [50]. They also report total system efficiency ($\eta_{sytem,total}$)

as the provided heat divided by the incident solar radiation upon the solar collectors, representative equations below in the following equations:

$$\eta_{system,internal} = \frac{\text{Heat Provided to Homes}}{\text{Heat Provided by Solar Collectors}}$$

$$\eta_{system,total} = \frac{\text{Heat Provided to Homes}}{\text{Total Incident Solar Radiation on Solar Collectors}}$$

Not surprisingly, both of these fraction amounts are considerably lower than other system metrics. While total system efficiency characterizes the overall performance of the system, solar fraction provides a better understanding of how well the system meets an energy goal, and BTES efficiency provides a better understanding of required energy storage size. This efficiency metric is discussed in Sweet and McLeskey [50].

The most common economic measure of BTES system effectiveness is cost savings, usually represented as a payback period contrasted with a conventional heating system [64,79]. This representative value is useful for retrofits and small-scale studies, discussed in [64,79]. Larger community scales systems are typically novel and difficult to compare against.

3.3. Examples of BTES Systems

Borehole Thermal Energy Storage makes a convincing case for effective STES based on multiple studies with diverse applications [35,43,64,72,92]. Previous studies acknowledge the push for centralized community thermal storage development, stating that the existing work on the performance of single family homes is insufficient when compared to community sized developments [50,56]. The greater development of community scale BTES technology is attributed to the scalable efficiency of solar assisted BTES technology with storage size [95]. Increasing thermal seasonal storage efficiency sponsors less grid energy draw from space heating loads because they are met with stored solar thermal energy [103]. Some numerical models validate with experimental data, but in large scale studies, often numerical studies are the norm because of construction costs.

3.3.1. Residential and Small Scale Demonstration of BTES

Contrasting the movement towards larger, community centric BTES installations there are several studies illustrating how coupled solar collectors can increase BTES efficiency [64,104–106]. These study focus on smaller scale, low-cost BTES system from a greenhouse study or single buildings. For example, Zhang et al. analyzed a retrofitted greenhouse possessing solar collectors, water tank, and a small borehole field using both TRNSYS and validated with experimental data. The system achieved an efficiency of 80% and 44% solar utilization and an expected payback of 14 years [64]. There are also instances of storage coupling with other mediums such as gravel water storage [63,107].

3.3.2. Community and Large Scale Demonstration of BTES

There are a number of successful large-scale BTES installations, especially in Europe. Sibbet et al. present a large scale and successful study at Drake Landing in Okotoks (AB, Canada) [43]. Drake Landing is an energy efficient community where each home meets the Canadian gold standard of building home efficiency [43,78]. Hugo and Zmeureanu confirm that improved home envelope thermal efficiency can significantly reduce heating loads in cool climates [108]. Figure 5 shows the schematic for the heating system at Drake Landing. The borehole field is insulated on the top, and runs through two buffer tanks filled with hot water (top) and cold water (bottom). The tanks provide the water needed for the BTES: heated water to store the energy in the ground from the hot water tank, and cool water for the cool water tank to extract the energy from the borehole. The heat transferred to the residential water lines provides DHW and space heating. Excess energy during the summer is stored in the borehole field during the winter for later use. Solar collectors mounted on building roofs provide the heat source. Boilers in fall and winter are backups should unusual occurrence avert the necessary energy. This system is an effective renewable energy storage system, where space and water heating consume approximately 80% of the energy supplied to residences [109].

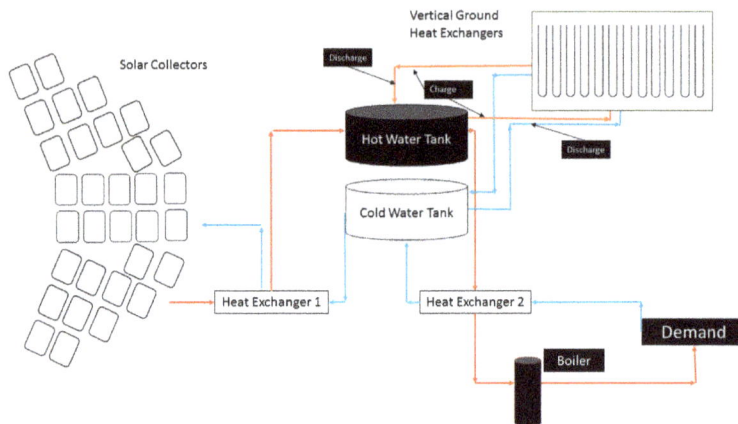

Figure 5. Design schematic of solar thermal energy system with short and long term storage [43].

Based on numerical and validated with experimental results, the overall performance for the fifth year show that the system achieved:

- 94–98.5% STTS Efficiency (short term thermal storage)—Annual average
- Efficiency $= \frac{Q_{out,tank}}{Q_{in,tank}}$ [98]
- 36–41% BTES Efficiency—5th Year
- 89–97% Utilized Solar Fraction (5th Year)

Reported BTES round-trip efficiency is relatively low at Drake Landing due to high groundwater flow [98]. However, other studies have reported BTES efficiencies of 80–90% [46,48,64,98]. Thus, proper site assessment regarding groundwater flow is important to promote higher efficiencies [98,99]. Seibertz et al. determined that monitoring of cooling behavior from thermal gradients makes it possible to identify high ground-water flow zones using a decay time comparison [59].

McDaniel et al. numerically analyzed in TRNSYS the annual energy cost reduction of combined heat and power coupled with a BTES system retrofit at the University of Massachusetts, Amherst campus [98]. The thermal energy comes from preexisting steam systems, which operate at maximum capacity during the summer months when demand and cost are lowest, storing the energy until the winter. The seasonal shift of energy results in a payback time of 9 years, with a BTES efficiency of 90%. This study from McDaniel delineates the relationship between increasing BTES efficiency and size, with a higher efficiency resulting from a borehole field consisting of 6000 boreholes.

BTES storage utilization features more prominently in Western Europe than other regions of the World [24,31,35,42,56,78,87,89,110]. Germany, Norway and Sweden, among the nations of Europe, boast the greatest number of STES systems. Switzerland is the world leader in BTES use, with annual geothermal heating of 1 TWh provided by installations [61]. Germany especially seeks to utilize BTES for solar energy storage in communities [35,39,48,57,88]. Table 2 show some of the more prominent examples.

Table 2. Community sized BTES constructions.

Location	Building(s)	Year Built	Energy Source	No. of Boreholes	Storage Volume [1000 m³]	Estimated Thermal Storage Capacity [MWh]
DLSC in Olbotoks, Canada [43]	52 Residential Homes	2007	Solar Thermal	144	34	780
Neckarsulm, Germany [88]	300 Homes & Shopping Center	1997	Soar Thermal	528	63	1000
Akreshus University Hospital, Norway [42]	Hospital	2007	GSHP	228	300	N/A
Nydalen Business Park Oslo, Norway [85]	Several Large Buildings	2004	GSHP	180	1800	N/A
Bræsdtrup District Heating, Denmark [111]	1500 households	2013	Solar Thermal	48	19	616
Crailshem, Germany [48]	School & Gymnasium	2007	Solar Thermal	80	39	1135
Attenkirchen, Germany [89]	30 Homes	2000	Solar Thermal	90	10	77
Anneberg, Stockholm [87]	50 homes	2001	Solar Thermal	99	60	1467
UMASS Campus, USA [98]	University Campus	Simulation	Existing Steam Turbines	6000	620	37,000
University of Ontario, Canada [112]	Four University Buildings	2004	Heating and Cooling GSHP	370	1400	9700
Groningen, The Netherlands [113]	96 residences	1985	Solar Thermal	20	23	595
Kerava, Finland ** [62]	44 Flats	1985	Solar Thermal			
Vaulruz Project [113]	Maintenance Center	1982	Solar Thermal	Horizontal Tubing	3.5	90
Kungsbacka, Sunclay, Sweden [62]	School Building	1980	Solar Thermal	N/A	85	86,000
Innsbruck, Kranebitten, Austria [113]	1200 MWh/year	1983	Solar Thermal	Horizontal Tubing	60	100

** Information lacking due to study year.

German practice appears to follow modularizing the construction of their BTES and accompanying solar collectors by adding additional solar collectors and ground heat exchangers after initial construction [48]. The Neckersulm borehole features the double U-pipe configuration validated by Zeng et al. to improve heat exchange with the surrounding earth, and has both numerical and experimental assessment of system performance [88,114]. The Crailsheim installation approaches the issue of separating diurnal and seasonal storage by isolating two solar arrays, one to seasonally service the BTES and another to charge hot water tanks for daily usage [48]. Attenkirchen provides a buried the water tank used for daily storage in the center of the BTES field, recycling some of the heat loss from the tank into the BTES [89]. Table 2 shows that solar assisted boreholes require less volumetric capacity, with injected energy supplementing the performance of GSHPs. This is supported by Rad et al. in a feasibility study of combined solar thermal and GSHP systems, determining that solar assistance leads to shorter required borehole lengths to meet the same loads [46]. Figure 6 below illustrates the relationship between increasing solar panel array area and increasing equivalent storage for a solar thermal system. This relationship demonstrates that more collected energy necessitates larger storage.

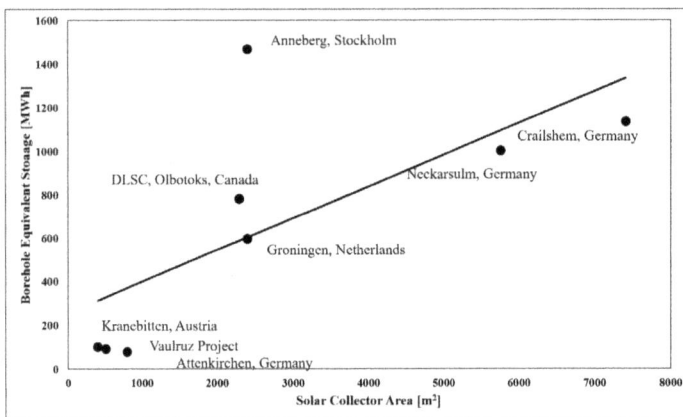

Figure 6. Increasing solar collector area and equivalent design storage of BTES.

4. Design and Modeling of BTES

Large costs for the construction of BTES tend to emphasis the importance in numerical simulations to ensure economic and thermodynamic feasibility. Three types of programs are present in building energy simulation: Building Energy Simulation (BES) tools such as EnergyPlus, building envelope heat and mass transfer (HAM) programs such as WINDOW (window and daylight modeling software from Lawrence Berkeley National Laboratory) and WUFI Pro (Wärme Und Feuchte Instationär, or transient heat and moisture) , and computational fluid dynamics (CFD) such as MODFLOW, COMSOL and TOUGH2 [115]. Modeling of BTES requires appropriate tool selection, which depends on the application and goal of the study: from whole building energy simulations, to more detailed heat and mass transfer programs. Accordingly, this study reviews various modeling techniques based on component level design, system level design, as well as a development of integration between these two.

4.1. Parameters to Consider in BTES Modeling and Development

Total storage capacity of the borehole depends on total volume of the borehole, porosity, and overall specific heat [30,40,99,116]. Catolico et al. assert that lower thermal conductivities allow for higher heat retention and thus better borehole efficiencies [99]. This is due to more concentrated thermal plumes which cause higher thermal gradients near the pipes and thus better heat transfer to and from the pipes during discharge periods [99].

Thermal conductivity depends on the material or soil temperature, but is often considered a constant property [116,117]. Moradi et al. utilize a model developed by Smits et al. adapted to the relevant geometry, treating the thermal conductivity as a constant [116,118]. Moradi finds that thermal conductivity increases proportionally to increasing water content because soil is a medium consisting of air, water, and organic matter in COMSOL, validated with experimental data [93,116,119,120]. Higher fractions of water lead to higher thermal conductivity and storage as the respective coefficients increase; these higher fractions of water are coined "water bridges" [119,120]. Greater measured porosity leads to higher levels of saturation with water present, resulting in higher thermal conductivity. In addition, thermal conductivity rises due to (1) increase in solid matter per unit volume; (2) less soil pores filled with air; (3) consistent contact for conductive heat transfer flow [120].

Soil saturation will lead to higher convection coefficients as well as higher thermal conductivity, which unfavorably impacts Borehole heat retention, and should be avoided [121]. Boundary layer models directly impact simulation accuracy and are critically important [99,114,116,119]. Convective heat losses induce more heat loss than conductive heat losses, however both lower the efficiency of BTES [99,116,122]. High permeability in soils, both unsaturated and saturated, leads to higher convection coefficients and subsequently higher heat losses [99,116]. This confirms McDaniel's observation that high groundwater flows lead to low BTES efficiency [98]. Li et al. establish that this applies only to BTES with solar or other heat injection, with GSHP installations lacking heat injection featuring higher thermal restoration in areas with higher groundwater flow [123]. The effects of convective boundary layers regarding heat transfer from soil to pipe have not been fully explored and should be studied further to fully understand the effect upon BTES efficiency [99]. The appropriate sizing of BTES and accompanying diurnal storage with ensuing codependence dictates system performance [41,44,49].

4.2. Component Design Level Modeling Software and Development

Numerical solutions providing a Multiphysics approach to defining ground heat transfer providing more accurate and robust solutions than a parameter fit white-box model approach taken by a larger whole building platform. Clio and Mirianhosseinabadi note that while numerical solutions present a high degree of versatility and accuracy, often they are prone to computational inefficiencies as a consequence of complicated hybrid coordinate systems, contrasted with analytical solutions [124]. Their study confirms that TRNSYS and EnergyPlus are highly utilized tools for residential homes in BTES simulation [124]. Simpler models like G-functions in GLHEpro (for EnergyPlus) or the Duct Storage Model developed by Hellstrom in TRNSYS may not accurately depict BTES heat transfer and fluid flow. BTES modules within larger simulations do not solve using a multiphysics approach, nor do they take ground water flow into account, to facilitate acceptable simulation times [125,126].

Studies examining more in-depth heat transfer in soil use COMSOL, GLHEpro and TOUGH2 [99,116,125,127,128]. COMSOL is a finite element analysis tool used for multiphysics approaches, such as fluid flow behavior [129]. TOUGH2 is a numerical solution program for heat and fluid flow in porous and fractured media [130]. GLHEpro uses a numerical simulation based upon "G-functions" which provide an accurate solution to the temperature profile of the earth envelope and do not have lengthy computational times [125]. Additionally, MODFLOW is the USGS program for modeling groundwater flow, or other flow in porous media [131]. Certain projects exist that utilize MODFLOW in [132].

In addition to commercially available software, Zeng et al. developed a quasi-three-dimensional model for the thermal network in a borehole field for a multitude of heat exchanger arrangements [114,133]. A more popular model often utilized in optimization case studies for BTES modeling is Ingersoll's infinite-line source model, and has been used in many studies as the underlying solution to the heat transfer in BHE [134–136]. Yet more optimization studies on system design use Hellstrom's DST model in Kjellsson et al. [137]. Finite line-source models by Zeng et al. and Molina et al. (with groundwater advection) illustrate the importance of consideration of axial heat transfer in BHE fields especially with shorter lengths [133,138]. Eskilson and Claesson present a detailed three dimensional computer model combining the interactions between convective heat flow and conductive ground process, which is incorporated into the GLHEpro software used by the EnergyPlus BES platform [139]. Catolico et al. establish that there are a number of existing models for modeling the behavior of heat and fluid flow in the ground, but there is a lack of effective property models for evaluating the pertinent heat and mass transfer parameters [99]. However, Shonder et al. present parameter estimation techniques coupled with Ingersoll's one-dimensional Borehole model giving an accurate solution for variable conductivities and heat capacities of grout and fluid of the Borehole [140].

4.3. System Design Level Modeling Tools and Efforts to Couple System Level and Component Level Models

TOUGH2 and COMSOL provide more robust analyses of BTES and can accurately model the heat storage. In contrast to high accuracy heat characteristic modeling, TRNSYS, EnergyPlus, and ESP-r are commonly used modeling platforms for system level analysis, discussed in Section 4. TRNSYS is the most prevalent modeling program when modeling BTES, using the Duct Storage Model (DST) to predict ground heat transfer [46,47,50,62,64,86,91,92,96,98,126]. There exists certain efforts to use open source EnergyPlus [41,49,124] which relies upon an outside program, GLHEpro, to perform the sizing parameters of a borehole [125,127,141]. Other studies have attempted to model vertical heat exchanger behavior in ESP-r [43,94].

The discussion then turns to coupling whole-building analysis tools (EnergyPlus, TRNSYS), with accurate multiphysics solutions (TOUGH2 and COMSOL). This approach, used for enhanced building envelope modeling, is known as "BES-HAM" or "BES-Hygrothermal" Coupling [115,142–145]. Co-simulation of software using MatLab and Simulink environment is not a novel process for HVAC application [142–144,146–149]. Ferroukhi et al. present a successful effort of TRNSYS and COMSOL co-simulation to model the hygrothermal effects in a multi-layer wall [143]. Huang et al. present a co-simulation of COMIS tool with EnergyPlus for hygrothermal effects of moisture transfer across multiple zones [150]. There are more efforts that seek to co-simulate TRNSYS and MODFLOW [151].

Catolico et al. present a BTES model utilizing TOUGH2, which at present has not been co-simulated with any BES analysis tool [99]. TOUGH2 can accurately model the effects of groundwater flow upon heat transfer in soil and thus would be a useful tool for effective modeling, especially in studies where high ground water flows are suspected [43]. The modular approach of the "types" built in TRNSYS facilitates co-simulation with programs like TOUGH2 and COMSOL, using Simulink "S-functions" as a linking mechanism to combine accurate building energy evaluation, and BTES heat transfer modeling More recently, Rad et al. developed an updated TRNSYS type for BTES simulation [100]. The type is based on the Ground Heat Exchanger Analysis Design and Simulation (GHEADS) developed by Leong and Tarnawski [152]. The advantage of the GHEADS model over the Duct Storage Model provided by Hellstrom is a coupled heat and moisture flow model, the presence of ground water table, and soil freezing and thawing cycles. When applied to the DLSC community design, Rad et al. found a 38% reduction in Borehole footprint a number could be achieved for similar system performance. This model by Rad et al. is a comprehensive model that combines complex coupled heat and mass transfer BTES modeling with system level modeling.

5. System Sizing and Integration of Diurnal and Seasonal Storage

BTES functions in either charging or discharging modes that are rather slow compare to diurnal storage systems such as water tanks or ice storage For example, the BTES system in DLSC took about 4–5 years to fully charge [43]. For this reason, some studies have proposed and/or implemented hybrid diurnal/seasonal systems illustrated in Figure 4. Solar panels heat water while solar irradiation energy is present, and pass the heat energy to the evaporator side of the heat pump. If system demand does not require solar energy, the heated fluid is stored in a water tank, or transported to the BTES to store for later use. The water thus heated from the boreholes is used to raise the temperature of the evaporator in a ground source heat pump and finally meet system demand [84,85].

System configuration is relatively consistent across design scenarios except in the case of Zhang et al. and DLSC when BTES temperatures are high enough to provide system heat by direct heat exchange, and GSHPs are not used [64,96]. However, the addition of short term storage is required to provide higher heat transfer rates from fluid medium rather than the earth medium of BTES [41]. Additionally, the sizing of diurnal water based storage, BTES, and solar collector area are dependent upon both each other, and the heating and cooling loads [41]. Sweet and McLeskey analyze the sizing of a borehole for a single family residence in TRNSYS [50]. Six different home sizes (from 75 m^2 to 220 m^2) were parametrically analyzed with six different BTES sizes ranging from 10 m^3 to 50 m^3 with solar collectors sized to 80% of the south facing roof on the home (Kalaiselvam notes that as a rule of thumb, solar collectors should be sized to 10% of the total floor area [33]). The results of the study show that for each home size, a borehole field of 15 m^3 provided optimal results, independently of the home size and solar panel area.

Effective storage system sizing is based upon both demand loads and supplied energy [43]. Hseih et al. performed a parametric analysis study in Rheinfelden, Switzerland of diurnal and STES configurations in a suburban area with 11 homes using EnergyPlus. With the addition of diurnal thermal storage the study concluded that system efficiencies rapidly increase, from 15% to 47% retention of solar fraction. The addition of long term storage increased the solar fraction used from 47% to 61%. Converting the system to centralized rather than decentralized storage reduced the system solar fraction utilization from 61% to 44% due to heat losses from transport piping. However, while the decentralized long term storage is effective for a smaller community, the potential to reduce installation costs outweighs the effectiveness of the decentralization in large applications. This is similar to the community in Okotoks (AB, Canada) [43,72]. The study by Hseih et al. acknowledges the need for water based diurnal storage integrated with STES for effective utilization of energy from solar collectors. Roth contributes a number of seasonal storage solutions that utilize a diurnal storage component in order to ensure the proper distribution of cooling energy [153]. Xu notes that water based storage is advantageous for faster response times, while seasonal thermal energy storage has comparatively lower discharge rates [36].

In addition to seasonal energy storage, diurnal thermal storage stores excess renewable energy generated during the day for later use at night, improving system efficiency [30,32,43,154]. Alternatively, some diurnal energy storage seeks to store energy purchased throughout the day during periods of off-peak loading [6,155,156]. Lee, Joo, and Baek demonstrate how thermal collector control strategies can be implemented to eliminate space heating during peak electric hours [18].

Rather than providing energy for the entire day, peak energy solar eliminates energy use during the most expensive energy cost periods associated with on-peak demand. Various combinations of these solar and off-peak heating purchasing schemes exist [88,155,156]. In addition to reducing consumer cost, peak-load generators often operate at much lower efficiencies than baseload generators, and implementation of solar thermal demand-side systems can reduce peak-load generator use [28]. Wholly grid-independent systems require significantly more storage than the aforementioned partially grid-reliant systems to meet 24 h demand, presenting difficulties and additional costs during installation [12]. Hyman details a thorough economic analysis for the installation details of thermal energy storage [25].

Diurnal Storage for Space Cooling Using Absorption Chilling or Ice Storage

Diurnal thermal storage can also provide space cooling by releasing stored heat to an absorption chiller, which produces cooling from a heat input via chemical process [155–157]. Chillers of this type have an expected performance (COP) of 0.15–0.6 [157]. The low COPs delineate that absorption chillers are more economical when used with waste or solar heat, rather than purchased heating [155–157].

Absorption chillers can increase solar thermal energy penetration during the cooling season if seasonal thermal energy storage is not available [155,156].

Most Cooling Thermal Energy Storage (CTES) is typically short term and provided by storing chilled water, or ice, chosen because of high heat capacity, due to stored latent heat. For example, chilled water with a 5 °C temperature difference has a density storage of 5.8 kWh of cooling energy per cubic meter of water [158], whereas ice storage boasts an energy storage density four to six times that of cold water storage [159]. Cooling capacity storage is an emerging technology utilizing natural resource to produce the intended affect [66,160–162]. The IKEA building, located in Centennial (CO, USA) utilizes combines an ice storage units in conjunction with a GSHP BTES system to supplement cooling season capacity [163]. After a review of the literature, it is apparent that no combined system for seasonal and diurnal cooling exists comparable to coupled diurnal and seasonal BTES with water tank systems.

6. Conclusions and Research Outlook

Energy storage is a critical component for future renewable energy grid performance [11,12,19–22]. The current United States grid relies heavily upon centralized generation and distribution transmission of energy in electrical form. This is not favorable for the implementation of intermittent renewable energy, which requires storage [11,12,19]. Additionally, thermal energy storage presents considerably lower capital cost than electrical energy storage [43,83]. This literature review considers seasonal energy storage mechanisms demonstrated in recent implementation. Acknowledging the importance of energy storage for renewable energy penetration, previous studies state that examination of various options for optimal energy management and system reliability remains the primary concern [164]. While BTES is the most universal STES method, other methods may be more effective depending upon geography and immediate hydraulic features [24,30,38,57]. The path to effective STES design is best navigated by thorough evaluation of environmental site characteristics and soil properties. Additionally, proper design practice regarding the integration of diurnal and seasonal storage yield higher system performance [37]. Integrated diurnal and seasonal energy storage provides a critical combination of extended storage periods (seasonal storage) and high discharge rates (diurnal storage) and promotes the highest levels of renewable energy penetration and efficiency, providing robust demand response. BTES modeling tools range from in-depth analysis allowing for subsystem design, to whole building simulations that incorporate simpler subsurface heat transfer models into energy design analysis [50,124,127]. Tool selection depends on the desired type of analysis, studies looking at whole building/community use analysis tools such as EnergyPlus and TRNSYS, while more detail Multiphysics tools such as COMSOL and TOUGH2 are used to model the heat transfer characteristics of BTES. Careful review on previous studies highlights that:

- Community scale BTES requires a "charging" period of few years for the design system temperature to be reached;
- Most single-residential scale BTES often do not require solar thermal panels because of low system demand, which thermal regeneration in the ground can recover;
- BTES is more commonly used for space heating and DHW applications than cooling applications;
- BTES is less geographically limited than ATES and requires lower installation costs than HWTES or GWTES;
- Coupled diurnal and seasonal storage increases the overall utilization of captured solar energy;
- Coupled diurnal and seasonal storage systems are much more prevalent for heating than cooling applications;
- Performance metrics for BTES systems and components can be inconsistent across the field, however BTES efficiency is always defined as the fraction of energy extracted divided by the energy injected;

- Although there are a handful of studies coupling BTES at the component model with a system level simulation, most previous studies have not bridged the modeling gap between the two levels of modeling.

Coupling integrated system and component level models is critical for modeling practice to improve system performance and lower capital costs. Jacobson et al. present a nation scale model to illustrate the effect of energy storage with renewable generation. However, no such model exists that segregates different energy end uses, undoubtedly for simplification. A comprehensive model that addresses different end uses in different sectors and regions could more accurately depict the role of BTES in the changing smart-grid.

Acknowledgments: The authors would like to thank Colorado School of Mines for the support to complete this literature review. Additionally, the authors thank Tim McDowell from Thermal Energy System Specialists for his critical input.

Author Contributions: Michael Lanahan was the principle author tasked with gathering the necessary studies and writing the review. Paulo Cesar Tabares-Velasco edited, provided guidance and some contributions to the studies and writing of the study. Paulo Cesar Tabares-Velasco was also responsible for obtaining the funding necessary for this work.

Conflicts of Interest: The authors declare no conflict of interest.

References

1. Monthly Energy Review June 2016. U.S. Energy Information Administration. Available online: https://www.eia.gov/totalenergy/data/monthly/pdf/sec2_3.pdf (accessed on 24 February 2017).
2. Hill, J.; David, T.; Stephan, P.; Douglas, T. Environmental, economic, and energetic costs and benefits of biodiesel and ethanol biofuels. *Proc. Natl. Acad. Sci. USA* **2006**, *103*, 11206–11210. [CrossRef] [PubMed]
3. Baldwin, C.; Cruickshank, C.A. A review of solar cooling technologies for residential applications in Canada. *Energy Procedia* **2012**, *30*, 495–504. [CrossRef]
4. Denholm, P.; Hand, M. Grid flexibility and storage required to achieve very high penetration of variable renewable electricity. *Energy Policy* **2011**, *39*, 1817–1830. [CrossRef]
5. Stadler, I. Power grid balancing of energy systems with high renewable energy penetration by demand response. *Util. Policy* **2008**, *16*, 90–98. [CrossRef]
6. Tweed, K. California's Fowl Problem: 10 Ways to Address the Renewable Duck Curve. 14 May 2014. Available online: http://www.greentechmedia.com/articles/read/10-ways-to-solve-the-renewable-duck-curve (accessed on 8 October 2016).
7. Denholm, P.; O'Connell, M.; Brinkman, G.; Jorgenson, J. *Overgeneration from Solar Energy in California: A Field Guide to the Duck Chart*; DE-AC36–08GO28308; National Renewable Energy Library: Golden, CO, USA, 2015.
8. Arteconi, A.; Ciarrocchi, E.; Pan, Q.; Carducci, F.; Comodi, G.; Polonara, F.; Wang, R. Thermal energy storage coupled with PV panels for demand side management of industrial building cooling loads. *Appl. Energy* **2017**, *185*, 1984–1993. [CrossRef]
9. Gujar, M.; Datta, A.; Mohanty, P. Smart mini grid: An innovative distributed generation based energy system. In Proceedings of the 2013 IEEE Innovative Smart Grid Technologies—Asia (ISGT Asia), Bangalore, India, 10–13 November 2013; pp. 1–5.
10. Lasseter, R.H.; Paigi, P. Microgrid: A conceptual solution. In Proceedings of the 2004 IEEE 35th Annual Power Electronics Specialists Conference (IEEE Cat. No.04CH37551), Aachen, Germany, 20–25 June 2004; Volume 6, pp. 4285–4290.
11. Sovacool, B.K. The intermittency of wind, solar, and renewable electricity generators: Technical barrier or rhetorical excuse? *Util. Policy* **2009**, *17*, 288–296. [CrossRef]
12. Jacobson, M.Z.; Delucchi, M.A.; Cameron, M.A.; Frew, B.A. Low-cost solution to the grid reliability problem with 100% penetration of intermittent wind, water, and solar for all purposes. *Proc. Natl. Acad. Sci. USA* **2015**, *112*, 15060–15065. [CrossRef] [PubMed]
13. Lew, D.; Brinkman, G.; Ibanez, E.; Florita, A.; Hummon, M.; Hodge, B.; Stark, G.; King, J.; Lefton, S.; Kumar, N.; et al. *The Western Wind and Solar Integration Study Phase 2*; Technical DE-AC36–08GO28308; NREL: Golden, CO, USA, 2013.

14. Driesen, J.; Katiraei, F. Design for distributed energy resources. *IEEE Power Energy Mag.* **2008**, *6*, 30–40.
15. Islam, M.R.; Gabbar, H.A. Study of micro grid safety & protection strategies with control system infrastructures. *Smart Grid Renew. Energy* **2012**, *3*, 1–9.
16. Bhandari, B.; Lee, K.-T.; Lee, G.-Y.; Cho, Y.-M.; Ahn, S.-H. Optimization of hybrid renewable energy power systems: A review. *Int. J. Precis. Eng. Manuf. Green Technol.* **2015**, *2*, 99–112. [CrossRef]
17. Proietti, S.; Sdringola, P.; Castellani, F.; Astolfi, D.; Vuillermoz, E. On the contribution of renewable energies for feeding a high altitude smart mini grid. *Appl. Energy* **2017**, *185*, 1694–1701. [CrossRef]
18. Lee, K.-H.; Joo, M.-C.; Baek, N.-C. Experimental evaluation of simple thermal storage control strategies in low-energy solar houses to reduce electricity consumption during grid on-peak periods. *Energies* **2015**, *8*, 9344–9364. [CrossRef]
19. Jacobson, M.Z.; Delucchi, M.A. Providing all global energy with wind, water, and solar power, Part I: Technologies, energy resources, quantities and areas of infrastructure, and materials. *Energy Policy* **2011**, *39*, 1154–1169. [CrossRef]
20. Mason, I.G.; Page, S.C.; Williamson, A.G. A 100% renewable electricity generation system for New Zealand utilising hydro, wind, geothermal and biomass resources. *Energy Policy* **2010**, *38*, 3973–3984. [CrossRef]
21. Becker, S.; Bethany, F.; Andersen, G.; Zeyer, T.; Schramm, S.; Greiner, M.; Jacobson, M. Features of a fully renewable US electricity system: Optimized mixes of wind and solar PV and transmission grid extensions. *Energy* **2014**, *72*, 443–458. [CrossRef]
22. Rasmussen, M.G.; Andresen, G.B.; Greiner, M. Storage and balancing synergies in a fully or highly renewable pan-European power system. *Energy Policy* **2012**, *51*, 642–651. [CrossRef]
23. Sabihuddin, S.; Kiprakis, A.E.; Mueller, M. A numerical and graphical review of energy storage technologies. *Energies* **2014**, *8*, 172–216. [CrossRef]
24. Nordell, B. *Large-Scale Thermal Energy Storage*; Division of Water Resources Engineering—Lulea University of Technology: Lulea, Sweden, 2000.
25. Hyman, L. *Sustainable Thermal Storage Systems Planning Design and Operations*; McGraw-Hill: New York, NY, USA, 2011.
26. Marnay, C.; Venkataramanan, G.; Stadler, M.; Siddiqui, A.S.; Firestone, R.; Chandran, B. Optimal technology selection and operation of commercial-building microgrids. *IEEE Trans. Power Syst.* **2008**, *23*, 975–982. [CrossRef]
27. Denholm, P.; Jorgenson, J.; Miller, M.; Zhou, E. *Methods for Analyzing the Economic Value of Concentrating Solar Power with Thermal Energy Storage*; NREL/TP-6A20–64256; National Renewable Energy Library: Golden, CO, USA, 2015.
28. Denholm, P.; Ong, S.; Booten, C. *Using Utility Load Data to Estimate Demand for Space Cooling and Potential for Shiftable Loads*; TP-6A20–54509; National Renewable Energy Library: Golden, CO, USA, 2012.
29. Evans, A.; Strezov, V.; Evans, T.J. Assessment of utility energy storage options for increased renewable energy penetration. *Renew. Sustain. Energy Rev.* **2012**, *16*, 4141–4147. [CrossRef]
30. Rad, F.M.; Fung, A.S. Solar community heating and cooling system with borehole thermal energy storage—Review of systems. *Renew. Sustain. Energy Rev.* **2016**, *60*, 1550–1561. [CrossRef]
31. Nordell, B.; Grein, M.; Kharseh, M. Large-scale utilisation of renewable energy requires energy storage. In Proceedings of the International Conference of Renewable Energys and Suitainable Development, Tlemcen, Algeria, 21–24 May 2007.
32. Dincer, I.; Rosen, M.A. *Thermal Energy Storage: Systems and Applications*; John Wiley & Sons: Hoboken, NJ, USA, 2011.
33. Kalaiselvam, S.; Parameshwaran, R. *Thermal Energy Storage Technologies for Sustainability: Systems Design, Assessment and Applications*; Elsevier: Amsterdam, The Netherlands, 2014.
34. Mangold, D.; Schmidt, T.; Müller-Steinhagen, H. Seasonal thermal energy storage in Germany. *Struct. Eng. Int.* **2004**, *14*, 230–232. [CrossRef]
35. Bauer, D.; Marx, R.; Nußbicker-Lux, J.; Ochs, F.; Heidemann, W.; Müller-Steinhagen, H. German central solar heating plants with seasonal heat storage. *Sol. Energy* **2010**, *84*, 612–623. [CrossRef]
36. Xu, J.; Wang, R.Z.; Li, Y. A review of available technologies for seasonal thermal energy storage. *Sol. Energy* **2014**, *103*, 610–638. [CrossRef]
37. ASHRAE Technical Committees, Task Groups, and Technical Resource Group. *ASHRAE Handbook*; American Society of Heating, Refrigerating and Air-Conditioning Engineers, Inc.: Atlanta, GA, USA, 2004.

38. Pinel, P.; Cruickshank, C.A.; Beausoleil-Morrison, I.; Wills, A. A review of available methods for seasonal storage of solar thermal energy in residential applications. *Renew. Sustain. Energy Rev.* **2011**, *15*, 3341–3359. [CrossRef]

39. Mangold, D. Seasonal storage—A German success story. *Sun Wind Energy* **2007**, *1*, 48–58.

40. Pavlov, G.K.; Olesen, B.W. Thermal energy storage-A review of concepts and systems for heating and cooling applications in buildings: Part 1—Seasonal storage in the ground. *HVAC R Res.* **2012**, *18*, 515–538.

41. Sibbet, B.; McClenahan, D. *Seasonal Borehole Thermal Energy Storage—Guidelines for Design & Construction*; Natural Resources: Hvalsoe, Denmark, 2015.

42. Midttomme, K.; Hauge, A.; Grini, R.S. *Underground Thermal Energy Storage with Heat Pumps in Norway*; Norwegian Geotechnical Institute (NGI): Trondheim, Norway, 2016.

43. Sibbet, B. The performance of a high solar fraction seasonal storage district heating system—Five years of operation. *Energy Procedia* **2012**, *30*, 856–865. [CrossRef]

44. Schmidt, T.; Mangold, D.; Müller-Steinhagen, H. Central solar heating plants with seasonal storage in Germany. *Sol. Energy* **2004**, *76*, 165–174. [CrossRef]

45. Skogsberg, K.; Nordell, B. The sundsvall hospital snow storage. *Cold Reg. Sci. Technol.* **2001**, *32*, 63–70. [CrossRef]

46. Rad, F.M.; Fung, A.S.; Leong, W.H. Feasibility of combined solar thermal and ground source heat pump systems in cold climate, Canada. *Energy Build.* **2013**, *61*, 224–232. [CrossRef]

47. Wang, X.; Zheng, M.; Zhang, W.; Zhang, S.; Yang, T. Experimental study of a solar-assisted ground-coupled heat pump system with solar seasonal thermal storage in severe cold areas. *Energy Build.* **2010**, *42*, 2104–2110. [CrossRef]

48. Nussbicker-Lux, J. The BTES project in Crailsheim (Germany)—Monitering results. In Proceedings of the 12th International Conference on Energy Storage, Lleida, Spain, 16–18 May 2012.

49. Hsieh, S.; Weber, R.; Dorer, V.; Orehounig, K. Integration of thermal energy storage at building and neighburhood scale. In Proceedings of the 14th International Conference of IBPSA Building Simulation, Hyderabad, India, 7–9 December 2015.

50. Sweet, M.L.; McLeskey, J.T., Jr. Numerical simulation of underground seasonal solar thermal energy storage (SSTES) for a single family dwelling using TRNSYS. *Sol. Energy* **2012**, *86*, 289–300. [CrossRef]

51. Kirkpatrick, D.L.; Masoero, M.; Rabl, A.; Roedder, C.E.; Socolow, R.H.; Taylor, T.B. The ice pond—Production and seasonal storage of ice for cooling. *Sol. Energy* **1985**, *35*, 435–445. [CrossRef]

52. Rutberg, M.; Hastbacka, M.; Cooperman, A.; Bouza, A. Saves energy, money thermal energy storage. *ASHRAE J.* **2013**, 62–66.

53. Silvetti, B. Application fundamentals of ice-based thermal storage. *ASHRAE J.* **2002**, *44*, 30–35.

54. Abhat, A. Low temperature latent heat thermal energy storage: Heat storage materials. *Sol. Energy* **1983**, *30*, 313–332. [CrossRef]

55. Novo, A.V.; Bayon, J.R.; Castro-Fresno, D.; Rodriguez-Hernandez, J. Review of seasonal heat storage in large basins: Water tanks and gravel-water pits. *Appl. Energy* **2010**, *87*, 390–397. [CrossRef]

56. Schmidt, T.; Mangold, D. New steps in seasonal thermal energy storage in Germany. In Proceedings of the Tenth International Conference on Thermal Energy Storage, Pomona, CA, USA, 2 June 2006.

57. Gabriela, L. Seasonal thermal energy storage concepts. *Acta Tech. Napoc.* **2012**, *55*, 775–784.

58. Lee, K.S. Simulation of aquifer thermal energy storage system under continuous flow regime using two-well model. *Energy Sources Part Recovery Util. Environ. Eff.* **2009**, *31*, 576–584. [CrossRef]

59. Seibertz, K.S.O.; Chirila, M.A.; Bumberger, J.; Dietrich, P.; Vienken, T. Development of in-aquifer heat testing for high resolution subsurface thermal-storage capability characterisation. *J. Hydrol.* **2016**, *534*, 113–123. [CrossRef]

60. Ganguly, S.; Kumar, M. A numerical model for transient temperature distribution in an aquifer thermal energy storage system with multiple wells. *Lowl. Technol. Int.* **2015**, *17*, 179–188. [CrossRef]

61. Lee, K.S. Underground thermal energy storage. In *Underground Thermal Energy Storage*; Springer: London, UK, 2013; pp. 15–26.

62. Hesaraki, A.; Holmberg, S.; Haghighat, F. Seasonal thermal energy storage with heat pumps and low temperatures in building projects—A comparative review. *Renew. Sustain. Energy Rev.* **2015**, *43*, 1199–1213. [CrossRef]

63. Rybakova, L.E.; Khairiddinov, B.E.; Khalimov, G.G. Investigation of heat-exchange processes of a subsurface pebble thermal-storage system in a solar greenhouse-dryer. *Appl. Sol. Energy* **1990**, *26*, 18–21.
64. Zhang, L.; Xu, P.; Mao, J.; Tang, X.; Li, Z.; Shi, J. A low cost seasonal solar soil heat storage system for greenhouse heating: Design and pilot study. *Appl. Energy* **2015**, *156*, 213–222. [CrossRef]
65. Skosberg, K.; Nordell, B. *Cold Storage Applications*; Lulea University of Technology: Lulea, Sweden, 2000.
66. Gorski, A.J. *Third International Workshop on Ice Storage for Cooling Applications*; U.S. Department of Energy-Energy and Environmental Systems Division, Argonne National Laboratory: Dukes County, MA, USA, 1986.
67. Yan, C.; Shi, W.; Li, X.; Zhao, Y. Optimal design and application of a compound cold storage system combining seasonal ice storage and chilled water storage. *Appl. Energy* **2016**, *171*, 1–11. [CrossRef]
68. Taylor, T.B. Ice ponds. In Proceedings of the AIP Conference, New York, NY, USA, September 1985.
69. Yan, C.; Shi, W.; Li, X.; Wang, S. A seasonal cold storage system based on separate type heat pipe for sustainable building cooling. *Renew. Energy* **2016**, *85*, 880–889. [CrossRef]
70. Reuss, M. Water pits and snow store: Technologies for seasonal heat and cold storage. In Proceedings of the 2008 ASHRAE Winter Meeting, Salt Lake City, UT, USA, 21–25 June 2008.
71. Seasonal Cold Storage Building and Process Applications: A Standard Design Option? Available online: http://docplayer.net/15828152-Seasonal-cold-storage-building-and-process-applications-a-standard-design-option.html (accessed on 21 July 2016).
72. Sanner, B. *High Temperature Underground Thermal Energy Storage State-of-the-Art and Prospects*; Lenz-Verlag: Neu-Isenburg, Germany, 1999.
73. Paksoy, H.O.; Andersson, O.; Abaci, S.; Evliya, H.; Turgut, B. Heating and cooling of a hospital using solar energy coupled with seasonal thermal energy storage in an aquifer. *Renew. Energy* **2000**, *19*, 117–122. [CrossRef]
74. Arnfield, A.J. Two decades of urban climate research: A review of turbulence, exchanges of energy and water, and the urban heat island. *Int. J. Climatol.* **2003**, *23*, 1–26. [CrossRef]
75. Vanhoudt, D.; Desmedt, J.; van Bael, J.; Robeyn, N.; Hoes, H. An aquifer thermal storage system in a Belgian hospital: Long-term experimental evaluation of energy and cost savings. *Energy Build.* **2011**, *43*, 3657–3665. [CrossRef]
76. Sommer, W.; Valstar, J.; Leusbrock, I.; Grotenhuis, T.; Rijnaarts, H. Optimization and spatial pattern of large-scale aquifer thermal energy storage. *Appl. Energy* **2015**, *137*, 322–337. [CrossRef]
77. Neilson, K. *Thermal Energy Storage: A State-of-the-Art*; NTNU and SINTEF: Trondheim, Norway, 2003.
78. McClenahan, D.; Gusdorf, J.; Kokko, J.; Thornton, J.; Wong, B. seasonal storage of solar energy for space heat in a new community. In Proceedings of the 2006 ACEEE Summer Study on Energy Efficiency in Buildings, Okotoks, AB, Canada, 13–18 August 2006; pp. 1–13.
79. Nam, Y.J.; Gao, X.Y.; Yoon, S.H.; Lee, K.H. Study on the performance of a ground source heat pump system assisted by solar thermal storage. *Energies* **2015**, *8*, 13378–13394. [CrossRef]
80. Geth, F.; Tant, J.; Six, D.; Tant, P.; de Rybel, T.; Driesen, J. Techno-economical and life expectancy modeling of battery energy storage systems. In Proceedings of the 21st International Conference on Electricity Distribution (CIRED), Frankfurt, Germany, 6–9 June 2011.
81. How Long Does It Take to Pay Off a Tesla Powerwall? IER, 5 January 2016. Available online: http://instituteforenergyresearch.org/analysis/payback-on-teslas-powerwall-battery/ (accessed on 17 September 2016).
82. Nykvist, B.; Nilsson, M. Rapidly falling costs of battery packs for electric vehicles. *Nat. Clim. Chang.* **2015**, *5*, 329–332. [CrossRef]
83. Gerssen-Gondelach, S.J.; Faaij, A.P.C. Performance of batteries for electric vehicles on short and longer term. *J. Power Sources* **2012**, *212*, 111–129. [CrossRef]
84. Mustafa Omer, A. Ground-source heat pumps systems and applications. *Renew. Sustain. Energy Rev.* **2008**, *12*, 344–371. [CrossRef]
85. Lund, J.; Sanner, B.; Rybach, L.; Curtis, R.; Hellstrom, G. *Geothermal (Ground-Source) Heat Pumps a World Overview*; Oregon Institute of Technology: Klamath Falls, OR, USA, 2004.
86. He, M.; Lam, H.N. *Study of Geothermal Seasonal Cooling Storage System with Energy Piles*; Department of Mechanical Engineering, University of Hong Kong: Hong Kong, China, 2017.

87. Lundh, M.; Dalenbäck, J.-O. Swedish solar heated residential area with seasonal storage in rock: Initial evaluation. *Renew. Energy* **2008**, *33*, 703–711. [CrossRef]
88. Nußbicker-Lux1, J.; Heidemann, W.; Müller-Steinhagen, H. Validation of a computer model for solar coupled district heating systems with borehole thermal energy store. In Proceedings of the EFFSTOCK 2009, Stockholm, Sweden, 14–17 June 2009; pp. 1–7.
89. Reuss, M.; Beuth, W.; Schmidt, M.; Schoelkopf, W. *Solar District Heating with Seasonal Storage in Attenkirchen*; Bavarian Center of Applied Energy Research: Garching, Germany, 2015.
90. Sliwa, T.; Rosen, M.A. Natural and artificial methods for regeneration of heat resources for borehole heat exchangers to enhance the sustainability of underground thermal storages: A review. *Sustainability* **2015**, *7*, 13104–13125. [CrossRef]
91. Wang, H.; Qi, C. Performance study of underground thermal storage in a solar-ground coupled heat pump system for residential buildings. *Energy Build.* **2008**, *40*, 1278–1286. [CrossRef]
92. Oliveti, G.; Arcuri, N. Prototype experimental plant for the interseasonal storage of solar energy for the winter heating of buildings: Description of plant and its functions. *Sol. Energy* **1995**, *54*, 85–97. [CrossRef]
93. Operational Response of a Soil-Borehole Thermal Energy Storage System. Available online: http://dx.doi.org/10.1061/(ASCE)GT.1943-5606.0001432#sthash.Via0Ht6e.dpuf (accessed on 24 February 2017).
94. Purdy, J.; Morrison, A. Ground-source heat pump simulation within a whole-building analysis. In Proceedings of the Eighth International IBPSA Conference, Eindhoven, The Netherlands, 11–14 August 2003.
95. Fisch, M.N.; Guigas, M.; Dalenbäck, J.O. A review of large-scale solar heating systems in Europe. *Sol. Energy* **1998**, *63*, 355–366. [CrossRef]
96. McDowell, T.P.; Thornton, J.W. Simulation and model calibration of a large-scale solar seasonal storage system. In Proceedings of the Third National Conference of IBPSA-USA, Berkeley, CA, USA, 30 July–1 August 2008.
97. The IEA TASK VII Swiss Project in Vaulruz; Design and First Experiences. Available online: http://eurekamag.com/research/020/246/020246506.php (accessed on 27 September 2016).
98. McDaniel, B.; Kosanovic, D. Modeling of combined heat and power plant performance with seasonal thermal energy storage. *J. Energy Storage* **2016**, *7*, 13–23. [CrossRef]
99. Catolico, N.; Ge, S.; McCartney, J.S. Numerical modeling of a soil-borehole thermal energy storage system. *Vadose Zone J.* **2016**, *15*. [CrossRef]
100. Rad, F.M.; Fung, A.S.; Rosen, M.A. An integrated model for designing a solar community heating system with borehole thermal storage. *Energy Sustain. Dev.* **2017**, *36*, 6–15. [CrossRef]
101. Xi, C.; Hongxing, Y.; Lin, L.; Jinggang, W.; Wei, L. Experimental studies on a ground coupled heat pump with solar thermal collectors for space heating. *Energy* **2011**, *36*, 5292–5300. [CrossRef]
102. Tiwari, A.K.; Tiwari, G.N. Thermal modeling based on solar fraction and experimental study of the annual and seasonal performance of a single slope passive solar still: The effect of water depths. *Desalination* **2007**, *207*, 184–204. [CrossRef]
103. Zhang, Y.; Zhou, G.; Lin, K.; Zhang, Q.; Di, H. Application of latent heat thermal energy storage in buildings: State-of-the-art and outlook. *Build. Environ.* **2007**, *42*, 2197–2209. [CrossRef]
104. Santamouris, M.; Lefas, C.C. Thermal analysis and computer control of hybrid greenhouses with subsurface heat storage. *Energy Agric.* **1986**, *5*, 161–173. [CrossRef]
105. Xu, J.; Li, Y.; Wang, R.Z.; Liu, W. Performance investigation of a solar heating system with underground seasonal energy storage for greenhouse application. *Energy* **2014**, *67*, 63–73. [CrossRef]
106. Gauthier, C.; Lacroix, M.; Bernier, H. Numerical simulation of soil heat exchanger-storage systems for greenhouses. *Sol. Energy* **1997**, *60*, 333–346. [CrossRef]
107. Zhang, L.H.; Liu, W.B.; Dong, R.; Chang, L.N. Experimental study on temperature improvement in solar greenhouse with underground pebble bed thermal storage. In Proceedings of the 2011 International Conference on Computer Distributed Control and Intelligent Environmental Monitoring (CDCIEM 2011), Changsha, China, 19–20 February 2011; pp. 797–799.
108. Hugo, A.; Zmeureanu, R. Residential solar-based seasonal thermal storage systems in cold climates: Building envelope and thermal storage. *Energies* **2012**, *5*, 3972–3985. [CrossRef]

109. Energy Publications. Survey of Household Energy Use; Office of Energy Efficiency—Natural Resources: Canada. 2003. Available online: https://oee.nrcan.gc.ca/publications/statistics/sheu/2011/pdf/sheu2011.pdf (accessed on 24 February 2017).

110. Guadalfajar, M.; Lozano, M.A.; Serra, L.M. Analysis of large thermal energy storage for solar district heating. In Proceedings of the Eurotherm Seminar 99, Lleida, Spain, 28–30 May 2014.

111. Tordrup, K.W.; Poulsen, S.E.; Bjørn, H. An improved method for upscaling borehole thermal energy storage using inverse finite element modelling. *Renew. Energy* **2017**, *105*, 13–21. [CrossRef]

112. Dincer, I.; Rosen, M.A. A Unique borehole thermal storage system at university of ontario institute of technology. In *Thermal Energy Storage for Sustainable Energy Consumption*; Springer: Dordrecht, The Netherlands, 2007; Volume 234, pp. 221–228.

113. Chuard, P.; Hadorn, J.-C. *Central Solar Heating Plants with Seasonal Storage-Heat Storage Systems; Concepts, Engineering Data and Compilation of Projects, Task VII*; International Energy Agency, Sorane SA: Lausanne, Switzerland, 1983.

114. Zeng, H.; Diao, N.; Fang, Z. Heat transfer analysis of boreholes in vertical ground heat exchangers. *Int. J. Heat Mass Transf.* **2003**, *46*, 4467–4481. [CrossRef]

115. Cóstola, D.; Blocken, B.; Hensen, J. External coupling between BES and HAM programs for whole-building simulation. In Proceedings of the Eleventh International IBPSA Conference, Glasgow, UK, 27–30 July 2009.

116. Moradi, A.; Smits, K.M.; Massey, J.; Cihan, A.; McCartney, J. Impact of coupled heat transfer and water flow on soil borehole thermal energy storage (SBTES) systems: Experimental and modeling investigation. *Geothermics* **2015**, *57*, 56–72. [CrossRef]

117. Thermal Conductivity of Soils. Available online: http://oai.dtic.mil/oai/oai?verb=getRecord&metadataPrefix=html&identifier=ADA044002 (accessed on 24 February 2017).

118. Smits, K.M.; Cihan, A.; Sakaki, T.; Illangasekare, T.H. Evaporation from soils under thermal boundary conditions: Experimental and modeling investigation to compare equilibrium- and nonequilibrium-based approaches. *Water Resour. Res.* **2011**, *47*. [CrossRef]

119. Bear, J.; Bensabat, J.; Nir, A. Heat and mass transfer in unsaturated porous media at a hot boundary: I. One-dimensional analytical model. *Transp. Porous Media* **1991**, *6*, 281–298. [CrossRef]

120. Farouki, O. *Thermal Properties of Soils*; US Army Corps of Engineers, Cold Regions Research and Engineering Laboratory: Hanover, NH, USA, 1981.

121. Givoni, B. Underground longterm storage of solar energy—An overview. *Sol. Energy* **1977**, *19*, 617–623. [CrossRef]

122. Gustafsson, A.-M.; Westerlund, L.; Hellström, G. CFD-modelling of natural convection in a groundwater-filled borehole heat exchanger. *Appl. Therm. Eng.* **2010**, *30*, 683–691. [CrossRef]

123. Li, C.; Mao, J.; Xing, Z.; Zhou, J.; Li, Y. Analysis of geo-temperature restoration performance under intermittent operation of borehole heat exchanger fields. *Sustainability* **2016**, *8*, 35. [CrossRef]

124. Clio, S.; Miriaiiliosseinabadi, S. Simulation modeling of ground source heat pump systems for the performance analysis of residential buildings. In Proceedings of the BS2013, Le Bourget-du-Lac, France, 25–28 August 2013; pp. 1960–1967.

125. Spitler, J.D.; Words, P.D.K. GLHEPRO—A design tool for commercial building ground loop heat exchangers. In Proceedings of the fourth international heat pumps in cold climates conference, Aylmer, QC, Canada, 17–18 August 2000; pp. 17–18.

126. Pahud, D.; Hellstrom, G.; Mazzerlla, L. Duct Ground Heat Storage Model for TRNSYS. Available online: http://repository.supsi.ch/3043/1/30-Pahud-1996-DST.pdf (accessed on 24 February 2017).

127. National Renewable Energy Laboratory. *EnergyPlus*; NREL: Golden, CO, USA, 2016.

128. Zhang, R.; Lu, N.; Wu, Y. Efficiency of a community-scale borehole thermal energy storage technique for solar thermal energy. In Proceedings of the GeoCongress 2012, American Society of Civil Engineers, Oakland, CA, USA, 25–29 March 2012; pp. 4386–4395.

129. COMSOL Multiphysics® Modeling Software. Available online: https://www.comsol.com/ (accessed on 27 December 2016).

130. Research-Projects-TOUGH-Software-TOUGH2. Available online: http://esd1.lbl.gov/research/projects/tough/software/tough2.html (accessed on 27 December 2016).

131. USGS MODFLOW and Related Programs. Available online: https://water.usgs.gov/ogw/modflow/ (accessed on 11 January 2017).

132. Angelotti, A.; Alberti, L.; La Licata, I.; Antelmi, M. Energy performance and thermal impact of a borehole heat exchanger in a sandy aquifer: Influence of the groundwater velocity. *Energy Convers. Manag.* **2014**, *77*, 700–708. [CrossRef]

133. Zeng, H.Y.; Diao, N.R.; Fang, Z.H. A finite line-source model for boreholes in geothermal heat exchangers. *Heat Transf. Asian Res.* **2002**, *31*, 558–567. [CrossRef]

134. Ingersoll, L.R.; Plass, H.J. Theory of the ground pipe heat source for the heat pump. *ASHVE Trans.* **1948**, *54*, 339–348.

135. De Paly, M.; Hecht-Méndez, J.; Beck, M.; Blum, P.; Zell, A.; Bayer, P. Optimization of energy extraction for closed shallow geothermal systems using linear programming. *Geothermics* **2012**, *43*, 57–65. [CrossRef]

136. Bayer, P.; de Paly, M.; Beck, M. Strategic optimization of borehole heat exchanger field for seasonal geothermal heating and cooling. *Appl. Energy* **2014**, *136*, 445–453. [CrossRef]

137. Kjellsson, E.; Hellström, G.; Perers, B. Optimization of systems with the combination of ground-source heat pump and solar collectors in dwellings. *Energy* **2010**, *35*, 2667–2673. [CrossRef]

138. Molina-Giraldo, N.; Blum, P.; Zhu, K.; Bayer, P.; Fang, Z. A moving finite line source model to simulate borehole heat exchangers with groundwater advection. *Int. J. Therm. Sci.* **2011**, *50*, 2506–2513. [CrossRef]

139. Eskilson, P.; Claesson, J. Simulation model for thermally interacting heat extraction boreholes. *Numer. Heat Transf.* **1988**, *13*, 149–165. [CrossRef]

140. Shonder, J.A.; Beck, J.V. Determining effective soil formation thermal properties from field data using a parameter estimation technique. *ASHRAE Trans.* **1999**, *105*, 458.

141. Yang, H.; Cui, P.; Fang, Z. Vertical-borehole ground-coupled heat pumps: A review of models and systems. *Appl. Energy* **2010**, *87*, 16–27. [CrossRef]

142. Tariku, F.; Kumaran, K.; Fazio, P. Integrated analysis of whole building heat, air and moisture transfer. *Int. J. Heat Mass Transf.* **2010**, *53*, 111–3120. [CrossRef]

143. Ferroukhi, M.Y.; Abahri, K.; Belarbi, R.; Limam, K. Integration of a hygrothermal transfer model for envelope in a building energy simulation model: Experimental validation of a HAM–BES co-simulation approach. *Heat Mass Transf.* **2016**. [CrossRef]

144. Ferroukhi, M.Y.; Djedjig, R.; Limam, K.; Belarbi, R. Hygrothermal behavior modeling of the hygroscopic envelopes of buildings: A dynamic co-simulation approach. *Build. Simul.* **2016**, *9*, 501–512. [CrossRef]

145. Woloszyn, M.; Rode, C. Tools for performance simulation of heat, air and moisture conditions of whole buildings. *Build. Simul.* **2008**, *1*, 5–24. [CrossRef]

146. Reiderer, P. Matlab/Simulink for building and HVAC simulation-state of the art. In Proceedings of the Ninth International IBPSA Conference, Montreal, QC, Canada, 15–18 August 2015.

147. Crawley, D.B.; Hand, J.W.; Kummert, M.; Griffith, B.T. Contrasting the capabilities of building energy performance simulation programs. *Build. Environ.* **2008**, *43*, 661–673. [CrossRef]

148. Van Schijndel, A.W.M. A review of the application of SimuLink S-functions to multi domain modelling and building simulation. *J. Build. Perform. Simul.* **2014**, *7*, 165–178. [CrossRef]

149. Cascetta, M.; Serra, F.; Arena, S.; Casti, E.; Cau, G.; Puddu, P. Experimental and numerical research activity on a packed bed TES system. *Energies* **2016**, *9*, 758. [CrossRef]

150. Huang, J. Linking the COMIS multi-zone airflow model with the energyplus building energy simulation program. In Proceedings of the 6-th IBPSA Conference, Kyoto, Japan, 13–15 September 1999; Volume 2, pp. 1065–1070.

151. MocDowell, T.; Vice President at Thermal Energy System Specialists, LLC, Madison, WI, USA. Personal communication, 9 January 2017.

152. Leong, W.H.; Tarnawski, V.R. Effects of simultaneous heat and moisture transfer in soils on the performance of a ground source heat pump system. In Proceedings of the ASME-ATI-UIT Conference on Thermal and Environmental Issues in Energy Systems, Sorrento, Italy, 16–19 May 2010.

153. Roth, K.; Brodrick, J. Seasonal energy storage. *ASHRAE J.* **2009**, *51*, 41–43.

154. Yu, M.G.; Nam, Y.; Yu, Y.; Seo, J. Study on the system design of a solar assisted ground heat pump system using dynamic simulation. *Energies* **2016**, *9*, 291. [CrossRef]

155. Solar Thermal Case Study: Shouldice Hospital (2012). Available online: http://www.solarthermalworld.org/content/solar-thermal-case-study-shouldice-hospital-2012 (accessed on 24 February 2017).

156. Case Study: Oxford Gardens Solar Cooling Project, Enerworks Solar Heating and Cooling. Available online: http://enerworks.com/wp-content/uploads/2013/04/casestudy-OxfordGardens.pdf?b3fdf2 (accessed on 24 February 2017).

157. Herold, K.E.; Radermacher, R.; Klein, S.A. *Absorption Chillers and Heat Pumps*, 2nd ed.; CRC Press: Boca Raton, FL, USA, 2016.

158. Hasnain, S.M. Review on sustainable thermal energy storage technologies, Part II: Cool thermal storage. *Energy Convers. Manag.* **1998**, *39*, 1139–1153. [CrossRef]

159. Saito, A. Recent advances in research on cold thermal energy storage. *Int. J. Refrig.* **2002**, *25*, 177–189. [CrossRef]

160. Silvetti, B. Application Fundamentals of Ice-Based Thermal Storage. Available online: http://www.calmac.com/stuff/contentmgr/files/0/9ee1a79e74c076ff2adac9661e9ef80f/pdf/020311_emjas_emailablearticle_ashraejournalfeb02.pdf (accessed on 9 November 2016).

161. Masoero, M. Refrigeration systems based on long-term storage of ice. *Int. J. Refrig.* **1984**, *7*, 93–100. [CrossRef]

162. Eames, I.W.; Adref, K.T. Freezing and melting of water in spherical enclosures of the type used in thermal (ice) storage systems. *Appl. Therm. Eng.* **2002**, *22*, 733–745. [CrossRef]

163. Jim, A.; Facilities Manager at IKEA, Cennetannial, CO, USA. Personal communication, 2 December 2016.

164. Denholm, P.; Ela, E.; Kirby, B.; Milligan, M. *The Role of Energy Storage with Renewable Electricity Generation*; National Renewable Energy Laboratory: Golden, CO, USA, 2010; pp. 1–61.

energies

MDPI

Review

Large-Scale Electrochemical Energy Storage in High Voltage Grids: Overview of the Italian Experience

Roberto Benato [1,*], Gianluca Bruno [2], Francesco Palone [2], Rosario M. Polito [2] and Massimo Rebolini [2]

1   Department of Industrial Engineering, University of Padova, 35100 Padova, Italy
2   Terna Rete Italia, 00156 Rome, Italy; gianluca.bruno@terna.it (G.B.); francesco.palone@terna.it (F.P.);
     rosario.polito@terna.it (R.M.P.); massimo.rebolini@terna.it (M.R.)
*   Correspondence: roberto.benato@unipd.it; Tel.: +39-049-8277532

Academic Editors: Rui Xiong, Hailong Li and Joe (Xuan) Zhou
Received: 24 October 2016; Accepted: 10 January 2017; Published: 17 January 2017

Abstract: This paper offers a wide overview on the large-scale electrochemical energy projects installed in the high voltage Italian grid. Detailed descriptions of energy (charge/discharge times of about 8 h) and power intensive (charge/discharge times ranging from 0.5 h to 4 h) installations are presented with some insights into the authorization procedures, safety features, and ancillary services. These different charge/discharge times reflect the different operation uses inside the electric grid. Energy intensive storage aims at decoupling generation and utilization since, in the southern part of Italy, there has been a great growth of wind farms: these areas are characterized by a surplus of generation with respect to load absorption and to the net transport capacity of the 150 kV high voltage backbones. Power intensive storage aims at providing ancillary services inside the electric grid as primary and secondary frequency regulation, synthetic rotational inertia, and further functionalities. The return on experience of Italian installations will be able to play a key role also for other countries and other transmission system operators.

Keywords: large-scale electrochemical storage; energy and power intensive; ancillary services

1. Introduction

This paper is an overview of the large scale electrochemical storage stationary installations in Italy. Many previous papers [1–24], which are briefly reported in the following, highlighted the role of Italy as a path-maker in the field of large scale electrochemical storage in the high voltage network. In [1–3], a detailed description of the Italian energy intensive installations can be found, whereas papers [4–10] offer some analyses of the main features of sodium-sulphur technology. Paper [15] thoroughly analyses the authorization procedures of Italian energy storage systems. In the papers [11,12], the reader can find scientific and technological details of the sodium nickel chloride batteries of the Italian installations whereas papers [13,19] are devoted to a steady-state electric model of this technology; in [20,21] this model has been enlarged to take transient behaviour into account. Safety tests performed on a Na-NiCl$_2$ battery are fully presented in [22,23]. Papers [16,17] consider the model of battery under faulted condition and their arc flash respectively. Paper [18] is devoted to the computation of battery efficiency including auxiliary equipment losses.

The main contribution of this paper is to thoroughly present all the features of these installations. In particular, the paper describes how the Italian transmission system operator (TSO in the following) has chosen two energy storage strategies in the high-voltage network. In the first one, the electrochemical energy storage systems (EESS) is conceived to release renewable generation from electric loads and to avoid overload conditions in the existing overhead lines. This use implies longer charge/discharge intervals (about 8 h) and a kind of "energy service" more than a "power service"; therefore, these installations have been called "energy intensive" installations. For Italian "energy intensive" installations [1–3], Terna was chosen because of its extensive history of successful installations, the Sodium-Sulphur (Na-S) electrochemistry [4–10], supplied by the Japanese NGK INSULATORS, LTD. There have been three installation sites located in the South of Italy (around Benevento city): two installations of 12 MW and one of 10.8 MW (wholly 34.8 MW Na-S storage has been installed). It is worth remembering that Na-S batteries belong to the Na-beta battery family (as Na-NiCl$_2$ [11–13]). The other direction has involved electrochemical technologies with short charge/discharge intervals (from 0.5 to 4 h). The tested technologies are in the Li-ion family and sodium-nickel chloride. The installation sites are Sardinia (9.15 MW installed power in Codrongianos) and Sicily islands (6.8 MW installed power in Ciminna). Due to their high use flexibility allowed by the Power Conversion System (PCS) [14], power intensive installations have been applied in the field of grid ancillary services. Moreover, a brief cost comparison is also presented in Section 5.

2. Energy Intensive Projects

The three energy intensive installations are very similar. A unit makes use of module series and parallel connection. A 12 MW installation extends over an area of 7000 m^2, and involves the use of medium (20 kV) and low (400 V) voltage levels, in detail (see Figure 1a):

- ➢ 10 units of 1.2 MW each;
- ➢ 10 Power Conversion Systems (PCS) of 1.2 MW (in other Italian installations, there are 2.4 MW PCS instead of 2 PCS of 1.2 MW);
- ➢ 2 shelters for MV switchboards (QMT1, QMT2);
- ➢ 2 shelters for LV switchboards (QBT1, QBT2);
- ➢ 2 shelters for emergency generators (GE1, GE2);
- ➢ Shelter for the control system.

The connection of the power systems to the national grid is performed by means of a MV/HV (20/150 kV) transformer. The HV level is 150 kV since all the energy driven installations are located in South Italy (in North Italy, the HV level is 132 kV). The Italian unit, also called "assembly", has a rated power of 1.2 MW with 40 modules of 30 kW (see Figure 1b). The structure is composed of a self-supporting latticed frame with shelves for module placement, a Battery Management System (BMS) container part, ventilation, and a cooling system. Each module is protected by a BMS. The BMS includes a disconnection relay that disconnects the battery from any load if the value of any battery parameter is outside the predefined operating range.

The controlled parameters are the module temperature, the dc-side current and voltage. The battery status is calculated from these basic parameters. Other BMS functions include: detection of alarms, warnings and battery capacity limitation signals; Control interface with PCS and data logging. The frame is of galvanized steel. The frame thickness is at least 2.3 mm both on lateral sides and on the covering. The dimensions (in meters) are 9.410 length × 4.800 depth × 4.820 height. It is worth noting that the standard NGK unit had five modules of 50 kW per rack (i.e., 250 kW per rack) whereas in the Italian installations five modules of 30 kW (150 kW per rack) have been employed: this gives greater spacing between modules. The Terna assembly is also equipped with a double redundant fire and gas detection system: the first is based on SO$_2$ detection (in extremely remote case of fire and cell breaking), the second is based on ventilation system continuous air analysis that can detect the possible presence of smoke. Figure 2 shows some photographs of the energy intensive installation in Ginestra.

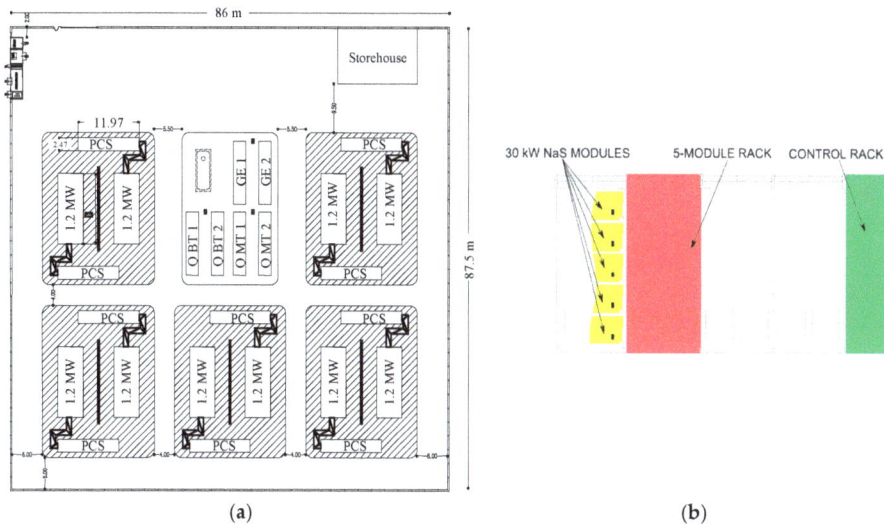

Figure 1. (**a**) 12 MW installation constituted of 10 units of 1.2 MW; (**b**) Frontal view of the 1.2 MW unit (the other four racks are back-to-back with the visible ones).

Figure 2. Some photos of Italian Na-S "energy intensive" installation in Ginestra.

2.1. Safety Features

In order to have a general overview of safety features, Table **??** reports all the safety levels which have been implemented in Italy. An additional protection for SO_2 confinement has been realized: it consists of an automatic system for ventilation block by means of ventilation grid shutting (based on SO_2 detection) with the dual effect of avoiding the access of oxidizing agents and the SO_2 spreading. It is worth noting that, once running, the heat produced by charging and discharging cycles is sufficient to maintain operating temperatures and no heaters are required. Heaters are in operation only when the battery is idle and the temperature falls below 305 °C. In any case, if the battery is not running and the heaters fail, the temperature tends to decrease. It is not an issue of safety but only a matter of battery performance degradation. The battery has a very good thermal insulation and the worst thing which can occur (if for a long time the heaters do not work) is that the molten substances solidify. The battery can withstand 10 cycles (the so-called freeze-thaw cycles) with temperature lower than 150 °C.

Table 1. Safety levels of the Na-S unit starting from the cell level.

Component	Function
Cell Level	
Safety tube	➢ Controls the quantities of sodium and sulphur which can combine in case of β″-alumina failure ➢ Avoids the rupture of cell case ➢ Limits the short-circuit current (blocking the sodium flow)
Corrosion protection layer in aluminium alloy Fe-Cr	Zeros the corrosion possibility due to sodium polysulphides during the discharge phase
Further thermal insulation and fire-resistant layers inside the cells	Prevents fire inside one cell from propagating to the adjoining ones
Module Level	
Fuses (equipped for each four-cell block)	Interrupt over-current in case of a short-circuit
Cell connections	Limit the over-voltages inside the module
Module dry sand filling	➢ Absorbs the active material in case of a cell rupture ➢ Avoids the fire spreading generated by a cell
Insulated double-walled stainless steel enclosure with thickness equal to $0.8 \div 1$ mm	➢ Avoids the material spill in the environment ➢ Avoids cell contact with oxygen and stops combustion
Control and monitoring	➢ Controls charge-discharge ➢ Failure detection and alarming ➢ Puts the equipment out of service if it fails
Electrically Insulated compartment	Prevents active material from leaking outside, hence short circuits can be avoided
Fire resistance panels in the upper and lower part of the module	Avoid the fire spreading between one module and the preceding or successive one for a given time
Unit level	
Galvanized steel cabinet walls with thickness ≥ 2.3	Good protection from direct lightning and to bullets due to vandalism or stray hunting shots

2.2. Tests Performed by NGK on the New Italian Module

In recent years, a lot of tests have been performed on the most used module, i.e., on NGK E50. As already mentioned, Terna has required a safety enhanced module with the same dimensions as the previous one but with fewer cells inside, with more sand between cells and with fire resistant carbon sheets inside the module. It is therefore inferable that all the results obtained in the tests on the "old" module would give equal or better results than the new one. This has been confirmed by some tests commissioned by Terna whose results are reported in Table **??**. In particular, in order to verify the effectiveness of sheets inside the cell a test has been performed: the cells adjoining to that fired are not damaged so that no fire propagation has occurred inside the module. Another test has been performed in order to verify the effectiveness of the sheets inside the module: the fire has not propagated outside the module.

Table 2. Safety Tests performed by NGK on the new module.

Test	Purpose	Figures	Results
External short circuit	Confirm safety against external short circuit		- No damages - No substance leakage - Sound connection status - Correct operation of protections

Table 2. *Cont.*

Test	Purpose	Figures	Results
Exogenous fire	Confirm safety against exogenous fire		- Exposed to fire for more than 60 min with outer temperature about 890 °C - No module fire or explosion - No substance leakage
Flooding	Confirm safety against flood		- Immersed in water for more than 12 h - No module fire or explosion - No substance leakage
Fall	Confirm safety against fall		- Collided part of module enclosure was deformed - No module fire or explosion - No substance leakage

2.3. The Authorization Procedure

The authorization procedures have been wholly presented in [15]. The most important European regulation which can be applied to the stationary application of sodium batteries is the Directive 96/82/CE, also known as "Seveso II", and the Directive of the European Parliament 2012/18/UE also known as "Seveso III", both concerning the major-accident hazards related to the presence of hazardous substances. It is worth noting that each EU member state had to incorporate the provisions of the "Seveso III" directive into national law by 31 May 2015. In order to evaluate the Seveso II implications, the entire amount of chemical substances (during charge and discharge and consequently sodium, sulphur, and polysulphides) in the energy storage installation project has to be determined. By considering the toxic substance amounts as the most restrictive ones, it is possible to demonstrate that up to a rated power of 12 MW the installation falls under article 6 prescriptions of Legislative Decree 334/99, which represents the Italian decree in force during the period when the evaluation of Terna projects was made by the competent authorities. These prescriptions foresaw to send a notification of the installation of the three sites to the competent authority for assessing the risk of major accidents (offices "relevant risks and integrated environmental authorization" of the Ministry of Environment, of the Region, of the Provinces and territorially competent Municipalities, Provincial Prefecture, the "Regional Technical Committee of the Fire Department") at least 180 days before the start of construction, together with:

➤ A detailed project information;
➤ A Risk Analysis Assessment (performed in collaboration with the Department of Industrial Engineering of Padova University), which showed that, in case of occurrence of given events (earthquake and vibration, flooding, mishandling, direct and indirect lightning strikes, endogenous or exogenous fire, sabotage and hunting, and external impacts), the safety/mitigation systems adopted in the plant would have reduced the risk of release of chemicals in the environment to negligible values. The tools used in the risk assessment have been the FMEA

(Failure Modes and Effects Analysis) and the FMECA (Failure Modes and Effects and Criticality Analysis). In a range between 1 and 25, the maximum computed risk priority number has been 9.

Furthermore, Terna, in accordance with the EU Directive "Seveso", has drafted an internal document called "Prevention Policy for Major Accidents", holding the management criteria to be undertaken for the prevention of "significant" accidents.

3. Power Intensive Projects

In the following, a detailed description of the power intensive installations and of their uses inside the high voltage network are presented. These installations have also been named "Storage Labs".

3.1. Ciminna (Sicily) and Codrongianos (Sardinia) Power Intensive Installations

Since the power system architecture of the two installations is very similar, only Codrongianos Storage Lab is described (see Figure 3). Its single-line diagram is shown in Figure 4. In Figure 5, some photographs of the HV/MV transformer and battery unit racks are shown. The different power ratings, and the different storage typologies are reported in Table ?? (in Table ?? for Ciminna). There are 10 different EESS branches, subdivided in two groups of five (three branches are foreseen for future storage technologies). Each EESS has its own PCS (composed by four 250 kVA rated power inverters), step-up LV/MV transformer (1.25 MVA, 15 kV/0.55 kV, Yd connected) and a dedicated MV cable line [16]. The 15 kV bus bar is then connected to the 150 kV HV ac sub-transmission grid through a 40 MVA, 150 kV/15.6 kV transformer shown in Figure 5 (Yy connected). A grounding transformer (GT) with a 385–770 Ω resistor (depending on the temperature) provides a ground path for the otherwise ungrounded 1.7 km long MV system connected to the 15 kV bus bar. PCSs are fully described in [14]: generally, PCS is constituted of a first stage made by a DC–DC converter and of a second stage made by a DC–AC converter to maintain the inverter dc side voltage. Moreover, this two-stage architecture avoids the PCS oversizing due to the EESS voltage variation during the charge/discharge operations. In fact, a $\Delta u_\%$ percentage EESS voltage variation with respect to the rated value requires an inverter component overrating of $1 + \Delta u_\%$ for both the voltage and the current (maximum current corresponding to minimum battery voltage), resulting in an inverter power oversizing of about $1 + 2\,\Delta u_\%$. For instance, by hypothesizing a maximum current and voltage variation of 20% (ΔV_{max} *and* ΔI_{max} respectively), the inverter rated power must be oversized of 40% as in the following:

$$P = \Delta V_{max} \cdot \Delta I_{max} = V_n(1 + 20\%) \cdot I_n(1 + 20\%) \approx 1.40\,P_n = P_n + \Delta P$$

Figure 3. Two views from above the Codrongianos substation.

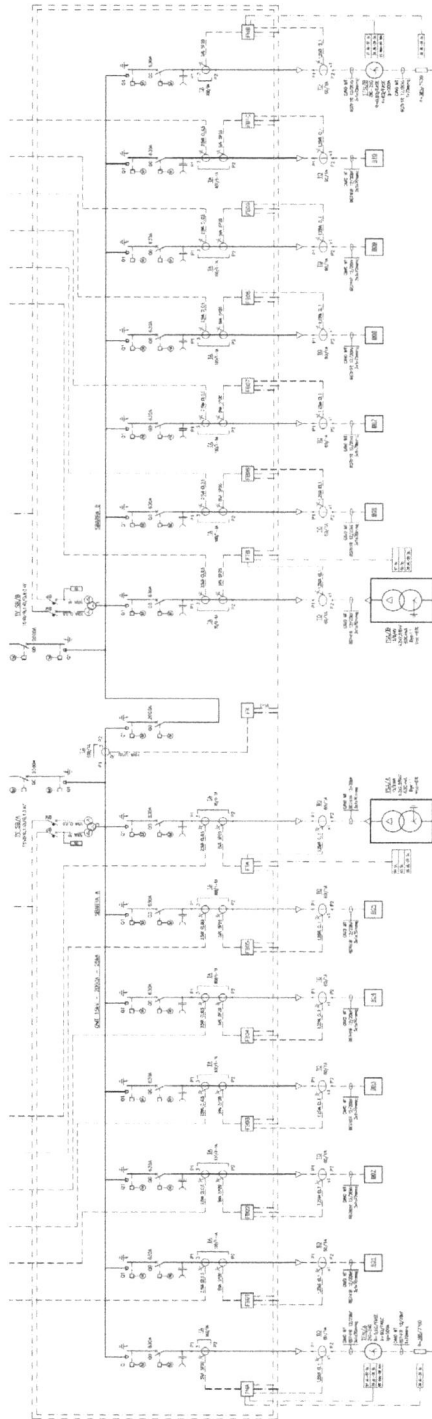

Figure 4. Single-line diagram of Italian "power intensive" installation (Storage Lab) in Codrongianos.

Table 3. Different storage technologies installed in Codrongianos (Sardinia).

Codrongianos (Sardinia)		
Power (MW)	Energy (MWh)	Electrochemistry
1	1.231	Lithium Iron Phosphate
1.2	0.928	Lithium Nickel Cobalt Aluminium
1	0.916	Lithium Manganese Oxide
1.08	0.540	Lithium Nickel Cobalt Manganese
1	1.016	Lithium Titanate
1.2	4.15	Sodium-Nickel Chloride
1	2	Sodium-Nickel Chloride

Table 4. Different storage technologies installed in Ciminna (Sicily).

Ciminna (Sicily)		
Power (MW)	Energy (MWh)	Electrochemistry
1	1.231	Lithium Iron Phosphate
0.9	0.570	Lithium Nickel Cobalt Aluminium
1	0.916	Lithium Manganese Oxide
1	1.016	Lithium Titanate
1.2	4.15	Sodium-Nickel Chloride

Figure 5. Some photos of the Italian "power intensive" installation (Storage Lab) in Codrongianos.

For these Storage Lab installations, some features have been studied and published; in particular:

➢ modelling of battery under faulted conditions and assessment of protection system behavior [16];
➢ the arc-flash in these energy storage systems [17];
➢ the efficiency calculations including auxiliary power losses [18];
➢ the steady-state and transient modelling of Na-NiCl$_2$ [13,19–21], including safety tests [22,23].

3.2. Power Intensive Ancillary Services and Advanced Functionalities

3.2.1. Primary Frequency Control: Provision of Frequency Containment Reserve (FCR)

This service [24] is delivered in accordance with the Italian Grid Code (Annex 15 of [25]) and requests EESS to modulate the active power output (ΔP, in MW) proportionally (depending on the droop parameter, σ, expressed in % and fully configurable) to the deviations of the grid frequency (ΔF, in Hz) around its nominal value of 50 Hz, as in (1):

$$\Delta P_{FCR} = -\frac{1}{\frac{\sigma}{100}} \cdot \frac{\Delta F}{50} \cdot P_{rated} \tag{1}$$

Figure 6 shows the FCR service regulation patterns.

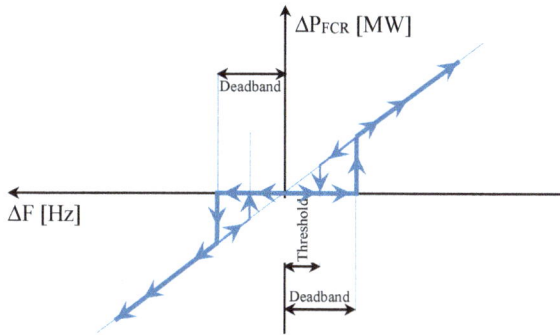

Figure 6. FCR ancillary service regulation patterns.

EESS may shift the active power flow direction continuously and take less than 100 ms for a complete inversion (full inversion from maximum discharge active power to maximum charge one) as shown in Section 4. The value of the droop parameter σ may be set to a small value in order to allow EESS to perform a greater active power contribution in case of wide frequency deviations. This service will play a key role since in the European grid there is an ever-growing decreasing of power frequency characteristic of primary control (also known as regulating energy) due to the increase of generation by renewable resources [26].

3.2.2. Secondary Frequency Control: Provision of Frequency Restoration Reserve (FRR)

In accordance with the secondary frequency control [24], this service consists in performing active power output variation ΔP_{FRP} as requested by an external control signal (L%, whose percentage range is 0–100%) fed to local control system from Terna Central Supervisory Control And Data Acquisition (SCADA). It is worth remembering that FRR regulation is a power plant service. A reserved regulation band (half-band, HB, expressed in MW) around the active power set-point (the actual balancing program or the stand-by mode) is dedicated in accordance with:

$$\Delta P_{FRP} = 2 \cdot HB \cdot \frac{(L\% - 50\%)}{100\%} \tag{2}$$

As already mentioned, the percentage set-point control signal *L*% is forwarded by Terna to the plant and it is defined in the range of 0–100%. Set-point signal is updated and forwarded every 8 s to the plant with a maximum variation within the total regulation band of 4%.

3.2.3. Provision of Synthetic Rotational Inertia (SRI)

In addition to FCR, EESSs could be equipped with SRI (operated independently from FCR) [24] in order to contribute in reducing the fastest frequency transient phenomena, since the beginning. The EESS high rapidity of varying the generated or absorbed active power *P* has made feasible several scenarios which were not possible with traditional power plants. In the specific, these EESS characteristics may help mitigate the reduction of European grid rotational inertia: therefore, EESS may be requested to deliver an active power proportionally to the measured derivative frequency, i.e., to the rate of change of frequency (df/dt). However, it is crucial to implement robust control blocks for a reliable frequency rate of change sampling. This could be fulfilled by feeding the control block numerical algorithm with a proper and adequate frequency sampling, insuring on one hand computational accuracy and on the other hand fast solution times. The result must be available within a time frame assessable in tens of milliseconds. The faster the P response, the more effective the mitigation of frequency variation is. Such a P response should not be confused with a fast FCR regulation set with low droop value because its contribution depends linearly on the instantaneous measured frequency deviation. By assuming both FCR and SRI committed together, each contribution must be distinguished: EESS may be engaged in FCR with droop not sufficient to let the plant fill the P capability though still guaranteeing fast response, and at the same time SRI may provide EESS full P capability access just in case of sudden frequency deviation. EESS has to deliver ΔP_{SRI} (in MW) proportionally (depending on a parameter, K_W, expressed in MW·s/Hz and fully configurable) to the filtered derivative frequency measurement ($\Delta f / \Delta t$, in Hz/s), as in (3):

$$\Delta P_{SRI} = -k_w \cdot \left(\frac{\Delta f}{\Delta t} \right)_{Butterworth-filtered\ value} \tag{3}$$

Figure 7 shows the SRI service regulation patterns.

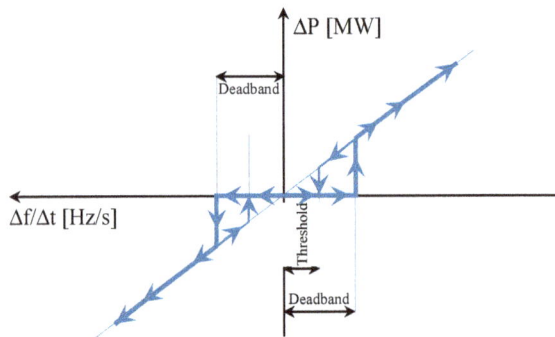

Figure 7. SRI ancillary service regulation patterns.

The necessity to implement such a service is to emulate the stabilizing contribution to frequency deviations guaranteed by the rotating masses, in prevision of an ever-growing passage from traditional power plants (with rotating synchronous generators) to static inverter-equipped power generation. For a prompt SRI response, it has been crucial to implement a reliable, accurate, and fully configurable digital filter of grid frequency for feeding the regulator: Infinite Impulse Response filters (IIR—e.g., Butterworth filters) have been selected as the best compromise between speed and accuracy of response. IIR filters are fully configurable in terms of filter order and cut-off frequency to adjust filtering to the effective harmonic content inside grid frequency and to achieve an adaptive filtering by means of high stopband attenuation and effective noise cancellation. In addition, IIR filters enable the achievement of low-rate phase delays and may be designed as inherently stable.

3.2.4. Congestions Mitigation and Balancing Program

HV line congestion mitigation/active power balancing service [24] is used to set a specific active power profile to EESS through XML file or with a manual set-point. As long as this service is activated, other active power regulations (FCR and FRR regulations), whenever turned on, must provide their contribution around this active power profile. In the case of active power balancing service deactivate, reference power is zero MW and all other services (in *P*) are requested to deliver their own output around this value.

3.2.5. Voltage Regulation

Two operating modes are available as mutually exclusive [24]:

(1) Local Bus Bar Voltage Regulator: the scope of this service is to adjust local substation voltage (HV bus bar). Measured voltage error will lead to a reactive power contribution (Q) in accordance

with a predetermined U-Q curve, fully configurable, so to reduce the deviation between actual voltage and its set value;

(2) Regional Voltage Regulator (RVR): the scope of this regulation is to adjust relevant Terna substation voltages (pilot substations with high fault levels) through a coordinated plant reactive power regulation. In this case, the remote controller is in charge of computing the exact amount of reactive power to be requested to the plant to reach Terna substation voltage set value. The remote controller aims at nullifying voltage deviation between the measured value and set value of each Terna HV pilot substation.

3.2.6. Further Functionalities

In addition, EESSs are equipped with advanced functionalities [24]. A functionality represents the capability of EESS to perform a specific service, in addition to the aforementioned ones or in a mutually exclusive way and it may be demanded through automatism or on request of the operator.

(1) Local Frequency Integrator (LFI): this functionality is operated in background and is automatically activated (fully configurable) in emergency conditions (high frequency transient) for restoring nominal frequency value through an integral control loop feedback. It is used when isolated grid conditions are detected.

(2) Defense plan (switch opening and active power modulation): the task is handling the shedding of load/production in order to keep the integrity of the grid, in case of abnormal conditions resulting from occurrence of extreme contingencies. This may be obtained through the following commands:

   a. Switch opening;
   b. Active power (P) modulation within 300 ms.

EESS rapidity of varying the active is also used for these additional commands:

- Instantaneous maximum P feeding into the grid;
- Instantaneous maximum P absorption from the grid;
- Instantaneous P exchange stop.

The extremely fast response recorded for active power modulation (less than 300 ms from the request to the full activation) leads to considering, when including EESS in system defense plan, the utilization of this command (maximum power feeding or maximum power absorption, depending on the desired direction) instead of switch opening, traditionally used also for pump storage. In this regard, in some operational conditions, the effectiveness of EESS as a defense plan is doubled.

4. Some Returns on Operational Experience

In order to have some measures on the real behaviour of the Italian energy intensive installations, Figure 8 shows the discharge/charge module power and current for a standard cycle. This involves:

➢ a discharge phase of 10 h where for 7 h the discharge constant power is 0.6 p.u. and for 3 h the constant discharge power is 1 p.u.;
➢ a charge phase of 10 h where a constant charge power of 1 p.u. for 8 h after which a supplementary charge (as already mentioned in Section 4) is needed in order to reach 100% of SoC.

In Figure 9, the voltage and temperature inside a module are shown with reference to the standard cycle shown in Figure 8. With regard to the transient conditions, Figure 10 shows the inversion of the power flow from discharge to charge. It is worth noting that power inversion occurs in about 150 ms which is fully suitable for energy intensive stationary installations in the HV grid. The inversion from charge to discharge has a very similar behaviour.

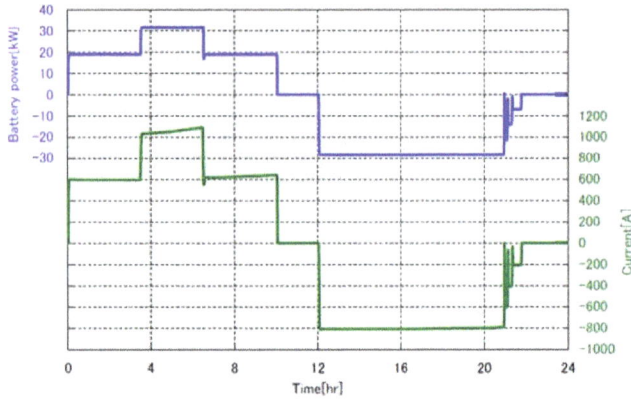

Figure 8. Current and DC power behaviour during a standard cycle in a module.

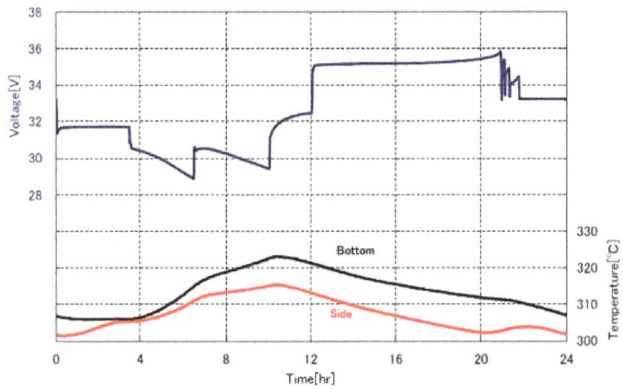

Figure 9. Voltage and temperature (at side and bottom of the module) during a standard cycle.

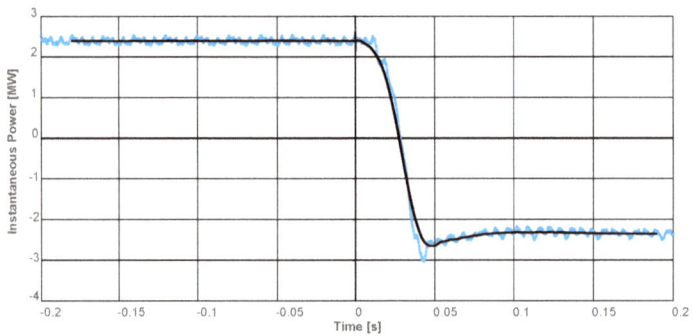

Figure 10. Rated discharge to rated charge power: active one (**dark** line) and instantaneous one (**azure** line).

With reference to the Storage Lab installed in Sardinia, an operational data diagram focusing on FCR regulation triggered by the occurrence of a deep frequency transient is shown (see Figure 11).

?? and Table 6 report the technical datasheet of the EESS devoted to FCR regulation and its configuration, respectively.

Table 5. Technical datasheet of EESS devoted to FCR in Codrongianos (Sardinia).

Capability	±1.0	MW
Nominal storage capacity	1.0	MWh
Overload peak	±1.3	MW
Overload peak sustainability	60	s
Battery technology	Li-ion	-

Table 6. FCR service configuration in the EESS of Table **??**.

Merit order	1	-
Frequency set-point	50	Hz
Deadband	20	mHz
Hysteresis deadband	50%	% of deadband
Droop	0.50%	%

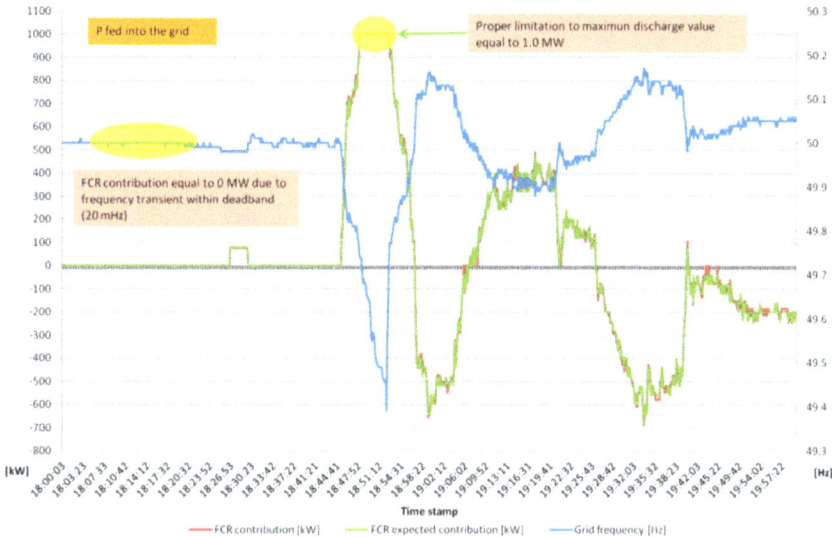

Figure 11. EESS GIG, total P output versus a real frequency transient (FCR contribution).

The frequency profile has experienced an initial under frequency phase, reaching the transient minimum value of 49.39 Hz, followed by an overshooting up to 50.17 Hz and consequent damped oscillations around 50 Hz.

Figure 11 clearly shows that the EESS can fully supply its rated power since the very beginning of the network transient, without noticeable delays. Furthermore, during the subsequent frequency variations, the EESS switches from discharge to charge according to the frequency error, contributing to oscillation damping. Notably, the actual EESS FCR behavior perfectly matches the expected one, confirming the efficacy of the control system.

5. A Brief Cost Comparison

In order to have a comparison between the different cost components of the battery installations [27], Figures 12–14 present the cost pie charts of Li-ion, Na-NiCl$_2$, and Na-S respectively.

The average total cost of Li-ion installations is 1.3 M€/MW and by considering an average discharge time for this technology equal to 1 h, the cost per MWh is 1.3 M€/MWh. The average total cost of Na-NiCl$_2$ installations is 3.0 M€/MW and by considering the nominal discharge time for this technology equal to 3 h, the cost per MWh is 1.0 M€/MWh. The average total cost of Na-S installations is 3.3 M€/MW and by considering the nominal discharge time for this technology equal to 7.2 h, the cost per MWh is 0.46 M€/MWh.

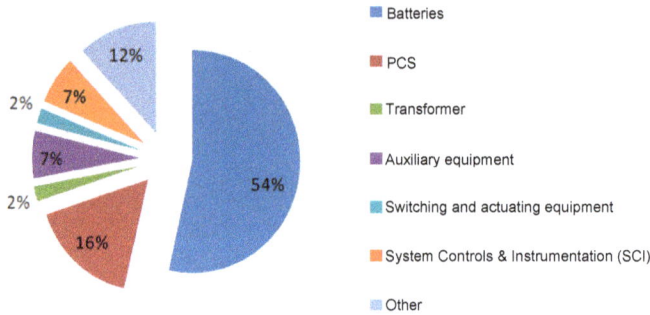

Figure 12. Li-ion percentage costs.

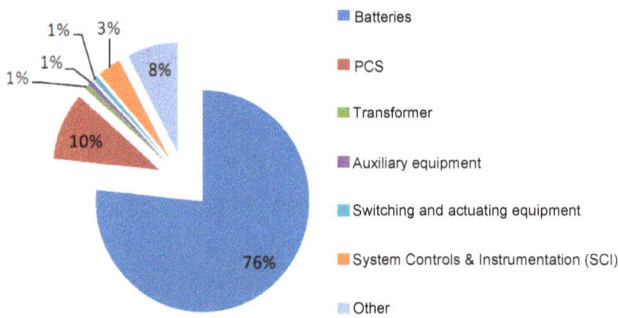

Figure 13. Na-NiCl$_2$ percentage costs.

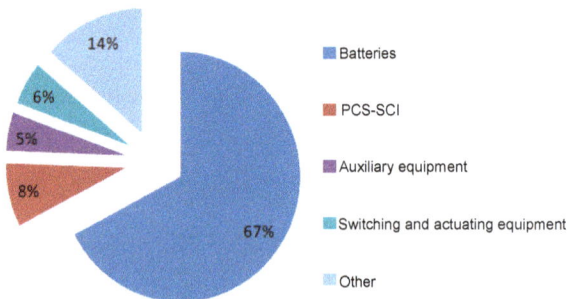

Figure 14. Na-S percentage costs.

The percentage ratio between PCS-SCI and battery costs is higher for Li-ion installations since the battery costs are lower.

6. Conclusions

This paper gives a wide overview of the energy storage projects installed in the Italian high voltage network. Safety issues, authorization procedures, and use applications of the energy and power intensive stationary electrochemical storage are throughout presented and developed. Li-ion (of different families), sodium-sulphur, sodium-nickel chloride electrochemistries have been tested with a total installed power of 50.75 MW.

In conclusion, it is possible to give some tendency lines: sodium-sulphur and sodium-nickel chloride with their long discharge times seem more suitable for energy intensive applications whereas Li-ion batteries seem more suitable for power intensive ones. Sodium-nickel chloride batteries show an attitude to be employed also in power intensive applications due to their intermediate discharge times.

Italian installations and their return on experience will play a key role in completely understanding the real battery behaviours in stationary applications including aging phenomena.

Author Contributions: Roberto Benato wrote the paper whereas the remaining co-authors collected the cost and installation data.

Conflicts of Interest: The authors declare no conflict of interest.

References

1. Andriollo, M.; Benato, R.; Dambone Sessa, S.; Di Pietro, N.; Hirai, N.; Nakanishi, Y.; Senatore, E. Energy intensive electrochemical storage in Italy: 34.8 MW sodium–sulphur secondary cells. *J. Energy Storage* **2015**, *5*, 146–155. [CrossRef]
2. Andriollo, M.; Benato, R.; Sessa, S.D. 34.8 MW di accumulo elettrochimico di tipo Energy Intensive mediante celle secondarie sodio-zolfo (Na-S). *L'Energia Elettr.* **2014**, *5*, 23–35.
3. Andriollo, M.; Benato, R.; Sessa, S.D.; Di Pietro, N.; Polito, R. *Large Scale Italian Energy Intensive Storage Installation: Safety Issues and Environmental Compatibility*; Paper C4–115; Cigré: Paris, France, 2016.
4. Sudworth, J.; Tilley, R. *The Sodium/Sulfur Battery*; Chapman and Hall: London, UK, 1985.
5. Linden, D.; Reddy, T.B. *Handbook of Batteries*, 3rd ed.; McGraw-Hill: New York, NY, USA, 2002.
6. Dustmann, C.-H.; Bito, A. Safety. In *Encyclopedia of Electrochemical Power Sources*; Garche, J., Dyer, C., Moseley, P., Ogumi, Z., Rand, D., Scrosati, B., Eds.; Elsevier: Amsterdam, The Netherlands, 2009; Volume 4, pp. 324–333.
7. Ohima, T.; Kajita, M.; Okuno, A. Development of sodium-sulfur Batteries. *Int. J. Appl. Ceram. Technol.* **2004**, *1*, 269–276. [CrossRef]
8. Wen, Z.; Cao, J.; Gu, Z.; Xu, X.; Zhang, F.; Lin, Z. Research on sodium sulfur battery for energy storage. *Solid State Ion.* **2008**, *179*, 1697–1701. [CrossRef]
9. Xiaochuan, L.U.; Xia, G.G.; Lemmon, J.P.; Yang, Z.G. Advanced materials for sodium-beta alumina batteries: Status, challenges and perspectives. *J. Power Sources* **2010**, *195*, 2431–2442. [CrossRef]
10. Iijima, Y.; Sakanaka, Y.; Kawakami, N.; Fukuhara, M.; Ogawa, K.; Bando, M.; Matsuda, T. Development and field experiences of NAS battery inverter for power stabilization of a 51 MW wind farm. In Proceedings of the 2010 International Power Electronics Conference (IPEC), Sapporo, Japan, 21–24 June 2010; pp. 1837–1841.
11. Benato, R.; Cosciani, N.; Crugnola, G.; Sessa, S.D.; Lodi, G.; Parmeggiani, C.; Todeschini, M. Sodium Nickel Chloride battery technology for Large-scale Stationary Storage in the High Voltage Network. *J. Power Sources* **2015**, *293*, 127–136. [CrossRef]
12. Benato, R.; Sessa, S.D.; Cosciani, N.; Lodi, G.; Parmeggiani, C.; Todeschini, M. La tecnologia sodio-cloruro di nichel (Na-NiCl$_2$) per l'accumulo elettrochimico stazionario sulla rete di trasmissione. *L'Energia Elettr.* **2014**, *4*, 71–84.
13. Sessa, S.D.; Crugnola, G.; Todeschini, M.; Zin, S.; Benato, R. Sodium nickel chloride battery steady-state regime model for stationary electrical energy storage. *J. Energy Storage* **2016**, *6*, 105–115. [CrossRef]
14. Andriollo, M.; Benato, R.; Bressan, M.; Sessa, S.D.; Palone, F.; Polito, R.M. Review of Power Conversion and Conditioning Systems for Stationary Electrochemical Storage. *Energies* **2015**, *8*, 960–975. [CrossRef]
15. Rebolini, M.; Tosi, S.; Vanadia, R.; Di Pietro, N.; Senatore, E.; Polito, R. *The Authorization Procedure for Energy Storage Systems Projects Installed on the Italian Transmission Grid*; Paper C3–103; Cigré: Paris, France, 2016.

16. Gatta, F.M.; Geri, A.; Lauria, S.; Maccioni, M.; Codino, A.; Gemelli, G.; Palone, F.; Rebolini, M. Modelling of battery energy storage systems under faulted conditions: Assessment of protection systems behavior. In Proceedings of the 2016 IEEE 16th International Conference on Environment and Electrical Engineering (EEEIC), Florence, Italy, 6–8 June 2016.

17. Gatta, F.M.; Geri, A.; Maccioni, M.; Lauria, S.; Palone, F. Arc-flash in large battery energy storage systems —Hazard calculation and mitigation. In Proceedings of the 2016 IEEE 16th International Conference on Environment and Electrical Engineering (EEEIC), Florence, Italy, 6–8 June 2016.

18. Gatta, F.M.; Geri, A.; Lauria, S.; Maccioni, M.; and Palone, F. Battery Energy Storage Efficiency Calculation Including Auxiliary Losses: Technology Comparison and Operating Strategies. In Proceedings of the IEEE PowerTech Conference, Eindhoven, The Netherlands, 29 June–2 July 2015; pp. 1–6.

19. Benato, R.; Sessa, S.D.; Crugnola, G.; Todeschini, M.; Zin, S. Sodium nickel chloride cell model for stationary electrical energy storage. In Proceedings of the 2015 AEIT International Annual Conference (AEIT), Naples, Italy, 14–16 October 2015; pp. 1–6.

20. Benato, R.; Sessa, S.D.; Necci, A.; Palone, F. Sodium-Nickel chloride ($NaNiCl_2$) Experimental Transient Modelling. In Proceedings of the 2016 AEIT International Annual Conference (AEIT), Capri, Italy, 5–7 October 2016; pp. 1–6.

21. Benato, R.; Sessa, S.D.; Necci, A.; Palone, F. Sodium-nickel chloride battery experimental transient modelling for energy stationary storage. *J. Energy Storage* **2016**. [CrossRef]

22. Benato, R.; Sessa, S.D.; Crugnola, G.; Todeschini, M.; Turconi, A.; Zanon, N.; Zin, S. Sodium-Nickel chloride ($Na-NiCl_2$) battery safety tests for stationary electrochemical energy storage. In Proceedings of the 2016 AEIT International Annual Conference (AEIT), Capri, Italy, 5–7 October 2016; pp. 1–6.

23. Benato, R.; Sessa, S.D.; Crugnola, G.; Todeschini, M.; Turconi, A.; Zanon, N.; Zin, S. Test Di Sicurezza Su Batterie Sodio-Cloruro Di Nichel Per L'accumulo Elettrochimico Stazionario. *L'Energia Elettr.* **2015**, *92*, 47–53.

24. Carlini, E.M.; Bruno, G.; Gionco, S.; Martarelli, C.; Ortoloano, L.; Petrini, M.; Zaretti, L.; Polito, R. *Electrochemical Energy Storage Systems and Ancillary Services: The Italian TSO's Experience*; Paper C4–116; Cigré: Paris, France, 2016.

25. TERNA Italian Grid Code. Annex 15: Participation in the Regulation of Frequency and Frequency/Power. Available online: www.terna.it (accessed on 1 December 2015).

26. Xu, B.; Oudalov, A.; Poland, J.; Ulbig, A.; Andersson, G. BESS Control Strategies for Participating in Grid Frequency Regulation. *IFAC Proc. Vol.* **2014**, *47*, 4024–4029. [CrossRef]

27. Tortora, A.C. Storage E Sicurezza Della Rete: I Progetti Di Terna. Available online: http://www.aeit-taa.org/Documenti/TERNA-2016-01-27-Storage-on-Grid-sicurezza-rete.pdf (accessed on 1 December 2016).

MDPI AG

St. Alban-Anlage 66

4052 Basel, Switzerland

Tel. +41 61 683 77 34

Fax +41 61 302 89 18

http://www.mdpi.com

Energies Editorial Office

E-mail: energies@mdpi.com

http://www.mdpi.com/journal/energies

www.ingramcontent.com/pod-product-compliance
Lightning Source LLC
Chambersburg PA
CBHW051705210326
41597CB00032B/5378